ISBN 978-88-8398-068-8

Florence, Italy
www.e-p-a-p.com
www.europeanpress.eu
Printed in USA, UK, Italy

The Meaning of Life, the Universe, and Nothing

Part I

By Kuenftigen Leuten[1]

EUROPEAN PRESS ACADEMIC PUBLISHING

[1] amurphy09@comcast.net

Preface: The Deeper Reality of Human Nature and Everything

Everything is composed of divided pieces of nothing that are attracted to their opposite halves in various combinations, which cause the macrocosmic feeling of mutually ecstatic love in the purest fractal form of this attractive process. The divisions of nothing by itself that continue all the time in a vacuum create the infinite variety and wonder in the universe.

Living organisms originated from various combinations of divided pieces of nothing existing in complex systems of attraction and movement that enabled their reproduction. To survive in the competition for the resources needed by these biochemical processes that were multiplying, the life-forms became ever more aware of their environment in order to learn to both cooperate with it and control it.

Human beings represent the current optimal harmony of attractive forces on earth today. That balance includes both an innate desire to cooperate between individuals in a diverse fashion as well as spiteful vindication against everything that is incoherent with that resonating harmony of diverse wills. This complex combination prompted an expansion of human brains and thinking capacities in order to enable optimizing relationships with other people who are also characterized by a powerful combination of ecstatic cooperation and brutal spite.

The expanding brain of human beings enabled them to figure out how to detect correlations between events from past cycles that allowed them to learn to forecast the future. Unlike other organisms that merely react instinctively to the environment, humans were thereby able to use their predictive knowledge to prepare for future situations. Such preparations enabled them to increasingly manipulate and control their environment, and their expanded impact on any future situation they faced motivated them to forecast themselves, as the future environment was increasingly affected by their future actions.

Predicting of oneself requires a self-awareness that enables thinking about what one would and could do in any situation. This self-awareness combined with the innate logic of all organisms' brains formed the consciousness people have today. Consciousness is merely this logical reflective self-awareness that expires at the death of the brain, although one's impact on everyone else while living can be long-lived.

The initial thoughtful awareness was a group consciousness, as might have been enhanced by group chanting that can create the powerful unity that the weak and slow human species needed in order to be able to survive an overwhelming environment. Such group consciousness naturally promoted equality in a diverse form that enabled each person to be responsible for the individual's own varied needs and desires.

However, about 35,000 years ago, matriarchal cultures apparently developed that created a hierarchy by revering femininity above masculinity. Males eventually revolted and spitefully replaced the matriarchies with patriarchies. Today's religions and societies largely reflect the patriarchal conquest that began 6000 years ago, although women have started to make some progress back toward equality in the last few hundred years.

Despite the recent imposition of inhumane cultures, human beings remain programmed by evolution to be highly motivated to serve humanity in an individualistic way. Evolved human nature has made feelings of happiness and self-fulfillment for each person be dependent on setting the individual's overriding purpose to be to promote mutual love among humans across time. Doing so maximizes not only the potency of the human group but also the power and thus the freedom of the individual.

Table of Contents **for**
The Meaning of Life, the Universe, and Nothing

Appendices

Preface to Everything
-1

This book provides what its title indicates. It has two parts, the first of which provides scientific facts needed to overcome the indoctrination to which most are subjected in today's societies. Such knowledge may be necessary for most to understand the second part, which applies logic to information compiled from hundreds of research publications to explain the meaning of life, the universe, and everything. In particular, the negative knowledge or lies, as can be symbolized by a negative number like –1, with which most people have been brainwashed from birth may have to be recognized with a reading of the first part before it is possible to comprehend the second part of the book that is actually based on the power of dividing by nothing or 0. A recognition of that fact enables discovering how the deeper reality of human nature and everything evolved.

Part I

The first part, *Beyond 666*, provides knowledge of what 666 really symbolizes, and it thereby enables people to move beyond the programming with which contemporary cultures indoctrinate the vast majority of human beings in the world from birth onward. It thus allows them to open up their minds to understand answers to the deepest philosophical questions of human beings. The essential aspects of Part I are summarized in pictorial form in Appendix E.

As explained in Part I, the mark 666 once denoted prehistoric human societies that existed before cruel patriarchs took control of the world a few thousand years ago. Those barbarians, whose descendants rule the world today, demonized the more advanced human civilizations that they conquered and later brainwashed humanity with an incredible set of lies to keep the resulting enslaved people from revolting against the inhumane vicious conquerors. A part of this deceitful process involved the vilification of the 666 insignia in some of the religions such as Christianity that were fabricated by the patriarchs.

Understanding the reason for the demonization of the 666 symbol can therefore facilitate elimination of infantile prejudices that are subconsciously ingrained in most people, especially Christians. The indoctrination that begins at a very early age inhibits a comprehension of the meaning of everything, and a knowledge of "What the Fundamentalists and Other Patriarchs Really Don't Want You to Know" may be a prerequisite for understanding the second part of the book. In particular, it's necessary to imbibe from the tree of knowledge in order to overcome the ignorance that is cruelly programmed into people by the inhumane societies of today. It's especially important to recognize the absurdity of the artificial, cruel definitions of "good" and "evil" with which most people have been indoctrinated by patriarchal societies from childhood.

The patriarchal system, upon which Christianity is based and whose basic form was molded thousands of years ago, was designed to be self-perpetuating. That fact can explain its lasting endurance in the face of human nature with which that system so seriously conflicts. The patriarchal system is akin to creating masses of brainwashed robots, who have been provided with strong motivation to repress both their own innate human nature and that of others to create hell on earth. The intrinisic human nature of these robots, however, has resulted in continuous modifications of the basic cruel system to be more consistent with human needs, desires, and tendencies.

An increasing percentage of the population in even the very christian USA continues to partially overcome their brainwashing and escape a belief in the fictitious patriarchal deity despite all the social and economic pressure that the USA system imposes on them. Although some of these escapees used drugs to find their way out of the hell on earth created by Christianity, drugs aren't necessary in order to experience the blissful freedom of uninhibited living according to one's own human nature alone. In fact, without enlightenment as to each human being's evolved purpose, drugs provide only temporary relief from the cruel slavery of the Christian concentration camp that becomes even more unendurable when the artificial effect wears off. On the other hand, escapes into the rather permanent ecstasy of simply resonating with one's own human nature are quite feasible with mere conscious knowledge about the meaning of life. Despite all the camouflaging of the way out of Christian hell that includes brainwashing from childhood on, a road map out of the concentration camp is incorporated into this book.

With an increasingly integrated world, the minority of Christian robots have little prospect for stopping escapes from their concentration camp of mechanical slaves, much less replacing them with new inmates, regardless of the social and economic incentives they provide. Even if some momentary relief from the torture suffered in the Christian concentration camp provides occasional short-term "highs" in various ways, including an infantile religious relationship with an imaginary deity, those moments of relative "highs" can't even come close to matching the bliss and self-fulfillment that comes from being truly human and free of the mental disease called christianity.

While recovery from the diseases of Christianity and similar religions can be difficult and may even be painful for some, the fact that so many have maintained themselves on a road to full recovery to a disease-free human nature indicates that the benefits of a cure far outweigh any costs. The people who have at least partially cured themselves provide useful proof to others seeking a cure that the disease isn't necessarily terminal. In addition, those on the road to recovery have paved the way for others to follow more easily. Moreover, as ever more people find an escape route, recovery is facilitated since there are more comrade escape prisoners with whom they can make true human connections. This book may serve as a useful highway to human nature that many more may choose to utilize, at least part of the way, to escape the living hell created by Christianity. It could even enable an even more rapid route back to human heaven on earth at spaceship speed.

By its very "intelligent design" of repressing human nature for the short-term benefit of a few cruel sadists, christianity does indeed cause true hell for those brainwashed into believing in that Christian deity, no matter how well off the concentration camp inmates are materially and no matter how many robotic slaves they can claim as "friends." While christian robots may perceive other robots to be their "friends", human nature makes many people experience better relationships with their animal pets than with other humans, as animals are at least natural living beings as opposed to inanimate brainwashing programs that have taken control of human bodies like a horrible disease. Robots, no matter the shape, are indeed less attractive than even live animals and are downright disgusting since they are akin to decaying dead bodies that have been programmed like zombies to do everything their exploitative brainwashing has indoctrinated them to do. In that sense, they are even somewhat similar to dung, which also may have once been part of a living human being.

The natural human tendency to seek ever more and better in any situation or game combined with the capitalist brainwashing of people to think that making money is the primary purpose of humanity enslaves humans into seeking ever more wealth in a futile effort to find happiness on earth. At the same time the Christian religion motivates them to worship and serve unconditionally a fabricated cruel deity to avoid the fictitious eternal damnation and hell after death that is preached to them. The result of the indoctrination into both slavery for the selfish rewards in life offered by capitalist materialism and after death promised by Christian "heaven" is an especially inhumane combination that motivates unnatural servitude to an elite in order to selfishly accumulate material goods and a ticket to a fictitious paradise in an imaginary afterlife. Both the capitalist indoctrination into materialist selfishness and the worship of a fictitious deity above all else that requires obedience to an inhumane set of rules diverts people away from their human nature and their natural love for each other. Christianity and capitalism thereby transform human beings into Pavlovian dogs that are merely trained to respond mechanically to stimuli that are associated with real or possible rewards and punishments.

The two different forms of slavery to money and a fictitious deity often result in conflicting demands on people by encouraging both pure selfishness and unquestioning servitude to a superior at the same time. Both pathological selfishness and any form of slavery are totally inconsistent with human nature. The conflicting demands of the cruel lies drilled into innocent children in places like the USA magnifies the torture of human beings there by fragmenting them in a frustratingly futile attempt to go into two different directions simultaneously that both artificially lead them far from their innate tendencies.

Human beings evolved in a culture of group consciousness and diverse but resonant equality, which permit virtually unlimited individual freedom for the good of humanity. Human nature therefore developed to enable creative individualism for the unified goal of serving all human beings in the way each individual feels is best. That system is a polar opposite to the capitalist Christian societies that replace the natural desire to help humanity as a whole with an unnatural competition to better serve a "superior" being in the way the "superior" being demands. The latter cruel system requires an unquestioning blind obedience that mandates regimented thinking and identical behavior. By establishing rewards and penalties based solely on one's performance as a robotic slave, christianity promotes a pathological indifference to the fate of anyone except the "superior" being.

Part II

Part II entitled "Scientific Answers to the Ultimate Questions about Everything and Nothing" is based on the logical and intuitive implications of documented scientific facts that are cited with references in parentheses. A full comprehension of Part II of the book may be difficult without recognizing the basic facts in Part I. In any event, the deduced "answers" do not represent any sort of doctrine that is unchangeable by new information and interpretations.

The very entertaining and thought-provoking movie "Hitchhiker's Guide to the Galaxy" provides part of the motivation for the titling of this book. In that film, the computer's answer to the ultimate question of "the meaning of life, the universe, and everything" was "42", but the significance of that response was left undiscovered because the ultimate question wasn't specified exactly enough.

While this book largely avoids number or word games, a brief analysis of a numeric answer to the ultimate question is useful here in the preface to provide a simple symbolic overview of the issues. For instance, with science indicating a mathematical pattern for the universe, the ultimate question might be: how many basic particles or forces are there in the universe? So far, science has discovered 40, but 2 more may very well exist somewhere in this universe to complete the puzzle. Nevertheless, both the answer and the question may of course be invalid, not only because the film itself is fictional, but also because answers developed by computers, science, or humans are, even if they utilize valid logic, based on assumptions that may not reflect reality very well.

Still, the symbolism of all phenomena, such as that of the author of the book upon which the film is based, may help one find one's own valid insights into the ultimate questions and answers. I and my children have interpreted the film's "42" answer as the identically sounding "For tea, two." The ultimate question with such an answer seems, in our subjective opinions, to be "What does it take for there to be happiness, meaning, and life?" The obvious answer is nourishment and a love pair that are symbolized by "tea" and "for two", respectively.

It could even be possible that the implied 42-40=2 basic particles remaining to be discovered by science are some sort of inseparable love pair, which can never be observed by any scientific instrumentation or method, which can only be experienced, and which may be more powerful than anything else in the universe. One possible pair might be the positron and electron which are irresistibly attracted to each other. Those two fermions, which are actually just the two opposite sides of the same divided nothingness moving toward each other in different directions in time, form the basis of all other particles and forces in the universe. The finding of this research indicates everything represents some combination of the attraction of electrons and positrons for each other that themselves are merely electromagnetic fields created by the division of nothing by itself.

However, for anyone believing that romantic love, or love in general, is everything, it's important to recognize that life also requires nourishment (or tea) and that there are at least 40 other known basic substances in existence that appear to serve that end. Those other substances in the universe may be formed by the two basic substances, but without those complex systems of the basic attraction between divided pieces of nothing, there would only be lifeless nothingness and a unified bland light. The positron and electron are actually identical to one another except that they travel in opposite directions in time, and so there is only one type of substance that is divided by movement in the temporal and spatial dimensions that were created as a result of their division. When those two divided substances meet at the same place in time and space, they create identical light photons that would be merely dull nothingness if there weren't the various combinations of divided nothing to observe and be affected by it.

The two substances or electromagnetic fields are drawn to their opposite halves across the spatial and temporal dimensions in a wave of infinite potential combinations, each of which will occur with a certain probability in parallel universes. The attractions of divided nothing for each other cause space and time to bend back on themselves to unify back into nothing after an eternity when the past meets the future. Thus, there is only one coherent whole consisting of nothing (0), which just is.

The logical conclusions of this book indicate that love isn't a force but just a macroscopic reflection of the attraction of the basic units of the universe, electrons and positrons, to each other that interact in massive numbers in an extremely complex manner to create the "mysterious magic" of love. The mutual attraction of electrons and positrons to each other in space can cause them to follow a long circular route through the time dimension where they unify after an eternal past and future that meet to add to zero time net. That attraction which is intrinsic to their nature represents the basic force or process that creates and drives all other microscopic and macroscopic forces and things in the universe.

The electrons and positrons themselves are merely separated pieces of nothing whose pull on each other results in electromagnetic force fields, which is all they and everything else are, and their separated tug on each other even creates space and time itself. A comprehension that this innate attraction just reflects the pull of divisions of nothing back into a unified whole nothingness can clarify the complexity and power of love as well as everything else.

In any event, it is important to understand the evolved purpose of love on a macroscopic level of perception. Love, even its romantic form, by itself can't fulfill a selfish hope that it alone will lead to happiness and fulfillment. To avoid disappointment, frustration, and unhappiness, and to enable the experience of love in the purely blissful form, it's necessary to recognize that romantic love evolved to bond people together for a unified goal to perpetuate humanity. The romantic form of human love developed to have two parents experience ecstasy while creating and raising offspring. Their reasons for being are co-determined jointly with all other human beings acting according to their evolved human nature, of which romantic love is only one characteristic.

Romantic love is essentially a mere narcotic that produces a temporary high, albeit one that is extremely ecstatic and that can last for some time. It has evolved as an internally generated blissful drug because it is highly useful in motivating evolutionarily productive behavior. However, like all narcotics, it can also be taken to extremes when an individual becomes selfishly addicted to it without using the drive for it, and the pleasure of it, for the higher purpose of serving humanity as a whole. Such selfish addictions are unnatural for human beings and counterproductive to their long-term happiness.

Human nature has evolved so that any addiction eventually becomes undesirable, and it is also possible to turn off an addiction to love. Addictions and obsessions can be discontinued simply through an understanding of what they are and knowledge that they can be terminated through the simple internal will power which all humans have evolved to have. Even after turning off an addiction to love, it can still be enjoyed, and enjoyed far more, when one controls it internally. These insights derived from recent research findings in psychology and neuroscience are just some of the revelations about the meaning of life, the universe, and everything that are provided in the book.

Actually, the best simple mathematical answer to the ultimate question of the meaning of everything probably isn't even 42. Instead, it's more likely 1/0, which represents a sort of inverted nihilism that has infinite value and meaning, as this book seeks to explain and document in great detail. This conclusion is partially derived from the fact that particles and antiparticles naturally pop out of vacuums of nothingness all the time, and it is those fermions that create everything.

However, the computation 1/0 more directly provides symbolic meaning by putting the whole (1) over the individual pieces of nothing (0). In particular, by putting the happiness of a group of others united in purpose over that of oneself, one creates infinity.

Scientific facts indicate that the infinite variety and wonder of the universe repeats itself forever, and so maybe the true answer might better be written as 1/0,1/0,1/0,.... Removing the line separating the one in the numerator from the zero in the denominator of each 1/0, and unifying the two figures, creates 666.... Thus, the prehistoric human societies perhaps had it right after all.

666 can also symbolize the Tree of Life. The upper part of the "6" here of course denotes the tree, and the circle below represents the earth. The Tree of Life symbolizes the naturally blissful human act of physical procreation via the erect male tree inserted into the nurturing female earth.

666 can also symbolize the Tree of Knowledge. The fruit on it represents the knowledge that comes from the growth resulting from the Tree of Life's union of the male tree with the female earth. The Tree of Life and the Tree of Knowledge are thus identical. The theory of some Eastern religions that sexual activity brings inspiration to the participants represents an even more direct integration of the Tree of Life and Knowledge.

The Tree of Life and Knowledge and its fruit, as well as 666 itself, are intensely vilified by patriarchal religions such as christianity in their books like the bible. The fictitious patriarchal deities of those religions can only survive on a persistence of ignorance and sexual frustration. They are as absurd and fictional as the religion on the planet in the film Hitchhiker's Guide to the Galaxy where conscious living beings seemed to believe they were snot to be wiped clean from a deity's nose by some cosmic deity, which doesn't sound that much different from the Christian god that created the world only to destroy it and all the evil human beings in it, as explained and prophesized in that religion's bible.

Such religions on earth thrive on sexual repression that diverts people's energies and attention into extreme and often bizarre efforts to satisfy a basic human drive in order to inhibit them from attaining the knowledge needed to figure out the entire religious system is built on lies. The hate-filled books like the bible are holy for fictitious patriarchal deities and their roboticized slaves, but they are sacrilegious and inhumane to human beings.

From a purely sexual perspective, the number 690 may represent a number beyond 666, insofar as it represents the deeper 69 component of human sexuality that can naturally lead to physical unity and procreation of the fertilized egg 0. The 690 permits differentiation of each within the context of a unified number that is far more than the sum of its parts (i.e., 690>6+9+0). That number therefore promotes diversity within the context of a resonant coherence and encourages egalitarianism since individualistic striving to meet the unique needs of each other is essential for the unity of the whole.

The symbol 666 may have originally been developed to emphasize the equality of all. However, the concept of egalitarianism is thereby perverted into a regimented system of unnatural sameness that disharmoniously discourages human diversity and freedom. Such a system that is incoherent with human nature needs a hierarchy to enforce it. Matriarchies were apparently the first of these inhumane cultures, and 666 came under those cruel systems to symbolize only the female in the form of the maiden, the mother, and the crone. The 666 matriarchies likely originated in the purely selfish desire of older women to receive the greater love and reverence that they received as maidens and mothers themselves when they were younger. The older women were able to use deception and manipulation to impose a religious, political hierarchy that enforced through their rule love and reverence for themselves that was equal to that of maidens and mothers.

For those seeking a purpose for everything as opposed to mere love, sex, and truth, which may all be related, the figure "28" could represent a more meaningful answer to the ultimate question(s). In particular, substituting the same English sounding of "to" for "two", and recognizing the infinity sign as being a sideways eight, "28" could also symbolize "two infinities" to indicate the infinite paths to the past and future that seem to move in opposite directions. Every point in perceived conscious time does indeed have an infinite number of different histories that could have led to that "now" and an infinite number of futures to which it will diverge.

The vertical position of the infinity sign in "28" could also symbolize "to infinity and beyond", insofar as it points up and down in space. This saying is identical to that of the animated toy Buzz Light Year in the movie "Toy Story", but it represents something far deeper for actual human beings than for owned toys.

Regardless, the computer's statement in the fictional movie "Hitchhiker's Guide to the Galaxy" that it is necessary to define the question better in order to understand any particular simplified answer may be insightful. Different numbers can possibly prompt via analogy a better insight into understanding the answers to the ultimate question depending on the particular aspect of the meaning of life, the universe, and everything being sought. They can at the very least thereby represent simple symbols for remembering and being consciously aware of complex concepts.

However, within the larger context of the ever-changing infinity of nothing or 0 that produces and contains everything including 690 and 42, love and the number of basic substances (whether 40, 42, or some other mathematical figure) aren't real. Mathematical figures are mere representations of a set of characteristics of predictable relationships, and love itself is just one of those processes. The implication for us humans is that we are more than mere flesh and blood, or atoms, or quantum chemicals, or electromagnetic fields, or energy, or vibrations, or love. We are of course all that, but much more. We are a part of a whole but partially fragmented and vibrating nothing that is stretched infinitely in what appears as time and space, and therein lies the incredible wonder and variety of all creation. And by putting the whole of all other combinations of split nothings above our own mix, we can reach the resonant nirvana of infinity, i.e., $1/0 > 666 > 0/1 = 0$ that makes humanity so powerful.

Within this context, it is imperative for maximization of 1/0, and the resulting happiness for oneself and others, to not use love for strictly selfish purposes. Love evolved as one means to the higher end of serving the human beings of the future, as one possible path to 1/0. On the other hand, the selfish use of love as an end in and of itself can enslave oneself into meaninglessness, inadequate fulfillment, and exploitation. The latter path leads only to the loveless unhappiness of 0/1.

Even for the maximum pleasure of each partner in a relationship of romantic love, mates must remember to serve each other only for the purpose of contributing to the development of future human beings. Following that reason for being requires subordination of all innate and conditioned tendencies to this higher meaning in life. It's important not to do so out of a desire for any personal happiness, pleasure, or reward, which may never ever come, and which provide minimal satisfaction if one acted for that purpose, but because of reverence for the natural human beings of the future. One's own pleasure and life is insignificant in comparison, as 1/0>0/1 indicates mathematically.

This book, however, is not about simple answers, numerical symbols, semantics, or fictional movies. For instance, the investigation into the nature of the matriarchies that are symbolized by 666 doesn't include any deep analysis of the number 666 other than brief references to its apparent general symbolic meaning in the prehistoric societies where it originated. In any event, this book goes way beyond 666, 690, 0/1, 42, 1/0, 28, and mere symbols.

Perhaps the most important benefit to me of this research is that I now understand what it means to be human. By deprogramming the indoctrination with which I and most people have been brainwashed in today's society from early childhood onwards permits one to be innately obsessed with human beauty, love, and sex spiritually, emotionally, and physically. Every minute of the obsession can be enjoyed no matter what, no matter how irrational, and this book can enable comprehending why. Being human is happiness, and being human only requires acting according to one's own human nature without inhibitions, resonating with other human beings and the environment for the benefit of oneself and all of humanity.

This book is actually designed to be very useful in de-programming oneself back into a human being, which we all can once again become. Part of that de-brainwashing may require eliminating internal repression of the naturally evolved human desires for companionship, sex, communal love, and equal happiness. Once that de-toxification process has been accomplished, it is possible to understand the scientific facts that lead to clear implications about the nature of life, death, consciousness, free will, will power, hypnosis, happiness, desire, sex, love, time, creation, existence, and reality.

The latter facts enable the final purging of the indoctrination that was designed to transform people into robotic slaves or machines. The indoctrinating brainwashing, which begins early in life with the perversion of innocent children's minds via vicious lies about an after-life of unbearable eternal punishment for refusing to enslave oneself to a fictitious god, represents extreme psychological cruelty. A moral, if not legal, case could be made that all those involved in spreading the deceptions of Christianity need to be prosecuted for torture, with especially harsh sentences for those abusing young children with the gruesome brainwashing of Christianity that is so psychologically painful to endure or break.

Existing in the very cruel and inhumane society created out of ignorance can sometimes seem difficult, but it is easier if one realizes that it is a question of fighting that sick system or perishing along with humanity as a whole. Each enlightened individual has 60 trillion powerful human cells on their side. The enemy has none; it has only very deceitful ideologies and religions that are relics created long ago by cruel patriarchs imposing slavery on human cells. Just as human ancestors have done for millions of years ago, human beings will adapt the environment to their wishes, as opposed to accepting it as is and merely reacting to it. As is normal human nature, humans will attempt to control the environment instead of being controlled by it. Even if one is unsuccessful in fighting the inhumane ideologies and religions of today, one's innate human nature enables happiness by just making the attempt, fighting the good fight, and recognizing that, in those many future parallel universes in which patriarchy dies and humanity survives, there will be human beings who will revere one's efforts.

As a human being, one is indeed free to both consciously and unconsciously fight human nature, whether that battle against oneself reflects religious indoctrination or not. For instance, due to capitalist brainwashing, one may fight human characteristics such as the evolved tendency to try to serve humanity as best as one can. However, battling human nature won't change it and will likely only make one very unhappy, if not totally depressed in a psychological hell on earth.

One pathway out of such hell may be accessed through a mere perception of all humanity as an omnipotent deity. Human nature innately causes people to act for the benefit of that collective god, and, as explained in this book, human will power can be considered to have created human beings and everything else in this universe out of nothing, which humans also control. An innate human Heaven in the mind derives from this religious belief in humanity as a whole, as everyone serving this higher purpose in an individualistically diverse fashion results in an ecstatic group consciousness that is individually experienced. That heavenly group conscious can exist initially in small groups, as it was in the earliest days of humanity when there were few humans on earth but many adversaries that included hominoids not resonating with humans' higher purpose. It will then inevitably spread quickly as others in contact with such people acting according to their innate human nature are drawn by their own desire to enter Heaven on earth. To reacquire this hypnotically blissful unity among individuals, it is only necessary to have them all focus all their efforts on serving humanity as a whole, including with respect to earning money for both their own comfortable survival and that of their heirs who are designated based on that higher purpose. This process will cascade all human beings into a communal Heaven on Earth. While it may take some time for the inspiring enlightenment to spread, paradise on earth could cause an avalanche in an incredibly short temporal horizon if the unified groups of people stay connected to the whole of humanity.

Regardless, one useful conclusion of this research is the fact that everything possible will happen in the future. What a person wants will be attained, at least in a future parallel universe.

It can be fun to fantasize about achieving a certain desired experience even if one doesn't focus enough to actually experience it. Moreover, fantasizing about a particular goal, such as experiencing the bliss of mutual love between two human beings, can actually help attain a particular goal, or at least recognize the effort needed to experience realization of it and to determine whether it is worth it or not.

Although everything is an illusion that is created by divided pieces of nothing, many complex forms of those illusions, such as live organisms, evolved to be aware of the illusions that are their reality. One type of organized system of the illusory divided nothing, namely human beings, has even evolved to be self-aware of its own illusion, partially even controlling the illusions they experience. Self-control of the illusion is maximized with knowledge about the nature of the illusion that itself indicates the maximum power of human beings exists when they cooperate equally in a unified will.

Human nature and its power developed to be most harmonious with egalitarian unity at least partially because the nothingness from which human beings evolved is characterized by this trait. Humanity is the most aware, macro form of divided nothing on earth. It may be the most enlightened system of electromagnetic fields in the universe, or at least it can be, as human nature seems to be driving it to that goal in a learning process that is especially enjoyable because of human humor.

However, there is far more to life, the universe, and everything than mere enlightenment, which exists in the form of a uniform light photon that is timelessly bland. Each person, all traits of human nature including spite, and consciousness itself exist to promote mutual love between people, and so it can be naturally concluded that this love, including its romantic narcotic form, is the meaning of life. Conscious life evolved in the universe which was thereby co-created by such life in a timeless unity, and the rational consciousness which developed out of that process leads to the inescapable deduction that the purpose of everything is the intrinsic attraction of nothing for itself in all its divided parts that construct reality. The mutual love between the individualized self-awarenesses that is the highest form of this attraction produces the most ecstatic macrocosmic form of diversely humorous processes to which each human being innately drives their self-reflective wills to create infinitely fascinating variety out of blandness.

Please recognize that I didn't want to write this book. My human nature drove me to find out the meaning of life, the universe, and nothing, and the only avenue open to me was to allow the knowledge of the future to flow through my writings. I therefore indicate "kuenftigen Leuten" as the author, as that means in German "future people", who inspired me with the knowledge contained in this document. I apologize in advance for any errors in my interpretation of that knowledge and any other errors I have made in logic, but I feel the basic tenets and conclusions of this book are irrefutable in any case.

Understanding the material in this book may change your life beyond comprehension. It has mine. Enjoy.

(Note also that a name of a person or periodical followed by a comma and a year in parenthesis after a statement is a research style indicating the source of the information that is given in the bibliography at the end of each of the two Parts of the book).

Introduction to Nothing
0

Nothing *(0)* **lasts forever. Even diverse coherence (*690*), matriarchy (*666*), patriarchy (*000*), or a balanced harmonious** *infinity* **(***1/0***) changes over time. Divisions of nothing by itself (***0/0***) mathematically equal** *0*, *1*, **or** *infinity*, **and the many divisions of nothing, when added back within the whole nothingness of** *1*, **sum to all the numbers in between that recur eternally in the continuous process of creation from divisions of nothing.**

Dividing anything by nothing or zero, including dividing zero by itself, can be shown algebraically to lead to not only infinity but an infinite set of mathematical possibilities, including irrational and imaginary ones. The laws of quantum physics that have been verified to be 100% consistent with the reality of the universe enable the same infinity of possibilities. Nothing comes from nothing net, but an infinity of possibilities is created via the divisions by nothing within nothing.

Creation from nothing has been verified to happen all the time. In particular, particles of matter and antimatter originate continuously out of empty space and then usually destroy each other in a quick spark that returns the vacuum into nothingness. There is nothing beyond the reality of nothing so splitting itself, vibrating thereby forever in a fragmented yet interconnected form, and then reuniting completely with itself in a spark of light. The vibrations of nothing create the reality we see and the love that we feel.

All perceived particles are just waves of various vibration frequencies or electric charges that result in different electromagnetic force fields. These force fields create the perception of solid matter because those normally unobservable barriers can't typically be penetrated by other force fields. However, an unusual electromagnetic environment can cause those force fields to collapse so that matter no longer acts like it is solid. For instance, a tornado can enable a piece of straw to get stuck inside a piece of glass without breaking it.

In any event, matter in any event is just a form of divided nothing that forms in different combinations within nothing. The apparent solidity of matter merely reflects electromagnetic fields of nothing that usually cannot be breached by other vibrations of nothing.

Quantum physics has shown that everything acts as a wave which explores every possible path. However, all points on a wave and all of the infinite wave paths into the many different potential futures will be followed in parallel universes. Within this set of infinite possibilities, the universe one experiences depends on one's own focused will. However, the environment and the will of others affect that focus as well as the set of potential experiences.

The universe in which human beings lives was probably created from the accumulation of matter in a black hole of an antimatter universe. Correspondingly, the black holes of our universe may result in the creation of further antimatter universes. There are likely an infinite number of such universes and anti-universes already in existence that originally developed when some of the matter and antimatter arising from the splitting of nothing didn't reunify and mutually annihilate. Everything possible will happen in such an infinite reality that also include an infinite number of parallel universes for each such universe at every moment in time.

There are two major levels of reality. At the core, there is the reality of potential vibrations that move in waves of different frequencies. Then there is the reality of perceptions that reflect categorized averages or points in waves of different systems of electric charge densities in the form of objects. The latter reality is what we humans normally perceive in macroscopic force.

Perception is merely a system of electric activation of brain neuron cells that permit complex analysis of many different pieces of information. These information bits are optimally categorized into groups in the brain for maximum efficiency in terms of storage space (i.e., brain size), energy usage (i.e., nutrition requirements), and speed of recall.

A process exists in mammalian brains that provides chemical pleasure rewards internal to the brain for evolutionarily useful behavior that itself is related to the firing of particular groups of brain neurons. This process may explain how the basic genetic code enables the brain to form efficient connections of neural firing activations from conception onwards. The connections are dependent on the environment and can be adapted as the environment changes in a process of Bayesian learning that uses probability analysis to evaluate the optimality of changes in the system of neural firings. At the same time, the genes that are responsible for brain development also have certain specific innate tendencies that categorize the effects of the various electromagnetic processes which affect one's own electromagnetic fields, including the neurons whose systems of interconnected electromagnetic activity are responsible for the categorization, learning, and reactions.

The human genome even permits a conscious self-awareness of the environment and enables self-aware analytical power. That power uniquely evolved over millions of years in people's ancestors to enhance the survival and reproduction of humanity as a whole in a process which also permitted substantial genetic variation in characteristics across individual human beings for maximum adaptability of the human species.

It is possible to develop some complex decision-making subconsciously, but consciousness, which seems to have been prompted by the development of self-forecasting, enhances that capacity greatly. Consciousness was preceded by forecasting, which itself may have been spurred by the urge for spite that evolved as a simple useful reflex at unequal treatment in organisms like monkeys that needed group cohesion in order to survive. In particular, the spiteful responses of others to selfish behavior at others' expense motivated prediction of such responses that then provided an important incentive to act for the benefit of others. This forecasting capacity expanded further to enable an analysis of spiteful urges before acting on them. That development was aided by further expansion of human intelligence in a mutually reinforcing process which increased both brain power and cooperation. The interactions between oneself and others, especially those involving empathy and spite, eventually led the growing brain to the self-forecasting that requires self-awareness or consciousness for maximum effectiveness.

Self-forecasting and consciousness become especially important once an organism reaches a stage where it begins to significantly affect the environment instead of merely reacting to it. This interactive situation naturally arose when spiteful actions inhibited uncooperative behavior among people who were motivated to forego selfish behavior out of a forecast of such spite. Empathy, which itself also inhibits selfish acts because of the empathized negative feelings that are naturally experienced in the human brain, may have at least partially developed out of such predicted behavior in others, as feeling what others were perceived to feel helped in forecasting their future actions.

Empathy may have co-evolved with spite to create the enormously complex human relationships that exist and that promoted the development of consciousness in human beings to evaluate them. While the working memory of human consciousness is rather limited for individualized pieces of information, it remains larger than that existing for other organisms. Conscious analysis is made much more powerful because the stimuli sources of information can be categorized into general concepts, such as through the use of words or symbolic language. Both the processing of information and their categorization are consciously perceived via a combination of neural firings in the human brain. This thinking system greatly enhances the conscious capacity of humans to evaluate very complex situations, such as the optimality of cooperative or spiteful behavior, better than other organisms.

Consciousness is based on a system of simple logic, consisting of "if A, then B", that is applied to the problem of forecasting to enable better decision-making. Information from the senses, emotions, and memories are fed into that logical processor via interconnected neural firings to create the sensation of self-awareness that is especially useful for self-forecasting and that is the core of consciousness.

The brain has developed a reward and penalty system for consciousness, and that system is at least partially controlled by conscious thoughts that evaluate the success in forecasting, analysis, decision-making, and goal setting in meeting human needs and desires. Conscious control of its own reinforcements and punishments, however, is limited by the evolution of the happy feeling of self-fulfillment from trying to serve humanity while suffering the hell of useless emptiness from at serving oneself alone. The success of the overall human brain processing system in perpetuating humanity caused it to expand via the process of evolution.

Maximum success in accomplishing the basic task of consciousness depends to a large extent on it programming the subconscious for many activities that can be more efficiently undertaken via simple rules that enable rapid action without conscious interference. The conscious has evolved to take complete charge of the human brain and hence the human body in its entirety for the purpose of perpetuating humanity, and it is highly motivated to follow this reason for being in a process that powerfully unifies the self within a greater group consciousness of all human beings.

Subconscious tendencies include some that are innate or purely genetic, but most have been developed at least in response to the past environment, and the opinions of and treatment by other human beings is an important component of that environment. Such inclinations continue to be affected by the environment, but the conscious is in full control at all times to overrule any and all subconscious tendencies, subject only to its strong motivation to serve humanity as a whole. Consciousness changes subconscious inclinations by having the system of coherent electrical firings by neurons that make up the conscious out-fire those that make up the system of neural activity that creates the subconscious tendencies. The conscious thereby also fully controls the filtering process that lets information, thoughts, and fantasies into consciousness. The conscious thus completely controls everything about one's brain, including emotions such as love, spite, and hate and desires such as hunger for food and sex. A natural human wish to surrender to some group consciousness can even be repressed, but the evolved human trait of only finding happy self-satisfaction through serving humanity as a whole can't be changed.

The conscious is an innate human program that evolved from a subconscious cause-and-effect analytical tool to help enable humanity as a whole to survive and reproduce. Unlike the minds of other organisms which regulate self-behavior in a process of linear decision-making to maximize the chances of satisfying needs and desires, humans learned to forecast their own actions in the future and to predict the corresponding effects those actions would have on their environment. Self-forecasting then resulted in an infinite feedback loop of cause and effect between the individual and the environment with an infinite number of outcomes that required much more complex multi-dimensional analysis. This process eventually became aware of itself as a component of its self-forecasting. The conscious mind is this self-awareness that became ever stronger as the environment was increasingly affected by the actions of humanity and as empathy combined with self-forecasting enhance the predictability of others and their own effects on the environment. The fact that humans increasingly changed their environment as their minds grew made self-forecasting and empathetically forecasting others ever more important for them with respect to evolutionary adaptation to that environment.

Empathy, which may have evolved in human beings at least partially to enhance their ability to forecast the actions of others by enabling prediction of what one might do oneself in the same perceived situation, enhanced their motivations and ability to cooperate. In addition, because of empathy, everyone naturally forms a perception of the feelings of others, and people's understanding and image of themselves follows from those perceptions of others' opinions. A group consciousness thereby results that creates ecstasy when there is unity of purpose in serving all.

As a result of empathy and various chemicals in each person's brain that evolved to promote human cooperation, each individual human originally wanted the happiness of all, not just of one's self. The evolution of these various unique human characteristics caused the original human consciousness to be a communal or group one as opposed to individual consciousness.

Group consciousness both facilitates and increases the importance of forecasting others. With group consciousness and empathy being the norm among humanity for millions of years, true human happiness requires individual striving for a mutual happiness of all, including oneself. This tendency promotes the cooperation that humanity needed to survive against overwhelming odds in a hostile environment that included faster and more powerful predators.

Self-forecasting evolved to enhance individual survivability and reproduction but only for the greater good of humanity as a whole. The evolved human brain system is designed to satisfy each person's current needs as well as to grow in preparation for better meeting the future needs of humanity that include the need to create and nurture offspring. While the root rational logic of the conscious can't be controlled by the environment or others, it is strongly motivated via pleasure chemicals in the brain that provide happiness to serve the goal of enhancing the survival and reproduction of all humans, including oneself. Informational inputs to the conscious mind about the environment and itself certainly can affect its deductions, and the logical conscious processor can rationally choose to surrender to subconscious tendencies to take control sometimes when it deems that optimal. For instance, a single goal that is consciously set can often be attained most effectively by uninhibited subconscious processing.

One of the means by which the conscious directs delegates control to the subconscious is via will power. A person's will can be maximized by a conscious intent that resembles a prediction with no conscious doubts or questioning. Without conscious interference, the subconscious is able to operate much more efficiently, and will power directs it toward a particular task or goal that is essentially predicted by oneself. Forecasting and will power are thus related, insofar as a prediction collapses the probability wave functions of current electromagnetic processes in a fashion to maximize the chance of the forecasted event happening in the future.

Even if the conscious mind tires of exercising its will, it remains in charge as long as it maintains its confidence that it is in charge and should be in charge. Doubting that fact automatically allocates control, by default, to the subconscious, which focuses on basic human needs and desires, such as for nourishment and sex. Giving up control to the subconscious without conscious direction enables easy manipulation by others and an inferior feeling of empty worthlessness. A lack of understanding of humans' evolved purpose in life can easily result in such an unsatisfyingly robotic existence.

Because the process of forecasting optimally requires determining the motives of the entity being forecast, it seems natural that humans would seek to figure out what motivates themselves. Humans' yearning to know the purpose of themselves and their lives may therefore follow directly from the self-forecasting process. The desire to know humans' purpose created the incentive for humans to determine or create a purpose for themselves. The optimal functioning of conscious activities had already created a need for better systems of categorizing things, and various religions or sets of "spiritual" assumptions about oneself and the environment developed that also led to implications about a purpose, which facilitated forecasting and decision-making.

Forecasting led to people trying to explain predictions to children. Such complex explanations need more than monkey-see-monkey-do communication for simple decision-making and tool usage. Humans therefore required the development of communication with words that symbolized complicated ideas about the future.

Some of the earliest human communication signals may have involved chanting or singing that enhanced group focus in a hypnotic effect and that concentrated the wills of all humans to affect the environment and the future for a particular purpose. As humans expanded their communication to educate children and others in complex concepts like forecasting, consciousness co-developed naturally out of the forecasting and communication process.

One possible trigger for self-forecasting and human consciousness was the perception that women's menstrual cycles are tied to the moon cycle. This connection may actually occur via electromagnetic or other waves emanating from the moon if the typical 28-day cycle for both isn't a mere coincidence. The menstruation cycles of a group of women living together are typically synchronized, perhaps resulting from evolution selecting human groups not exposed to the extreme danger of predators constantly being able to smell their blood, and group forecasting based on those common cycles that could have been at least partially motivated by empathy may have actually preceded individual predicting. Group consciousness would have even grown out of such unified self-forecasting that would have also been further motivated by empathy. At any rate, predicting for oneself, or for a group that includes oneself, enabled consciousness or self-awareness, or group self-awareness. As consciousness developed further, it required ever-stronger self-forecasting and cognitive abilities, especially because those very abilities increased the degree of control human groups had over their environment that therefore increasingly depended on future human actions.

The group consciousness caused people to see all other humans as equal and to want to act for a unified purpose of benefiting both oneself and other humans. However, each person acted individually in pursuit of that goal, thereby resulting in a very diverse, adaptable, creative, and unified group, especially since each person selected their own optimal balance of rational conscious decision-making and subconscious processing.

Nonetheless, less than 40,000 years ago, some individuals began to be perceived as superior, possibly due to some special knowledge, magic, or technology like selfishly motivated hypnosis that they kept secret. Instead of teaching others, as would be natural in a purely group consciousness, they exploited their powers to selfishly fulfill personal desires that were not being met by the existing society. Because of their unique creative and forecasting powers, women were likely the first such "superior" beings. Older, fatter ones may have led the way because they might have been the people most frustrated in realizing their desires due to the evolved attraction of males to younger, more slender women. Older women so motivated to return to a more revered situation might have grown wiser by experience. In any event, the "superbeings" began to receive special reverence or worship, and an artificial division into "superior" and "inferior" classes resulted that fragmented group empathy and consciousness. This unnatural process intensified individual consciousness and selfishness.

While the first worshipped entities or deities were likely female ones, male deities have predominated over the last few thousand years as a result of the conquest of matriarchal civilization by patriarchal barbarians. That conquest appears to have begun as the most recent Ice Age was ending about 10,000 years ago. These barbarians may have originally been exiled from coastal or fertile societies for selfish behavior detrimental to others, but they might also have left due to dissatisfaction with a second-class status of males in matriarchal cultures. Their descendants later invaded the lands occupied by matriarchies from desolate areas like central Russia and Middle Eastern deserts after the massive flooding and destruction of those "civilizations". That environmental catastrophe, which was caused by the glacier melting that continued for thousands of years that took place as the last Ice Age was ending, enabled conquest by the less culturally developed barbarians, who imposed patriarchal religions on the vanquished that persist throughout the world in various forms today.

Most of today's patriarchal religions exploit the human tendency to seek a belief system that provides them with greater chances of satisfying the human desire for love and a purpose in life, as is enhanced by sets of assumptions that enable better forecasting. These indoctrinating systems brainwash people into believing in the love of a fictitious deity, servitude to that deity as their purpose, and ridiculous set of related lies that facilitate exploitation by powerful elites. In this current artificial environment which encourages, promotes, and indoctrinates people with slavery to selfishness, people are motivated into servitude to "superiors" without question via a Pavlovian reward-punishment scheme fit for dogs. As a result, many human beings are unaware of their true human nature that gives the complete power their conscious has to totally control their own brains, and they merely react to the manipulated direction like robotic slaves. Through this process, they allow their perceived "superiors" to control them.

Contrary to the widespread lies and propaganda to the contrary, humans still have a desire to return to the group consciousness that innately maximizes group happiness and power. People are becoming increasingly aware that there is no such thing as a superior human being, and the fragmentation of communal consciousness may become ever more tenuous as people also recognize that belief in and servitude to a fabricated superior deity is even more absurd.

There is no longer any reason for unequal treatment of people, and attempts to manipulate people into believing otherwise for exploitative purposes can only lead to human spite. Such spite may have been one of the motivations for the replacement of a worldwide matriarchy with a global patriarchy a few thousand years ago. In particular, males had been mistreated during the unnatural matriarchal system of female worship.

Despite the recent development of selfish matriarchies, patriarchies, and any other inhumane system, however, human groups remain connected through mutual love, which creates happiness chemicals in the human brain. Mutual love requires human empathy that enables a shared consciousness and a unity of purpose that is ever changing for mutual happiness that maximizes satisfaction of individual human needs and desires. Romantic love between members of the opposite sex creates an especially ecstatic, hypnotic high and likely preceded the artificial worship of fictitious deities by millions of years. The reverence involved in romantic love borders on worship that prehistorically involved the male partner becoming obsessively desirous of a woman who remains naturally turned on by such attention. Mutual love and simultaneous orgasm innately drive both sexes toward pleasing each other in a harmonious union that represents a fractal form of the basic force of the universe consisting of electromagnetic attraction. However, the trusting surrender to each other necessary for such ecstasy is difficult today because of brainwashing by exploitative "superiors" that transforms humans into robotic zombies selfishly seeking to satisfy only their own needs and desires.

Free will exists for those who want it to, i.e., for those human beings who consciously choose not to let others control their actions and thoughts, and so control by the elite is inevitable only for those who aren't aware of, or don't want to exercise, the power of their free will. Similarly, life after death exists for those making the selection to have their psychic forces continue to affect matters even after the extinguishments of their physical life.

Human happiness comes from one's conscious self that is its own judge and jury. The conscious self has a great deal of latitude in dispensing its own rewards of happiness for deeds that are consciously perceived to have been positive ones, while unhappiness is levied for failure. Decisions to act consistently with evolved human nature can most easily bring the maximum happiness, as self-contentment appears to have evolved to be dependent on serving a higher goal for humanity as a whole.

While human happiness may be dependent on trying to satisfy the basic needs and desires of oneself and other humans, human nature has given each individual unlimited freedom in carrying out that task and even in trying to do so. Psychic powers may even be accessed that may enable one's self to be integrated across time to positively influence the past, the present, and the future in some parallel universes. The vibrations of each part of the creation from nothing affect others, including across time, and conscious focus can therefore influence and direct others as well as oneself, for maximum entertainment and bliss.

After the death of the human body, the conscious self is eliminated, except in other people's memories. While each individual's emissions of radiation and electro-magnetic waves may persist forever, as will the quantum entanglement of people with whom a person has come into contact in some fashion, those effects are fragmented representations of a whole self and also disjointed by changes over time. The electromagnetic waves, which constantly emanate from all people while they are alive and which are affected by conscious will, therefore do not represent a continuation of the whole conscious self or soul that expires at death.

Even though human beings themselves evolved to function for the good of the group as a whole, there may actually be some higher universal meaning to life and everything. An infinitely vibrating nothing represents the essence of everything and this vibration process represents a sort of inverted nihilism that creates infinite variety, beauty, and meaning, just as dividing 1 by 0 equals infinity. These insights remain undefined and incomprehensible to many, just as dividing a whole nothing by itself is. However, they create the opportunity for a truly carefree enthusiasm for life that permits a large degree of freedom in experiencing what one wishes in an environment which one not only partially controls but with which one is integrated.

The drive of living beings to be ever more aware of, and coherent with, their environment may be the purpose of life. In particular, evolution appears to be pushing everything to be more coherent with the characteristics of the divided pieces of nothing that make up the universe, including living beings. The purpose of human beings and everything that formed out of nothing constantly dividing itself may be to explore, experience, and know the resulting creation better in a harmonious unified but individualistic fashion that even allows individualized setting of purpose for humanity. While nothing in the aggregate lacks any control over anything in particular, it may be possible for life-forms coherent enough with the rest of the divided pieces of nothing to control future creation and nothing itself.

Nevertheless, it is important to recognize that human happiness in the universe is maximized by having goals that one has complete control over, such as merely trying as well as one can to achieve something without necessarily having success in doing so. The focus on doing the best one can is especially important with respect to the natural human desire to be at least as "good" as others. The innate human desire for equality invariably results in frustration in the unnatural hierarchal cultures of today. A natural society that is consistent with innate human nature has everyone equally trying to make everyone else as happy as possible because such a culture is the type in which humans evolved. However, attempting to replace the current sick hierarchical societies with such a healthy one is also likely to be very frustrating unless the goal is merely to try as best as one can.

Humans have evolved to feel most self-fulfilled when they seek to help humanity in their own particular way, but the innate altruism of human beings has little to do with reciprocity between individuals. Altruism merely requires mutuality based on unified purpose of action as a group that also seeks to help the group's individuals, who are primarily responsible for their own needs and wishes but whom the group seeks to help when optimal or necessary.

However, in today's cruel society based on exploitation, it may be more productive to focus more on trying to make oneself and one's loved ones as happy as possible. The golden rule of "love thy neighbor as thyself" that predates Christianity by thousands of years is applicable here, but only for "neighbors" who are human and therefore love those who love them. Those individuals can include the people of the future, as is especially rewarding psychologically in today's societies with only a limited number of actual human beings obeying their innate nature.

A Heaven in the mythological Christian "afterlife" doesn't exist for those following the golden rule. Instead, Christian believers, who artificially force themselves to love and servitude of others even when they are being exploited as part of their slavery to serve a fictitious god fabricated by cruel patriarchs to manipulate and control people, will suffer only eternal frustration and hell forever.

On the other hand, mutual love can bring Heaven on Earth to humanity because it's innate to human nature. Human nature is subconsciously driving to free human consciousness from the shackles of ignorance, and no amount of repression seems able to repress that drive. The vibrational frequencies of human nature and nature as a whole overwhelm those of artificially programmed ones.

The return of human nature is thus ensured, unless humanity is completely destroyed first as a result of a continuation of the hierarchical culture that enslaves people. If humanity is destroyed, it likely will be replaced by an even more self-aware form of life somewhere in the universe that can carry on humanity's purpose.

The world perceived by human beings may actually be a fractal macrocosm of the micro worlds of atoms, quantum particles, and light that derived from nothing. As an entangled part of that system, which also integrates time in similar but not identical cycles, humans could conceivably travel into the past, just as quantum particles of antimatter do. They may already access some of the holistic knowledge of light photons, and human consciousness, which seems to focus on the future, may permit an awareness of the virtually unlimited possibilities which exist at the more macro level of human perceptions.

While the universe has few boundaries, coherence with the nature of the universe and the human nature that evolved within it is likely to maximize human self-fulfillment. A harmonious playful balance between opposites which originated from nothing and which created everything seems to be the optimal means to human happiness that evolved to reward actions which resonately serve humanity as a whole.

Since people are subject to the evolved tendencies of human nature and all other things, it is important for true human happiness to place the individual beneath a higher goal, such as to serve the group of human beings united in purpose. Human nature also motivates people to pay more attention to the future than the present, thereby enabling the focus to be on helping humanity persist long-term. Living for the future involves consideration of both contemporary humans and future descendants.

Humans living in the future makes the present the past. Perceiving the present as the past enables seeing contemporary actions as past ones that have already happened. Such a belief system enables more efficient subconscious processing for the moment to enable attaining a future that one has consciously chosen in the past. Maximum will power is thereby effected to facilitate the ability to choose the future possible parallel universe one wishes to experience.

The future is likely to be even more fascinating than the present, as there is always something new in the environment continuously created out of nothing that integrates everything. The wholeness of time itself and the quantum nature of causality working in both temporal directions can even enable the future to affect the present. The latter possibility may help explain many otherwise mysterious psychic phenomena, as well as the innate spirituality and drive of human beings to serve a higher goal.

The incredible human abilities of today developed within the context of a prehistoric mutual hypnosis, which was a component of a group consciousness. That unification of wills enabled otherwise weak human beings to become the most powerful by far on the environment of earth. Putting the whole of the future of humanity above the self can lead to infinite possibilities and happiness (via 1/0). On the other hand, placing oneself first leads back to nothingness (via 0/1) just as placing some subgroup of humans artificially placed in hierarchal positions leads to little better (as in 666). Once one has found the purpose of serving the whole future of human beings, following one's human nature will pave a path toward meaning that is coherent with but different from the path of other human beings.

The conscious developed as a tool to serve this higher goal. While the conscious has a great deal of latitude in determining how best to advance humanity, setting goals or taking actions inconsistent with the higher purpose that has been programmed genetically by evolution into all human beings results in unpleasant feelings of unhappiness, emptiness, discontentment, and even guilt that discourage such behavior.

With the power created by inner knowledge, belief, and focused emotions, one can make oneself and the external environment coherent with what one wants. One can then reawaken human spirituality in a simple process of seeking harmonious human companions and sexual mates, thereby increasing the opportunities open to both oneself and them.

If one doesn't know what one is looking for, one won't find it. If one dosn't derive one's own power from one's own confidence and emotional focus built on knowledge of the characteristics of 1/0, one won't derive it from other individuals (0s/1). Instead, one will be controlled by others (0sx0/1=0) and find only what the others are looking for (0/1), or futilely search aimlessly with them in a sea of meaninglessness (0/1+0/1+0/1+…+0/1=0).

Any individual today can exert more powerful forces than all the past humans in existence. The more such a 1/0 is coherent with other 1/0s from the present and future, the more powerful, as long as one doesn't make the mistake of trying to be at all coherent with the brainwashed 0s/1.

While an infinite set of possibilities is allowed by the nature of divided nothing as illustrated by the laws of quantum mechanics, the chance of anything based on pure selfishness approximates zero. As a result, the mere phenomenon of the development of the Christian religion having been motivated by a selfish patriarchal elite seeking to manipulate others into servitude to them makes its ultimate elimination as a belief system inevitable. The fact that christianity and other patriarchal religions are based on lies that violate the laws of quantum physics makes it ultimately impotent in the face of the infinite power of human nature. Despite attempts to make patriarchal religions like Christianity more compatible with the quantum laws of divided nothing, they have no chance against belief systems based on the facts of quantum mechanics that are devoid of selfish manipulative lies. Divided nothing in its essence consists of positrons and electrons that are selflessly driven inevitably toward their opposites, as is reflected in fractal macroscopic form with human love driving human beings toward mates of the opposite sex, and that force will overcome all else.

Even without the existence of divisions of nothing and the resulting creation of the dimensions of time and space in this universe, there are and always were only the infinite possibilities in nothing (1/0) that included the chance of such divisions leading to the evolution of human conscious emotions. The mere potential of that very evolution enabled the collapsing of that possibility into its occurrence. The "magic" of the creation of a universe which spawned the evolution of human beings was therefore inevitable within the context of the infinite possibilities in nothing and its division into parts of nothing that include the evolution of consciousness. No matter how improbable, that evolution is proof of the infinite possibilities that exist, thereby facilitating the seemingly impossible with mere recognition of that fact.

Within that context of human beings having willed humanity into existence, it is possible to better understand human nature, including both its power and its constraints. While the power of humans in the aggregate is virtually limitless, that power is dependent on the level of resonance among human wills of both the present and future. Merely trying to make one's will coherent with that of human beings of the future can create enormous happiness as well as enable harnessing their incredible power.

All human beings have an innate desire and capacity for feeling unified with humanity and the universe as a whole. The human urge for love and sex are just components of that drive. We are all intrinsically connected via electromagnetic fields, light photons, and quantum entanglement, and people naturally feel spite for anyone or anything that disrupts human harmony. Even the individual person experiences unhappiness when actions are taken that are incoherent with the good of humanity as a whole.

Humanity and human nature itself continue to evolve to permit ever greater awareness and responsive control of the environment in which human beings exist. However, putting the interest of the group of human beings above the individual is the evolved tendency of humans that innately creates happiness. It's not an externally imposed moral code but an innate internal one that can't be permanently suppressed any more than the drive for sexual intercourse. Living in harmony with such natural human tendencies creates ecstasy, which is the natural state of human beings.

On the other hand, unhappiness is the typical experience for selfish robotic monsters. While these machines can reprogram themselves back into human beings, they currently are not human, and it is an insult to humanity to consider their selfish interests in any way. Despite the innate drive to serve humanity's long-term interest the best way that one can, each human being is dependent on no one, not even oneself.

With the knowledge that one evolved to serve humanity as a whole, one can always act according to the simple guideline of having that innate purpose be primary, thereby enabling efficient focus in a manner most coherent with human nature. In addition, if one is ever unsure how to act or react in a given situation, one can ask oneself what a true human being would do. One can thereby access all of the tremendous power of human nature, resonating powerfully with the human nature of other people according to the laws of that universe. Carrying out the quest of serving humanity by following the path of human nature enables one to maximize the bliss of experiencing all the infinite wonders of the universe created by vibrations of nothing that one thereby co-vibrates harmoniously.

Humanity itself appears to have evolved to make itself the superspecies of the universe, or at least one of the super-species that maximizes the self-controlling, self-awareness of coherent systems of divided pieces of nothing. The characteristics of human beings seem to have evolved to create a self-satisfying and self-fulfilling heavenly paradise for those who coherently serve that goal of humanity in harmony with the characteristics of divided nothingness. The people of the future will resonate more with each other and humanity as a whole, thereby enabling them greater will power that can also influence the present. The more a person seeks to harmonize with that collective coherence, such as by trying to act like an imagined human being of the future would in the service of humanity, the happier and more powerful the individual will be. In fact, as long as one resonates with the people of the future, any possible future one chooses will be experienced.

Understanding that one is a part of the collective human deity, it is possible to comprehend better the purpose of everything. For instance, within the context of the reality of causality working in both directions in time, group consciousness can be understood as each person mutually willed into existence by those who mutually love the individual, including one's own future potential offspring. Those individuals willed into existence had to be different in order to avoid the blandness and impotency of pure self-love that leads to empty nothingness. Through that group love or consciousness, a joint determination of reality occurs that creates infinite beauty because of the uniqueness of each human being.

The purpose of each human being can therefore be deduced to be to love and be loved in a mutually selfless but individualized fashion that resonates the human wills with the environment across space and time. Carrying out that purpose requires resonating with the loving wills of the future human beings who do likewise in an individualized fashion. This harmonious process, which involves the collapse of the wave functions of the past to enable control of the future, necessitated the evolution of forecasting that differentiates humanity from all else in the world. The power of prediction also enables each individual human being to select the particular set of temporally connected universes which they wish to experience, thereby allowing maximum individualized free choice in a diversely common coherence. They can even choose their own individual interpretation of the purpose of humanity as a whole and then live in their own separate set of parallel universes to act for that purpose. Establishing a higher purpose of loving and living according to one's own unique human nature while harmoniously serving the wave functions of humanity as a whole is all that is necessary for attaining Heaven Now and Paradise Forever.

No individual human, no matter how directionless and empty of purpose, can collapse the whole back into bland nothingness because of the wave functions of other humans' wills seeking individualistic realities to create the infinite diversity that exists and that are given more coherent but diverse structure by willing harmonious resonance between the personal human wills. In addition, no matter how destructive and hellishly empty a person's pure slavery to selfish desires can drive a person to be, and no matter how personally unsatisfying strict selflessness an individual may be driven pointlessly into toward the hell of nihilism and robotic control by the environment and the selfish wills of others, the collective waves of past, present, and future human wills ensure sufficient balance of selfishness and selflessness to cause an eternal infinity of diversity that continued divisions of nothing enable. The diversity of the many different wave functions of human wills thus make it possible to avoid the emptiness of nothingness and permit the ecstasy that human life can be reached through enlightened resonance with the infinite combination of potential created by those wills.

Merely acting according to one's human nature may be all that is required in order to allow a blissful existence regardless of circumstances. Doing so only requires making the overriding goal for all actions to be the purpose for which human consciousness evolved, i.e., to serve humanity as an equal but different member of that species. Establishing this purpose in life isn't an external moral obligation in any sense, but it is instead an internal mental code to which all human beings are driven through the evolved desire for feelings of self-fulfillment.

Acting as one thinks a human being of the future can help in carrying out this task. However, it is important to remember that future people, while more powerful and knowledgable than present ones on average, are merely systems of divided nothing that are far from omnipotent. Their very existence depends on existing people both creating them and developing the knowledge upon which they can build. The power of future people to even will backwards to the present also requires a resonance with the wills of human beings today. Both contemporary and future humans can maximize the potency of their wills through a cooperative intent that harmonizes the actions of all for the benefit of everyone.

Putting the whole (1) of diverse pieces of nothing (0) above individual combinations of split nothingness (0+0+0=0) is a naturally evolved inclination in human beings because they developed into the most powerful system of divided pieces of nothing on earth. An infinity of possibilities in divisions of nothing by itself (0/0) that often enough result in new creations in the form of two light photons out of two divided pieces of nothing (0/0+0/0=2x0/0) make free will possible, including the freedom to place an individual's own self-interest above that of the whole of humanity (0/1). However, individual selection of the particular temporally connected set of *infinite* ***potential parallel the individual wishes to experience is only feasible through resonance with the whole (1/0), as the individual will is otherwise swamped by the wave functions of all other human wills. In addition, human happiness and self-fulfillment have evolved to be dependent on such coherence, providing further motivation for living in harmony with the waves of possibilities existing in other human beings acting according to their innate human nature. Because of the infinite power of placing the individual below the whole (1/0), evolution caused human beings as the most powerful form of divided pieces of nothing to become ever more aware of and resonate with the wonderful reality of the interconnected split nothingness that is the basic essence of everything. Such resonance, which exists in mutual love in its highest form, creates infinite power, as well as ecstatic happiness, seems to be the meaning of life, the universe, and everything (1/0).***

Part I

Beyond 666

This Part I surveys research and evidence on the existence and characteristics of the matriarchies that permeated human societies for so long prior to six thousand years ago. In the process, new insights and hypotheses on unresolved issues are logically derived.

By providing a review and integrative analysis of the literature on the nature of prehistorical and subsequent human societies, the current state of the world can be understood better. In contrast, the contemporary perspective on ancient history that is marred by patriarchal propaganda and lies is employed by fundamentalist religions to inhibit investigation of facts relating to intrinsic human nature.

An unbiased view of the world today requires some knowledge of the distant past in order to understand and overcome the patriarchal brainwashing with which virtually all humans today have been indoctrinated. It may be difficult for many to overcome their infantile subconscious misconceptions and sexist biases to recognize the likely reality that human societies have, through tens of thousand of years of prehistory, been ruled by women. However, no other known hypothesis fits the aggregated anthropological and other facts even remotely as well. While each piece of evidence alone is insufficient to conclude that matriarchies were the norm until 6000 years ago, the combined collection of facts is rather overwhelming. The last few thousand years of cruel patriarchal tyranny and destruction have been unable to eliminate the evidence that include innate human tendencies.

Chapter 1

Evidence on the Existence and Nature of Matriarchies in the Distant Past

{reprinted with minor changes from Austin Murphy's published article "History Without the Standard Patriarchal Bias" that appeared in 2008 on pages 75-84 of Volume 3 of the *Journal of Interdisciplinary Social Sciences*}

Human societies were ruled by women before they were dominated by men. These matriarchal civilizations were overthrown by patriarchal barbarians a few thousand years ago, but the evidence of their prior existence is overwhelming. The evidence also indicates that they were likely cooperative and peaceful, and, despite today's rule by violent patriarchs, human nature continues to be affected by that heritage. Unlike the remainder of the book, this first chapter is largely reprinted from a published article (Murphy, 2008) that may be a prerequisite to understanding the meaning of life, the universe, and everything.

A. Evidence on the Existence of Matriarchies

There is substantial evidence that humans lived in a matriarchal society for thousands of years (Baring and Cashford, 1993). The evidence includes archaeological findings indicating special prehistoric reverence for women and the predominant existence of goddess worship and female priestesses in prehistoric times when religion was even more integrated into the social framework. Also supportive of the matriarchy theory of prehistory is the extensive evidence of an overthrow of matriarchal civilization. The evidence includes not only archaeological artifacts but also human traits which exist today that seem to only be consistent with greater female power in the past. Widespread stories or myths in patriarchal religions of a prehistoric replacement of the original rule by females, as well as written records explicitly indicating ever-decreasing women's rights contribute further proof. The existence of matriarchal forms of marriage among common people even after the patriarchal conquest would be especially difficult to explain if there hadn't earlier been earlier rule by women. Evolutionary evidence, which show women to be smarter, and which imply female choice from the size of the human male reproductive organ, among other things, indicates matriarchies extended into the far distant past.

Prehistoric Goddess Worship and Female Rule

Some of the evidence on the prehistoric existence of matriarchies is provided by the widespread existence of ancient symbols of fertility cults, goddess worship, and honoring femininity (Butler, 2004). Goddess worship is well documented by Gimbutas (1989) and Eisler (1988) for Europe and the Middle East and by Jiayin (1995) for China, as well as to at least some extent by Hancock (2002) for other parts of Asia and by Malmstoem (1997) and Henderson (1997) for the Latin American Indians. Humans tend to create deities and myths to reflect their societies and beliefs, and the process is self-reinforcing, with people tending to follow the lessons of their religious stories and even act like their deities (Sjoo and Mor, 1987).

Given that ancient societies integrated religion with the social and political system as well as with knowledge or science much more than they do today (Howells, 1990), the prehistoric worship of a goddess provides significant support alone for the hypothesis that prehistoric human societies were characterized by matriarchies. The exclusion of uncastrated males from representing goddesses even in more recent times, such as in the periods of patriarchal Greek and Roman domination (Baring and Cashford, 1993) represents more evidence that the prehistoric times of goddess worship were largely characterized by females, such as priestesses, ruling society.

While there is evidence of gods as well as goddesses tens of thousands of years ago, many of the apparent male deities may have represented mere male consorts for the goddesses, as goddess worship and a matriarchy seemed to predominate (Gooch, 2006). About 90% of all human sculptures found to have been created between 7000 and 32,000 years ago are female (Cruttenden, 2006), despite widespread evidence of destruction of goddess symbols by later patriarchal conquerors (Stone, 1976).

Widespread surviving evidence of vulva drawings and sculptures for tens of thousands of years in the past is also consistent with the existence of a matriarchy in prehistoric times and may even reflect a worship of women's bodies (Gadon, 1989). Such art could represent a reverence for where humans began independent life and where most men return (Baring and Cashford, 1993). Although the images might also be construed to be mere prehistoric pornography (Goodison and Morris, 1999), the discovery from the ancient past of so many fat female figurines (Gimbutas, 1989) that are sexually unattractive to males (due to a big belly signaling motherhood and hence no potential for impregnation) implies such ancient art more likely represents reverence for motherhood. It's also possible, however, that females may have manipulated males into accepting a mystification of women's sexual and reproductive processes, possibly even through "magic" like hypnosis that originated far back in time, and at least the vulva symbols may have reflected, at some point, female body worship or other fetishes indicative of an unbalanced society under matriarchal rule.

The procreative and breast-nurturing powers of women would have likely contributed to ancient humans honoring them for their life-giving "magic" and looking to them for leadership if not outright worship, where the predominance of female deities would merely reflect that fact of life (Sjoo and Mor, 1987). The prehistoric tracing of every person's birth through the females, is natural when the true identity of the father is never known for sure, would have added to the power of and loyalty to women (Eisler, 1988). The potentially realistic analogy of the universe being a Mother Nature that doesn't create anything separate from itself but delivers parts of its whole is the basis for most ancient religions and mystery rites (Baring and Cashflord, 1993) and probably also contributed to a worship of a goddess and consent for a matriarchy.

Some archaeological evidence of women being buried with more possessions in seemingly more sacred places about 10,000 years ago (Baring and Cashford, 1993) is also consistent with a matriarchy. The rituals apparently reversed a few thousand years ago to provide substantially more possessions for at least some males (Stone, 1976).

Widespread Evidence Indicating Overthrow of Matriarchal Civilization

There is extensive archaeological evidence that patriarchal barbarians from desolate areas of the world like central Russia succeeded via relatively recent warfare in destroying more advanced, civilized, egalitarian, and peaceful matriarchies over a period of a few thousand years starting around 6000 years ago in Europe (Gimbutas, 1977) and the Middle East (Baring and Cashford, 1993). The archaeological findings are supported by ancient myths in many patriarchal religions throughout the world indicating an original matriarchal rule by a goddess who permeated everything but who was later replaced/subjugated by male god(s), often through violence (Husain, 2003).

The newest evidence that horse domestication for transport and milking began over 5500 years ago in the steppes of Kazakhstan (*NewScientist,* 2009a) supports these findings insofar as horses would have been extremely useful for the barbarian conquerors for both mobility and battle attacks. Although sheep were domesticated about 11,000 years ago and cattle subsequently, domestication of horses started only 5500 years ago, being initially concentrated in what is now Russia, the Ukraine, and Romania and providing "unmatched military might, speed and mobility" (Coghlan and Callaway, 2009).

Barbaric warlike male-dominated societies are also alleged to have been imposed in other areas of the world in more recent times. For instance, about 1500 years ago, barbarian warlike people appear to have conquered highly civilized but purely egalitarian societies previously living in peace in South America (Polich, 2001), even though vestiges of matriarchy remained in many tribes in North America into the modern era (Underwood, 1965).

Evidence on a transformation of prehistoric society in Australia doesn't seem to exist, and that continent may very well have been originally settled by patriarchal barbarians. It's possible that the desolate areas like those of Australia, as well as central Russia and the Middle East, were the destinations for those exiled from matriarchal civilization, and those refugees spitefully developed patriarchal societies that later conquered and replaced the matriarchies through most of the world.

On the other hand, much of Asia has been documented to have transformed itself somewhat more peacefully than Europe and the Middle East from matriarchy to patriarchy about 5000 years ago (Jiayin, 1995). Matriarchies may have persisted even longer in Africa (Jackson, 1972). The less violent overthrow of the matriarchies in Asia and parts of Africa may have caused those cultures to maintain more of the collectivist aspects of the prehistoric matriarchies than the more individualistic peoples of Europe and the Middle East that were conquered by violence.

In any event, widespread written records indicating changes in laws and memories of prior times when women had more power provide additional evidence of the matriarchies being overthrown around that time (Stone, 1976). The declining appearance of goddess figures after 6000 years ago that accompanied a decline in cultural achievements in general with the increasing sign of barbarian conquest is also very indicative of patriarchal conquest and destruction (Gimbutas, 1977). The change a few thousand years ago to treating some males with much more reverence in gravesites is also consistent with the replacement of an egalitarian matriarchal culture with a hierarchical patriarchal one (Stone, 1976).

During the replacement of matriarchies with patriarchies, female goddesses were substituted over time with male ones that culminated in one dominating patriarchal god in the Middle East a few thousand years ago (Sjoo and Mor, 1987). While goddesses were still allowed in some conquered matriarchal societies in the Middle East, such as in Sumer 5000 years ago (Baring and Cashflord, 1993), they were eventually replaced by more patriarchal religious systems that recorded the defeat of the resisting matriarchal peoples (Sjoo and Mor, 1987).

Although the rewriting of history and myths was the norm for the patriarchal conquerors (Malkowski, 2006), some element of truth was likely to remain despite the distortions (Graves, 1960). With much of civilization having been destroyed as opposed to being adopted by the barbarians, setting human culture back thousands of years (Eisler, 1988), later patriarchal rulers were not aware of the true history of civilization that may have gone back as much as 70,000 years (Malkowski, 2006), as anything relating to the matriarchies that were so hated by the patriarchal conquerors was destroyed (Stone, 1976).

Continuation of Matriarchal Marriages Even Within Patriarchies

Even after the matriarchies were replaced, widespread special honoring of women and goddesses continued (Walker, 2000). The persistence of the reverence provides additional evidence of how deeply entrenched matriarchal cultural has been implanted into human nature by long periods of ancient societies being led by women. For instance, writings by actual historians from patriarchal Greece from less than three thousand years ago indicate that, in Egypt, which was widely recognized as the most advanced civilization at the time, matriarchal customs continued so strongly among the common people that Egyptian marriages stipulated "the man shall obey the woman in all things" (Jackson, 1972).[1] The Germans who stopped the Romans from conquering the world (Murdoch, 2006) still regarded their women as goddesses two thousand years ago (Davidson, 1990).

Marriage rites that were prevalent in rural Europe until the patriarchal Christian Church intervened actually required the male to promise to "worship" his future wife forever while the women merely had to agree to share her future husband's bed (Walker, 2000). Such unequal marital pledges provide some indication of the relative reverence for and power of women until "modern" patriarchal times. Millennia of patriarchal burning of books and other forms of censorship, as well as the patriarchal persecution and mass murder of millions of women and male sympathizers, have been unable to eliminate the honoring of women, providing an indication of how long-term and deeply rooted a matriarchy really is in human nature (Stone, 1976).

Forms of matriarchies persisted in some societies into more modern times. For instance, in some Native American tribes such as the Iroquois and Hopi, the women owned the property and ran the households, and the women even chose and censored male figurehead chiefs (Underhill, 1965). Aboriginal peoples with similar matriarchal systems existed in China until recently that had the male partners sleep in the homes of the female and required the man to obey the rules of woman's clan while there (Jiayin, 1995).

Evidence of Matriarchy in the Very Distant Past

There are various different types of evolutionary, genetic, and physical evidence that, prior to the violent patriarchal conquest a few thousand years ago, females exerted a great deal of power in human societies for a very long period, perhaps spanning hundreds of thousands of years (Sjoo and Mor, 1987). For instance, female choice resulted in male faces being selected that are handsomer and look less violent or dominant (Coghlan, 2004), and such preferred characteristics would not have been allowed to evolve in an alpha male hierarchy of dominating males.

More powerful evidence of female choice in mate selection exists in the size of the male human penises. The fact that they are so large compared to humans' closest ape relatives clearly implies free female choice (Miller, 2001). In particular, male dominated societies result in tiny penises like those of alpha male apes, which females have little power to refuse (Etcoff, 1999), and whose size does not contribute directly to enhanced fertilization (*NewScientist,* 2007a). Thus, today's human male anatomy is inconsistent with an extremely long period of male domination that would have prohibited female choice in mates.

Alpha male societies can permit some female selection in terms of seeking the strongest patriarch who can protect them and their offspring in a harem (*NewScientist,* 2006m). However, only in a society without the need for alpha male protection, such as in an egalitarian or communal matriarchal one where male violence is not exercised to control women's reproductive activities, do females have the luxury of choosing males based on the size of their penises.

In the rare species where rape sometimes occurs, the female genitals have evolved adaptations that inhibit the success of that strategy (*NewScientist*, 2007f). For instance, with female water strider bugs have developed genital shields that they raise only after a 15-minute tap dance courtship (*NewScientist*, 2009b). The fact that such adaptations to prohibit reproduction by force don't exist in human females implies that human males rarely if ever committed rape in prehistory.

The female ancestors of humans had evolved by several million years ago to be approximately the same size as the males (Jurmain, Kilgore, Trevathan, and Nelson, 2004), and such a situation is uncharacteristic of an alpha male society (Stringer and

Andrews, 2005). Male dominating systems are optimized by overwhelmingly dominant male strength. With approximate equal strength between the sexes in human ancestors, women were most likely treated as at least equals.

Perhaps most important is the role that female choice can play in selecting mates who are able to adapt to varied environments (West-Eberhard, 2003), as the evidence indicates human females have (Miller, 2001). It actually would have been nearly impossible for humans to have evolved from ancestral apes without judicious female selection, and humans likely would have remained ignorant apes if females weren't given the right to choice. In fact, human intelligence itself could not have evolved in a society of brutal alpha male apes where the strongest ruled and chose the prettiest servile sex objects/slaves.

The fact that a component of the cortex thinking area of the human brain is inherited strictly through the female, whereas only portions of the pre-human midbrain section is inherited solely through the male (Rose, 2005), provides some evidence that the intelligence of humans originated in women. While the development of traits originating in one sex that naturally spread to the other are usually difficult to trace (West-Eberhard, 2003), the genetic evidence implies that intelligence originated in women. Greater thinking power would have further enhanced women's ability to rule in prehistoric times.

Female children tend to do better in school than males in Western countries like the U.S., at least until puberty (Tyre, 2005). Since female teachers, who outnumber male ones, have not been found to be biased in favor of students of their own sex (Kramarz, Machin, and Quazad, 2008), the evidence is indicative of females being smarter than males on average (Tyre, 2005). The decision-making and problem-solving area of the brain (the frontal lobes) continues to be larger for women than men on average, as is the short-term memory section (the hippocampus) of women's brains, while male brains actually have larger brain areas only for non-analytical processes like sensory perceptions and emotions (Hoag, 2008). These facts provide confirming evidence of superior females intelligence on average, and the difference may have been even larger before the patriarchal conquest.

As indicated by the overwhelming destruction of the matriarchal civilizations the they conquered (Gimbutas, 1977), the patriarchs clearly preferred servility over intelligence in the vanquished women and, through both sexual selection and outright murder, likely advanced the relative number of dumber females who were more docile. For instance, patriarchal religious leaders mass murdered millions of the most intelligent women during the Inquisition in the Middle Ages in Europe (Walker, 1988), Since traits that are life-threatening tend to be de-selected in only one sex of a species, as with feather plumage and bright colors that exist in only male birds because such characteristics would endanger an immobile mother nurturing her eggs and offspring (West-Eberhardt, 2003), it is probable that women's advantage in intelligence was even greater than before the mass purges of the patriarchs over the last six thousand years,

Within this context, it's interesting that one Oriental religion called Tantric Buddhism, which seeks enlightenment through sexual ecstasy, actually regards wisdom as the female side of human nature at the same time that active love and compassion are considered male attributes (Capra, 1999). Since the development of the very important human tendency to cooperate actually seems to also be chemically related to women and their nurturing (Begley, 2007b), which were essential for the survival of humanity, it is likely that the Tantric view derives from the sexual fact that greater love and compassion is generally required of males to ensure mutual orgasms and ecstasy. Such a precondition for the useful bonding effect of a mutual orgasm would seem to only arise if females had the power to select their sexual mates, and they developed greater intelligence as a result of that right to choice that was vital in providing for care during pregnancy and offspring nurturing periods. In any event, all this evidence supports Sjoo and Mor's (1987) assertion that females were largely responsible for the earliest development of human intelligence, culture, and technology/inventions, as would be consistent with matriarchies existing in prehistoric times.

B. Characteristics of the Prehistoric Matriarchies

The rule by women prior to 6000 years ago seems to have been of a fairly egalitarian sort, and nonviolent forms of persuasion apparently were the norm (Eisler, 1988). For example, the widespread existence of communal graves without any evidence of special objects like tools and pottery being given to any individuals until about six thousand years ago would seem to indicate communism and collective ownership of material possessions without any special materialist treatment for any individual (Goodison and Morris, 1998). While many scientists continue to believe in the selfish gene theory that inhibits cooperative or communal systems from developing, models have been developed to explain the increasing evidence of evolutionary pressures that exist for organisms to develop for the good of a group, species, or entire ecological system (Holmes, 2009). As will be explained in Part II, the human traits of empathy, feeling good when cooperating, exhibiting shame when not cooperating, and spite all apparently evolved to help humanity survive an overwhelming environment, and these traits contribute to human cultures being naturally communal. Since there is evidence that cooperation will win out over selfishness in evolutionary simulations (Holderness, 2007), it would be difficult to believe humans could have evolved otherwise.

In addition to humans being innately cooperative organisms, prehistoric paintings on caves showing no evidence of war, rape, or human violence against humans supports the hypothesis that violent behavior was neither the norm nor revered, thus making wars during the last few millennia of patriarchal domination the exception (Eisler, 1995). The use of social pressure and exile as opposed to violence to maintain group cooperation that is normal for Native American societies, many of which have matriarchal tendencies (Underwood, 1965), was probably typical until patriarchal cultures based on obedience to an imposed hierarchy were imposed. The paucity of fortifications and other evidence of warfare prior to 6000 years ago provide further evidence that humans have historically been peaceful (Baring and Cashford, 1993).

While there is some evidence of battle-related injuries occurring 10,000 years ago, including an arrow in the back of a female buried with her baby and a healed skull that appears to have been hit with an axe, any fighting that caused such injuries may have been spurred by competition for scarcer resources in the environment of flooding catastrophes that are associated with glacier melting after the end of the most recent ice age (Spinney, 2008). Such evidence could also stem from accidental injuries, as in hunting accidents, especially since the evidence of any violent combat before six thousand years ago is sparse and provides little indication of actual battles or wars. In addition, the fact that many aboriginal cultures observed in recent times have been characterized by some violent tendencies doesn't imply violence is intrinsic to human nature, since much of such violence might have been spurred by contact with violent Europeans (Mithen, 2006). Moreover, since most "modern" aboriginal societies have patriarchal characteristics that may be unlike prehistoric human cultures and that may unnaturally spur violence.

The finding that women's greatest fear is being killed by men while men's deepest fear from women is to be laughed at (Shlain, 2003) is consistent with a hypothesis that the matriarchies enforced cooperation or female rule in a nonviolent manner. Ostracization may have been the typical natural punishment for selfishness or violence. The ancient matriarchal belief in the unity of all things that contrasts so strongly with the patriarchal worldview, which fragments all aspects of life and promotes divisive judgments, feelings of superiority, and conflicts, would have created an environment that discouraged violence (Baring and Cashford, 1993).

The patriarchal observation of men in ancient matriarchal systems indicating that "children are their oppressors" (Stone, 1976) supports this hypothesis. In particular, submission to children implies a voluntary cooperation for the good of others, society, and humanity that could not have been maintained by force, since children cannot physically dominate male adults nor control them sexually. The theory of non-violent cooperation is also consistent with the typical feminine traits that exist today and that assumedly would have characterized matriarchies.

Prehistoric art indicates an exclusive reverence for creation, nature, and pleasure, as opposed to the destruction, killing, and pain that is today considered "divine" (Eisler, 1995). A contemporary matriarchal culture, that of the Hopi Indians, exhibits similar signs of a combination of peacefulness and reverence for nature, creation, and sex (Underhill, 1965).

Engels (1972) long ago documented the fact that humans have historically lived worldwide in matrilineal societies which were not only egalitarian and communal but also permitted free love and sex except between known relatives. Studies of ancient matrilineal societies that persisted into the modern era of China have confirmed the evidence of free-love communes. There, matriarchal societies freely permitted short-term and long-term sexual partners, as well as easy partner switching and complete freedom of choice for both sexes (Jiayin, 1995). Because of communal consumption, economic equality, and jobs in separate clans, the sexual relationships were not artificially influenced by economic factors. There were sometimes exchanges of trinkets reflection mutual reverence among partners in long-term relationships, but there was never any payment for love or sex in those societies. Since the co-habitation among the sexes occurred in a home of the woman's clan, the male was a guest and would have been subjected to the customs and social order there, thereby making it a matriarchy, albeit without a binding hierarchical relationship.

The fact that "rural and peripheral areas of Europe removed from the turbulence of European history" continued to exhibit more peaceful, non-patriarchal systems and beliefs even into more recent historical times (Walker, 1988) provides additional evidence on the issue. The widespread belief that humans are innately selfish and violent therefore seems to be a product of patriarchal propaganda.

C. Relevant Evidence on Chimps

There is even more basic evidence that patriarchies, capitalism, and warfare among humans may very well be very foreign to human nature. In particular, a species called bonobos that resemble small chimpanzees are "our closet living relative" (Graham-Rowe, 2005), with both genetic studies and communication abilities indicating even a closer relationship than with the larger chimpanzees (*NewScientist*, 2007d).[2] Evidence on characteristics that humans share with bonobos can provide some indication of human characteristics that are more intrinsic as opposed to culturally influenced.

Wrangham and Peterson (1996) have described some of the characteristics of bonobo culture. While larger chimpanzees have many similarities to today's human culture in terms of being patriarchal and promoting both domination of females and violence, bonobos in contrast are very peaceful creatures both intra- and intergroup. Bonobo societies also tend to be matriarchies where females dominate. Sex is an integral part of their life and culture, with genital stimulation being virtually constant (Savage, 2010).

Like human females, bonobo females do not display the times when they can be sexually impregnated and can have sex at any time. As a result, it would be necessary for a male to dominate them continuously to ensure fatherhood, making it more difficult for genes of an alpha male to be rewarded in an evolutionary sense with wider propagation. The development of a matriarchy was thereby facilitated in both humans and bonobos.

Bonobo females bond together in groups that permit defense against testosterone-induced male aggression, domination, or rape. Despite the current patriarchal marriages among humans, human females today also demonstrate such tendencies to bond with the same sex, as shown by Spinney (2004a) even in a capitalist setting. Such bonding tendencies would have been more necessary and natural historically when ancient humans or hominids on the ground had to remain in groups to defend against predators. In contrast, chimps living in trees remain freer from the risks of predators and can therefore afford the risk of traveling in groups of a few or even of just one that inhibits defensive female bonding and promotes male aggression.

The tendency for women living in a group to have their menses biologically synchronized would also promote female bonding and alliances that would help inhibit any dominator alpha male system (Shlain, 2003). Since the difference in size and strength between males and females has actually been less in ancient humans in the last two million years than in bonobos who can prevent male domination anyway (Wrangham and Peterson, 1996), it seems very unlikely that humanity was characterized by cultures of female submission prior to six thousand years ago. Bonobos actually have about the same relative sex size ratio as for the larger chimps existing in alpha male societies (Stringer and Andrews, 2005), but other characteristics of bonobos that are similar to humans apparently enable them to impose matriarchal cultures, and the relatively larger human females armed would seem quite capable of doing the same at some point(s) in the past, especially if they were armed with the weapons that apparently did develop about two million years ago.

Humans may have started evolving separately from alpha male types only 5+ million years ago (Holmes, 2006). However, the multi-million-year existence of weapons such as rocks sharpened on two sides (Miller, 2001) could easily have been used by female human ancestors to defend against male aggression and domination, especially within the context of cohesive female groups. Thus, human cultures probably evolved more like those of the bonobo than the chimp in terms of being more matriarchal than patrriarchal.

Like humans, bonobos also exist on a wider variety of foods and live in larger societies than larger chimpanzees (Stringer and Andrews, 2005). Humans are therefore more like bonobos in those respects as well. Such a lifestyle would seem to require cooperation that is more easily achievable and maintainable through the sexual bonding and peaceful behavior of bonobos than through violent male-alpha patriarchal systems. Peaceful cooperation therefore seems to be the more evolved form of behavior that is more compatible with the larger groups in which humans evolved to live. The inferior communication abilities of the more violent and hierarchical chimpanzees compared to bonobos (*NewScientist*, 2007e) may be understandable within this context, thereby providing more evidence of humans having evolved to be more peaceful. Bonobos, which can actually be trained to understand human speech (Burling, 2005), share and work together much more effectively than chimpanzees (*NewScientist*, 2007d), and the expanded language abilities of humans would likely have reflected, and be reflected in, cooperative tendencies that were even greater than in bonobos.

Kaplan (2006) has reported that the meat-eating bonobos, which generally permit females first choice of meat, evolved in an environment with plenty of non-meat food. Bonobos haven't been observed to use tools in the wild because they don't need them in a land of plenty, but the author indicates that they are capable of easily learning to use them in captivity. That research further reported that even a more violent form of primate, the baboon, develops a pacifist society when there is adequate food and when aggressive role models are removed.

The capacity of both bonobos and humans in prehistory to have peaceful societies would have been enhanced by their ability to have sex at any time, since it would have served to mitigate frustration and aggression and thereby enabled useful cooperation to enhance survival of the species. Frequent sex has been found to provide extensive health benefits to both the human mind and body (Farnham, 2003), providing further advantages for which evolution selected. Bonobos themselves have been found to engage in free love and sex to reduce stress (Eisler, 1995), and there is even evidence that sex originated as a means to relieve tension while providing at the same time for the production of offspring with new gene combinations to adapt to a stressful environment (*NewScientist,* 2005c). Eisler (1995) has hypothesized that violent sex and violence itself would disappear in a return to a culture of cooperation and pleasure as opposed to the current patriarchal one that emphasizes domination and pain.

Bonobos engage in sex about every hour and a half (Parvin, 2006). Although this frequency is currently much higher than for most humans in our society, it certainly is not beyond the abilities and wishes of many humans today. People having experienced group living and a free love that includes stable relationships as well as flings and partner changes find human nature still best attuned to the bonobos' free love style (Newitz, 2006). Many native Hawaiians today continue to practice the ancient way of life of naturally preferring long-term monogamous relationships but having no rules or penalties associated with breaking up or having multiple partners (Morrell, 2005). However, the human patriarchal societies and religions that glorify suffering and death, as well as vilify sex and women, promote violence over pleasure, leading to human experience over the past few thousand years being closer to that of a violent primate patriarchy (Eisler, 1995).

Bonobos provide evidence that is potentially illustrative of the characteristics of matriarchies for primates. This information may be especially relevant to humans, who appear to be like bonobos in many ways despite the patriarchal culture in which we now live.

D. Evidence on Human Nature Today

The scientific evidence indicates that the peaceful and cooperative nature of humanity has not been changed significantly by patriarchal rule over the last few thousand years. For instance, although women may have recently learned to adapt to the patriarchy and male violence by selecting mates who can protect them, their basic nature remains largely unchanged.[3]

In addition, Spinney (2004a) has shown women's natural instinct to band together has continued, and Pinker (2002) has noted women's strong intellectual capacities, which Israel (2002) mentions would seem to be superfluous in a patriarchy. In fact, there is evidence that, despite the past disparaging of women's mental abilities, females actually remain smarter than males in many ways and on average (Tyre, 2005). Females may lose some of their relative intelligence advantages as their self-confidence is reduced at puberty (Motluk, 2006c), possibly because their sexuality, creative powers, and very existence are treated as evil and dirty under patriarchal rule (Eisler, 1995). As cited by Walker (2000), "At her first menstruation the young 'modern' girl is abandoned by her culture. She is made to feel that her body and its rhythms are a biological impediment to 'freedom' and 'fun'. The subliminal message she receives from her culture is that a properly functioning body is male and non-cyclic."

Some evidence on the incredibly negative effect of patriarchal propaganda on females' measured intelligence is the finding that women's scores in math are improved by 50% when they are informed before a test that any differences in intrinsic mathematical ability can be offset by practice (Adler, 2007). Given that men's mathematical superiority has been found to disappear in societies that have just moved more toward gender equality (*NewScientist*, 2008e), one can only imagine the effect if females were informed that their abilities equal or exceed those of males, and if such facts were disseminated instead of the patriarchal lies. The fact that the area of the brain involved in decision-making and problem-solving remains larger on average in women than in men (Hong, 2008) provides some indication of women's potential.

The inability of the military to make soldiers into psychopathic killers in a large number of cases (*NewScientist,* 2005d) provides further corroborative evidence of the persistence of human nature. Even extremely brave and motivated resistance fighters, like Czechs who witnessed the loss of a quarter of a million of their people at the hands of cruel Nazi occupiers in World War II, have a strong aversion to killing humans (Luza, 2002). It has even been found that the desire to kill other people is so foreign to human nature that as many as 85% of all soldiers trained to kill deliberately shoot to miss when in actual combat with observable enemy soldiers (*NewScientist,* 2005d). A large percentage of soldiers who are forced to engage in "close-quarters fighting" suffer long-term psychological diseases (Mackenzie, 2005). It is possible that many human beings can only be motivated to kill other people by demonizing the enemy and making people think the enemy is less than human.

Patriarchal rule over the last few thousand years has permitted powerful males to have more offspring and may have begun an evolutionary process that increases the tendencies of male humans toward seeking power (Wrangham and Peterson, 1996). Nevertheless, while many believe that selfish violence is innate human nature (Wrangham, 2007), and while success in capitalism is negatively correlated with caring for others (Nettle, 2007), only about 1% of the population even in the capitalist USA which encourages such behavior is truly psychopathic (Spinney, 2004b).

There is also more general evidence that the human traits of peaceful and egalitarian democracy have persisted through the darkest ages of patriarchal feudalism and are returning (Butler, 2004). For instance, in controlled environments where an individual can neither be exploited for altruism nor penalized for selfishness, there continues to be a strong innate tendency for humans to choose equality in income and wealth (Gigerenzer, 2007). In addition, it has been documented that men have the same biological tendency toward caring and personal bonding, especially for children, that women have when society permits these characteristics (Morgan, 2005), as is natural in a balanced culture (Sjoo and Mor, 1987). Moreover, people still make better decisions as a group than as part of a hierarchy (Surowiecki, 2004), as is rationally explained by Gigerenzer (2007) for cases where the public is not totally ignorant or manipulated.

Thus, the effect that the patriarchal suppression of sexual drives has on pushing people's energies toward selfishly and violently seeking power and money (Reich, 1946) has apparently not deeply affected the innate benevolence of human nature. Since there is evidence that mutual love and choice creates healthier offspring (Purdy, 2005), it is possible that the past rape and other forms of patriarchal sex by psychopaths will not create enough survivable offspring to significantly affect the humanity of people as a whole.

E. Conclusion

This chapter provides overwhelming cross-disciplinary evidence that much of human prehistory was characterized by greater female power than has existed over the last 6000 years, and that outright matriarchal cultures existed for thousands of years. Is overwhelming that It is unclear whether an alternative hypothesis could be created that would imply prehistory was characterized by something other than a matriarchy and yet be consistent with all the foregoing phenomena cited here. It would be interesting if some patriarchal fundamentalists at least tried to do so instead of continuing to regurgitate their tiresome lies that are so inconsistent with the facts. Perhaps the revival of fundamentalist Christianity (Sjoo and Mor, 1987) that attempts to disavow and discredit evolution (*NewScientist*, 2007l) and make people believe the biblical fairy tale that time began 6000 years ago (Sjoo and Mor, 1987) represents an act of desperation by the patriarchs to inhibit the truth about the prehistoric matriarchies being told. In particular, insecure males worried about losing their undeserved higher status could already recognize the impossibility of rationally explaining away all the evidence that human society was characterized by greater female power and even matriarchies until 6000 years ago.

Chapter 2
Hypotheses on the Origin of Matriarchies

From being one of the smallest of the primate females relative to its male counterpart, the females of the earliest human species (called hominids) began growing relative to the human male a few million years ago (Jurmain, Kilgore, Trevathan, and Nelson, 2004). Brain growth also appeared to begin from ape sizes about 2.5 million years ago, although initially the growth may have been related only to one sex (Stringer and Andrews, 2005). Albert and Otto (2005) have shown theoretically how traits can develop differently from those of the other sex via genetic inheritance. In the case of humans several million years ago, larger human brains necessitated rising head sizes that increased the pain and mortality of bearing children (Shlain, 2003), and the growth in the relative size of the females may have been spurred by this head growth. A related increase in the time for female dependency during pregnancy and nurtured infancy may have made intelligent female selection of a caring mate more important, thereby prompting evolution to cause special increases in the size of the women thinking area of the brain to better evaluate the all-important mating decision. Research indicating that a portion of the primary thinking area of the brain is inherited strictly through the mother but that no portion of that thinking component is inherited strictly through the father (Rose, 2005) is consistent with this hypothesis.

The increasing size of the human brain in turn helped humans become smart enough to create tools (or weapons) millions of years ago (Jurmain, Kilgore, Trevathan, and Nelson, 2004). Some of the first tools were possibly used to throw, just as other primates have learned how to do, and hurling rocks to scare off competing meat scavengers like hyenas may have originated long ago (Shlain, 1998). Since humans have a biological need for meat to meet basic B12 vitamin requirements and to facilitate other nutritional demands of the human body (Sizer and Whitney, 2003), it is likely that human consumption of meat through scavenging or hunting began far in the past. Human ancestors may have started consuming some small mammals as long as 6 million years ago (Wilford, 2009). Ungar (2007) provides evidence that hominoids began significant hunting and scavenging between 1.8 and 2.6 million years ago, and that some of the earliest tools could have been used to butcher meat.

Throwing objects at predators might have also entered early in human history. Increased intelligence enabled humans to figure out that pointed objects were more effective and to thereafter learn to make sharp bones and stones even sharper until the humans became the predators and hunters. Research does indicate that higher intelligence can assist both in avoiding predators and capturing carnivorous dinners (*NewScientist*, 2006f). Given the difficulty of fleeing predators when pregnant or defending infants, given maternal instincts, and given the ferocity of many animals in defending their young, it is likely that women were the first to work together to successfully fight off predators with weapons. The fact that there are ancient goddesses that were huntresses (Gadon, 1989) and the evidence that increasing mental capacity may have originated with women would be consistent with women taking the lead in hunting.

Although there is a widespread belief even among so-called anthropological experts that men were always the hunters (Greer, 2007), women do hunt in some aboriginal hunting societies (Peat, 2005). Females likely took the lead in both scavenging and hunting originally (Sjoo and Mor, 1987), especially since their bodies need meat more (Shlain, 2003).

Evolutionary developments occurred several millions years ago in the brain promote cooperation among humans via the internal chemical oxytocin, thereby enhancing humans' abilities to both defend against predators and later hunt, and that brain chemical also just happens to be instrumental for essential female functions like lactation and labor (Begley, 2007b). Oxytocin has also been found to significantly improve abilities "to read other people's emotions" (*NewScientist,* 2007c) and to be important social relationships between humans (Bernstein and Nash, 2008). This characteristic is crucial in permitting people to co-exist in social groups (Szalavitz, 2008), as was necessary for prehistoric humans in order to intelligently defend against predators as well as hunt animals. Oxytocin's development therefore also seems to be consistent with a hypothesis that cooperating women formed both the first human societies for defensive and hunting purposes. The usefulness of cooperation with animals like dogs in avoiding predators (*NewScientist,* 2007i) and in hunting may also have appeared at a fairly early point in time.

Although archaeologists and anthropologists tend not to speak on issues with non-patriarchal tendencies without overwhelming evidence (Gadon, 1989), my own hypothesis is that both the increase in intelligence and the relatively faster growth in the size of the females may very well have contributed to ending any dominating alpha male or patriarchal system. Armed with larger brains and tools or even weapons such as the double-bladed instruments found to be as 1.8 millions years old (Miller, 2001) that appeared about the time women started to grow closer to the height of men (Jurmain et al., 2004) might have enabled females to fight off not only predators but also alpha males. In fact, the initial use of a weapon to defeat a human predator may have actually caused females to figure out that a weapon, which can stop an attacker like a lion or crocodile, can kill the alpha males and could also be used to stop the alpha males from abusing them.

One hypothesis exists that the non-dominating or submissive males may also have formed into alliances for the overthrow of the alpha males (Jackson, 1972). However, that conjecture does not explain the reason for the increased relative size of the females nor why another form of patriarchy did not replace the alpha male one. Alpha male cultures do permit subordinate males to have sex with females in the group (*NewScientist,* 2010b), but non-alpha male chimps are known, after grooming female chimps, to violently dominate them if the females don't go out for sex with them outside the view of the alpha male (Wrangham and Peterson, 1996).

Since even chimpanzee societies that tend to have a ruling alpha male are often heavily influenced by female chimps through female alliances that can result in an alpha female (de Waal, 2008), it is more likely that the female humans, who tend to have more of the cooperation-enhancing oxytocin chemical (Begley, 2007b), worked together as a group to replace the patriarchy with a matriarchy. Both humans and bonobos, which are the smaller matriarchal peaceful species of chimp most genetically like human beings (Graham-Rowe, 2005), could have evolved from the cooperation of non-alpha males and females, either as exiles from groups run by alpha males or after the overthrow of the alpha males. Evidence that human societies were characterized by the matriarchal trait of compassion as many as 1.8 million years ago (Cruttenden, 2006) provides some evidence against the cruelest alpha-male patriarchies continuing to dominate after that time.

The fact that human control of fire such as for cooking may have begun as much as 1.9 million years ago (Kenneally, 2008b) may also have something to do with the end of the alpha male patriarchy. In particular, control of fire could be used as a weapon against alpha males, as well as against other predators. Given that most animals panic in the presence of fire whereas primates tend to understand it sufficiently to merely remain cautious about it while communicating knowledge of flames to others (*NewScientist,* 2010a), humans with developed enough brains could have easily learned that predators were afraid of fire. For instance, just one human ancestor running out of desperation toward an existing forest fire to escape a predator would have been enough for learning the usefulness of fire.

Armed with that knowledge, a logical deduction to harness fire by keeping it going for ready use at any time in campfires might have followed rather quickly. Figuring out how to start a fire once it went out might have taken some additional lucky experimentation, but the fact that tools and weapons were already at that time being formed or sharpened by flaking or rubbing two substances could have easily led to sparks that made such "luck" inevitable. Recognizing the usefulness of fire against both predators and alpha males certainly motivated its discovery, and women being less mobile and more vulnerable during pregnancy.might have had both the time and incentive to draw the necessary logical conclusions from trial and error.

The reversal of the earth's magnetic poles 1.8 million years ago (Stringer and Andrews, 2005) that would have likely been accompanied by catastrophic events that led to widespread existence of many and frequent natural fires on earth (White, 1980) seemed to have created an optimal environment of surplus fire for primitive thinkers to learn to harness. That environment also provided strong motivation to figure out how to use fire in the rapidly fluctuating situations of those disastrous times that were no doubt full of life-and-death emergencies for humanity's ancestors. The pole reversal on earth (Stringer and Andrews, 2005) and a change to a colder climate around 1.8 million years ago (Jones, 2009) may therefore have contributed to the harnessing of fire as a weapon against predators and then alpha males. The catastrophic events around a pole reversal can include both extremes of cold and fire (White, 1980), and human ancestors would have been highly motivated to learn to use the fire in some form just to try to survive the temperature changes. A generally cooler climate subsequent to then would have provided the incentives to maintain this skill.

Alternatively, the earliest use of fire could have been initially motivated strictly by a desire to stay warm in colder times that humans might have stumbled upon in their early experimental efforts to develop new tools, for which extremely variable climates would have provided great incentives. It still could have then been instrumental in humans later conquering both predators and alpha males. Within the latter context, it's especially interesting that the harnessing of fire occurred about the same time of about 1.8 million years ago (Kenneally, 2008b) that the size difference between the sexes for human ancestors became roughly equal (Jurmain et al., 2004), that human society showed evidence of compassion (Cruttenden, 2006), and that humans began eating meat (Ungar, 2007). Some of the earliest such meals were potentially the kill of predators which were scarred off by fire-wielding humans, and predators themselves could have been killed in such battles that enabled them to also serve as human food.

The first appearance 1.8 million years ago of the species homo erectus, which developed out of homo habilus and represents the more modern ancestor of today's human beings that genetically evolved to walk more upright (Stringer and Andrews, 2005), may be no accident within this context. In particular, an ability to harness fire enabled human beings to overcome their predators on the ground and therefore exist more permanently outside a tree environment. The fact that human speech began to develop about this time (Fisher, 2004) may also be an effect of the harnessing of fire that motivated communication at night around a campfire lit for both protection and warmth as well as cooking.

In any event, the widespread existence of ancient religions that reflect the harnessing of fire as a very significant event and even made fire sacred (Das, 2009) provides more than circumstantial evidence in support of a hypothesis that the various developments around 1.8 million years ago may have been related to fire. The widespread creation of double-bladed instruments that also began about the same time and continued for over a million years with little evidence of actually using them (Miller, 2001) for anything but symbols honoring females (Gimbutas, 1989) and possibly ceremonies involving labyrinths (Conty, 2002) may have been related to these circumstantial events. Those double-axes may have served not only symbolically, perhaps hinting at female ingenuity being responsible for the earth-shattering event of harnessing the power of fire by rubbing them on wood and stones. They might thereafter have been utilized for emergency defensive purposes in case of attack by predators even after more effective weapons and means of producing fire were created.

One hypothesis that might be drawn from the circumstantial evidence of 1.8 million years ago is that females not only initiated the use of fire with the aid of double-edged axes but also used those instruments to either dominate males in their group or enforce a cooperative culture. Such a fire-wielding, weapon-armed group could have easily dominated other groups of hominids or at least easily persuaded them to adopt the new culture that seemed to have produced homo erectus. The weapon thereafter served as a symbol of the new humanity dedicated to equality and cooperation that rarely had to be enforced with violence as evolution quickly selected human beings with characteristics that motivated them to prefer such peaceful new societies that maximized human harmony and power.

The evidence explained later in Chapter 12 that a truly dominating matriarchal human culture didn't appear until about 35,000 years ago indicates these early advanced cultures were based on cooperation as opposed to domination. The fact that women's sizes relative to males changed very little in the intervening years would seem very improbable if there had been over 1 million years of matriarchy. Given that males remain slightly larger and stronger than women on average, it seems virtually certain that there was no female domination based on superior physical force, and an overtly unequal two-caste system for any large portion of the last two million years can also be ruled out given that no significant inequality in male and female intelligence or other powers evolved over that time.

However, the existence of peaceful, cooperative matriarchal cultures among bonobos, whose females remain much smaller than male bonobos, indicates that some sort of matriarchal tendency within a largely egalitarian society could have existed for over a million years among human ancestors. If fire were the original source of greater female power, it could help explain the vilification of fire as the basis for hell in patriarchal religions that attempt to push humanity back toward alpha male societies. On the other hand, fire as a symbol of hell in some patriarchal religions like Christianity could more likely symbolize the violent volcanic destruction resulting from the ending of the ice age about 10,000 years ago that enabled conquest by patriarchs who believed a male deity caused such a fiery hell for believers in female images. In addition, if there had been a matriarchy for over a million years that included the worship of older women, the innate lust for older females that exists in other primates would exist in men today, and the fact that older women today have sexual desires that isn't complemented by male lust for them seems to be impossible if much of the last 1.8 million years had been characterized by a hierarchical matriarchy that would have selected for such lust. In addition, men's typically complete lack of sexual interest in older women today seems to make human matriarchies impossible without some development of special technology that appears to have first been harnessed by women about 35,000 years ago, as explained in Chapter 12.

On the other hand, the capacity of both female and non-alpha male chimps to submit is consistent with the development of human societies based on submitting to a higher power in the form of a group. In particular, such submissive traits in human beings would facilitate cooperation that would lead to the development of cooperative cultures, which would cause the further development of characteristics to increase the motivation to exist for the good of the group as a whole. Such tendencies do seem to exist in human beings today, as explained in Part II of this book, that would be intrinsically inconsistent with any very extended period of inequality or hierarchy between individuals or the sexes in general.

The stimuli that spur women's sexual excitement may provide a link to the past that enables understanding the sort of equality with matriarchal tendencies that might have existed in the prehistoric past. In particular, Bergner (2009) has found that women are sexually turned on by the mere sight of an erect penis of even an ape, even though they won't admit to it. This finding is consistent with human females still having remnants of the sexual excitement associated with alpha male domination. However, for millions of years, women have had some choice in who mates with them. In particular, given the relative unattractiveness of older females among humans, there is often competition among males for an individual fertile female. As a result, instead of mating with any sexually excited male, women can choose among prospective mates based on their own preferences for intelligence, looks, personality, humor, and caring. The resulting competitive selection process naturally results in younger females being treated special and therefore likely contributed to the change from patriarchal cultures to societies with mild matriarchal tendencies with respect to fertile women. Since chimps aren't subject to menopause, and since male chimps are at least as attracted to older females as to younger ones (Hooper, 2007), it can even be inferred that the ending of male dominance was tied in with the development of menopause that provided younger women with the power of choice. Such a natural tendency was far from hierarchical, as the extra reverence younger women received would have only compensated for their greater discomfort relating to menses, pregnancy, childbirth, and breast-feeding that also represented an especially important contribution to humanity that could be carried out best with special treatment.

The very pain of bearing children with larger brains may have also contributed to the decline of a patriarchal society. In particular, women may have seen the connection between sex and the pain and mortality of bearing children, thereby providing them with strong motivation to fend off unwanted patriarchal sexual advances. Because of this fact, female ancestors without sex drives would have largely died out without leaving any offspring. Nevertheless, the painful pressures of childbirth may have contributed to a weakening of the nerve endings inside the vagina to prevent unbearable pain, thereby causing women sexual excitement to be concentrated in the nearby clitoris (Reich, 1936). The sexual desires and pleasures of today's women may have evolved from this situation that required male submission to women's desire for clitoral stimulation for sexual intercourse to be allowed.

In particular, confronted with females adverse to the consequences of sex but having the capability to enjoy the pleasures thereof, males seeking to have sex would have had to try to use persuasion, since the women were of approximately the same size as men and also had the tools, intelligence, and cooperative ability to fend off rape. Women can typically be brought to sexual excitement through rubbing the clitoris without vaginal penetration by the male, as that female part acts similar to the male penis itself (Geddes, 2008), and men may have learned to massage that parts as a means of enticing women to provide themselves with at least oral sex. While such satisfaction of females desires haven't been necessary in the harshest form of patriarchy, there is a trend in sexual relationships toward ensuring female sexual excitement prior to males enjoying sexual pleasure (Reich, 1936). The fact that women's vaginal juices are natural aphrodisiacs for men (Chia and Wei, 1994) implies that males have evolved to be innately addicted to providing oral stimulation to women's genitals. In particular, men developed a liking for sucking women's genitals because it led to more offspring for males with such preferences since women preferred them. Evidence that even male fruit flies are smart enough to figure out how to do this (Purdy, 2005), as are the males of other organisms such as flying foxes that perform oral sex on females of their species (Gefter, 2008), and other indications that mutually cooperative sex is more productive (*NewScientist,* 2005e) would seem to make the hypothesis feasible.

Certainly, the human females would have had to develop some sort of strong motivation to overcome their aversion to the pain of giving birth, and the unique sexual pleasure that female humans enjoy (Shlain, 2003) may be one of the motivators. The portion of women's brains associated with anxiety (the amygdala) and alertness (the hippocampus) is greatly deactivated during clitoral stimulation and orgasm, far more so than for men (Page, 2005). This fact is consistent with the need for women to evolve a mechanism for overcoming the inhibitions associated with sexual intercourse causing the discomfort of pregnancy and childbirth. The "puzzle" of why women have orgasms at all and usually only during non-penetrative stimulation (Marshall, 2009) is also thereby solved. In particular, an orgasm prior to penetration reduced women's inhibitions against penetration by the penis and actually creates so much mind-numbing pleasure that it makes such penetration desired by the female in a mutual surrender to each others' innate sexual desires for procreation.

Since even monkey males have been found to vary the extent of their fondling depending on the situation that is affected by factors such as the degree of male competition (Barras, 2008), it is likely that male humans would also have adapted to female desires, especially in the face of more men than sexually desirable fertile women. Given that male Barbary macaques ejaculate upon hearing the call of a female while having sex.(*NewScientist,* 2008a), it is quite possible that male humans evolved to become sexually excited by providing oral sex to fertile women, perhaps partially through men becoming especially excited sexually upon hearing a woman's moans while being sexually stimulated.

Some potential ancient evidence on this issue is provided by the fact that the Egyptian myth on the creation of life has a goddess breathing life into a human with an ankh (Baring and Cashford, 1993). The female power sign, which symbolizes the female genitals (Walker, 2000), has a striking resemblance to the ankh whose breathing to create life may be symbolic of oral sex as the initiator of the procreative act. My own experiences indicate that it is naturally enjoyable for males to provide oral stimulation to women and perceive signs of their sexual excitement as especially exciting, even if some cultures today discourage that tendency.

However, females can also engage only in oral or manual sex, masturbation, or intercourse out of estrus, and those females who did so exclusively would have left no offspring. The widespread female desire for mutual orgasm after other stimulation (Rathaus, Nevid, and Fichner-Rathaus, 2008) may have developed as a result to save humanity from non-procreating extinction. It is also likely that the signs of estrus must have disappeared in some women around that time, so that women no longer knew for sure when they would be susceptible to the pain and danger of pregnancy.

The greater dependency of women upon others around giving birth and nursing may have also resulted in evolution selecting for humans who can have sex outside estrus. The pleasure at having sex with any fertile woman would have provided greater motivation for men to help mothers in need, as well as help motivate care for pregnant women because of the forecasted sexual pleasures available thereafter, with romantic love developing and being integrated with that care and prediction.

Such social bonding between men and women could not have taken the form of the patriarchally biased hypothesis mentioned by Shlain (1998) of females selling sex to males for the meat of hunted animals. In particular, with a communal distribution of resources, trading food or other goods for sex and/or love would not have been possible, as evidence from recent cooperative societies that persisted into the modern era demonstrates (Jiayin, 1995). Instead women likely hooked up with men who are swayed emotionally to help without an unnatural patriarchal trade of resources for sex.

Sexual and emotional bonding can also occur among women without men to motivate helping dependents, but humanity would have died out if there hadn't been an innate irresistible female desire for penetration by erect male penises, especially at the point of a woman's orgasm. Even if they often don't admit it, women today have been found to be sexually excited by the mere sight of an erect penis as well as the sight of young fertile women (Bergner, 2009), and they also experience more mind-numbing pleasure at orgasm (Page, 2005) that lasts longer (*NewScientist,* 2005b), that can freely occur multiple times (Bernstein and Nash, 2008), and that is preferred in a mutual form (Rathaus, Nevid, and Fichner-Rathaus, 2008), as all seems to be associated with an innate motivation for penetration and impregnation by males sexually excited by their youthful fertility. At the same time, males seeking sex with otherwise unwilling women would have evolved to satisfy female desires for sexual stimulation and long-term emotional care. As opposed to material gifts, horny men may have had to offer just sexual pleasure that served as a sign of submission to the long-term care needs of women in return for the satisfaction of males' own stronger sex drives. Offerings of knowledge and entertainment, as well as the demonstration of collectively productive skills, like hunting, could also have been used to woo females (Miller, 2001). In addition, special honoring of fertile women might have also helped persuade women to continue to allow procreative sex.

Women's right to choose their partners, as is evidenced in ancient times (Gadon, 1989), and as exists even on a more symbolic level in terms of the scientific fact of the egg catching the sperm inside the womb (Eisler, 1995), would have also contributed to their desire to have sex. The increased chance of impregnation that exists when the female has an orgasm to a level above the average 5% rate per mating (Gynn and Wright, 2008) would also contribute to the temptation to have sex, especially since there always exists some uncertainty with respect to pregnancy that enables the pleasure without the pain. The greater psychological benefits that people enjoy from traditional sex (*NewScientist*, 2006a) no doubt also contributed to the procreative acts that had to continue in order for the human species to survive.

Shlain (2003) has mentioned that women's sexual enjoyment, which is not found in most species, serves in another way to encourage procreation, insofar as the G-spot can provide women with some compensatory pleasure while giving birth, along with maternal instincts of wanting and loving children, thereby also contributing to the desire of women to bear children. Moreover, the discomfort of menses, along with the dangers it once brought from predators, as with lions smelling menstrual blood, would have certainly provided additional motivation to women to get pregnant and thereby avoid the bleeding of menses via pregnancy and subsequent child-nursing.

In addition, there apparently are psychological benefits to women from in-taking semen, insofar as it appears to increase their happiness (Persaud, 2002). The fact that there are significantly more hormones in men's sperm than in chimps' that can spur female ovulation and thereby increase the chance of pregnancy even outside a "normal" menstrual period, especially in conjunction with a female orgasm that raises the probability of fertilizing a ovulated egg cell, provides a further evolved connection between pleasure and procreation (Gynn and Wright, 2008). Such a connection would have been necessary to continue the human race despite the evolution of a "rational" consciousness that may have selfishly focused only on the happiness of the individual or mating pair alone without regard for the future of humanity as a whole.

A. Natural Sexual Bonding Among Prehistoric Humans

Fortunately, human sex now leads to human emotions that bond more naturally and chemically (Fisher, 2004) than via Shlain's (1998) hypothesized sale of human flesh for animal flesh. Evolution likely selected for the development of those emotional bonds. The normal lifespan for the chemical proteins that create the feeling of romantic love in the brain, tends to last less than two years according to scientific research (Emanuele et al., 2006). The latter time span is evolutionarily convenient, since it approximately matches the interval that females are pregnant and might be at least partially dependent upon external care. The fact that scientific evidence indicates a natural chemical reduction in the feeling of romantic love after about a year would also make evolutionary sense, as it would enable a shifting of love of both partners to the offspring, thereby ensuring primary emphasis on the care of those dependents.

In addition, in groups, the dissipation of romantic love would facilitate frequent switching of mates for a more diverse gene pool mix as well as enable recovery from romantic loss due to death and other causes. Evidence such as a recent aboriginal custom of all people of child-bearing age calling each other "husband" or "wife" would seem to indicate peaceful group marriages in ancient times (Jackson, 1972) that would be consistent with uninhibited sexual activities existing at the same time that everyone desire to bond for life and some pairs do (Morrell, 2005). The aboriginal custom of children calling all elders "father" or "mother" (Jackson, 1972) is consistent with communal raising of children occurring in prehistoric times (Sjoo and Mor, 1987), as would have been especially useful in a culture of uninhibited sexual activity. Group childrearing, which helps avoid some neurotic tendencies created by the isolated nuclear families of the modern era (Thimm, 2006), might have become the norm after humans bonded together in larger groups, but the continued existence of romantic love that bonds mates for the term of pregnancy may have continued because of its usefulness in providing special care by a male mate to each pregnant woman to optimize fetal development when there would be no innate male sexual motivation to do so.

Regardless of sexual and childbearing practices, humans have been bonding for love for life for millions of years (Fisher, 2004), and there remains a human tendency to prefer bonding for life when there are no social, legal, and economic pressures to do otherwise, as evidenced by the sexually uninhibited Polynesian culture that largely persists (albeit with some patriarchal corruption) even today (Morrell, 2005). Despite many marriages today being influenced by money matters, love today remains the most important determinant of human bonding when each mate is economically independent (Fisher, 2004).

Males have evolved a preference for younger females because of their greater number of years of childbearing (Miller, 2001). When they can, older males seem to prefer to mate with those 2-3 decades younger (Buss, 2003). In contrast, female chimps do not have barren menopause years, and male chimps tend to prefer to mate with older females, probably because the older females had proven their "genetic fitness" (Hooper, 2007b). A preference for younger female mates among human males may also stem from the fact that younger women are more likely to create healthier offspring that live longer (*NewScientist*, 2006n) and at least partially because "nubile" female "teenagers" were the ones most likely to be fertile since they were not either pregnant or nursing (Etcoff, 1999).

However, with younger women being less sexually driven than older women and vice-versus for men (Shlain, 2003), it would seem logical that there was often a pairing of young with old of both sexes. Some modern evidence of a natural attraction/bonding between older women and younger men does exist (Sjoo and Mor, 1987). Here, ancestor worship and honoring one's elders, at least partially because of their greater wisdom and experience, may have played a part in, or resulted from, such a pairing. The evolutionary factor of older men having had more time to develop a reputation for "good" genes, relating to brains, kindness, bravery, strength, etc., may have also contributed to the mating of young with old, as it makes it easier for younger women, who would have more children, to be able to select the ones with the best genes (Glantz and Pearce, 1989). The greater knowledge and experience of older men in many respects would also have given them an advantage here. The continued prominence of many older men in aboriginal societies (Burling, 2004) provides some indication of traits that would attract younger women to older men.

The younger males like all other men are "hard-wired to find nubile teenagers" (Etcoff, 1999), but they had to earn women's respect over time by hunting, etc. for that privilege. They may have settled in the meantime for older women who were less sought after but who were better matches with respect to the all-important sex drives and the emotional and learning needs of the younger males who are still dependent and developing intellectually. Older women would have also had the sexual experience to make sex more enjoyable and also to teach the younger males how to eventually please younger mates. In addition, since older women were less likely or unable to get pregnant, sex with them might have been preferred by those males who weren't predisposed to help take care of a pregnant mother and offspring for over 10 years. Such a cultural and emotional commitment has probably long been expected by females and human society as a whole, and younger carefree males may have preferred older women for that very reason before they matured into desiring the innately evolved bonding pleasures associated with such long-term caring responsibilities.

Moreover, the experience and connections of older females may have once provided them with greater power as in alpha female chimp societies (de Waal, 2008). Older female humans would have been motivated to wield that power to attain the sexual attention of the younger males who have the greater sexual appetites that match those of older women better. Masks and hypnosis might also have been utilized, especially by older women seeking to at least initially disguise their looks, with the human capacity to love at first sight potentially allowing such relationships to last.

Regardless, the extraordinarily high divorce rate of teenagers in the U.S. of 85% (eHarmony, 2008) provides an indication of the disaster mating young with young is. Besides sexual incompatibility, the reason for same-age relationships being so unsuccessful may be because both young males and young females want and expect to continue to be taken care of or spoiled, as they were as younger children. They may not be prepared for the partner's expectations, whereas older mates might be, at least partially because they would appreciate the opposite partner's youth more. Under those circumstances, it appears that humans evolved to match young with old in sexual relationships, with the young typically being significantly older after the death of their original mates and then prepared to take care of a young mate of their own.

Since older women would have been unable to create offspring if in menopause, this mating of young with old may have enhanced the efficiency and speed of evolution in choosing the genes of men who had already proven them to be of optimal benefit to human society. Perhaps human menopause and the resulting young-old mating patterns even significantly contributed to the evolution of humans into more advanced creatures than apes for this reason. It is difficult to explain how or why the sexual drive of women in menopause could have evolved without such a productive system. Freed from birthing needs, women in menopause may have also contributed significantly to advances in human technology and civilization, as Sjor and Mor (1987) have indicated. Given the inspiration that is obtained via orgasms (Daeda, 2005), as well as the motivation, that contribution could have been enhanced by older women having stronger sexual drives.

Regardless, current societal pressures for mating among those of approximately the same age may therefore represent a perversion against human nature. Societal attempts to inhibit natural and irresistible mating desires inevitably encourage truly perverted sexual activity, including incest. The natural sexual bonding between the young and old that apparently has existed for millions of years is described in more detail in Part II, as are the effects of today's cultural indoctrination into the innate perversion of same-age matings.

The widespread prohibition on sexual relations within a clan in human societies no doubt helped reduce matings between relatives in any sense (Jackson, 1972), thus reducing the chance of genetic defects (Cummings, 2006). In addition, there is a natural human aversion to, or at least lack of interest in, incest and sexual relations with those with whom one has grown up. These facts, combined with an otherwise free sexual environment that, in contrast to today, didn't label virtually all human sex evil, would likely have helped our ancestors to avoid incest and the widespread sexual child abuse that exists today.

With the males having, on average, a stronger sex drive than females, the pairing of old with young may have further contributed to the greater intelligence of women, as the older males had to offer wisdom to the young girls as part of the sex bonding, whereas the older women may have felt less obligation to the horny young men. The older women may even have eventually been prohibited by a budding matriarchal religion from revealing all their secrets in order to maintain a female mystique. Sex for knowledge and reverence therefore would have also contributed to an increase in women's wisdom and self-confidence.

A sex-for-knowledge trade seems a little more plausible than the sex-for-food, money, and wealth hypothesis that the patriarchy promotes but could not have happened in the communism of the prehistoric past. In any case, the pairing of young and old also enhances group strength by combining youthful strength and energy with elderly experience and wisdom, thus enhancing the synergy of such partnerships.

B. The Origins of Religion

The human nature of falling in romantic love leads to adoration and a virtual worship, which blindly and unconditionally sanctifies the opposite partner, and which is often triggered by love at first sight in men who derive pleasure from just seeing female beauty signaling likely fertility (Etcoff, 1999). Given that human communication probably began via the visual senses with gestures and facial expressions that are understood subconsciously and are often more meaningful than actual words (Shlain, 1998), given that people make lasting impressions about people's attractiveness and personality in the first second of seeing a new face (*NewScientist,* 2007g), and given that love preceded language in humans, it is likely that falling in a deep, lasting, and unconditional love upon mere visual sight is deeply entrenched in the human psyche. Romantic love that deludes and maintains deceptions about one's partners is natural and actually helps keep couples happier together (Jones, 2007). Total mutual adoration of one's partner also leads to more exciting emotions and sex (Deida, 2005).

The characteristics of "blind" romantic love so resemble those of virtual worship that it is likely religion began as the mutual adoration of partners. The "emotional" nature of religious belief that is "akin to longing" (Motluk, 2006a) is consistent with this hypothesis. The blind sanctification and "mythologization" of partners in romantic love (Gazzaniga, 1998) which leads to an idealization of the partner that ignores contrary evidence (Etcoff, 1999) also resembles religious worship. The willingness of two partners to sacrifice themselves for over 10 years for the sake of raising the offspring product of their love union provides an indication of the reason why such intense feelings evolved.

There is evidence of sex itself also being revered, including the female reproductive organs as well as to a lesser extent the male one (Gadon, 1989). Such reverence can create both pleasure and power in submission (Lai, 2001). Many traditional Polynesians still find sex sacred, especially within a self- and mutual- respecting relationship, and enjoy the bliss of adoring each other with complete knowledge of imperfections in the partner and with complete freedom to end a relationship if desired by either partner (Morrell, 2005).

Women's feeling of being adored and even worshipped when experiencing oral sex (Deida, 2005) may have also contributed to the development of religion, which apparently did encompass the deification of females before the worship of males (Graves, 1960). The magic and mystery of birth and creation by females of course may represent another reason for the special reverence for women (Gadon, 1989). However, although mutual romantic adoration/love evokes emotions similar to that of worship of a deity, it probably didn't originally deteriorate into a fetish worship of a non-human object such as a deceased partner or ancestor or artificially deified human, except possibly upon the loss of a loved one as a temporary offset to the negative feelings of loss suffered thereby.

Religious experiences themselves seem to be related to a chemical DMT that is created in the pineal gland (Strassman, 2001), and such religious experiences are also promoted by loving sex between two mutually surrendering human partners (Deida, 2005). Since religious experiences can also be induced with DMT itself, including through drinking a natural South American tea that has been in use for many centuries for that purpose (Strassman, 2001), the first discoverers of this "magic" in ancient times might have been able to utilize their knowledge to promote greater reverence for themselves. Such greater reverence and power to bestow ecstatic feelings of love might have easily been transformed into a worship of a deity ideal with some manipulation or exploitation.

Regardless of whether natural DMT sources were used in the process of developing human religion, the initial manipulation to worship probably was originally carried out by older women who sought a return to the reverence they received in their earlier days of more youthful beauty. The biological creation and nurturing powers of females could very well have been used to create an aura of mystique that would have resulted in them being able to manipulate some males into believing they were not only different but more advanced and more important than men. The greater sex drive of the male might also have been manipulated to increase the level of reverence for females. The first human hierarchy may have resulted, with females eventually being held to be not only superior but also actual deities that deserved special treatment throughout life and not just during their special procreative years.

If human intelligence originated with women, as some evidence indicates (Sjoo and Mor, 1987), wisdom might have provided an additional reason for greater reverence for females. Such reverence could have decreased as male intelligence began to catch up, but women might have utilized prior secret knowledge of some technology, hypnosis, magic, and other tricks to maintain the aura. Some older women might have found such selfish manipulation to be especially useful for purposes of unnaturally obtaining male sexual partners.

However, worship is not an innate human tendency in any event. For instance, the Native Americans don't worship anything (Underhill, 1965). Worship represents an unnatural fetish that inhibits procreation, as it's unnatural to want to have sex with something that isn't human, whether it be an animal, a deceased person, or a deity. Worship also inhibits creative activities, as it results in blind one-way obedience that doesn't encourage the striving that mutual loving reverence does. Over very long periods of time, religious worship of female deities or blind obedience to them would have resulted in human cultures turning into bee-like societies that would never have developed the striving and creativity to survive given the inadequacies of human strength and speed in comparison to predators. In addition, since any live worshipped people, such as "holy" priestesses for a universal goddess, have an incentive to deceive the worshippers into continually believing that they are better than human, counterproductive ignorance is promoted, and if anyone figures out the lies, destructive spite results. Human societies that encouraged ignorance or spite could not have lasted long in the face of predators who would have overwhelmed them until the development of more modern technology..

Since the worshippers could never be worthy of the worshipped by definition of worshipping superior beings, it results in a desire for self-sacrifice that inevitably leads to human frustration, which is then typically directed into either masochism or spite. Both the slavery of masochism and the conflict associated with spite are inconsistent with the harmonious mutuality needed to spur the creative cooperation that human individuals had to develop in order to survive their environment that would have otherwise dominated them. However, as explained in Part II, human beings evolved to be happy only upon a mutual love for other beings perceived as equal that can't exist with a "superior" fictitious deity, and that trait along with a feeling of self-fulfillment that comes with an individualistic commitment to serve humanity as a whole enabled the species to survive.

Humanity had to develop a healthy balanced mutual adoration among love pairs that was combined with a tendency for selfless service to the group as long as the group acted for the benefit of the individual love pairs. The major sacrifice for others was likely in the earliest cultures mostly for the children, with love pairs likely obligating themselves to take good care of the fetus and the new born for 10+ years, in return for the pleasure and the eternity that children bring. Such service to only children, however, wasn't sufficient to survive overwhelming predators in the treeless environments humans found themselves, and so the self-sacrifice had to evolve into one related to the human group as a whole, as was probably promoted by a group hypnosis or consciousness that probably began over a million years ago, as explained in Part II.

Any worship of females or female deities that arose in the last 35,000 years or so, as well as any worship of males or male deities in more recent centuries, is only possible because human mental capacities and cultures had advanced by that time to a point where they could survive such counterproductive activities. Such worship hasn't lasted long enough to be innate, however. The goddess religion that did develop around 35,000 years ago was so spiritually and materially integrated that it was necessary for the later patriarchal conquerors to separate sexual or sensual pleasures from spiritual worship in order to permit the replacement of the worship of a goddess with a forced blind spiritual worship of a threatening, vindictive male god that forbids the earthly pleasures of that goddess (Husain, 2003). While much of the goddess religion is so deeply ingrained into the human psyche that Jung and Freud have developed theories of psychology that reflect such tendencies (Graves, 1960), the last few thousand years of patriarchy have prevented that religious fetish from continuing long enough to become innate, as human beings continue to be most motivated by a mutual, egalitarian, harmonious love and unity with all in a non-hierarchical framework.

The goddess religion, even if more natural than patriarchal repression of sex drives, remains far from the natural human nature of mutual reverence between love pairs that is combined with the innate human tendency for all to serve the group or humanity as a whole as they individually see fit or optimal. In an even more inhumane manner, the possessions and self-interest forced upon people by the artificial capitalist patriarchy today naturally creates inhibitions and fears that prevent most people from ever attaining the bliss of mutual human adoration today. As a result, people learn to cope as best they can with something less, since someone taking advantage of a person's love is too probable and frequent in a capitalist patriarchy that deifies self-interest and money and that actually glorifies deception in love for cruel egoistic purposes/gain. Human adoration within this context is often focused on objects or fictitious deities that represent an unnatural worship fetish that leads to perversions in human nature.

It has been asserted that religion began as magic, as human predecessors believed in a connectivity between objects that a magician specialist at some point was selected or found to be best at harnessing to achieve desired effects or make good forecasts, and the magician(s) eventually became so often consulted that the magician(s) became divine ruler(s) by consent (Jackson, 1972). This hypothesis is consistent with goddess worship given the evidence indicating women were the first magicians, rulers, and deities, and given that evolution may have contributed to women being smarter and better predictors of the future, since the dumb ones and their dependent children would likely not have survived (Walker, 1988).

In addition, the prehistoric use of sex for healing (Achayra, 1999) and spiritual inspiration (Eisler, 1995), along with the magic of childbirth and menstruation in women (Walker, 1988), would have contributed to a belief in the sacredness of women that later patriarchs felt necessary to demonize (Acharya, 1999). Some of the priestesses in the matriarchy may have even utilized hypnosis, which can have powerful and seemingly magical effects, and which is especially useful in facilitating childbirth (McClenon, 2002). The magic of the labyrinth spiral (Johnson, 2003) and its documented use in ancient times (Conty, 2002) may be more understandable in the context of the labyrinth/spiral being symbolic of both hypnosis and of the vagina, as explained in more detail in Part II. Given that older people tend to have greater ability to synthesize across all parts of the brain and thereby efficiently integrate the intuition and emotions with the logical thinking and knowledge (Cohen, 2006), it would have been natural for the wisdom of older women to be especially revered.

A belief in magic, superstition, the paranormal, and a deity permits greater access to subconscious knowledge and can therefore enhance the ability to see useful patterns in the environment that may more than offset any false perceptions and gullibility associated with such belief (Wilson, 2006), at least in the face of ignorance. A greater will to success, thereby increasing the ability to integrate all knowledge and methods of analysis and intuition, can also result from a religious belief, and the positive psychic benefits of having beliefs in common with others (Motluk, 2006a) may also be obtained thereby for ignoramuses. These facts can help explain how evolution selected in prehistoric times for the continuing widespread human trait of religious belief.

The social bonding that comes with group religious beliefs/rituals (Dunbar, 2006) would have also contributed to the evolution of human religious belief. In the face of hierarchical cultures of the last 35,000 years that encourage exploitation as opposed to natural bonding, religions may have seemed to offer some respite via a unified belief in the religious lies. The psychological pleasure and health benefits actually derived from religion (Motluk, 2006a) exist only because the unified feeling created by those belief systems actually deceives humans into a false feeling of more social bonding than really exists in the exploitative hierarchies that have existed in the most recent 35,000 years of human history. A feeling of oneness with nature may have indeed assisted in matriarchal rulers in maintaining their cruel hierarchies as they explained their "mysterious" powers of analysis, intuition, forecasting, and "magical" technology by their connections to nature goddesses, even though most of their "mysterious magic" may have originated solely from selfish applications of hypnosis.

The ancient reverence for animals (Gadon, 1989) and the identification of individual clans with individual animals in ancient cultures (Jackson, 1972) may have merely reflected the magician's own identification and spiritual unity with the particular animal that was carried over to their clan, as is evidenced by many ancient pictures of a goddess figure with animals and even figurines that are part animal and part woman (Baring and Cashford, 1993). The switching between clans by males seeking sexual mates (Jackson, 1972) might have begun as a natural result of males worshipping non-relative females and the especially "mysterious" other women, as would have been reinforced by the advantages of diversity caused by such mixing of genes. The natural tendency of people raised with each other not to be attracted to those with whom they grow up, as indicated by unrelated children raised communally in the Israeli kibbutz, may have preceded this cultural phenomena, with such naturally evolving characteristics being reflected later in formal traditions.

The worship of ancestors (Howells, 1986) probably developed earlier than nature worship as a result of a continued reverence for mates and other live humans, such as parents or guardians, after their death. Since the dead can't provide mutual reverence, such reverence would have eventually dissipated when the benefits provided by the deceased while alive had been compensated with adequate reverence. Only if a person were perceived as a "superior" being whose powers persisted even after death would the fetish of ancestor worship have evolved

The later religious reverence for an earth goddess (Gadon, 1989) and incorporation into religion of astrological bodies (Jackson, 1972) may have reflected a further development and integration of such belief systems that perceive a true unified connection between the earth, the entire universe, and everything in it (Peat, 2005). The relationship between the menstruation cycle and the lunar cycle may have contributed to the deification of astrological bodies in matriarchies (Gadon, 1989), and the procreative capacities of women would have naturally led to a feminization of the earth and an original greater deification of females as opposed to males (Sjoo and Mor, 1987).

Evidence of vulva reverence among prehistoric peoples within the last 35,000 years (Gadon, 1989) provides further reason to believe there was initially a greater, if not eventual exclusive, worship of female deities, who not only were directly and easily perceived as life creators via birth and nourishers via breast feeding, but who also provided for the stronger sexual desires of men. The reverence for animals hunted for food and clothing that was especially strong among Native Americans in recent times (Underhill, 1965) would be consistent with a deification of satisfiers of human needs and suppliers of human pleasure.

C. The Contribution of Women's Menstruation to Human Development

As previously mentioned, menstruation is one of the sources of motivation for women to have procreative sex. Menses bleeding may once have been tied more closely to the degree of moonlight (Gadon, 1989) but may be distorted in contemporary times by artificial night lighting (Eisler, 1995). The cycle is also delayed by stress (Gooch, 2006), may be affected by sexual activity through a hastening of the ovulation process by the presence of sperm (Motluk, 2006b), and is strongly affected by the release of pheromones in other women's sweat that tends to tie the menstrual cycles of women living together (Bernstein and Nash, 2008). Menstruation appears to have begun 25 million years ago among human ape predecessors, although the human line of apes and their especially strong menstruation characteristics seem to have branched off about 20 million years ago (Gooch, 2006).

Menses may originally have served to promote bonding among human mates. In particular, menstruation provides a sign to men that a woman is fertile, i.e., is healthy and isn't too young or old to procreate, isn't currently pregnant, and is therefore available for possible impregnation in the near future (Zahavi and Zahavi, 1997). However, given that a menstruating woman isn't at her maximum level of fertility until a couple of weeks after menstruation (Bernstein and Nash, 2008), evolution would have rewarded the males who bonded with a menstruating woman for at least that long. To motivate male support of women during their vulnerable time when offspring was in their bellies, bonding had to evolve to continue through that interval of pregnancy. Such bonding is motivated today in the form of the chemical pleasures of romantic love that last about a year (*Wall Street Journal*, 2005) and that likely developed long ago due to that evolutionary pressure to bond human mates.

Women living together naturally synchronize their menses periods (Schlain, 2003), and it is possible the synchronization was an early development that evolved probably to protect the women from the alpha males. An alpha male may not have liked the bleeding, especially since it might attract predators, but he couldn't kill or exile all women. He also couldn't impregnate all females, thereby enabling the development of less hierarchical situations, either through the evolution of alpha male societies into egalitarian ones or through the dying out of those inefficient systems that were replaced over time with harmoniously powerful ones based on cooperation that developed among exiles from the alpha male cultures. Less breeding for size and a more diverse gene pool that is therefore healthier and more adaptable thus resulted in any event. Synchronization of menses would indeed have doubtlessly led to female bonding that would have assisted in a revolt against alpha males and led to their demise.

Gibbons, which also have menses synchronized by the moon so that they have sex at any time and not just while a female is in estrus, often bond in pairs of individual males and females, and so these characteristics may have led humans to do the same, at least when they are faced with scarcity or small populations (Gooch, 2006). Unlike domineering alpha male apes, male human ancestors bonded to women might have learned to protect females from blood-sniffing predators when they were bleeding heavily. Humans might have even figured out how to use menses blood on the ground as bait to attract both scavengers and predators, which could thereby be ambushed for killing and eating.

A synchronization of menses among women certainly helped decrease the frequency of the dangers associated with attracting predators that may have been especially large before those risks were reduced by greater human harnessing of fire and stone weapons, as well as ingenuity like grabbing the predator by the tail while strangling them by the neck. In particular, the coherent timing of menses by women living together would have made the extremely dangerous and stressful situation associated with female bleeding, which no doubt once attracted dangerous predators, happen only once a month. Human beings thereby had some needed rest and relief between the menstruation cycles that enabled greater freedom of movement during those pauses in the danger. Some variation in menses periods across human groups and time because of varied clouding of moonlight, stress caused by changing environmental conditions, and different sexual activity might have provided sufficient uncertainty to inhibit predators from evolving a distaste for human prey only during a particular time of the month when they were prepared with planned traps. The many years of such ambushes may very well could have promoted the innovative weapon-making and cooperation skills of humans that promoted their own hunting abilities. Shlain's (2003) analysis of women's need for iron due to menstrual bleeding that prompted humans to hunt for meat, might very well have occurred in conjunction with such predator ambushes that likely made predators eventually evolve to dislike human blood and meat altogether.

Prior to predators developing a complete aversion for human blood, pubic coverings or clothes for women could also have been partially motivated to inhibit predator detection of the bleeding females and their scent, thereby enabling greater freedom of movement for them during their special period without needing to retreat to caves, swamps, and trees, or maintain campfires. Although evidence for permanent human clothing exists only within the last 75,000 years (Holmes, 2008), more primitive temporary covering such as leaves could have been utilized much earlier. Whether only temporary clothing was motivated by menses or not, the development of permanent clothing might have not only facilitated migration to colder climates for the humans living in Africa with limited hair covering but also contributed to a greater mystique and male fascination for the covered female genitals that led to matriarchies.

The fact that the supreme deity of the legendary hunt was, according to the ancient Greeks, a lunar huntress who preceded the Greek and Roman huntress goddesses Demeter and Diana (Walker, 1988) would be consistent with the hypothesis that women's menstrual bleeding used to attract predators who later became prey. Only later as a result did predators evolve an aversion to the scent of humans and their blood. The greater success of larger groups of humans in fighting predators, especially with bleeding females, and in hunting, along with the female bonding caused by synchronized menses, would have contributed to the development of the human tendency to exist in groups as opposed to single families. The greater variety in diet that is possible in larger groups of hunters and gatherers may have also helped humans develop since they would provide more of the diverse nutrients needed by the complex human creatures.

As hypothesized by Gooch (2006) and Shlain (2003), human menstruation may have also prompted other evolutionary developments with respect to the further advancement of human mental abilities, especially those related to knowledge of astronomy, prediction, and preparing for the future.[4] The menstruation cycle that corresponds so closely to that of the moon would easily have led to humans, or at least the female ones seeing cyclical patterns tied to an astrological body, to foresee the future. The disadvantage of menstruation attracting dangerous predators and causing distracting pain to women would have likely helped evolution select for only those able to forecast and plan for hiding from or ambushing menstruation-attracted predators. Humans later learned more generally to take advantage of an ability to plan for a future they could forecast with ever greater accuracy in more situations as their brains grew in size and power with evolution.

In contrast, gibbons with their menstrual cycle also corresponding to lunar phases don't excrete nearly as much menstrual blood as humans and are therefore less endangered by its scent. They also spend far less time on the ground. because their jungle environments never turned into the savannahs to which humans had to adapt. Moreover, gibbons can escape from ground predators more easily anyway because they are faster than humans, and because they are also far smaller, they don't have the same potential to defeat predators with crude weapons and cooperative skills. Gibbons therefore never had the same need and use for the mental abilities of humans.

The minds of other apes may also have remained more primitive than those of humans for similar reasons. For instance, other apes, like bonobos which have estrus about every 47.7 days, have menstrual periods that don't match the moon cycle (Schwartz, 2005).

In any event, the intellectual powers of female humans may have evolved especially strongly to help them during menstruation. In particular, females during menstruation today exhibit greater creative capacity including through dreaming, and their aggressiveness also increases, as remains useful for them in winning and retention of sexual mates, who assumedly would assist in propagating the genes responsible (Gooch, 2006). Such creative assertiveness relating to menstruation also might have once helped women with their ambushing and hunting skills.

D. Human Society and Conflicts Between Selfish and Cooperative Tendencies

As Burling (2004) has explained, human intelligence could not have been selected by evolution for the advancement of human use of tools and related technology alone. Instead, sexual selection must have contributed, where intelligence represents a trait that attracts women because it provides an indication of the health and strength of a mate given the massive energy requirements of the brain, just as a male bird's plumage and colors do (Miller, 2001). The dangers of menses for a slow species on the open savannah created an evolutionary advantage for greater male intelligence, which would have been much more useful immediately than bright colors that would have inhibited hiding from or ambushing predators. In any event, female selection of mates in human ancestors likely contributed significantly to the development of human intelligence

Males could have also participated productively in the selection process (Burling, 2004). Given that a single female, including older ones, can sexually satisfy several males, as allegedly often occurred in the past (Sjoo and Mor, 1987), male selection could certainly have left some females without procreative mates and therefore resulted in some females without offspring if they were not attractive to males, or if they were outcast from society for any reason, However, given the innate male fixation on female youthful beauty that evolved to maximize the number and health of offspring (Miller, 2001), it is unclear how much male selection would have contributed to intelligence.

While traits developed in one sex are naturally transferred to the other sex unless there are some special disadvantages of that trait for that sex (West-Eberhard, 2003), the foregoing forecasting and mate selection processes spurred its development further in humans. In addition, the fact that humans had to cooperate in groups to survive against more powerful predators created social complexities which could make a larger brain useful (Miller, 2001).

One of the most important impetuses for the development of human intelligence may be related to the only truly unique human characteristic of having feelings of both altruism and spite that maximized creative individualistic cooperation, as explained in Part II. In particular, to avoid costly injuries, losses, and frictions related to fighting among humans, whose relative weakness in strength and mobility prohibited incurrence of such costs, human ancestors had to develop a tendency to cooperate with each other, and so male competition for female mates had to occur without violence. Such a peaceful society, where violence might lead to banishment and a genetic end for both males and females, may have caused any competition or disputes to be fought on a social or verbal level. Verbal and social competition spurred innovative creativity in a form that had the additional advantage of enabling women to see which males were the best entertainers, educators, and fathers for their offspring (Burling, 2004).

In any event, the most successful societies probably had competition for female mates influenced by successes in protection against predators as well as in food gathering, scavenging, and hunting, as opposed to dominating violence. Such mutually beneficial competition would naturally lead to the further advancements in intelligence, forecasting, and cooperation that are exhibited today in human beings as opposed to larger size. The complexities associated with this unique form of competitive cooperation certainly required incredible mental powers to resolve. In fact, since the conflict between competition and cooperation continues to perplex humans today, human minds may need further evolution to truly resolve it.

E. Violent Aspects of the Matriarchies

Although there is no evidence of matriarchies being characterized by the wars and sorts of violence that are typical in patriarchies (Sjoo and Mor, 1987), there was at least some deliberate killings involved. For instance, one widespread example of matriarchal violence is the annual killing of a new young husband or king consort of the goddess or queen (Husain, 2003). Such sacrifices seem to have originated in ancient rites of goddess worshipping cultures that involved sacrificing males in the spring for the purpose of fertilizing the earth with their remains to encourage plant growth (Baring and Cashford, 1993). Some of the sacrifices may have began over 10,000 years ago and seemed to involve extreme torture of males just to symbolize the harvesting and threshing of grain seeds in the fall (Gimbutas, 1989).

Offering animal or human sacrifices to deities is a widespread ancient religious custom to bribe or give thanks to some power perceived to be responsible for the fates of humans and to affect such factors as the abundance of food (Howells, 1986). The matriarchal sacrifices may have reflected such tendencies with respect to the matriarchal belief in an earth goddess that was perceived as providing for humanity (Baring and Cashford, 1993). Such sacrifices of males may have originally been motivated by local overpopulation or food shortages and then continued when they became superstitiously perceived to be related to good fortune. Given the greater sexual desires of males, it may have been possible to bribe them to be the sacrificial lambs in return for sex with the most desirable females, especially since the matriarchal religion of deifying femininity may have led to masochism because of their perceived inferiority.

Humans have an innate tendency to make sacrifices for their loved mates that originated in their willingness to mutually unite with each others' wishes. Combining this tendency with superstition and a religious fetish of worship leads to the bizarre acts like sacrifices that characterize religions. Sacrifices continued under the patriarchy, and some extreme patriarchal cultures, such as early Semite ones, even killed their first born for a patriarchal god, although the sacrifices may have been more symbolic in the more mixed religions that had male and female deities like the Greek and Roman ones (Baring and Cashford, 1993). The actual killings of males in matriarchies (Sjoo and Mor, 1987) clearly evolved into nothing more than a symbolic killing of initiates after a holy marriage in the "mystery" rites that were allowed in more recent Greek and Roman patriarchal cultures and that were designed to enable spiritual rebirth and reawakening to "forgotten knowledge" (Baring and Cashflord, 1993).

The latter "forgotten knowledge" that was remembered in the ancient mystery rites may very well have included some aspects of the prior matriarchal religions, however. For instance, the matriarchal tendency to focus on female sexual pleasure (Sjoo and Mor, 1983) was no doubt emphasized in the Greek and Roman "mystery rites", since most of the rites were for women who were said to learn true happiness thereby and since the breaking off or harvesting of an ear of corn apparently was the high point (Baring and Cashford, 1993). The actual grim treatment of males under matriarchy during harvest (Gimbutas, 1989) may have been symbolized by this symbolic phallic act that could be used for female sexual orgasms and masturbation, information about which wasn't widely disseminated under the patriarchal rule existing during those times. Although some true insights into the nature of the universe could also have been attained thereby (Baring and Cashford, 1993), the rites more likely represented an outlet for release of female frustration and sexual repression under patriarchy, since matriarchal propaganda and lies certainly can't provide the spiritual rebirth, insights, and related pleasurable high that simply replacing the ignorance of patriarchal brainwashing with true scientific knowledge can.

In any event, since male sacrifices under matriarchies appear to have long been of consorts of an older priestess or of an older queen who sometimes ruled jointly with the consort as a temporary king, and since there were males who "voluntarily" had themselves castrated and periodically suffer wounds to cause bleeding like menstruating females in order to enable them to become priests in the matriarchal religion (Baring and Cashford, 1993), it is possible that sufficient men volunteered to be sacrificed in order to be a consort with a ruling female. In particular, sexual frustration, second-class status, and denial of knowledge other had might have motivated some males to "freely" choose castration or sacrificial death.

Given ancient myths of ruling Goddesses having male consorts who sometimes act for them (Gadon, 1989), male kings, consorts, and partners may have been appointed by ruling women to represent them at times, especially when they were pregnant, even before they began to be annually rotated and then sacrificed in some matriarchal societies. However, the second-class status of males in all matriarchies may have significantly contributed to a male desire to overthrow the matriarchy.

Chapter 3
The Barbarian Revolt Against the Matriarchies

After tens of thousands of years of matriarch, women were transformed into male property a few thousand years ago (Stone, 1976). There is substantial physical and other evidence of a conquest of matriarchal civilization by destructive barbarians who originated in the harsh environments of Central Russia and the desert west of Mesopotamia, and both groups came with religions based on feelings of guilt and hatred of earth and nature (Baring and Cashford, 1993).

The latter facts are consistent with a hypothesis that the invading barbarians had originally been exiles from the matriarchy. Exiling occurred in Native American Indians societies for violation of group norms, with each exiling determined democratically by the group (Underwood, 1965) that reflected the seriousness of the infraction, such as violence, deceit, etc. However, other reasons for exile could have existed, including even males leaving voluntarily because they didn't want to live in a society where they were ruled by women. The exile of some men may have even been related to sexual frustration that stemmed from their inability to provide adequate sexual satisfaction for women, perhaps because they didn't want to massage female genitals, because they could stay sexually excited long enough, or because their penises were too small.

Whatever the cause, exile in the ancient past was typically fatal (Nettle, 2007). However, some exiles no doubt may have survived by moving to desolate unsettled areas such as central Russia and starting various alternative societies there. Since there were likely societies still in existence from pre-matriarchy times that had balanced relations between the sexes, some exiles, such as those who might have even voluntarily left matriarchal civilization for more balanced systems, may have used such cultures as examples or even been adopted into such existing cultures. The Native Americans, who tend to have balanced relations between the sexes (Peat, 2002) as well as the Mongolians, who had virtual equality between the sexes up through the time of Ghenghis Khan (Jiayin, 1995), would be examples of the latter. On the other hand, those exiled for sexual inadequacy and/or violent crimes committed as a result under the nature-loving matriarchy probably selected a cruel system of patriarchy as spite for their banishment to the desolate areas of the world.

That theory would explain the existence of many of the barbarians living in harsh areas tending to have a cruel hatred of women and the nature-loving matriarchal civilization. In particular, those who were banished may have reacted with spite against women, nature, and matriarchy and set up hierarchies where status was determined by violent domination. The resulting patriarchies would have provided the strongest and most violent males to take power by force that thereby resulted in the survival of the strongest and most violent. Those exiles who preferred more balanced societies, such as the Native Americans, may have fled farther to more distant lands like North America.

Some of the exiles from matriarchy may have been young males who revolted from the matriarchal system and willingly left to start or join patriarchal cultures in areas undesired by more civilized, matriarchal societies. Without females, however, the barbarians would have died out, and so it possible that the patriarchies in the waste lands were created, maintained, and reinforced by violent males who kidnapped women from the matriarchies.

As in the case of ancient Hawaiian cultures where both males and females are exiled for acts of incest (Morrell, 2005), which can be extremely cruel to offspring (Cummings, 2006), many of both the men and women of the patriarchies could have originated from involuntary banishment. However, the intense hatred and cruelty these patriarchs later exhibited toward women, along with their tendency toward homosexuality and contempt for heterosexual relationships (Acharya, 1999), would seem to indicate some sort of very strong resentment among the patriarchs that might not be expected to exist among groups of male and female exiles who might be more likely to bond together. The intensity of the barbarian despising of women would therefore seem to rule out the most typical source of their females being fellow exiles. Violent male groups could have seized many of their women not only via kidnappings from matriarchies but also by seizing them from peoples who fled matriarchies to live in more balanced and peaceful societies before they were attacked and destroyed by the violent barbarians, or before they left for unsettled lands like North America.

One likely cause of the patriarchs' resentment may have been related to sexual rejection by the civilized women, maybe even due to the patriarchs having small penises and/or small brains, or just plain being unable or unwilling to satisfy women. The great fear that the patriarchs through the ages have had of the sexual appetites of women, as partially symbolized by the patriarchal religious vilification of one ancient goddess (Lilith) for rejecting the alleged original man (Adam) for being sexually inadequate (Walker, 2000), provides evidence in favor of this hypothesis. This conjecture may at least apply to males who continue to support the patriarchy even when knowledgeable of the past.

In the ancient human societies that preceded matriarchies and patriarchies, it is likely that the sexual needs of males with small brains and penises were balanced with those of unattractive older women beyond menopause. The sexual frustration of both sexually unattractive men and older women would have thereby been avoided at the same time that genes for small penises and small brains weren't perpetuated. While some mating resulting in offspring might have occurred between the less attractive of each sex and the more attractive when the less attractive accomplished tasks that were so beneficial to the existing human societal grouping that their sexual desirability increased, unattractive traits would have tended to be de-selected by evolution over time. As a result, at some point in prehistory, there might have been an insufficient number of unattractive males for the older women, whose sexual frustration contributed, as explained in Part II, to the development of matriarchy. That matriarchy was very unfair and cruel to at least some men (Gimbutas, 1989), and the least attractive males might have suffered the most under matriarchy, resulting in many exiles and refugees from matriarchy despising women, as explained in more detail in Part II.

At any rate, global flooding of the widespread coastal civilizations occurred about 10,000 years ago and continued for thousands of years because of the ending of the ice age that was accompanied by increased volcanic activity (Hancock, 2002). Since melting ice caps affect underground pressure that also reduces the melting temperature of magma (*NewScientist*, 2008b), the end of the ice age not only raised sea levels but also caused catastrophic volcanic eruptions (Hancock, 2002) that greatly weakened the matriarchal civilizations. The past connections between the melting of the earth's ice sheets and catastrophic events have been verified, possibly providing some indication of the environmental changes that will come with global warming in the future (Fisher, 2009) and that may likewise lead to future revolutionary social upheavals and a new balance of power between the sexes. In any event, there is evidence of people losing faith in their goddess at some point (Baring and Cashford, 1993), as would be expected in the face of incredible natural catastrophes.

The surviving matriarchies fell prey to invading patriarchal barbarians, who conquered them over a few thousand years time (Eisler, 1988). The barbarian conquest of matriarchal civilization a few thousand years ago is especially well documented for Europe (Stone, 1976). However, substantial evidence also exists of a patriarchal takeover of a civilization in ancient China that was goddess-worshipping and exhibited the matriarchal characteristics of a partnership culture as opposed to the hierarchical society that is typical of patriarchies (Jiayin, 1995). Similarly, the invasion of an advanced civilization in India by Indo-European Aryans a few thousand years ago (Hancock, 2002) seems to have led to a caste system that rigidly left the invaders and their descendants in a dominant ruling position (Dubois, 1965). That caste system left its genetic mark on the population of India, with the dominant class having genes more consistent with those of Europeans and the lower castes being more genetically tied to those of East Asians (*NewScientist*, 2009d). While Joseph (2004, 2006) has provided evidence that some of the people of the old civilized world were also Caucasian, the ancient matriarchal civilization seems to have originated in Africa and been widespread there as well (Gooch, 2006).

Although powerful groups of patriarchal barbarians also emanated from the Arabian desert (Baring and Cashford, 1993), the main force of patriarchal barbarians apparently originated in central Russia (Stone, 1976), which until about 10,000 years ago was covered with ice (Gooch, 2006). It has been thought that the increasing desertification of central Russia caused by the nomadic patriarchal barbarians raising cattle may have helped prompt them to invade and seize the rich agricultural lands of the matriarchies (Eisler, 1995). The patriarchs' move from violent hunting to more peaceful cattle shepherding may very well have also redirected some of their lust for searching for animals to kill for meat into a desire to seek humans to battle for cattle and other booty (Shlain, 1998). Their violence could have been magnified by them continuing to be sexually frustrated due to a patriarchal hierarchy that made women the exclusive possessions of only the most alpha-type males, who likely further magnified the shortage of females by increasing women's death rates via harsh treatment.[5]

Within this context, it is interesting that women are likely to have been responsible for the original development of cattle domestication (Goodison and Morris, 1998), with cattle herders first drinking milk about 7500 years ago in the Balkans and Central Europe (*NewScientist,* 2009c). The science of raising cattle and horses may have been obtained by the uncivilized barbarians through tricking or kidnapping knowledgeable women as is consistent with the patriarchs' own myths (Husain, 2003).

The first domesticated animal was the wolf, which started its transformation into the dog 10,000 years ago, and which was especially honored by women because wolves and dogs howled at the moon, which itself was believed to have spiritual connections to women's reproductive cycles, and the domestication of goats for nutritional purposes followed only afterwards (Gimbutas, 1989). Since domesticated wolves were also likely used for protection, including by matriarchs against rebellious males, the art of animal domestication seems to have begun for religious and defensive reasons, as may be reflective of a dominating use of advances in technology under the matriarchy for purposes of control as opposed to resource production. These technologies that protected matriarchal power were then turned on them by the patriarchal barbarians. For instance, the patriarchal harnessing of domesticated warhorses was instrumental in the successful conquest by the barbarians (Eisler, 1995), especially with respect to war chariots (Baring and Cashford, 1993).

However, continued flooding and global catastrophes may have also contributed significantly to the decline of the older, more civilized, matriarchal world (Joseph, 2004). The surviving people of this civilization, which originally consisted of peaceful, nature-honoring goddess-worshippers until the natural catastrophes weakened it materially and spiritually (Joseph, 2006), likely were very badly demoralized by the fact that their civilization was being destroyed by the very nature they honored. The extensive evidence of civilized settlements across the world that are thousands of years old but now under water (Hancock, 2002) implies that much of their civilization perished, along with perhaps many of their people as well. The coastal refugees would have had to flee further inward and increase competition for scarce resources, possibly leading to frictions and even conflicts with the barbarians that may have also helped spur the barbarian invasion.

Regardless of the causes for and the success of the barbarian invasion and of the origin of the "race" of the barbarian conquerers,[6] substantial evidence exists of a patriarchal conquest (Stone, 1976). For instance, there is substantial evidence that the older matriarchal customs and goddess religion were replaced by patriarchy and worship of a male deity or deities (Stone, 1976). Given that there is extensive physical evidence of material destruction and long-term technology declines after the patriarchal conquest (Gimbutas, 1977), it seems that the patriarchs were indeed barbarians in comparison to the matriarchal societies.

Some Druid myths even indicate that abundance and plenty came from the Vanir (Paxson, 2005), which were Nordic matriarchal deities who were defeated by a set of patriarchal gods (Davidson, 1990), implying that, despite their destructive conquest, the patriarchal barbarians eventually were able to advance their level of prosperity and wealth from the earlier, more advanced matriarchal civilization. However, the fact that the peoples of Asia Minor and Egypt were more "sophisticated" and better educated than those of patriarchal areas like Greece even three thousand years ago (Strauss, 2006), and the fact that Asia Minor and Egypt contained some of the last vestiges of matriarchal civilization long after Greece was taken over by patriarchs much earlier (Baring and Cashford, 1993) provide some indication of how long it took the barbarians to assimilate the advanced culture of matriarchal civilization.

Many of the insights and technological advances of prehistoric civilization have only recently been discovered (Childress, 2000), although it is likely that much of that extensive knowledge originated from earlier human cultures that were more balanced than matriarchies. For instance, Achayra (2004) has cited evidence that much of the earlier belief systems that may go back millions of years were astrologically based and purely symbolic of luminous bodies in space. Since the ancient myth of the earth-goddess procreating all people through her son (Grieves, 1960) seems to indicate prehistoric people recognized their earthly origins in matings between people of opposite sexes and ages, the extraordinary knowledge and fascination of prehistoric people with the cosmos external to their worldly environment may stem from the forecasting power humans gained shortly after they learned to forecast women's menstrual periods from the lunar cycle. Other cycles in the heavens were also detected, along with attempts to relate them via astrology to earthly events. The power of the celestial bodies to influence phenomena in human beings and on earth would have helped people recognize both their dependence on and integration with the greater cosmic environment in their spiritual beliefs. Since the matriarchal goddesses were generally most associated with earthly phenomena like animals as opposed to stellar objects (Gimbutas, 1989), it would seem that the spiritual reverence for the heavens wasn't related to any particular religious worship,

Some stars were later transformed by some patriarchal religions into identifying symbols of "real" people who eventually became patriarchal deities (Acharya, 1999), perhaps because of the dominating appearance of the lights in the sky at the time of such patriarchal leaders who could have been identified with them, and perhaps because of the pre-matriarchal reverence for the celestial bodies that were therefore never tightly tied in with worship of femininity. However, the heavens in most patriarchal religions today like Christianity seem to retain their roots in being the location for human bliss that may be associated with the spiritual oneness that once was felt with them over millions of years of human prehistory because of the integration of human and cosmic cycles.

Most of the religious myths seem to be associated with worldly events even if they might be related to astronomical conditions that might have been partially associated with deities in both matriarchies and patriarchies. For instance, the ancient Greek myths and legends of male giants and monsters being defeated by Zeus and the Olympian gods may reflect the unsuccessful fight of the males of the goddess culture, just as may ancient stories of the mixing of the defeated women in the goddess culture with the conquerors (Graves, 1960). The Christian bible's book of Revelations may also reflect the historic conquest of matriarchal civilization by the patriarchs, although it could also have been used by the Romans to predict a disastrous defeat for any attempt to return the matriarchy (Achayra, 1999).

The Revelations chapter of the bible is so similar to the Hebrew book Enoch, which was written over 2100 years ago, that it has been hypothesized to be largely based on it (Stray, 2005), and so its adoption by the Roman Empire when it imposed Catholicism on their subjects after 313AD may merely reflect the Romans' desire to also use the story of the past patriarchal defeat of matriarchy as an indication of the results of future attempts to impose such satanic systems on the world. Many aspects of the book of Revelations are indeed based on ancient myths going back thousands of years, although it adds a special vilification of women who are symbolized by the prehistoric female genital sign of "6" (Acharya, 1999). The worship of the ancient triple goddess of maiden, mother, and crone that has been documented by Walker (1988) seems to imply that the feared "666" sign of the beast mentioned in Revelations probably represents the triple goddess.

Women's powers have been feared by the patriarchies long after their conquests of the matriarchies (Baring and Cashford, 1993), and the patriarchal vilification of attempts to return to matriarchal religions has taken the form of masculinizating the female deities in Judaism (Acharya, 1999) that has culminated in the male Satan symbol of evil in christianity (Baring and Cashford, 1993). Perhaps such de-feminization of the triple goddesses was intended to take away the sex appeal and mystery of a religion based on worshipping a goddess, as well as attempt to portray even female power to be under the rule of a male. Perhaps such male rule of femininity may even reflect a pre-matriarchal balanced culture where males provided so much delightful sexual and emotional bliss to women that they were considered deities who controlled them, just as the females controlled their own deities through the ecstasy they provided in a mutual love that is explained in more detail in Part II of this book.

In any event, the entire story of a final doomsday in the biblical Revelations could be partially related to a widespread threat of destruction that goddess-worshipping religions forecast would occur after the imposition of a patriarchy, as the patriarchs themselves perhaps adapted and reversed the Armageddon-like threat so that the ending destruction was only applicable in a future war to the enemies of the patriarchal religion (Walker, 1988). The goddess religion has managed to persevere to this day in some form, consciously in some and possibly subconsciously in everyone else, despite all the efforts of the patriarchs to destroy it (Sjoo and Mor, 1987), and the patriarchal adoption of Christianity represents yet another attempt to at least redirect many of the recurring indications of it into the service of the patriarchy (Baring and Cashford, 1993). As will be explained in Part II of this book, most of the recurring aspects of the goddess religion themselves are merely innate human tendencies that were naturally adopted by the matriarchies and then treated as evil by the conquering patriarchs because of their identification with femininity whose power and return they feared.

Chapter 4
During and After the Barbarian Conquest

During and after the barbarian conquest, many negative traits were attributed by the victorious patriarchs to the goddesses and their culture (Husain, 2003). While there may have been some exaggeration and distortion in the patriarchs' subsequent vilification of women (Baring and Cashford, 1993), there may have been some truth to the patriarchal myths.

For example, Jewish mythology has the traditional Goddess Inanna transformed into the wild and mischievous Lilith as a result of her refusing to submit to patriarchal sex with the "original" man Adam (Gadon, 1989). Adam may have actually symbolized here the original barbarian conquerors, who liked only "missionary" sex with submissive women who needed only via threats of physical violence for alpha male penetration as opposed to female sexual stimulation that they didn't understand.

In addition, as is consistent with a recent finding that women "enjoy" killing monsters in computer games as much as men as long as there is a story to justify it (*NewScientist,* 2006d), the women of the goddess culture probably did display more violent traits than women are allowed to portray in today's patriarchal societies. For instance, matriarchal rituals involving murderous sacrifices of kingly consorts of ruling matriarchs did indeed exist (Sjoo and Mor, 1987), and stories or rites associated with them did persist in patriarchal religions (Baring and Cashford, 1993).

Subsequent to the armed conflicts with the patriarchal barbarians, the annual sacrifices to the matriarchal earth goddess may have been of selected barbarian prisoners, who may have been allowed to temporarily marry a matriarchal queen or priestess. The latter hypothesis would also be consistent with the stories of beautiful Amazons who mated with males and then killed them (Husain, 2003) that existed in Africa and Europe (Sjoo and Mor, 1987). There are tales of societies of such female warriors existing only 3500 years ago, and archaeological evidence of female warriors just over 2000 years ago has been found in southern Russia and the Ukraine (Strauss, 2006).

Nevertheless, the historical incorporation of prisoners as equals into less patriarchal societies, such as in Africa (Howard, 1982) and Native America (Melzer, 1993), implies that many barbarian prisoners may have also been treated as equals at times. Mating between the two cultures might have been allowed, with the offspring of such intercourse being used to continue to fight the patriarchs. Mixing genes with the barbarians may have been at least partially motivated by war strategy. In particular, there may have been some short-term increase in population growth, as the potentially smaller brains and heads of the barbarians could have reduced child-birth deaths. In addition, the likely larger body size for the barbarians that had evolved for fighting could have increased the physical stature of the matriarchy's warriors, especially if females chose the largest brutes.

In contrast, the cruel treatment of prisoners by patriarchal whites (Melzer, 1993) would be consistent with much less inter-mating with matriarchal captives in the patriarchies. The patriarchs who originated the concept of "holy wars" during their conquest of the Palestine by Hebrew patriarchs, even conducted mass extermination of conquered peoples that, in the case of the Hebrew conquests, were religiously "justified" because of the people's failure to worship the patriarchal Hebrew god which was alleged in the biblical fairy tale to mandate their conquest and mass murder (Sjoo and Mor, 1987). The Christian/Jewish Bible itself admits to the Hebrews slaughter of over a million people in the conquest of Palestine alone just for worshipping an alternative deity (Walker, 2000), which seemed to symbolize a prehistoric goddess (Acharya, 1999).

Myths have indicated that, prior to the patriarchy's imposition of male gods, the worshipped goddess was actually three in one, consisting of a young maiden, a mother, and an older woman (Sjoo and Mor, 1987). The triple goddess may have been represented by 3 female priestesses or queens, with the maiden being selected from the fastest or youngest nubile girl (Walker, 1988). If alpha female chimp systems, in which females leaders use their experience and connections to enforce obedience and unity (de Waal, 2008) are any indication, the crone would have been the primary ruler and decision-maker in this trio. The maiden probably served mostly as a figurehead to induce natural male reverence for youthful fertile femininity, while the mother could have been used to solicit the reverence of children. The widespread vilification of older unattractive women as witches might be understandable within the context of such manipulation of males and younger females by crones. Regardless of the degree of power of women of different ages across the world at varying times, the matriarchal system was replaced in different ways across the globe.

A. Evidence in Europe and the Middle East from Ancient Myths and Religions

Myths in Europe and the Middle East indicate that rule by triple goddesses, priestesses, and queens was replaced with rule by a male king and god that was initially selected by a nymph goddess (Graves, 1960). A son of the king/god was sacrificed in many myths to correspond with a pre-patriarchy practice of killing a young boy to replace the annual sacrifice of the young "consort" of the matriarchal queen that occurred prior to the implementation of the patriarchy (Walker, 1988). The ritual killing of the young boy who symbolizes the son of the king/god probably led to the belief in a son of god being sacrificed that is so widespread in many religions like Christianity. In Greece, the sacrificed son of the king/god was Dionysus, who had a philosophy and story very similar to the later Jesus (Freke and Gandy, 1999), was alleged to have sprung from the thigh of the god without any mother, and essentially replaced an erotic nymph goddess (Graves, 1960). Most of the stories of the sacrificed sons of god, including Dionysus, Mithra, and Krishna, included a belief in their resurrection and apparently were adopted almost exactly into the later version in the form of Jesus Christ, whom the earliest Christians recognized was a mere symbolic myth that reflected astrological events in many respects as did the other sons (Achayra, 1999).

There may actually have been a Jewish Jesus who initiated an early Gnostic form of the Christian cult and whose brother James may have indeed been martyred by orthodox Jews, even if Jesus himself was never actually killed by anyone (Baring and Cashford, 1993). The entire idea of a sacrifice of a son of god appears to have originated from an attempt to adopt the "mystery" religions of other ethnic groups that were in existence at the time, such as that of the Roman Bacchus, the Asian Attis, and the Persian Mithra, for the Jewish people (Stray, 2005).

The earlier annual killing of the queen goddess's "consort" who was considered to be the "son' of the queen (Sjoo and Mor, 1987) may have actually provided the cultural basis for the sacrifices that were considered religiously necessary even by the patriarchs to renew the cosmos after each winter, and the patriarchal rulers merely adapted it to their liking. Since the sacrificed "son" under the matriarchies may have really reflected not only the young age of the king consort for the ruling queen but also an unrelated boy who was selectively adopted by her at puberty, the patriarchal sacrifice of a first son may have also involved an adopted child in some form. However, any patriarchs killing of "adopted" sons may have actually reflected the execution of their most respected or most homosexually desired enemy combatants, whereas the adoptions in matriarchy may actually have reflected a common prehistoric practice of the old to adopt youth of the opposite sex from other clans at puberty that had been the norm for hundreds of thousands of years before there was any matriarchy, patriarchy, or battle between the sexes, as will be explained in more detail in Part II of this book.

Given the prior widespread belief in reincarnation in the matriarchy, the myth of a sacrificed son of god being resurrected probably originated in the earlier belief in reincarnation (Husain, 2003). In particular, the death and resurrection story was originally related to the sun and its life-giving light being born anew on the shortest day of the year in the northern hemisphere that was December 25 a few thousand years ago (Acharya, 1999). The fact that the barbarian patriarchs emerged from central Russia where some may have encountered the midnight sun in the summer and the 24-hour-a-day darkness for times in the winter (Sjoo and Mor, 1987), the death and rebirth of the sun or "son" might have held special significance to them.

In particular, the practice of killing an actual son of a leader may have been considered to be especially essential in a northern area of the world where the sun actually did disappear for several days in the winter, and so the matriarchal superstition requiring a human sacrifice to enable the birth of a new sun or "son" may have easily persisted there despite the patriarchs hatred of matriarchies. Such sacrifices might have also helped control population growth at times and thereby minimize competition for scarce resources in the harsh environment of patriarchies. Given a patriarchal tradition of killing the first-born son of a leader reported even in the Christian bible (Walker, 1988) that otherwise makes little sense, such a hypothesis is not terribly far-fetched.

Given some traditions of having a younger male replace the king when the king was too old to defend himself from the stronger youth (Howells, 1986), such killings of a first son might have also been developed by the patriarchal leaders themselves to reduce the chance of them being replaced before their deaths by their oldest sons. Such acts also served as a rather serious warning for subsequent offspring, as well as of the cold, bloodthirsty nature of the patriarchal leaders whose own power required them to draw more blood than even the ancient goddesses with their bloody births, menstruation, and sacrificial rites that were once considered magical and holy.

Jackson (1972) has indicated that the sacrifice and death of a son was preceded by religions calling for the sacrifice of an animal to the earth for resurrecting new plant life after every winter. Such nonhuman sacrifices may have reflected more humane religions that developed greater balance in the relationships between the sexes than existed in either patriarchies or matriarchies. While such mixed religions may have plagiarized the morals, rituals, and concepts behind the original goddess-worshipping religion (Acharya, 1999), they were indeed more humane than the patriarchal or matriarchal ones based on human sacrifices. The entire concept of giving blood sacrifices to the earth may actually stem from an ancient tradition to consecrate the earth and make it more fertile via the sprinkling with women's menstrual blood (Walker, 1988), with such potentially productive traditions (Walker, 2000) may have actually existing prior to matriarchies or patriarchies, but the inhumane hierarchical nature of matriarchies and patriarchies caused such practices to be perverted into inhumane acts.

Drawings over 10,000 years old indicating bound men seemingly fertilizing the ground with erect penises and at least symbolic traditions of horrible sacrifices of males annually (Gimbutas, 1989) imply that males may not only have contributed their sperm to the fertilization of the fields but may have also been subjected to ritual sacrifice under the prehistoric matriarchies. Such atrocities may have preceded those of the barbarians, thereby providing further grounds for their spite against women. However, since a form of trials, death, and rebirth occurred in ancient religious mystery rites that were focused around a deity (Baring and Cashford, 1993), it is also possible that many tortures and sacrifices were merely psychological, as with U.S. water-boarding that simulates the extreme agony of drowning. As with the water-boarding that the U.S. applied rather widely along with the mass murder of nearly a million civilians over a hundred years ago in that country's conquest of the Philippines (Schirmer and Salon, 1987), the tortures applied by the matriarchs were designed to force submission to a "higher" authority both for the tortured victims and others inferior male subjects who were supposed to learn by example.

In any event, the old goddess religion was not completely destroyed after the patriarchal conquest, and the patriarchal conquerors may have felt compelled to continue the matriarchal rituals in some form to appease the vanquished peoples and motivate them into more voluntary submission to the patriarchs by adopting many of their superstitious beliefs and rituals. More mixed religions, which included creative and powerful goddesses like Ishtar, were widespread shortly after the initial patriarchal conquest (Baring and Cashford, 1993), and a belief in the need for some actual sacrifice to the earth may have actually continued among the conquered after they became patriarchal subjects.

The use of the goddess temples to sell sex (Walker, 2000) was probably one of the main reasons for the persistence of goddess worship under patriarchal rule. Such temples provided a source of income to support the temples, priestesses, and women in general, and they were no doubt very popular among many of the males who were sexually frustrated in the patriarchies. The fact that, even three thousand years ago, the temples of the goddesses of the enemy were typically spared even after the conquest and destruction of an enemy city (Strauss, 2006) provides some evidence of the popularity of such temples, as well as of the continued power of women and the fear of their deities long after patriarchal conquest.

Despite some plagiarizing of the ancient goddess worship, the mixed religions (i.e., the ones that permitted some female deities as well as male gods) themselves remained far removed from the integration of spirituality with the physical world that characterized matriarchal religions (Husain, 2003). The sale of sex for physical goods like money or meat itself appears to be a recent development among humans that doesn't characterize other primates like chimps (*NewScientist,* 2010e), and the combination of physical and spiritual unification that occurs with natural human sex (Daeda, 2005) would seem to be inhibited by payments for just the sexual pleasure or release. Prostitution may have actually first begun after women were put into an artificially inferior position that motivated them to exploit the ruling males' sex drives instead of naturally enjoying it. In any event, such a replacement of an innately mutual human pleasure, which can create such bliss as to be spiritually moving for both partners, with a physical trade represented one aspect of the separation of human spirituality from the physical world that the imposition of the patriarchal hierarchy brought with it.

One purpose of the division of the spiritual world from the physical one, especially with respect to eroticism, was to separate religion from the worship of nature, whose creative powers were naturally associated with the procreating females (Gadon, 1989). Rule-based religious/legal systems were implemented, replacing the spirituality of the female rule (Shlain, 1999), in order to control the natural human emotions and sexual lust that had supported the matriarchy (Gadon, 1989). The patriarchy may have been fully aware of the connections between sexual pleasure, religious experience, and enlightenment (Deida, 2005) and the special mysterious neural connections between women's sexual pleasures and the "high brain center" (Sjoo and Mor, 1987). The fact that being more interested in women than war was considered a grave insult to men a few thousand years ago (Strauss, 2006) provides some indication of the artificial morals that the patriarchies attempted to impose on males in order to inhibit female power.

In the partriarchal Roman Empire of Western Europe, there remained as many goddesses as gods (Goddison and Morris, 1998), but all traces of conquered matriarchal cultures like the Carthagian and Etruscan ones were destroyed (Sjoo and Mor, 1987). The patriarchal nature of Roman culture existed despite the fact that gender was more clearly incorporated into images of the local deities than existed in matriarchies (Goodison and Morris, 1998). Those deities were earlier symbolized under matriarchal religions by shapes like those of the chevron and of animals such as the serpent that have been more subtly associated with the goddess (Gimbutas, 1989). The more symbolic form for goddesses in matriarchies may reflect their desire to maintain their deifying aura by having themselves appear as something other than human, as might have been especially important for ugly old hags who didn't even have the mystique of sexual attractiveness to males.

In any event, the Roman deities were very similar to the system of Greek gods and reflected the dominant role of the males and the subjugation of females (Baring and Cashford, 1993). The ancient Greeks themselves had an extremely patriarchal society despite the existence of some female deities that reflected a continued recognition of some special female powers (Sjoo and Mor, 1987) such as those relating to love, forecasting, agricultural, and even hunting (Baring and Cashford, 1993). Male domination in the Roman Empire was intensified by a later Emperor's adoption of Christianity, which is a modified form of the extremely patriarchal Judaism, and which even vilifies the ancient matriarchal snake (Sjoo and Mor, 1987). While the snake has remained to many a symbol of the regeneration of life and, together with the egg, rebirth (Randall, 1993), it represents the cause of evil and human suffering in the christian bible that mandates an ignorant faith in and worship of a fictitious cruel male deity as the source of good and heaven.

The concept of a possible afterlife in Hades, which existed in the Roman Empire and to which steaming caldera terrain in Italy was thought to represent the entrance (Bojanowski, 2009), may have also been merged into the Judeo-Christian religion to create today's fantasy of Hell. The fiery volcanic god of Judaism's god that destroyed its enemies with fire (Acharya, 1999) may be one source of the Christian belief in a fiery fate for the enemies of their god which is basically just a Hebrew religious fabrication that itself evolved to meet the religious needs of patriarchal conquerors who may have been as ignorant of the religion's lies as their vanquished subjects.

Some cultures managed to maintain or return more power back to women after the patriarchal takeover. For instance, Sparta actually permitted women fairly equal rights, along with some characteristics of communism and democracy, despite a formal patriarchy, albeit at the cost of the enslavement of neighboring people (Cartledge, 2003). In Germany, the fertility goddess actually existed simultaneously with the merciless but mortal god of war (Wotan) up through the invasion and defeat of the Romans in 9AD, and Germans regarded women as goddesses (Davidson, 1990).

After the Romans' defeat by the Germans in 9AD, there was actually a widespread trend within the Roman Empire toward more goddess-oriented religions, or at least toward those based more on the innate human characteristics of caring as opposed to selfishness (Baring and Cashford, 1993). The myth of the birth of a christ around that time seems to reflect the trend which itself was based on a renewed hope that the cruel patriarchal rule could be overthrown by a more humane system. The victory by the German Druids in 9AD certainly stopped the Romans from further significant conquests and indicated that the Roman system wasn't invincible (Murdoch, 2006), thereby spurring renewed interest in other belief systems.

While the Rome maintained its old patriarchal religion and rule for several hundred years after the defeat by the Germans, it finally co-opted the widespread movement for more humane morals within the Empire under the overlay of patriarchal Judaism that mandated obedience to the hierarchical rulers (Baring and Cashford, 1993). In 312AD, the Roman Emperor Constantine apparently mistook a large meteorite or other extraterrestrial object that is documented to have fallen from the sky and crashed into the earth in that year as a sign to officially convert to Christianity (Chandler, 2008). Regardless of Constantine's actual convictions, he apparently tried to harness the empire-wide matriarchal tendencies through imposition of Christianity as the new state religion, which, despite Christianity's adoption of some aspects of the matriarchal religion was actually far more inhumane than other patriarchal religions in existence at the time because of the severity of its hierarchical overlay (Acharya, 1999).

In particular, Christianity represented a sprinkling of some aspects of humane philosophies that existed even under the matriarchies on the top of the foundation of Judaism, which had largely been purged of matriarchal influences (Baring and Cashford, 1993). Jews and their special adaptations of Judaism had been spread in many forms throughout the Roman Empire after the exiling of the Jews from Palestine following one of their many revolts that coincided with the founding of the Christian form of that particular religion many decades after the alleged birth of Christ (Shlain, 1998). Christ himself was originally a mere Jewish copy of a symbolic male character like Buddha, Krishna, and Mithra which advocated benevolent matriarchal characteristics and subservience for common people in various Asian religions (Acharya, 1999). Christianity merely adapted those belief systems, including "pagan" rituals like Christmas and Easter, from the prevalent ones that had spread throughout the Roman Empire (Baring and Cashford, 1993), and the Roman Catholic guise of that adaptation enabled it to take a less "foreign" Asian form as well as maintain some harsh aspects of patriarchy that already existed in the Hebrew religion that had been widely spread after the Roman exiling of the Jews. In particular, the concept of Christ, who like Mithra and other similar gods like Osiris and Dionysius was mythologized in annual rituals to have been sacrificed (Stray, 2005), may have appeared useful to the Romans for teaching docility at the same time that the Jewish Old Testament.provided needed militancy to fight off invaders.

The more mixed religions tend to be less patriarchally harsh and dictatorial than others such as Judaism, on which Christianity is based, that don't have goddesses (Achayra, 1999). The Hebrew religion Judaism was actually among the first to eliminate the goddess temples and make prostitution illegal a little over two thousand years ago (Baring and Cashford, 1993), and it thereby put women in an even lower position than previously.

Judaism itself was founded on an older tradition involving the sacrificial offering of a first son to a cruel patriarchal god (Walker, 2000) and even mandated a torturous mutilation of newborn sons' penises in an attempt to imitate the blood-letting of menstruation that was perceived to be holy and creative (Gadon, 1989). The killing of the first son in patriarchies may also have reflected a shortage of resources, a shortage of women, and a desire by patriarchal rulers to have less competition for power, resources, and women, as well as a superstitious attempt to resurrect the sun that actually disappeared for days near some points of possible original exile near the North Pole. It may also have been an attempt to patriarchally emulate older matriarchal rituals that might have at one time entailed actual physical torture and killing of a young male consort or son of a goddess or priestess, as well as psychological torment and symbolic death in later times. Such matriarchal beliefs may have been so strongly indoctrinated into people's overall belief systems since infancy that even people exiled from matriarchies continued to believe them even after they returned as conquerors.

It's interesting that the later Christian religion, which built upon Judaism, didn't have such cruelties to males except with respect to a fictitious crucifixion of a mythical Christ, which provided a symbol for why no further bloody sacrifices had to be made to renew the earth and cosmos. The later Moslem religion, which built on the two former ones, actually imposed cruel mutilation of women's genitals at puberty instead. Female genital mutilation continues today in parts of northern Africa and the Middle East, with an estimated 3 million victims per year, and with 85% of all girls suffering this torture in Egypt despite government laws against it (*NewScientist*, 2008f). It's possible that the widespread extreme cruelty to women in contemporary Egypt relates to its later movement away from matriarchal rule that persisted for commoner males, who had to agree at marriage to "obey the woman in all things" until about 2000 years ago, insofar as the male spiteful revenge for prior matriarchal cruelty hasn't yet totally run its course.

B. Evidence from Myths and Religions Elsewhere

The assumption of power by the barbarians may have taken a different path in other areas of the world, however. For example, Western Africa was characterized by matriarchies until the Islamic invasion many hundreds years after the imposition of the Christian religion in Europe (Sjoo and Mor, 1987). Non-patriarchal beliefs may have been purged quickly there without intermediary religions and battles, and a crueler form of patriarchy therefore exists there that likely will be moderated as it has been elsewhere in the world over time.

In Egypt, the resurrection myth is slightly different, with the husband/brother of the goddess Isis brutally being killed by his evil brother and then resurrected through her son (Achayra, 1999). This myth is symbolic of female regeneration powers, with the goddess being the mother of her husband by bearing the son who is the husband reincarnated (Husain, 2003). While somewhat different from the myths of Europe and the Middle East, the same themes of a murdered but reborn male who is also the son of the ruling deity seemed to also exist in ancient Egypt even after patriarchal conquest by violence.

Ancient Egyptian civilization, which, like ancient African civilization in general, may have been far more advanced than previously thought, was originally organized on a matrilineal basis (Malkowski, 2006). Women's right to choose their mates may have even contributed to the advancement of the culture if they chose wise ones. However, there is evidence that some cattle-raising human "brothers" seized the Nile area of Egypt about 7000 years ago after a Sahara drought, gave the Nile River a male name with nurturing female characteristics, and labeled their kings to be sons of Isis (Goodison and Morris, 1998).

Given that there was a story of murder of father gods by their sons and stories of father gods laying eggs in the Egypt of the pharaohs (Jackson, 1972), it is likely that the patriarchal barbarians who seized power in Egypt adopted the matriarchal belief in the necessity of sacrificing kings or men, if not boys or sons. Instead, the Egyptian myths seem to indicate that the son of a goddess eventually took power by violence, perhaps after exile by a matriarchal queen mother or after fleeing from her intended execution of him, and the powers of the goddess that include even giving birth, laying eggs, and creating fluid productivity were then given to male gods after the son's conquest.

Some vestiges of goddess/female power remained in Egypt after patriarchal conquest even in symbols, albeit under the control of male gods as in the famous "eye" of wisdom and chaos existing in the Egyptian godhead Horus (Sjoo and Mor, 1987). The matriarchal religion also persisted for a long time with the Egyptian common people after patriarchal priests spread their ideology (Baring and Cashford, 1993). However, the level of technology actually deteriorated significantly after imposition of rule by the male pharaohs began (Malkowski, 2006), just as it did in Europe after conquest by the patriarchal barbarians there (Gimbutas, 1977). The cruelty of Moslem mutilation of female genitals in Africa may not only reflect the more recent spiteful overthrow of matriarchy but also a more modern recognition that males don't have to suffer or be sacrificed as they did in the matriarchies.

In China, the patriarchs took over after massive flooding that led to reverence for men who implemented procedures to control the devastating effects of natural disasters and who later engaged in wars, including some successful ones against barbarian invaders from Russia and Central Asia, that led to the glorification of men, an oppressive patriarchy, and even widespread killing of female babies (Jiayin, 1995). While Chinese religions retain a few goddesses (Husain, 2003), their philosophy as stated by Confucius was based at the core on women's "sacrifice to men's genitals", which led to binding women's feet so they couldn't dance, and which formally preached total "ignorance" and subservience for women (Jiayin, 1995). The conquest of China much later by the Ghenghis Khan's Mongolian invaders (Hartog, 1999), who had actually a more balanced relationship between the sexes, resulted in some interruption in the harsher forms of

Asian patriarchies with some periods of female rights and sexual freedom (Jiayin, 1995). The greater reverence for the elderly in Eastern cultures than in Western ones may actually stem from the fact that the wisdom of older males led to patriarchy in the Asia as opposed to the brute force of young barbarian conquerors in Europe.

The rest of Asia had some myths that are similar to those of the Christian religion, except that they precede them. For instance, Buddha, who was very similar to Christ in philosophy but originated before Jesus, was believed to be a reincarnation as opposed to a resurrected sacrificed son of god (Achayra, 1999) This slightly different myth in Asia might reflect an even different path toward patriarchal rule, perhaps because the matriarchies of Asia never murderously sacrificed males as they apparently did in Europe and Africa, even if the Asian matriarchies may have sacrificed men's genitals to the desires of their ruling women.

In the Americas, there doesn't appear to be a similar continent-wide takeover by patriarchal conquerors. Since the northern Native American societies were characterized by a reasonable balance between the sexes (Underhill, 1965), and since their religion displayed more balanced in reverence among the sexes despite a reverence for an earth mother (Shlain, 1998), it is possible that the patriarchal takeover in the rest of the world was at least partially motivated by imbalances in the matriarchal systems that were perceived to be unjust by some males, as explained in Part II. However, there is evidence of at least one powerful conquering tribe in North America (the Aztecs) overthrowing a matriarchal moon-goddess society (that of the Mayans) in today's Mexico simply out of a lust for power a few hundred years ago (Sjoo and Mor, 1987). The subjugated Mayan people of Mexico had developed a somewhat cruel hierarchy themselves at least a thousand years prior to conquest, despite having a religion that included patriarchal deities as well as goddesses, which symbolized various forces of Nature, and that was largely focused on astrology grounded in some fairly accurate astronomical forecasts (Henderson, 1997). The Aztecs imposed a particularly cruel hierarchy on the vanquished that made the purpose of humans to provide blood sacrifices to a sadistic patriarchal Sun-god and that may have been close to as terroristic as the Christian and other Old World religions which were later imposed on the Native American survivors of the European conquest (Sjoo and Mor, 1987). Whereas the violence of both the patriarchal Europeans and Aztecs may have originated out of spiteful hatred for the matriarchal cruelties many centuries earlier, the ignorance of the people under patriarchies seems to result in violence being spread against everyone else, including other patriarchies.

While many Native Americans originated in parts of Asia and crossed the Pacific in boats, many others crossed through Alaska and originated in Siberia (Adovasio, 2002), and the patriarchal leaders of the Aztecs may have been among the latter. Those Native Americans, who came by boat as opposed to a more continuous coastal route via Siberia and Alaska, apparently included some that appeared directly from Asia in Oregon over 14,000 years ago (*NewScientist*, 2008c), and these immigrants may have had the more balanced society that characterizes most Native Americans in what is now the USA. It is possible that the Native American emigrants were themselves exiles or refugees from less balanced matriarchal or patriarchal systems that existed their places of origin in Asia over 10,000 years ago.

C. Resistance to Patriarchy

Even after the victory of the patriarchs, there was a great deal of resistance to the elimination of the matriarchy and to the replacement of the worship of a goddess with a god (Stone, 1976). Despite extreme efforts that included mass extermination policies, some of which are recorded in the Old Testament of the Christian bible, the patriarchal rulers often had to make many compromises with human nature to manage their subjects, such as permitting the continued worship of divine females in at least some respects (Sjoo and Mor, 1987).

For instance, the patriarchal Greeks/Romans had to tolerate a religious festival honoring a goddess Demester/Diana who rescued her daughter from patriarchal rapists (Gadon, 1989). In addition, before the religious system in Western Europe that included goddesses was finally replaced by the single patriarchal god of Christianity in the final stages of the Roman Empire, many values characteristic of matriarchies were adopted such as kindness that were widely advocated by many, including early Christians called Gnostics (Shlain, 1998) and others like the other strong rival religion based on a Persian Mithra upon which the Jewish Christ was based (Baring and Cashford, 1993). Many of the humane morals incorporated into the holy books of the Gnostics, who were the only christians actually persecuted by the Romans, remained in the main Christian religious text called the New Testament that had been appended to a patriarchal Old Testament of Judaism by early Gnostic Jews like Paul, who succeeded in converting many Gentiles to his religion, but the Roman Catholic version of christianity actually emphasized the Old Testament values of obedience to a patriarchal elite and even discouraged sex (Shlain, 1998).

Perhaps because of a less violent seizure of power by the patriarchs in Asia (Jiayin, 1995), Goddess worship continued even longer in Asia (Husain, 2003). However, the killing of female babies in China, the harsh life for women as indicated by the shorter lifespan of women who were allowed to grow up, the early portrayal of a woman working on her knees in the pictorial character for the word "woman" in the Chinese language, and the use of the same word for "wife" and "slave" in ancient Chinese (Jiayin, 1995) are indicative of cruelty similar to that at times in Europe and the Middle East. Such apparent barbarism in Asia may actually reflect spite for the suffering that males had endured under the prior matriarchies.

In addition, in some areas of Asia, society values, such as symbolized by Buddha, were less viciously patriarchal, and the most loved deity in China was a benevolent goddess, who was a human female resurrected after being killed for wanting to marry a physician (Husain, 2003). The latter myth may symbolize some regret at persecution of Asian women who may have tried to help men through their knowledge of ancient herbal medicine. As in the Middle East, where Judaism had long severely persecuted believers in matriarchal deities (Acharya, 1999), many lost their lives across the world in their attempts to retain at least the humane aspects of pre-patriarchal belief systems and morals.

In many places, there was a complete repudiation of patriarchal religious morals by the masses who retained their matriarchal customs and rituals. For instance, the actual matriarchal relationships within families among the common folk of Egypt that treated men as virtual slaves to their wives and that were observed by people in Greek patriarchies just a few thousand years ago (Sjoo and Mor, 1987) existed long after patriarchal conquest. Patriarchal people apparently invaded Egypt 5000-7000 years ago and took formal control at that time by establishing an official kingdom there (Stone, 1976). The formal Egyptian religion that was imposed was quite similar to the ancient patriarchal Greek and Roman ones, which may have adopted it in part (Shlain, 1998), but the common people of Egypt continued their matriarchal culture for some time (Stone, 1976). The patriarchal conquest and patriarchal relationships didn't permeate the entire African continent until thousands of years later when Islam made its appearance in Northern Africa and Christianity entered the rest of the continent even later in history.

Even in Europe, Walker (2000) has reported that, up until the Inquisition in late medieval times, common men had to pledge to worship their wives upon marriage. Such matriarchal family relationships were overshadowed by patriarchal rulers or lords who were considered to own the common folk, which included both male and female serfs (Nicole, 2007). At least in Germany, a less patriarchal situation among the masses may have reflected the last vestiges of their Druid religion. The Druids included as one of their few immortal Nordic deities a goddess Nerthus, who had been associated with the matriarchal belief in a triple goddess and the eternal cycle of regeneration, who was the main surviving goddess in a set of "serpent" deities whose worshippers had venerated ancestors and Nature without good or evil judgments, and who had been defeated by patriarchal "eagle" gods (Davidson, 1990). The return of some aspects of special honoring of women and chivalry after the fall of the Roman Empire in the rest of Europe (Shlain, 1998) might be easily explainable if the peoples who came through Germany to conquer Rome had or adopted religions similar to those of the German and other European Druids, as seems quite likely.

Not too long after the fall of Rome, there is some evidence that the leaders of Europe were persuaded by the Catholic Church to adopt Christianity (Davidson, 1990). Hypnosis, including the group hypnosis that continues to be used by Christianity by having worshippers adopt the hypnotic position of closed eyes and surrender during prayer (McClenon, 2002), was also likely employed. The commoners were not initially forced to adopt Christianity, but their fertility goddess was eventually replaced by a fertility god after the goddess was effectively "put into a brothel" before Christianity was finally imposed by feudal lords upon the formal abolishment of their communal society (Davidson, 1990).

It is interesting that the elite rulers of the Middle Ages in Europe were called lords, just as their Christian god was. They may have been identified as being one and the same. The lord often had first right to bed a new commoner's bride (Walker, 1988), as might even have been welcomed in this bizarre medieval culture as a means of "sanctifying" the woman. Any resulting offspring would have potentially been considered to be like jesus, and the virgin mary, which wasn't perceived to be ":violated" after insemination by the christian deity or the medieval lord, could have represented the ideal for such women in the European feudal society.

The Christian lord system might even have been developed earlier by the romans after their official adoption of that religion in 312AD as a means of helping them rule more effectively the more matriarchal peoples like the Druids throughout Europe. The later conquest of the Roman Empire around 500 AD by the barbarians from Eastern Europe (such as Hungary) itself might have resulted from more balanced cultures even farther east exiling violent patriarchal individuals, who could have quickly seen the advantages of Christianity in imposing their own hierarchical rule on the European

Druids. The changing of the Druid German gods prior to the conquest of Rome (Davidson, 1990) may have been related to the barbarians initially trying to impose a patriarchal system on the conquered Germans, whose defeat of Rome in 9AD had enabled them to avoid having them replace their more matriarchal religious culture with the patriarchal one of the Holy Roman Empire.

Regardless, the matriarchal marital relationships within the context of an unnatural patriarchal superstructure that persisted among the commoners in Europe long after Christianity was imposed (Walker, 1988) may have developed as a means to transform the natural mutual obedience between human mates into a trade of male service for female sex. Only much later did the Christian leaders eventually eliminate the reverence and rights for women that existed throughout much of Medieval Europe through violent persecutions that included the Inquisition (Shlain, 1998).

Some remnant of goddess worship is still allowed today under Catholicism via the reverence for the Virgin Mary, the mother of Jesus in the Christ myth, as may reflect the ancient symbolism of the Goddess giving birth to the "sun" (Gadon, 1989). The fact that it is a virgin figure which is revered permits continued adoration of women's bodies but as something too sacred to be violated as opposed to being an object of intense pleasure and creation. Women are therefore encouraged by the revered figure's example and Christianity's vilification of sex to be prudish, thereby frustrating both men and women sexually and making them more easily exploitable and manipulated by the patriarchal rulers (Walker, 2000). Sex becomes a commodity for sale within the patriarchal capitalist framework, with women being motivated and willing to sell their bodies in this environment (Sjoo and Mor, 1987), as the perception of females as innately evil in both Judaism and christianity by their very existence that promotes the sexual desire which is vilified as sinful in those religions (Walker, 2000) motivates them to get the highest price for their soulless selves that act as mere vessels for the production of male offspring.

Chapter 5
Possible Genetic Effects of Rule by the Patriarchal Barbarians

A recent discovery of a new gene in about 25% of people that controls human brain size and that is most prevalently in those of European and Middle Eastern descent began to spread about 6000 years ago and (Inman, 2005), and it may have arisen from the patriarchal invasion of the barbarians from the north. In particular, the patriarchal barbarian hunters and warriors, who left few signs of ever creating an intelligent civilization in the harsh environment of their central Russian origin (Eisler, 1988), would have been able to survive better in their harsh environment with a smaller brain, given that a smaller thinking capacity needed less energy and nutrition.[7] Given the high death rate in childbirth for bigger-brained humans (Shlain, 2003), especially for those mothers who are suffering from poor health and malnutrition that results in maternal mortality rates much higher than an estimated 1-1.5% "normal" rate and causes infant death rates as high as 10% in some developing countries at childbirth (Gosline, 2007), a smaller brain might also have enabled the barbarian population to grow more rapidly than otherwise. A high level of fertility among barbarian women might have been especially difficult without a reduction in brain size given both the harsh environment and the limited resources that the barbarian patriarchies would likely have provided to childcare and overexploited women.

With the coastal matriarchal civilizations beginning to be decimated by worldwide flooding and related calamities caused by the end of the ice age about 10,000 years ago (Hancock, 2002), with those natural catastrophes lasting thousands of years (Joseph, 2004), continued growth in the inland barbarian population may have given them both the motivation and local numerical superiority to gradually conquer the civilized humans in more temperate coastal zones in a few thousands years. The concentration of the flooding destruction on the civilized coastal areas might have demoralized the matriarchal people and even caused them to doubt the goddess religion since the barbarians who didn't believe in a nature goddess had escaped the natural disasters unscathed. There is indeed some evidence of increased building of forts during that period in at least one area (*NewScientist,* 2005f), as would be expected during barbarian invasions taking advantage of flooding. This phenomenon is also consistent with a hypothesis that flooding caused population displacement and territorial encroachments that may have spurred warfare between the barbaric and civilized peoples.

After the patriarchal conquest, rape and other methods of patriarchal procreation would have resulted in massive self-selection for their pea-brain genes. Just for instance, in patriarchal aboriginal societies characterized by barbaric patriarchy, "leaders" procreate twice as many offspring as other males (Burling, 2004). In addition, it's quite possible that the barbarians slaughtered many of the conquered peoples, keeping around just enough slaves skilled enough to produce for the conquerors. Such a scenario would be consistent with the policies imposed by Ghenghis Khan thousands of years later (Hartog, 1999). Thereafter, the barbarians may have imposed restrictions on sexual reproduction of the survivors in order to hold down their population, thereby enabling the conquerors to consume more of their resources, at least short-term, at the same time that the number of slaves who might revolt was kept to a minimum, thereby further increasing the prevalence of the conquerors genes. The genetic results of separation of castes of people that is still present today in India, with the dominant class being descended from a small pool of ancestors (*NewScientist*, 2009d) itself probably reflects the conquest of India by a group of barbarians a few thousand years ago, who maintained racial "purity" through that caste system (Hancock, 2002). However, the genes of at least the conquered classes of India might have been mixed by the patriarchal barbarian patriarchs often forcing sex with their servants, just as the aristocracy of the old South in the USA did to their slaves to create a mixed gene pool for the African Americans of today.

However, the purity of the pea-brain gene pool would have been diluted somewhat because of a need to permit some mating of the conquered slaves with normal-sized brains in order to motivate them to serve the barbarian "elite" better and keep their sexual frustration from rising to a level that might lead to revolts. Much of that mating may actually have been between the conquerors and the conquered, as the conquerors may very well have found the conquered sexually attractive. The latter hypothesis is consistent with ancient Greek and Jewish religious myths telling the story of such prehistoric conquests of goddess-worshippers who were sexually desired and impregnated (Baring and Cashford, 1993). When these sexual relationships became affectionate and resulted in loved offspring, the barbarity and sexual repression of the conquerors would likely have moderated, thereby enabling the advance of culture and a lower increase in the sexual selection for barbarian genes. The documented history of patriarchy in China in the last few thousands years during which time its harshness was moderated over various intervals due to such attachments (Jiayin, 1995) might be typical of the history of the world for the last 6000 years.

While some of the rapid evolution of human genes over the last 40,000-80,000 years has been related to cultural changes such as the development of agriculture and urbanization (Wills, 2009), some of the changes in the human genome may have stemmed from the rapid mixing of barbarian genes into the human gene pool. Because there is evidence of genes evolving in the last 6000-10,000 years that increase susceptibility to Alzheimer's and brain inhibition functions (*NewScientist,* 2006b), it would make evolutionary sense that the source of that evolution was the barbarians who might benefit from the decrease in the amount of energy needed by such brains in harsh environments with scarce resources.[8]

Inman (2005) also reports having found another gene that controls brain size to be about 40,000 years old that is prevalent among non-African peoples. There are myths of the barbarian hunters and warriors living in harsh environments even before the massive worldwide flooding over 10,000 years ago (Goodwin, 1996), and those barbarians could very well have originated from people leaving or exiled from matriarchies. However, many of these earliest settlers of desolate lands may have come from balanced societies prior to the worldwide imposition of imbalance matriarchies, perhaps being exiled for violent behavior that might have stemmed from frustration out of being physically strong but mentally feeble. In any event, a smaller brain size would have been self-selected in the harsh environments where the energy for thinking had to be conserved, especially when energy for physical strength might have been very important for both hunting and warfare needed to defend against more violent and stronger exiles. One gene existing in half of all Europeans and West Africans has been found to be associated with a higher probability of obesity as well as smaller occipital and frontal lobes (*NewScientist*, 2010c) that are associated with thinking, as would be consistent with invaders from barbarian violent cultures in harsh environments like northern Europe spreading their genes through southern areas through the rape and other forms of patriarchal procreation that was enabled by conquest

Given that there appears to be some Neanderthal genes among human beings today (Gibbons, 2010), given that the Neanderthals had smaller thinking areas of the brain (Gooch, 2007), and given that the Neanderthal form of homo sapiens tended to be pushed to more desolate areas by the expanding smaller humans as they died out about 35,000 years ago (*NewScientist*, 2007h), it's possible that the spreading of the genes of the Neanderthals occurred through their matings with the exiles from more modern human societies. While some of the mixing between the two forms of the human species might have resulted in reasonably balanced cultures, such as that of the Mongolians and the Native Americans who emanated from Asia, it's also possible that the Neanderthal genes were absorbed into the modern human gene pool through alpha male forms of procreation like rape after the more violent male exiles from more civilized societies in more fertile areas conquered some of the desolate lands. The fact that a category of mitochrondrial DNA called Haplogroup Group N appears to have originated from the modern humans who settled farther inland into Eurasia after emanating out of Africa about 75,000 years ago and can today be found in Europeans, Middle Easterners, Asians, and Native Americans (Wikipedia, 2010) seems to be consistent with this hypothesis.

Many of the pre-matriarchal exiles from the coastal civilizations of Eurasia, may have stemmed from a civilized preference of women for older men who had proven themselves. As explained in Part II, such a bias might have created a problem for impatient, impetuous males, who may have been frustrated because of their preference for younger as opposed to older women with whom they were actually more compatible. Some may have turned into complete perverts by initiating intra-human violence as a means of imposing satisfaction of their disgustingly inhumane desires on real human beings.

These subhuman barbarians may have mixed with both Neanderthals and later with human exiles from cruel matriarchies. Such mixings could very well have resulted in more psychopathically violent and selfish natures evolving in many of the inland societies that would inevitably self-destruct eventually without some sort of system to deter excesses. Systems deterring violence in psychopaths can include formal laws that are enforced by punishment for breaches of the defined rules. However, since barbarians with small thinking capacities might have difficulty in developing such forms of control, and since such systems can never completely eliminate the violent tendencies of psychopaths to circumvent the system, societies of these subhumans have to often be diverted by wars of conquest, pillage, and rape that enable some mixing of the psychopaths' genes, or they eventually self-destruct out of pathological selfishness. The more mixed breeds tend to predominate today, although the purer form of psychopathic subhumans seems to exist in 1% of contemporary populations, thriving by exploiting others and thereby endangering human civilization and humanity as a whole.

In any event, other evidence exists that seems to imply a special small-brained group of barbarians having conquered matriarchal civilizations in the last few thousand years. For instance, since infertility is related to the existence of narrow hips relative to the waist of women (Miller, 2001), and since large hips in women relative to their waistlines naturally enhance the brains and intelligence of children through increased nutrient intake such as during breast-feeding (*NewScientist,* 2007m, 2007n), a matriarchal tendency to honor, worship, and glorify women with large hips (Gimbutas, 1989) may actually stem from an innate trait of human beings that helped them evolve greater thinking capacities in their offspring. However, barbarians having or evolving smaller brains in their harsh environments that required maximum conservation of energy for physical exertion might have also led to hip sizes evolving in their women that were only large enough to enable giving birth to offspring with those smaller brains. The genes for smaller hips were then passed on to many of the mixed descendants of the barbarians who procreated with civilized women with larger brains after their conquest of civilization. The existence of an unusually painful and dangerous childbirth for humans (Schlain, 2003) may at least partially be caused by the resulting mixing of patriarchal genes with those of humane humans over the last 6000 years that has resulted in many cases of larger brains having to come out of smaller hips.

People like Polynesians who have not mixed their genes much with white barbarians are indeed known to have far less difficulty in childbirth (Morrell, 2005). The biblical myth of Eve eating of the Tree of Knowledge may possibly symbolize the higher level of intelligence associated with the larger brains of humans in the matriarchal societies being the cause of the pain of childbirth. In particular, after the matriarchal Eve mated with the patriarchal Adam, who may have symbolized the first patriarchal leader, an unfortunate genetic mixing in many cases resulted in a mismatching of brain and hip sizes. Human hips did indeed begin to grow to accommodate larger brain sizes as long as 1.2 million years ago (*NewScientist*, 2008h), and so the existence today of small hips and child birth difficulties might most likely be explained by the mixing of humans with barbarians with small hips and brains.

The association between narrow hips and infertility (Miller, 2001) might also be explained by the hypothesis of patriarchs having small brains and hips. In particular, small-hipped white barbarian women might have tended to be less fertile because the more fertile ones would have died out with all their children from overwork and inadequate resources in a harsh environment that could only have been made worse by patriarchal rule which forced childcare and many other tasks exclusively upon their female property. This pure conjecture might help explain why African women on average seem to have wider hips than white women, which itself, however, is a casual observation that may not be substantiated.

In addition, there is evidence in the form of female figurines with exaggerated hips that women with larger hips were found to be more attractive, if not outright worshipped, tens of thousands of years ago (*NewScientist,* 2007b). As a result, it seems quite plausible that there was also a tendency for women with smaller hips to be less happy, and there might have been a few of those women who had some motivation to join the pea-brained barbarians. They themselves might have been exiled themselves for violent or undesirable behavior that could be induced by their unhappiness and frustration related to the perception that they were unattractive. If true, they would have contributed to the barbarian women having smaller hips. However, this conjecture, which is especially speculative, isn't necessary to help explain an evolution of smaller hips in patriarchal barbarian societies that may explain the intense labor of child birth today resulting from mismatches in female hip and infant head sizes.

Because the head size of whites approximates that of blacks on average (Pinker, 2003), those whites without the barbarian pea-brain genes might be the most intelligent of all. However, it is unclear as yet whether that particular gene which reduces brain size has an effect only in certain environmental circumstances or not, and so this hypothesis is also speculative. The tendency of humans today to be more inclined to copy behavior than chimps, who are more likely to figure things out and "think for themselves" (Macleod, 2006), may also reflect some association with the pea-brain gene, but this hypothesis, like some of the other ones presented in this chapter, requires empirical testing for verification or refutation. In fact, the greater capacity of humans to copy may be more deeply entrenched in human nature, since copycat behavior avoids having to reinvent the wheel and can therefore facilitate the spreading of innovation and culture (Gigerenzer, 2007).

A gene in about 1/3 of Americans predisposing them to a greater tendency to violence (Phillips, 2006) could also be related to the barbarians' spreading their genes. The tendency for human spite which doesn't exist in apes (*NewScientist*, 2007k) might also be related to the existence of human violence. It is quite possible, however, that any tendencies for violence due to spite encompassed in genes are very limited, and violence may largely be caused by the nature of patriarchal society (Baring and Cashford, 1993). That inhumane culture itself appears to have evolved from the long-ago banishment of some people who may only have created a cruel system as a way to survive and fight the matriarchy, and who may have seen violent subjugation of women as a spiteful human necessity..

A puzzling gene that has recently evolved and is associated with attention deficit disorders might also be explained by making humans "fitter" for some environments (Douglas, 2006). The barbarian invaders from the north are once again likely to be the source of this gene, given that colder climates require more physical energy expenditures, and less mental ones, to stay warm. The same may be said for autism, as social relationships may have been less important in an environment of extreme violence and pure survival of the strongest. A gene for intensity of desire that motivates more extraverted aggressive behavior and is favored by evolution in unstable environments has also developed and increased in frequency in the last few thousand years (Nettle, 2007), and the barbarians may be the ones who spread this gene as well.

Given that there were likely some evolutionary advantages at some time somewhere for most genes existing today, including for early aging and death, such as Huntington's disease (Cummings, 2006), it is possible that the harsh environment of central Russia, which might not have been able to support many older people, also helped evolve such early-death genes among the barbarians. Cultural barbarism like a formal sacrifice of a king, when he was too old to be of use for hunting and was superfluous in a primitive society that valued only strength, may also have naturally evolved in that setting. Later old patriarchal rulers may have actually replaced such a custom with the sacrifice of his son that preemptively prevented the king's undesirable fate.

Overall, the rate of genetic change over the last few thousand years has been 100 times the norm for evolution, with a peak in the rate of change about 6000 years ago, and while some of the changes (including a greater resistance to malaria for African peoples) are clearly identified to be adaptive (such as to new diets that evolved a greater tolerance to milk from domesticated cattle for most European people, and such as to climatic changes such as greater intake of vitamin D via paler skin for some), many are related in unclear ways to the brain (Kaplan, 2007). However, there remains no really great systematic genetic differences across nation and ethnic group (Robins, 2005), and so it would appear that the patriarchal barbarians succeeded only in widely mixing their genes with those of other humans but not in totally supplanting basic human nature among any ethnic group. The genes of the more peaceful and benevolent humans remain widespread, as will be documented in greater detail in Part II. Although there continues to be a small number of people with genes that promote extremes in personality behaviors, most of the various personality types that exist in today's humans can be adapted to non-destructive behavior (Nettle, 2007), at least in a humane society.

Perhaps the mixing of the special barbarian, Neanderthal, and matriarchal genes within the basic framework of the pre-matriarchal human genome may have even created a more adaptable species. In particular, the barbarians may have contributed to a more powerful spite and strong will for humanity, just as the matriarchal influence could have provided a more integrated and potent intuition. If nothing else, the past 35,000 years should provide a very strong motivation to return human culture back to a harmony between the sexes, as those tens of thousands of years of hell should provide an example of the extremely unpleasant results of continuing the inhumane societies of today that promote the senseless battle of the sexes. People may only need to be made aware of their prehistory in order to enable bringing Heaven back to earth through a simple recognition of the suicidal futility of fighting their innate human nature.

Chapter 6
Human Nature Perseveres

Despite long-term, cruel attempts to silence women and force them to be weak slaves, especially in places like "democratic" Athens of ancient Greece (Eisler, 1995), women remain at least as intelligent as men and are nearly as strong (Pinker, 2003). Although some traits have been found to develop in only a few hundred generations of a species in the right environment, it can also take up to 10,000 generations (*NewScientist,* 2010d). Apparently, the mass purges of strong non-submissive women in the Inquisition and earlier (Achaya, 1999) haven't been enough to force a rapid evolution of human beings that are characterized by male biological dominance.

One reason for the failure of the patriarchs over thousands of years to eliminate strong, smart women and impose a genetically unassailable system of dominating women and creating an apelike alpha male society for all time may be related to the stupidity of the males in charge. For instance, although the Christian Inquisition from the 15th to the 17th century wiped out millions of women in Europe, many were killed due to a desire for their property and that of their friends or relatives, while many of the rest were murdered for superstitious reasons (Gadon, 1989). Although Sjoo and Mor (1987) have hypothesized that some of the bravest and brightest women were killed in the Inquisition, it's quite likely that many of the smartest women were able to survive, or at least left offspring, because they were able to outwit the idiotic patriarchal cruelty. Many women may have merely been discarded when they were too old to be of any sexual use for common men, many of whom could have welcomed the extermination of "old hags" because of their perceived "uselessness" and potentially undesirable "magical" influence and control.

As an example of the absurdity of the Inquisition, there was a mass murder of many mainland Europeans with freckles simply because they were superstitiously construed as being a sign of a witch by the Inquisition mass murderers (Ellerbe, 1995). Since some of the barbarian tribes that invaded the matriarchies had red hair that is often associated with freckles (Stone, 1976), it seems unlikely that freckles provided any indication of matriarchal ancestry, strength and intelligence in women, or goddess-worshipping tendencies.

In contrast, though blue eyes have been alleged in ancient myths to belong to the people of an advanced culture (Joseph, 2004) and though ancient images of the Goddess have had blue eyes (Gadon, 1989), blue eyes were not persecuted by the patriarchal Christians. Blue eyes actually developed only about 10,000 years ago, which is 10,000 years after the white skin color pigment evolved in people of northern areas to help them access more vitamin D from the sun (Kaplan, 2007). Perhaps the biological tendency of blue eyed people to be less aggressive (Etcoff, 1999) spared this trait from being expunged from the people of continental Europe, as the ruling males with apelike brains would have felt less threatened by less aggressive women. A natural inhibition of male primates to being violent against people with blue eyes (Gooch, 2006) may have also enabled people with this characteristic to survive despite any possible association with matriarchal cultures.

The long-term senseless randomness of the Inquisition slaughter, which was conducted by both the Protestant and Catholic Churches and may have been prompted by the black plague and the resulting fear it inspired (Sjoo and Mor, 1987), may represent one reason it did not affect the characteristics of today's women more. However, one of the motivations of the mass murder by Christian males may have been to torture the superior knowledge of women out of them (Gadon, 1989). It may be no accident that the scientific revolution began shortly after the several hundred years of mass interrogation/murder of females (Sjoo and Mor, 1987), as is also hypothesized in the Donna Read documentary film, "Women and Spirituality: The Burning Times".

Many of the ancient secrets of magic, science, and religion, such as electrical lighting caused by friction (Childress, 2000), may have, even after the patriarchal conquest, remained unknown to all but women initiated in the mysteries (Sjoo and Mor, 1987). The Inquisition could have only partially unlocked these female insights known to at least the elite priestesses in the matriarchy. The evidence that the stereotype of witches' brooms may have been related to older unattractive women on the fringe of society using long sticks for both mixing hallucinogenic herbs and massaging their genitals that can enhance psychic powers, and the archaeological evidence that witches' pointed hats that may have resembled the fabled pointed hat of Merlin the magician were merely related to ancient astronomical time keeping activities (Stray, 2005). Given the history of matriarchal and mystery rites imposing cruel sacrifices on men (Baring and Cashford, 1993), it's possible that many of the Inquisition's cruelties were directed against older frustrated women whose bizarre activities and seemingly mysterious knowledge of the cosmos may have been regarded with great suspicion. Patriarchs may have had some justification in suspecting some of these women's rites were symbolic reenactments of cruel matriarchal ones.

The scientific revolution may have also been spurred by the Europeans' seizure of the advanced civilization in African and Moorish Spain in the fifteenth century that had not moved so far from matriarchy due to a more recent conquest by the patriarchs (Jackson, 1972). Out of a need to survive the wars with the threatening Moslems, the Christian patriarchs may have sought to equalize the Moslems' advantage in knowledge that they came to recognize their own women might have. The discoveries of the Knights of Templar in the only Crusade to succeed in seizing Jerusalem for a short time certainly helped spark an opening in European thought in their own secret societies that spread widely thereafter and was even reflected in the female shape of the Gothic Churches they built (Butler, 2004), and the ruling Catholic patriarchs may used the Inquisition as a means to obtain some of that secret information.

However, the original motivation to employ mass torture to innocent people may have originated in a christian attempt to suppress the more open sexuality that was encouraged by the contact with the Moslems, many of whom hadn't yet developed the extreme sexual inhibitions of the more fundamentalist patriarchal religions (Sjoo and Mor, 1987). Many of the European secret societies such as the Cathars, Troubadors, Templars, and Kabbalists were indeed characterized by much more open sexuality (Baring and Cashford, 1993) that may have been seen as threatening to the patriarchal Catholic rulers who were struggling against invading Moslems with a similar human culture that was more popular with their own people than the inhumane sexual repression of the Catholic Church. Acquisition of knowledge could have originally been a mere byproduct of the torture that may have initially been designed to just obtain information on any collaboration between Moslems and the secret societies. While the choices of victims of the Inquisition seemed in many cased to be either totally senseless or based more on selfish motives like greed for the victims' property choice of victims in that system of knowledge acquisition included people just for having freckles (Ellerbe, 1995), the cruel nature of the christian bible itself certainly makes it feasible that the Christian torturers themselves may have actually believed they were thereby really serving a deity in which they truly believed..

Frances Bacon's instruction to "torture" Nature's "secrets from her" may have indeed had more than figurative meaning with respect to the Inquisition's torturing of women (Baring and Cashford, 1993). In particular, many patriarchs perceived females to be soulless animals lacking in the males' spirituality that was divorced from Nature, but these "soulless" beings were recognized to be very knowledgeable about Nature, especially with respect to herbal medicine that was more effective than Christian faith and obedience (Sjoo and Mor, 1987). Accessing such knowledge for adoption by the patriarchs themselves was perceived to be useful to the maintenance of the christian system.

Chinese myths of a daughter of a king being killed for promoting medicine (Husain, 2003) may reflect events similar to the European Inquisition in Asia at some more distant point in time that enabled patriarchal elites to access women's secret knowledge about medicine and other technology. Because it happened much further in the past with potentially less violence, Asian religions permitted incorporation of the ancient goddesses into their beliefs that also enabled a more widespread acceptance of natural remedies and other prehistoric technologies. The Chinese did indeed have a higher level of technology until Europeans were able to access it through travelers such as Marco Polo, the Inquisition's torture of their own women, and knowledge obtained from the Middle East. This hypothesis might also help explain why Asian religions tend to be more attuned to integrated thought and being that more approaches the reality of quantum physics and modern science than the anti-Nature fragmented Western ones (Capra, 1999).

Women were the likely inventors of culture and civilization (Sjoo and Mor, 1987) and were often identified as the creators of wisdom even by the early patriarchies (Husain, 2003). The existence of only somewhat superior female mental capacities today may reflect a relative deterioration for women. However, the sheer senselessness of much of the Inquisition killings, as reflected in the bias against freckles, may have resulted in many of the smartest women surviving by hiding their intelligence and knowledge. There may therefore have actually been little change in the relative balance of brains during patriarchy.

The rapid evolution of genes over the last 5000-10,000 years (*NewScientist*, 2006b) may have actually had a greater effect on the males. In particular, given that females would have had no right to select their mates in a patriarchy over the last 5000 years, the attributes that women prefer in men, such as larger penises and more intelligence,) might have deteriorated. For instance, if the barbarians were truly the descendants of individuals exiled from civilization, it would make some sense that many started out with the smaller penises and less intelligence that might have increased sexual frustration and even ostracization in the matriarchy, thereby possibly prompting desperate acts like violence that resulted in exile. The evidence that males today are genetically inferior to women in terms of a higher propensity to die at almost any age (*Wall Street*

Journal, 2007) may at least partially reflect the widespread forced procreation with the inferior barbarian males who left more their inferior genes to the males of the mixed offspring.

The fact that women weren't even allowed to see men's penises prior to marriage or mating for life during most of the thousands of years of patriarchal rule would be consistent with many patriarchal barbarians being sensitive and embarrassed about their small penises, or with their general sexual incompetence that can include a refusal to try to please mates such as with oral or manual stimulation of the clitoris. Since men fear being laughed at by females more than anything else (Shlain, 2003), this interpretation of using abstinence before marriage as a way of preventing women from knowing how small the penises of the males in power really were and how incompetent they were sexually seems to make sense. Labeling of sexually frustrated women, including ones with partners or husbands, as diseased with "hysteria" before 1952 (Maines, 1999), along with the prudery that prohibited women from enjoying or even talking about sex, would have helped disguise the ineptness of the ruling descendants of the sexually inadequate pea-brained barbarians. A similar conclusion can be reached about the current USA Republicans and fundamentalists, who disgustingly glorify violence over sex and seem to be intent on re-imposing this cruelty.

The patriarchy has long attempted to suppress female sexual pleasure (Eisler, 1995), possibly in order to inhibit women from seeking men who are less sexually inept than the pea-brained barbarians. Since the capacity of women to experience orgasm, which may have evolved to promote bonding just as it does in bonobos, has been found to vary based on genetic factors (Hooper, 2005), it is possible that the patriarchal society has begun to select for women who experience orgasm less frequently or not at all.

Prior to the invasion of the patriarchal barbarians, virginity referred to women unviolated by men but did not imply an absence of sex or procreation, both of which were once sanctified in goddess-worshipping societies in stark contrast to being made profane under patriarchal rule (Gadon, 1989). This concept of a sexually active and procreating virgin would be consistent with the other evidence that women used to naturally choose their mates without any economic, social, or other pressure like fear of violence. Many of the exiles from pre-patriarchal societies may indeed have been banished or left because they were unable to attract females naturally or were unable to satisfy them sexually, and these exiles might have come to believe that any sexual act violated women because they didn't know how to "do it". Regardless, sex within the patriarchal system may represent a violation of women because the typical unnatural pressures that do affect their willingness to engage in sexual intercourse make it so they enjoy it less, if at all, and because sex under patriarchy typically represents something closer to violating rape than natural human love making that innately turns women on and can hardly be considered a violation of them.

However, the patriarchs, especially those who didn't know how to turn women on and only had sex via rape, would have been motivated to turn this view of virginity among matriarchal peoples on its head. In particular, it's possible to speculate that the mythical procreation of a Christ son of god through a virgin could reflect an ancient patriarchal tradition of women having sexual intercourse only with the patriarchal elites to be more desired by those elites. In fact, the patriarchal conquerors might have been considered such hallowed deities soon after their conquest that sex with them was considered to be holy. Women therefore might have been considered virgins even after such holy acts, whereas sex with mere common mortal men might have been considered as "violating" the females so "sanctified". The conquerors and the people they brainwashed into believing in them may have then perpetuated the myth in the lie of the virgin birth of Jesus. It's even possible that the sacrifice of the sons of such "sanctified virgins" was a tradition which originated in the superstition associated with needing to sacrifice a person each year to regenerate the earth that the hated matriarchies themselves had subconsciously brainwashed the stupid barbarian patriarchs into actually believing. Obtaining vanquished women for such hideous rituals in the patriarchal cults might have actually been one of the motivations of their kings designed to encourage conquest or be sacrificed.

In any event, a continuation of the patriarchal system of sex and relationships between the sexes would tend to drive the human species back in the direction of apes. However, the trends in women's liberation over the last hundred years (Israel, 2002) imply that women will keep humanity from evolving back into violent monkeys. The fact that violence in women is extremely rare and is often motivated by male friends when it does happen (Klein, 2003) implies that they remain peaceful, despite a natural feeling of spite for all that has been done to them in the last 6000 years and despite the passing through of the male genes for violence. They and all human beings in general who can overcome patriarchal brainwashing can return humanity to its peaceful roots.

Chapter 7
Prehistory May Lead to an End of the Patriarchal Hell

Despite the possibility of patriarchal barbarians spreading genes for pea-sized brains and clit-sized penises through rape and other forms of patriarchal sex, it does seem highly probable that the egoistic barbarian genes will be overcome. In particular, there is widespread evidence of a continued human tendency to cooperate that is more likely to show itself in a cooperative environment which discourages pathological traits, and there are evolutionary forces that are selecting for that cooperative tendency (Pinker, 2003). A new field of social neuroscience has opened up based on the "idea that human brains are 'wired to connect'" (Underwood, 2004). There even remains a specialized region of the brain that is related to "maternal love", is associated with "compassion", and appears to have extreme power to control the rest of the brain and create feelings of happiness (Begley, 2007a). The natural bonding effect of sex (Eisler, 1995) and the continued close biological connections between wisdom and sex in the brain (Sjoo and Mor, 1987) that have overcome millennia of unnatural patriarchal repression (Stone, 1976) would seem to ensure that the current triumph of patriarchal evil was only temporary.

As information spreads via the enhanced communication among people that is enabled via the internet and travel for example, more and more people are likely to find out that the patriarchal religions effectively worship a penis that is reflective of an unnatural and perverse homoeroticism which essentially managed to do a sex-change on a Goddess (Achayra, 1999). Even the Goddess's pleasure- and life-giving vagina, which was previously worshipped (Gimbutas, 1989), was assumed by the fabricated omnipotent god, which created all by merely dictating, including the pleasure of Heaven that existed after death through mere worshipping obedience to the god. This fairy tale has to be seen to be increasingly absurd as people are merely confronted with the facts. While exact reasons for the fabrication of every aspect of the origin of the Christian myth aren't completely known, many have been fairly well documented in earlier chapters.

An increasing number of humans may also become aware of the disgusting torture and other acts of the patriarchy over the last few thousand years (Ellerbe, 1995) and of its inhumane senselessness in light of recognizing whey such cruelty occurred. As a result of such recognition of the hypocritical cruelty Christian religion as well as of the manipulative nature of its lies, it is possible that the patriarchy will collapse out of the patriarchs' contempt for their own system.

As males figure out that the patriarchy has evolved into a system that does not satisfy their needs for emotional and sexual release nearly as well as the ancient one that sanctified love and sex as opposed to vilified it, men will likely support the revolution in social relations at least as much as women.[9] The deceit and artificialness that has marred relations between the sexes for millennia (Gibson, 2006) should quickly disappear once natural feelings are allowed to replace the perversions of the patriarchal system (Eisler, 1995).

Regardless, the patriarchal mechanistic science, which replaced magic and religion in many respects, may not be the final stage of knowledge and enlightenment (Jackson, 1972). Instead, a return to an unfragmented religion may help advance technological developments. For instance, the fragmentation of the universe by traditional science is being proven to be less reflective of reality than more unified theories (Capra, 1999).

Since the very torturing of knowledge out of women during the Inquisition would seem to indicate that even the modern but fragmented scientific revolution would have been impossible without stealing information from women, it becomes doubly clear that future progress depends on accessing women's brains in a natural, cooperative environment. At the very least, contrary to Harvard University's ex-President Lawrence Summers' assertion that women "may lack the 'intrinsic aptitude' to succeed as top scientists" (McGinn, 2006), it should become increasingly apparent that society needs women in equal leadership positions (Gadon, 1989), especially in modern science where integrated analysis and cooperative communication are badly needed (Capra, 1999). Given the scientific evidence that women tend to be more patient, more caring, less violent, more capable of using their intuition, better able to integrate human emotions into decision-making, and more capable of weighing options (Tyre and Scelfo, 2006), it would seem important that women assume more leadership roles in all areas, as in prehistoric times.

In fact, given females' superiority in reading and writing today (Tyre, 2005), Shlain's (1999) assertion that the introduction of alphabetic linear reading and writing corresponded with the introduction of the patriarchy because such writing was more suited than holistically imaged writing for males, who tend to have relatively stronger left brains that are heavily used for the processing of reading and writing. Women's greater ability in learning even the more fragmented sound symbols of more modern languages provides evidence that females are so much more intelligent than males that they are actually mentally superior to males even in areas where they have been excluded for thousands of years of the patriarchal hell on earth.[10] Women's problem-solving and decision-making sections of the brain have also been found to be larger than that of males on average (Hong, 2008). However, the patriarchy's indoctrination of women into thinking they are futilely inferior in mathematics has been found to decrease their performance on math tests by 50% (Adler, 2007). There is also especially strong peer pressure against girls doing well in mathematics in the U.S. to avoid being labeled "nerdy", and that fact also contributes to the extremely low number of American females who excel in math (Grayling, 2008).

The status and situation of women in many countries like the U.S. has improved substantially in the last hundred years (Israel, 2002). Although women still do not have equal rights/power and are badly underrepresented at the top of society everywhere in the world,[11] the trend toward greater equality might be expected to continue.[12]

However, given what the patriarchal rulers have done in the past to attain and maintain power, there remains a danger that they will try to reverse the trend. The Republican fundamentalists appear to be trying to do just that. Besides justifying their crimes with an alleged war on terrorism that provides cover for their fascist imperialism, they also frequently make themselves out to be crusaders serving a Christian god. If their power is especially threatened, they may even accurately quote their patriarchal bible to insist again that women are evil, and, invoking the book of Revelations, assert that feminist leaders are antichrists.

During the millennia of goddess worship, the "6" was actually an ancient symbol for the female genitalia (Acharya, 1999). Since goddess worship typically involved a 3-

deity set of maiden, mother, and crone (Walker, 1988), the "666" in the biblical book of Revelations could easily be construed to represent anyone advocating any movement toward greater power for women as beasts serving the "devilish" matriarchal goddess.

As previously mentioned, the book of Revelations may very well relate to past events as opposed to predicted future ones (White, 1980). The battles against the antichrist having the mark of the beast ("666") on the forehead reported in Revelations actually seems to reflect the patriarchal conquest in ancient times of the old matriarchies. The latter societies worshipped nature, mother earth, and earthly pleasures and hence could be considered an evil earthly beast by the patriarchal barbarians, who abhorred their physical dependence on women so much that many became religious asexuals, masochists, homosexuals, and/or rapists (Eisler, 1995).

Ancient myths indicate that the prehistoric system of matriarchal communism was a heavenly paradise (Gooch, 2006). For instance, there are ancient Celtic myths of a distant civilization that worshipped goddesses, promoted free love, was devoid of personal possessions, and was considered a heaven (Gooch, 2006). Since there is also extensive evidence of advanced civilizations being destroyed worldwide by natural disasters over the last 12,000 years (Joseph, 2006), it is quite likely that such paradises once permeated much of the globe.

It is true that the nature and astrological worship of these former cultures (Gadon, 1989) probably contributed to their undoing, as it was nature and extraterrestrial bodies themselves that devastated them with earthquakes, floods, and comets (Joseph, 2006). Suffering from a long stream of such demoralizing events that also greatly weakened the matriarchy materially, it is no wonder that they eventually succumbed and adapted to a patriarchal religion that saw nature as the enemy. The superior ability of men to deal with the extreme problems probably also helped advance the patriarchal takeover (Jiayin, 1995), since the matriarchy may very well have given men insufficient power and equality, as explained in more detail in Part II.

It is that latter problem which reflected the sexual hierarchy existing in matriarchies that actually made the matriarchal societies something less than paradises, at least in comparison to pre-matriarchal human cultures. The matriarchies could have been perceived as paradises merely in comparison to the hell patriarchies have imposed on earth. In addition, it's quite possible that the inequities and cruelties under matriarchy were only moderately increased over time, so that they were easily differentiated from the true heaven of the balanced human societies that preceded them, especially because of rapidly improving technology over time that made human life easier and that potentially resulting partially from matriarchal knowledge obtained from the Neanderthals or insights gained while female exiles from the pre-matriarchal cultures themselves had used drugs or other means of seeing the world differently.

Nevertheless, the deepness with which a communal matriarchal paradise has been impressed in humans' behavior and genes is indicated by the fact that this concept has not been totally driven out of human thought despite thousands of years of cruel patriarchal rule. A human desire for that paradise is very real, although it has merely been manipulated by the patriarchal religions via an unfullfillable promise after death for obeying the cruel patriarchy. The mere enlightenment of enough people may permit a return to a real heaven on earth that the patriarchs so vehemently forbid and fear.[13] At the very least, knowledge of where humans have been can help determine humanity's future potential (Sjoo and Mor, 1987).

However, it must be emphasized that the overthrow of the matriarchies thousands of years ago was not only useful but also necessary, as even enlightened feminists admit (Baring and Cashford, 1993). As indicated by Jiayin (1995), it was men who save humanity from the calamity and destruction (during the flooding and catastrophes at the ending of the last Ice Age) that was brought on coldly and unjustly by the previously worshipped Nature, which women could not control even with all their worship, tricks, hypnotism, and human sexuality. Women need men and vice versus, and not just for sex. Some women may even have a tendency to prefer, due to millions of years of both cultural and genetic history, a return in the direction of the alpha male ape society, just as some frustrated males may want a return in the direction of matriarchy in order to have the opportunity for more sex. However, the excitement, pleasure, and happiness of both sexes remains maximized by mutual love without any worship, artificial sexual inhibitions, or other manipulation, since such mutually revering relationships are most deeply embedded into human genes via millions of years of such, as is explained in more detail in Part II of this book.

Chapter 8
Is the Evidence on Matriarchies Conclusive?

The foregoing facts are generally ignored by "mainstream" anthropologists. One of their widely accepted views admits the egalitarian nature of humanity for most of the last 2.5 million years but asserts that the development of agriculture after the end of the ice age about 13,000 years ago resulted in a surplus of wealth that resulted in conflict and a need for a hierarchy to own and distribute it until about 250 years ago when those at the bottom of the hierarchy began to achieve greater freedom (Vugt, 2008). Even if some evidence could be provided that prosperity, as opposed to scarcity, causes conflict and inequality among humans today, it would still be necessary to provide some facts supporting a hypothesis that the existence of additional wealth could cause humans to deviate from their natural egalitarian ways that had been ingrained into human nature over millions of years. There actually doesn't appear to be any evidence whatsoever supporting the wild conjecture. In addition, that speculative hypothesis seems to ignore the effect of the catastrophes that occurred as the ice continued to melt over the last 13,000 years, as well as all the extensive evidence that matriarchal rule existed between 35,000 and 6000 years ago before it was replaced by a hierarchical patriarchy.

When confronted with the theories and hypotheses presented in this paper, many anthropologists state that the widespread myths telling of original female deities being overthrown by male deities are insufficient alone to imply conclusively there was once a matriarchy. They then typically refuse to consider or discuss the matter further. While it is certainly true that the mere widespread belief in that common story in ancient religions doesn't prove the existence of prehistoric matriarchies, it is indeed consistent with the hypothesis. In addition, as mentioned in the prior chapters, there is such a substantial amount of other evidence that is so consistent with the hypothesis that it merits it being categorized as valid if not outright accepted as true beyond a shadow of a doubt. Instead of bigotedly refusing to even consider the matter, it would be useful if anthropologists came up with an alternative explanation for such evidence that matriarchies were widespread up until at least 6000 years ago. It would be even more useful if they were able to acknowledge at least to themselves that their prior refusal to even consider the evidence derives strictly from their brainwashing starting as infants by today's societies that are propped up by an indoctrination of alpha male superiority.

All archaeological evidence going back two million years seems inconsistent with human cultures being characterized by patriarchies for more than the last few thousand years. The large number of female symbols emanating from time periods between 35,000 and 6000 years ago far outnumber male symbols and seem alone to imply greater reverence for females than males then, as does the more extensive archaeological evidence at Crete and Catal Huyuk. Gimbutas's (1977) findings of a large reduction in the female symbols after the Kurgan/Aryan invasions from central Russia and the decline in evidence of civilization is consistent with a conquest by destructive patriarchal barbarians that happened alongside an ever lessening importance for female goddesses in older religions and ever lessening rights for females in written laws, etc.

While some "investigators" have suggested that the widespread existence of female figurines were just prehistoric pornography (Goodison and Morrison, 1998), perhaps these "experts" could also explain why that "pornography" so widely displays women with big bellies that are a sexual turn-off to males who evolved to copulate with fertile women who didn't have the bloated stomachs of pregnancy. Maybe they could also invent an explanation for why such "pornography" displaying women at all disappeared after the Kurgan/Aryan invasions and imposition of their patriarchal gods? Since some forms of matriarchal societies clearly even existed into modern times such as in China (Jiayin, 1995), it would be helpful if such "experts" were able to explain why the mere possibility of the existence of matriarchies doesn't exist in their own infantile minds.

There is evidence that matriarchal tendencies tended to persist in even some parts of Europe until the Roman conquest, such as with the Druids. The Romans even reported that the German Druids treated their women like goddesses (Davidson, 1990), and matriarchal marriages persisted in Europe as well as in Egypt long after patriarchal conquest that itself occurred only a few thousand years ago (Walker, 1988). If matriarchies never did exist, as many "experts" insist, it would be wonderful if these brainwashed individuals could explain the evidence here also.

There is no question that an explanation for some individual pieces of evidence can be fabricated by anthropologists indoctrinated with a patriarchal religion. However, if there were some way to rationally explain all such evidence in a patriarchal manner, why hasn't it been done yet? Why instead are there attempts to belittle and silence those who even mention the hypothesis of the existence of prehistoric matriarchies?

It's interesting that fundamentalism has been making a comeback since the initial advancement of the matriarchal hypotheses. Fundamentalists, who strive so hard to dispute evolution and have even succeeded in convincing a majority in the U.S. (Hecht, 2006), may recognize that the only way to fight facts is with fiction that makes people think they will go to hell if they lose their "faith" in the fiction. It's probably no coincidence that the fundamentalists deny the world and the universe even existed before 6000 years ago (Stray, 2005), which just happens to be the time before the patriarchal conquest began that overthrew the prevailing matriarchies. The fundamentalists' allegation of time beginning 6000 years ago derives from the patriarchal bible, which of course was written long ago in an attempt to erase all memory or knowledge of the matriarchies (Walker, 1988).

"Recorded history" essentially is "nothing more or less than a catalogue of causes and campaigns that were justified on the basis of beliefs and values that fit a unique time and place and were designed for specific ends" (Shealy, 2009). The evidence of destruction of all culture upon the initial conquest by the patriarchal barbarians thousands of years ago easily explains the lack of written records, even where there had been written languages prior to conquest. The nearly successful attempt by the Spanish Conquistadors to destroy all the vast written materials of the Mayan natives of Mexico, whom they subjugated in the 16th century, and whose very language symbols or letters were prohibited because they were intrinsically "evil" (Henderson, 1997), may be indicative of the effectiveness of the destruction of such evidence that endangered the brainwashing system of patriarchal fundamentalism. Because it remains difficult to publish facts greatly inconsistent with theories prevalent in the existing system (LeDoux, 2002), knowledge of the facts of human prehistory continue to be only sparsely disseminated.

Chapter 9
Summary of What the Fundamentalists Really Don't Want You to Know

Part I reviews the evidence on prehistoric social relations and discovers a matriarchal civilization which patriarchal barbarians destroyed and replaced with cruel patriarchal religions. Although patriarchal fundamentalists have long sought to erase all memory of. the prior human societies, an overwhelming body of evidence still exists that irrefutably demonstrates the fact that the lies of the patriarchal religions were created to manipulate and subdue others, especially women. Religious fundamentalists have long used their intolerant religious beliefs to "justify' inflicting incredible cruelties on other people (Gynn and Wright, 2008), and this first Part of the book seeks to reveal them for what they are.

Because it has been documented that the patriarchal religions like Christianity were indeed fabricated long ago to make obedience to a ruling elite more acceptable to the masses, there is no chance that the Heaven and Hell or reward and punishment in any form after death as promised by those religions has any chance whatsoever of happening. It is more likely that my youngest son's dream of living on a planet with Pokemon will be the final destination for his soul (as long as he "believes in" Pokemon of course, thereby avoiding the punishment and extreme torture after death reserved for anti-Pokemon heretics and skeptics!). Given today's scientific method of accepting theories that appear based on evidence and Bayesian probability to be true (Matthews, 2008), true scientists and anyone not brainwashed by the Christian cult would have to believe that Pokemon world to be far more likely to be true, and therefore more acceptable as a belief system, than Christianity.

However, in either the case of a belief in a Christian or Pokemon deity and universe, it is clear that each such system was invented by humans as opposed to the invented deity or universe having created the humans themselves. Because the latter was designed largely for imaginary entertainment that could reflect some subconscious insights on life, whereas christianity has been clearly established to be a set of fabricated lies designed to cruelly manipulate people into submission to ignorant violent patriarchs, the Pokemon world might actually represent reality better than the Christian fairy tale, but both remain pure fabrications without any foundation in the true nature of things. The sole fact that both a belief in the Christian god and a belief in Pokemon are known to have been invented for a purpose other than to actually describe reality makes the likelihood of them having any significant degree of truth compared infinitely small, even in comparison to any of the infinite number random belief systems that a person could think up instantaneously without thought.

In contrast, the theory of everything presented in the second part of this book is grounded in scientific facts and intuitive logic. It reviews a massive amount of research across many disciplines to provide likely answers to the true meaning of life, the universe, and everything, but it may require overcoming infantile beliefs in religious lies for full comprehension, as is made possible by this first part of the book.

Much of the information provided here may assist those seeking the truth, oneness, and nirvana that mystics and so many others have long been seeking (Eisler, 1995). Most religions regard such heavenly insights as the Holy Grail (Baring and Cashford, 1993), including for many religions that have little else in common otherwise

(McNamara, 2006). In matriarchal religions, the Holy Grail itself was the Goddess's Cauldron or womb where all life was created (Walker, 1988), but that holey phenomenon was later vilified in patriarchal religions to be a fictitious Hell (Acharya, 1999). Beliefs in a fictitious goddess are just as unlikely to lead to an understanding of the nature of the universe that so many seek as a belief in a fabricated god. In contrast, knowledge of the distant past can help free people from the actual hell on earth that the patriarchal and fundamentalist religious lies create in the minds and spirits of its naïve slave believers and exploited robots in human bodies.[14]

The evidence in this article may supply sufficient rational proof to overcome many people's preconceived notions that patriarchy is the norm of human relations and that increased women's rights and equality is unnatural. Unjustified prejudices against women may thereby be eliminated. A better understanding of evidence on past relationships between the sexes can also provide women with sufficient knowledge to eliminate their own self-doubts that inhibit their capacity to achieve their true potential.

Given that most women and many men may be unlikely to accept today's unnatural situation,[15] and given that the end of humanity is virtually certain under patriarchy via violence, environmental destruction, etc., a return to the naturally balanced human society before there was a matriarchy or patriarchy seems optimal. Such a return to paradise doesn't require abandoning any useful modern technologies, many of which were originally designed in prehistoric times and only more recently rediscovered. While some might fear such a society could transform itself again into a matriarchy and start the vicious cycle of the battle between the sexes anew, such a retransformation seems impossible if men are made aware of women's nature and the past. In the meantime, while awaiting a better society, or trying to spur it on, one can adapt to the existing system best when one understands it is based on pure lies at the core.

Footnotes

1. Since there was no mention of any enforcement of the male obedience vow, it is likely that it was based on emotional bonding as opposed to a form of slavery. In addition, because marriages were not forced, men still had some say/choice in the women they married and pledged to obey. Moreover, since relationships and marriages could be broken up at any time, men could always just leave and pick a new mate if a female became unreasonable. Over time, this situation might have been expected to result in only the nicer women being selected for marriage and procreation. If this phenomenon reflected in any way the ancient past, it would have led to the evolution of the nicer women, as is consistent with typical good nature of women today. The past few thousand years of patriarchy may not have affected this characteristic of women, since females in an extremely oppressive patriarchy are largely chosen strictly on physical beauty (with any inherent "meanness" in some women merely repressed by male violence). The physical beauty of women was actually most important in a society (Sparta) which provided for such an equal relationship between the sexes (including equal education and property rights for women and open marriages atypical of patriarchies) that it was widely perceived by the patriarchal world at the time to be a matriarchy (Cartledge, 2003). Some honoring of women has long existed even in the patriarchal U.S., with some of the historical U.S. customs of putting women on a pedestal even causing one Japanese visitor to conclude in 1872 that American men "Follow the Way of Serving their Wives" (Huffman, 2004). The relationships of the future may very well have both partners pledging love-blinded obedience to each other, as may indeed have also been characteristic of ancient relationships to some extent. In fact, the matriarchies that persisted in some parts of China until recently indicate that any one-way obedience vow may have been related to sexual relationships occurring within the boundaries of the female's clan, thereby making the male guest subject to the female clan's rules, but only as long as the male chose to remain in the relationship, which could be freely terminated at any time by either party (Jiayin, 1995). It is instructive to realize that actual or formal slavery does not appear in societies until after the collapse of a prehistoric communal economy (Rodney, 1982). Communism and a free cooperation were innate to humans for millions of years until the patriarchal conquest a few thousand years ago that has led to a dominator culture whose hierarchies enforce unwilling obedience via material punishments (whether via formal slavery, capitalist "incentives", or other means) and create sexual perversions as a means of adapting to this artificial system that also inhibits sexual release (Eisler, 1995). It seems very unlikely that any form of economic slavery actually existed under the matriarchy except related to mutual voluntary adoration and cooperation, given no evidence of it, given that even differential grave cite goods for men (hunting weapons) and women (jewelry for beauty and children) in one area (Sjoo and Mor, 1987) likely represented differential importance attached to such goods across sexes (and possibly some division of labor between the sexes) as opposed to economic inequality or hierarchy, given that mutual love and adoration remain the most attractive and enjoyable for

humans (Deida, 2002) and likely the most useful from an evolutionary point of view, and given that the quotes on male servitude are from a time when the patriarchs had largely conquered the world (and may have represented some female defense against the one-sided surrender/subservience of women in the patriarchal system that had been at least partially imposed at the top of society, as in Egypt under the pharaohs, but that still attempted to retain some vestige of the bliss of mutual adoration for common folks). In addition, since the patriarchal Greeks also made statements about women ruling in Sparta that were related more to some equal rights for women there (that were extremely unusual at the time in the other parts of Greece) as opposed to actual female rule or matriarchy (Cartledge, 2003), the allegations about male servitude were likely incredibly exaggerated in Egypt (since they were made by Greeks living in an extremely patriarchal society at the time), as is readily apparent from the statement on the USA in the nineteenth century.

2. There is some evidence that humans are more similar to orangutans in some respects, such as the ability to walk upright (Schwartz, 2005), but humans appear to have diverged from orangutans over 10 million years ago (Stringer and Andrews, 2005). Some evidence exists that humans' ability to walk upright developed that long ago while both humans and orangutans were still largely living in trees (Hooper, 2007a), but both species could have also developed that ability separately. It's also possible that chimps and bonobos also once had the ability but lost it (as bonobos and chimps may have always lived in environments where the ability to walk upright was not necessary or even counterproductive). Evolution often can cause unnecessary traits to disappear in some but not in others for whom the traits turn out to be more important (Spinney, 2007). Since a large number of genetic changes spread laterally via viral infections as opposed to through strict heredity of random mutations (*NewScientist*, 2008g), the most likely cause of humans being most similar to bonobos but also have many characteristics akin to orangutans may be a direct hereditary relationship of humans with orangutans over ten million years ago before the dispersion of the humans to Africa (while the orangutans remained in Asia) where the physical proximity to bonobos caused humans to be affected by their genes through viruses over millions of years that were selected by evolution because they were adaptive to the African environment they shared.
3. Although there was a massive purge of strong, intelligent women in Europe during the Inquisition (Ellerbe, 1995), it is unlikely that evolution could have changed women's basic preferences very much in the few hundred years of those atrocities (especially since the strongest and most intelligent women may have strong and smart enough to make men believe they were weak, stupid, and pro-patriarchy out of self-preservation). For instance, while women with financial problems value men's wealth more than their appearance, affluent women still find men's physical appearance more important than their finances (*NewScientist*, 2006c). While a majority of females might be attracted by macho men like Arnold Schwarzennegar in the *Terminator* and Rhett Butler in *Gone With the Wind* (Wrangham and Peterson, 1996), it should be pointed out that those men are perceived to be protecting women from even worse, more violent men. In

addition, although both males and females tend to be attracted to more "aggressive" mates (*NewScientist*, 2005a), it is quite possible that this tendency is strictly because an attractive physical appearance might be correlated with "aggressive" behavior (perhaps because the good looks promote self-confidence and strong will), and the "aggressive" behavior itself may not be attractive. More violent males may be more popular with other males, and females may feel pressured to be nice to attract their nicer friends (Wilkerson, 2005). Some women may actually like violent males if they are popular and held in high esteem by society (Wilson, 2005), as is likely in a violent patriarchy where males rule. However, a conjecture that many women are attracted to violence and violent men (Bond, 2007) is unsupported by any evidence and is utterly absurd, especially since any female love of more violent males is probably dependent on those men being nice to their women. Moreover, while women like emotionally strong men, they still far prefer men who in continuous small ways make them feel like individual princesses (Louis and Copeland, 1998), as would be consistent with an ingrained goddess culture that has not yet been purged. Similarly, although patriarchs like Hitler and the Nazis were attractive to some women, even those sadistic patriarchs felt the need to both honor and protect women (Knopp, 2000). The continued natural inclination of males to serve a pure, beautiful woman (Reich, 1946) provides evidence of the goddess culture still being innately buried in the male psyche as well. Engels (1972) has hypothesized that relationships between the sexes in the future might very well be characterized by voluntary egalitarian partnerships between loving non-relatives that could easily be dissolved by any partner.

4. Astronomy was originally integrated with astrology under ancient matriarchies but was condemned by patriarchs (Achayra, 1999). Research has indicated that personalities (and even mental diseases) do demonstrate some systematic variation depending on the month of birth, although weather effects on the mother and fetus are thought to be the cause (Motluk, 2007). Given that the universe has very strong electromagnetic fields (including from other stars and galaxies) that have effects from enormous distances that are not entirely understood (Ananthaswamy, 2006), and given that organisms such as bats actually use electromagnetic fields to guide themselves (*NewScietnist,* 2006o), it is possible that there are more subtle causes for systematic astrological traits, especially given the effect that cosmic rays might have in determining the initial pairing of genes during conception. There are many factors which go into the pairing of genes at conception that are far from understood, including with respect to the "simple" ones that determine the sex of the child (Douglas, 2002) that have even been alleged/reported to somehow be affected by the attractiveness of the parents (*NewScientist*, 2006e). Although the finding that eating more during pregnancy increases the chances of creating boys due their greater nutrient needs (*NewScientist*, 2008d) might explain the latter phenomenon if attractive women tend to eat less in order to maintain their attractiveness, the finding could also be at least partially related to evolution rewarding specialization in sex attributes, where beauty is more important in female humans to win the love/worship/support of males. Just for example, male monarch deer that are most

attractive to females have been found to create female offspring that have less sexual success and create less offspring when they do mate than other female deer (*NewScientist*, 2007j).

5. Some have hypothesized that the domestication of cattle helped men figure out that they had a part in the magical powers of life creation (Sjoo and Mor, 1987), hence reducing justification for deification of women (Baring and Cashford, 1993). If women had previously figured out the connection between sex, pregnancy, and birth, as might be assumed to be likely from an evolutionary point of view in order to enable them to minimize the risks of child deformities via incest, men might have felt an additional element of spite against women for being deceived on this issue for untold millennia. However, the inability of women to do anything to stop the horrible and widespread natural calamities over thousands of years, combined with male successes in reducing their impact (Jiayin, 1995) was probably a much more important factor in causing the decline in the worship of goddesses, who began to be perceived as objects to be controlled by men especially since they represented the forces of nature whose infliction of damage needed to be inhibited..
6. Race in the traditional sense refers to groupings such as Asian, African, and Caucasian that differ most merely in white blood vessel genes (Coglan, 2007) and therefore immunity to different diseases, as well as skin color to adjust for sunlight (the fact that South Africa with a mild climate didn't have whiter inhabitants prior to European colonization might imply a more tropical past there, perhaps due to different positioning of the poles, which definitely do change at least slightly, if not drastically, over time). The "race" of barbarians being referred to here may have represented a mix of the traditional races but have been differentially characterized by a more widespread existence of a smaller penis, a smaller brain, and genetic mutations that increase tendencies to violence and inhibit natural tendencies toward cooperation and maternal/female love. Given the capacity of the barbarians to mate with the true human races (and many of today's patriarchal dictators could be descendants of such matings/rape), it is not possible to label them subhumans or another species in spite possible huge difference in the aforementioned traits.
7. Given that autism is 4 times more likely for boys than girls (*NewScientist,* 2006l), and given that older men are several times more likely to produce autistic children than younger men (Cummings, 2006), it is conceivable that the genes for autism developed amongst barbarians who did not honor their elders and instead relied on brute force to determine mating and other privileges. In such an environment, an autistic child would be more dependent upon a parent and be easily motivated to use any extraordinary intelligence or youthful strength (that might be relatively greater in the males) to guarantee the reproductive access of the older parent, thereby making the existence of the gene rational from an evolutionary perspective. However, women's desire for older men who had proven themselves may have continued to create a problem for the younger barbarians, who may have been the ones who were originally exiled (or exiled themselves) from the matriarchy because of a frustrated preference for younger as opposed to older women, and who continued to view violence as a means of satisfying that

preference (thereby resulting in more psychopathically violent and selfish natures evolving in those societies that would inevitably self-destruct eventually without some sort of system to deter excesses such as through formal control or laws and punishment for breaches of the defined rules—since such rules can never eliminate natural tendencies by psychopaths to circumvent the system or).

8. Having developed a test for determining the existence of this brain mutation (and even applying for a patent on the test), researchers with a reputation for racism have been unable to discover a positive relationship between IQ scores and the existence of the mutation and even stated, "[We] had no luck … no matter which way we analyzed the data" (Balter, 2006). This result is far from surprising, and it's interesting that no test for a negative correlation was mentioned. It's also important to point out that IQ test results have been found to be positively correlated with family wealth levels for reasons that have nothing to do with intelligence (Pizzigati, 2004), and so even a finding of insignificant results might imply a negative correlation if there were a control for family wealth level if it is assumed that the descendants of the patriarchs have greater wealth through the patriarchal system of inheritance, patriarchal networking, and power transfers (but given the patent application for the genetic test, it's unlikely that an unbiased scientist will be able to conduct the necessary research). Regardless, the mutated gene's effects on brain size may have been turned on only in special circumstances (such as causing a very small brain size a harsh environment with limited food resources during brain development) and therefore only rarely display effects on intelligence in modern populations.
9. The finding that even male possums (which normally prefer polygamy) have a natural tendency to turn to monogamy when they become sexually frustrated through artificial changes in their environment (*NewScientist*, 2006h) implies that there is nothing intrinsically civilized/advanced/human about a formal monogamy imposed through sexual repression and other pressure. In addition, since women tend to have a reduced sexual drive in a committed relationship (*NewScientist*, 2006g), men can easily be sexually frustrated (and controlled) by women even in a monogamous relationship, and this characteristic may have helped permit the transition from the ape alpha male society to the bonobo-type and Hawaiian-style cultures that were blissful for all humans for millions of years (as well as exceptionally productive/creative/progressive) until recently (as well as enabled women to persevere and reverse abnormal cases/times of patriarchal rule that may have resembled the era of the most recent few thousand years that natural instincts/reason should end soon, unless the males destroy the earth and themselves first). As indicated by Reich (1946), many psychological problems and diseases among humans stem from sexual repression (that leads to anger, violence, and war), and these problems would be expected to become less important in a society that is less inhibited sexually. However, sexual liberation is only one component of the solution (Eisler, 1995). Addictions and compulsive behavior that are counterproductive to the individual and society (Phillips, 2006) may still occur unless we humans return to our most basic electrical/chemical instincts/pleasure of unconditional love/worship of our sexual mates.

10. Moreover, since the imposition of the patriarchy and the oppression of women set civilization back thousands of years (Stone, 1976), even simple patriarchal logic implies that culture, technology, and other aspects of civilization can be expected to advance more rapidly in general as women move closer to equal rights and are better motivated to contribute to progress.
11. Many of the women who do actually make it to the top of the pyramid may feel forced (even in the patriarchy "lites" of countries like the U.S.) to essentially become "males" (effectively trapped inside women's bodies) in order to rise and maintain their high position. Even then, the mere fact that there is a reasonably high probability of a female taking time out from a career to go through pregnancy and child-rearing results in women managers and entrepreneurs being prejudiced by the capitalist system even if they don't ever become pregnant. For instance, unless there is some sort of government quota or compensation scheme that somehow offsets any costs/risks associated with the greater possibility of women (in comparison to men) leaving the workforce temporarily or permanently (and thereby making a costly and potential inadequate replacement necessary), capitalist participants (including employers and financiers) might be biased against women out of self-interest alone. Thus, even if the old-boys network and male prejudices that continue to bias opportunities against women (Walker, 2000) were somehow eliminated, and even if the government-financed maternity leave and child-care existed that is necessary (but not sufficient) for women's equality as well as vital for the long-term health of society (Albrecht, 2002), the capitalist system would remain stacked against women.
12. Patriarchal religions and their leaders represent an attempted reversal toward alpha male ape societies. There are indeed frequent genetic reversals, such as fish born with limbs and people born with tails or webbed limbs, but such backward mutations generally do not prevail unless the environment changes to make the reverted species a better adaptation (Page, 2007). There is nothing to indicate that the environment would make alpha male apes better adaptations to it, and instead the evidence seems to indicate a greater need for a return to humans.
13. Cruttenden (2006) provides some fascinating astronomical and archaeological evidence that there may actually be cosmic forces at work that are helping for a return to an enlightened paradise. According to that author, much of the evidence has been destroyed, but far more has been preserved in "myths", which along with seemingly overwhelming astronomical evidence, are largely ignored/distorted by the patriarchal rulers.
14. Given that many ancient religions indicate that humanity is about to enter a new period of enlightenment (Perkins, 2004), perhaps facilitated by cosmic events/cycles (Cruttenden, 2006), enhanced knowledge of humans' past may represent an important component of that process. Note that this paper is not intended to motivate a return to a matriarchy but to our own human roots of balance. By contributing to knowledge on human history, some of the ignorance that enabled women to trick men into accepting a matriarchy with its cruelties to males can be reduced, thereby encouraging both sexes to seek a balanced society that does not repeat the deception and subsequent violence on a small or large scale, as explained in more detail in Part II. In any event, a return to the

prehistoric matriarchal religion of worshipping a single goddess completely responsible for creation, life and pleasure, and death persisted for tens of thousands of years (Sjoo and Mor, 1987) is unlikely, given that men have figured out that they are partially responsible for creation, for life and pleasure, and especially for death. Any new religion is likely to be consistent with modern science (and quantum physics), perhaps even returning to a balanced worship of the entire cosmos that contains both male and female forces, as in the religions of many native Americans (Peat, 2002)

15. Actually, many women may be quite happy with less anti-female forms of patriarchy. For instance, in much of the Western world, women are treated with lots of special male attention when they are young, and they are somewhat protected by marriage morals and divorce laws from being abandoned as they age and become less popular. However, a return to a system that facilitates freer, more frequent, and more healthy sex for all should be desirable to all who haven't been totally brainwashed by the patriarchal system. An example of the power of women to enact change is provided by strike in Russia in 1917 by women that shortly thereafter led to the overthrow of the dictatorial czar and the eventual replacement of capitalism with communism later that year (Durkin, 2008).

References

Achayra S. *The Christ Conspiracy: the Greatest Story Ever Sold.* Adventures Unlimited Press: Kempton (1999).
Adler, R. "The Curse of Being Different." *NewScientist* (January 13, 2007), 17.
Adovasio, J. *The First Americans.* Random House: New York (2002).
Albert, A. and S. Otto. "Sexual Selection Can Resolve Sex-Linked Antagonism." *Science* 310 (2005), 119-121.
Albrecht, G. *Hitting Home: Feminist Ethics, Women's Work, and the Betrayal of "Family Values*. Continuum: New York (2002).
Ananthaswamy, A. "North of the Big Bang." *NewScientist* (September 2, 2006), 28-31.
Balter, M. "Brain Man Makes Waves With Claims of Recent Human Evolution." *Science* 314 (December 22, 2006), 1871-1873.
Baring, A. and J. Cashford. *The Myth of the Goddess.* Penguin: New York (1993).
Barras, C. "Macaques Prepared to Pay for Sex." *NewScientist* (January 5, 2008), 6.
Begley, S. "How Thinking Can Change the Brain." *Wall Street Journal* (January 19, 2007a), B1, B2.
Begley, S. "Beyond Stones & Bones." *Newsweek* (March 19, 2007b), 53-58
Bergner, D. "What Women Want." *New York Times Magazine* (January 22, 2009).
Bernstein, D. and P. Nash. *Essentials of Psychology.* Houghton Mifflin: Boston (2008).
Bond, M. "Fighting the Great Delusion." *NewScientist* (September 1, 2007), 51.
Bowjanowski, A. "Piercing the Heart of a Fiery Colossus." *NewScientist* (November 7, 2009), 18-19.
Burling, R. *The Talking Ape.* Oxford University Press: Oxford (2005).
Buss, D. *The Evolution of Desire.* Basic Books: New York (2003).
Butler, A. *The Goddess, the Grail & the Lodge.* O Books: Winchester (2004).
Capra, F. *The Tao of Physics.* Shambhala: Boston (1999).
Cartledge, P. *The Spartans.* Overlook Press: Woodstock (2003).
Chandler, D. "Extraterrestirial Impacts in Human History." *NewScientist* (February 2, 2008), 42-43.
Chia, M. and W. Wei. *Sexual Reflectology.* Destiny Books: Rochester (2003).
Childress, D. *The Technology of the Gods: the Incredible Sciences of the Ancients.* Adventures Unlimited Press: Kempton (2000).
Coglan, A. "Handsome Men Evolved Thanks to Picky Females." *NewScientist* (May 12, 2004)
Coghlan, A. "A Subtle Key to Human Diversity." *NewScientist* (January 13, 2007), 8.
Coghlan, A. and E. Callaway. "Intelligent Design of Domestic Breeds." *NewScientist* (May 2, 2009), 10.
Cohen, G. "The Myth of the Midlife Crisis." *Newsweek* (January 16, 2006), 82-86.
Conty, P. *The Genesis and Geometry of the Labyrnth.* Inner Traditions: Rochester (2002).
Cruttenden, W. *Lost Star.* St. Lynn's: Pittsburgh (2006).
Cummings, M. *Human Heredity.* Thomson: Belmont (2006).
Das, S. "Creatures of the Flame." *NewScientist* (June 8, 2009), 47.
Davidson, H. *Gods and Myths of Northern Europe.* Penguin Books: New York (1990).
Deida, D. *Finding God Through Sex.* Sounds True: Boulder (2005).
De waal. "Midnight Tips for the Clinton Camp." *NewScientist* (March 1, 2008), 50-51.

Douglas, K. "Boy or Girl?" *NewScientist* (September 14, 2002), 42.
DuBois, B. *The World and Africa: an Inquiry into the Part which Africa has Played in World History*. International Publishers: New York (1965).
Dunbar, R. "We Believe." *NewScientist* (January 28, 2006), 30-33.
Durkin, K. "International Women's Day." *Workers World* (March 13, 2008), 9.
eHarmony. "7 Fatal Flaws of Selection." *eHarmony Advice* (2008), http://advice.eharmony.com/?page=articles/view&AID=51&cid=23600&aid=70351
Eisler, R. *The Chalice & the Blade*. Harper & Row: New York (1988).
Eisler, R. *Sacred Pleasure*. HarperCollins: New York (1995).
Eisler, R. and R. Miller (editors). *Educating for a Culture of Peace*. Heinemann: Portsmouth (2004).
Ellerbe, H. *The Dark Side of Christian History*. Morningstar and Lark: Windermere (1995).
Emanuele, E., P. Politi, M. Bianchi, P. Minoretti, M. Bertona, and D. Geroldi. "Raised Plasma Nerve Growth Factor Levels Associated With Early-Stage Romantic Love." *Psychoneuroendourinology* (2006), 288-294.
Engels, F. *The Origin of the Family, Private Property and the State*. Progress Publishers: Moscow (1972).
Etcoff, N. *Survival of the Prettiest*. Anchor Books: New York (1999).
Farnham, A. "Is Sex Necessary?" Forbes.com (October 8, 2003).
Fisher, H. *Why We Love*. Holt: New York (2004).
Fisher, R. "Don't Provoke the Planet." *NewScientist* (September 26, 2009), 8-9.
Freke, T. and P. Gandy. *The Jesus Mysteries*. Three Rivers Press: New York (1999).
Gadon, E. *The Once and Future Goddess*. Harper: San Francisco (1989).
Gazzaniga, M. *The Mind's Past*. University of California Press: Berkeley (1998).
Geddes, L. "Fabled G Spot Traced at Last." *NewScientist* (February 23, 2008), 6-7.
Gefter, A. "Doing What Comes Naturally." *NewScientist* (August 30, 2008), 44.
Gibbons, A. "Close Encounters of the Prehistoric Kind." *Science* (May 7, 2010), 680-684.
Gigerenzer, G. *Gut Feelings: The Intelligence of the Unconscious*. Viking: New York (2007).
Gimbutas, M. "The First Wave of Eurasian Steppe Pastoralists into Copper Age Europe." *Journal of Indo-European Studies* 5 (1977), 277-338.
Gimbutas, M. *The Language of the Goddess*. Thames and Hudson: New York (1989).
K. Glantz and J.. Pearce. *Exiles from Eden*. Norton: New York (1989).
Gooch, S. *The Dream Culture of the Neanderthals*. Inner Traditions: Rochester (2006).
Gooch, S. *Psychic Phenomena*. Inner Traditions: Rochester (2007).
Goodison, L. and C. Morris (Editors). *Ancient Goddesses*. University of Wisconsin Press: Madison (1999).
Goodwin, J. *The Polar Myth*. Adventures Unlimited Press: Kempton (1996).
Gosline, A. "Extreme Childbirth." *NewScientist* (January 6, 2007), 40-43.
Graham-Rowe, D. "Bonobos Dying as They Flee Hunters." *NewScientist* (February 5, 2005), 16.
Graves, R. *The Greek Myths*. Penguin: New York (1960).
Grayling, M. "Math in School Should be Cool." *NewScientist* (October 25, 2008), 48.
Greer, G. "Not So Simple, Not So Strange." *NewScientist* (March 31, 2007), 49.

Gynn, G. and T. Wright. *Left in the Dark.* Kaleidos Press: Morrisville (2008).
Hancock, G. *Underworld.* Three Rivers Press: New York (2002).
Hartog, L. *Ghenghis Khan.* Victoria House: London (1999).
Hecht, J. "Why Doesn't America Believe in Evolution." *NewScientist* (August 19, 2006), 11.
Henderson, J. *The World of the Ancient Maya.* Cornell University Press: Ithaca (1997).
Hoag, H. "Sex on the Brain." *NewScientist* (July 19, 2008), 28-31.
Holderness. M. "In the End, We Are All Part of One Another." *NewScientist* (March 10, 2007), 46-48.
Holmes, B. "Civilization Left its Mark on our Genes." *NewScientist* (December 24, 2005), 8.
Holmes, B. "Beastly Tales." *NewScientist* (January 19, 2008), 30-33.
Holmes, B. "The Not-So-Selfish Gene." *NewScientist* (March 7, 2009), 36-39.
Hong, H. "Sex on the Brain." *NewScientist* (July 19, 2008), 28-31.
Hooper, R. "Genes Blamed for Fickle Female Orgasm." *NewSicentist* (June 11, 2005), 6.
Hooper, R. "Four Legs New, Two Legs Much Older." *NewScientist* (June 9, 2007a), 18-19.
Hooper, R. "Menopause Set Us Apart From Chimps." *NewScientist* (2007b), 13.
Howells, W. *The Heathens.* Sheffield Publishing Company: Salem (1986).
Huffman, J. *Modern Japan.* Oxford University Press: Oxford (2004).
Husain, S. *The Goddess.* University of Michigan Press: Ann Arbor (2003).
Inman, M. "Our Brains They Are A-Changing." *NewScientist* (September 17, 2005), 8-9.
Israel, B. *Bachelor Girl.* HarperCollins: New York (2002).
Jackson, J. *Man, God, and Civilization.* Citadel Press: New York (1972).
Jiayin, M. (editor). *The Chalice and the Blade in Chinese Culture.* China Social Sciences Publishing: Beijing (1995).
Jones, D. "The Love Delusion." *NewScientist* (March 31, 2007), 42-45.
Jones, D. "A Window on the Past." *NewScientist* (April 25, 2009), 29-32.
Joseph, F. *Survivors of Atlantis.* Bear & Co.: Rochester (2004).
Joseph, F. *The Lost Civilization of Lemuria.* Bear & Co.: Rochester (2006).
Jurmain, R., I. Kilgore, W. Trevathan, and H. Nelson. *Essentials of Physical Anthropology.* Thomson: Belmont (2004).
Kaplan, M. "Make Love, Not War." *NewScientist* (December 2, 2006), 40-43.
Kaplan, K. "Study Finds Humans Still Evolving, and Quickly." *Los Angeles Times* (December 11, 2007).
Kenneally, C. "So You think you're Unique." *NewScientist* (May 24, 2008), 29-34.
Klein, S. *The Most Evil Women in History.* Barnes & Noble: New York (2003).
Knopp, G. (translated by A. McGeoch). *Hitler's Children.* Sutton: Gloucestershire (2000).
Kramarz, F., S. Machin, and A. Quazad. "Assessed by a Teacher like Me: Race, Gender, and Subjective Evaluations." INSEAD Working Paper 2008/58/EPS.
Lai, H. *The Sexual Teachings of the White Tigress.* Destiny Books: Rochester (2001).
LeDoux, J. *Synaptic Self.* Viking Penguin: New York (2002).
Louis, R. and D. Copeland. *How to Succeed With Women.* Prentice Hall: New York (1998).
Luza, R. *The Hitler Kiss.* Louisiana State University Press: Baton Rouge (2002).

Mackenzie, D. "Trauma of War Hits Troops Years Later." *NewScientist* (August 27, 2005), 6-7.
Maines, R. *The Technology of Orgasm.* John Hopkins University Pres: Baltimore (1999).
Malkowski, E. *Before the Pharaohs.* Bear & Company: Rochester (2006).
Malmstroem, V. *Cycles of the Sun; Mysteries of the Moon.* University of Texas Press: Austin (1997).
Marshall, M. "Six Things Science Has Revealed About the Female Orgasm." *NewScientist* (May 30, 2009), 31-33.
Matthews, R. "Some Swans are Grey." *NewScientist* (May 10, 2008), 44-47.
McClenon, J *Wondrous Healing*. Northern Illinois University Press: DeKalb (2002).
McGinn, D. "Larry, We Hardly Knew Ye." *Newsweek* (March 6, 2006), 48-49.
McNamara, P. (editor). *Where God and Science Meet.* Praeger: Westport (2006).
Meltzer, M. *Slavery: A World History.* Da Capo Press: New York (1993).
Miller, G. *The Mating Mind.* Anchor Books: New York (2001).
Mithen, S. "On the Origins of Warfare." *New Scientist* (July 22, 2006), 54-55.
Morgan, E. "Father Nature." *NewScientist* (August 27, 2005), 38-41.
Morrell, R *The Sacred Power of Huna.* Inner Traditions: Rochester (2005).
Motluk, A. "Particles of Faith." *NewScientist* (January 28, 2006a), 34-36.
Motluk, A. "The Secret Life of Semen." *NewScientist* (August 5, 2006b), 8.
Motluk, A. "The Teen Gene." *NewScientist* (July 22, 2006c), 35-38.
Motluk, A. "Born Under a Bad Sign." *NewScientist* (January 27, 2007), 41-43.
Murdoch, A. *Rome's Greatest Defeat.* Sutton: Phoenix Mill (2006).
Murphy, A. "History Without the Standard Patriarchal Bias." *Journal of Interdisciplinary Social Sciences* (2008 forthcoming).

Nettle, D. *Personality*. Oxford University Press: Oxford (2007).
Newitz, A. "Love Unlimited." *NewScientist* (July 8, 2006), 44-47.
NewScientist.. "Bully Boys." March 5 (2005a), 49.
NewScientist. "Women's Orgasms Are a Pure Turn-off for the Brain." June 25 (2005b), 14.
NewScientist. "Sex and Stress: Intimately Linked Right From the Start." August 13 (2005c), 18.
NewScientist. "In Harm's Way." August 27 (2005d), 3.
NewScientist. "Slugs Show Benefit of Sperm Trade." October 15 (2005e), 19.
NewScientist. "Weather's Role in Early Irish Building." November 26 (2005f), 19.
NewScientist. "To Stay Calm, Engage Fully." (January 28, 2006a), 17.
NewScientist. "Species Angst." March 11 (2006b), 5.
NewScientist. "Why Women are Changing their Minds about Men." April 8 (2006c), 21.
NewScientist. "A Smarter Kind of Massacre." July 22 (2006d), 29.
NewScientist. "Soundbites." August 5 (2006e), 9.
NewScientist. "Hmm, I wonder Who is Daft Enough to Be Eaten…" August 12 (2006f), 15.
NewScientist. "Soundbites." August 19 (2006g), 12.
NewScientist. "Polygamous Possums Turn to Pairing for Life." August 19 (2006k), 15.
NewScientist. "X Marks the Spot in Hunt for Autism Genes." September 9 (2006l), 20.
NewScientist. "Beat you to the Fittest Mate." October 28 (2006m), 17.
NewScientist. "How to Live Long and Prosper." November 25 (2006n), 19.

NewScientist. "Bats Have Magnetic Super-Sense." December 9 (2006o), 21.
NewScientist. "It Pays to be Well Hung." February 26 (2007a), 17.
NewScientist. "Curves Were Best for Stone Age Men." March 10 (2007b), 19.
NewScientist. "Insight Hormone." March 17 (2007c), 19.
NewScientist. "Bonobos Join Forces to Outdo Chimps." April 7 (2007d), 17.
NewScientist. "Hands Up Who Wants to Talk." May 5 (2007e), 6-7.
NewScientist. "The Twists and Turns of Duck Sex." May 5 (2007f), 21.
NewScientist. "How to Make Better Choices." May 5 (2007g), 35-43.
NewScientist. "Secret of Neanderthals' Last Stand." May 19 (2007h), 21.
NewScientist. "Tough Times for China's Tigers." June 23 (2007i), 4.
NewScientist. "Monarchs of the Gien End Up With Underachiever Daughters." June 30 (2007j), 19.
NewScientist. "Spite, a Uniquely Human Emotion." July 21 (2007k), 16.
NewScientist. "Fifty Years on…" November 3 (2007l), 4.
NewScientist. "Plump Mums Have Brainy Kids." November 10 (2007m), 21.
NewScientist. "Breastfeeding Gives IQ Boost—for Some." November 10 (2007n), 22.
NewScientist. "Macaques Are Keen to Listen in on Monkey Business." April 5 (2008a), 15.
NewScientist. "Volcanic Boom." April 12 (2008b), 15.
NewScientist. "Faeces Ids Early North Americans." April 12 (2008c), 15.
NewScientist. "Baby Boy Food." April 26 (2008d), 6.
NewScientist. "Equality Makes Maths a Girl Thing." June 7 (2008e), 16.
NewScientist. "The Abuse Goes on." September 20 (2008f), 4.
NewScientist. "Jumping Gene' Leapt into Ancient Mammals' DNA." October 25 (2008g), 15.
NewScientist. "Big Hips, Big Brain." November 22 (2008h), 17.
NewScientist. "Earliest Evidence on Humans on Horses." March 14 (2009a), 14.
NewScientist. "The Best Singer Gets the Girl." June 20 (2009b), 15.
NewScientist. "Adults First Drank Milk in the Balkans." September 5 (2009c), 16.
NewScientist. "Years of Separation Belie Indians' Shared Ancestry." September 26 (2009d), 17.
NewScientist. "Chimp's Dance Suggests a Mental Grasp of Fire." January 9 (2010a), 17.
NewScientist. "Who's the Daddy? Uncertainty Pays Off." January 16 (2010b), 13.
NewScientist. "'Fatness Gene' May Thin Your Brain." April 24 (2010c), 15.
NewScientist. "Fat Lips Evolve at Record Speed." May 22 (2010d), 19.
NewScientist. "Exchange Meat for Sex? No Thanks." June 5 (2010e), 15.
Nettle, D. *Personality*. Oxford University Press: Oxford (2007).
Nicole, D. *Fighting for the Faith*. Pen & Sword: South Yorkshire (2007).
Page, M. "Women's Orgasms Are a Turn-off for the Brain." *NewScientist* (June 25, 2005), 14.
Paige, M. "The Ancestor Within." *NewScientist* (January 13, 2007), 28-31.
Parvin, P. "Of Primates and People." *Emory Magazine* (Spring 2006), 20-24.
Paxson, D. *Talking Up the Runes*. Red Wheel: San Francisco (2005).
Peat, D. *Blackfoot Physics*. Weiser: Boston (2005).
Persaud, R. "Feeling Good? You'll Never Guess Why." *NewScientist* (Jnne 29, 2002), 5.
Phillips, H. "Just Can't Get Enough." *NewScientist* (August 26, 2006), 30-35.

Pinker, S. *The Blank Slate.* Penguin: New York (2003).
Polich, J. *Return of the Children of Light.* Bear & Co.: Rochester, Vermont (2001).
Purdy, J. "Love is the Answer." *UGA Research Magazine* (Summer 2005), 33-43.
Randall, E. *The Serpent Mound.* Arthur W. McGraw (1993).
Rathaus, S., J. Nevid, and L. Fichner-Rathaus. *Human Sexuality.* Pearson: Boston (2008).
Reich, W. *Die sexuelle Revolution.* Sexpol-Verlag: Kopenhagen (1936).
Reich, W. *The Mass Psychology of Fascism.* Farrar, Straus, and Giraux: New York (1946).
Robins, R. "The Nautre of Personality: Genes, Culture, and National Culture." *Science* (October 7, 2005), 62-63.
Rodney, W. *How Europe UnderdevelopedAfrica.* Howard University Press: Washington (1982).
Rose, S. *The Future of the Brain.* Oxford University Press: Oxford (2005).
Savage, S. "The Bonobo's Secret." NewScientist (June 19, 2010), 48.
Schirmer, D.and S. Shalom. (editors). *The Philippines Reader.* South End Press: Boston (1987).
Schwartz, J. *The Red Ape.* Westview: Cambridge (2005)
Shealy, C. "Beliefs and Values: Understanding Global Implications of Human Nature." *Beliefs and Values* 1 (2009), 3-5.
Shlain, L. *The Alphabet Versus the Goddess.* Penguin: New York (1998).
Shlain, L. *Sex, Time, and Power.* Penguin: New York (2003).
Sizer, F. and E. Whitney. *Nutrition Concepts and Controversies.* Wadsworth: ninth edition (2003).
Sjoo, M. and B. Mor. *The Great Cosmic Mother.* Harper & Row: New York (1987).
Spinney, L. "Why We Do What We Do." *NewScientist* (July 31, 2004a), 32-35.
Spinney, L. "Snakes in Suits." *NewScientist* (August 21, 2004b), 40-43.
Spinney, L. "Back to their Roots." *NewScientist* (June 16, 2007), 48-51.
Spinney, L. "Where are the Bodies?" *NewScientist* (November 8, 2008), 40-43.
Stone, M. *When God Was a Woman.* Harcourt Brace: Sand Diego (1976).
Strassman, R. *DMT: the Spirit Molecule.* Park Street Press: Rochester (2001).
Strauss, B. *The Trojan War.* Simon and Schuster: New York (2006).
Stray, G. *Beyond 2012.* Vital Signs Publishing: East Sussex (2005).
Stringer, C. and P. Andrews. *The Complete World of Human Evolution.* Thames & Hudson: London (2005).
Surowiecki, J. *The Wisdom of Crowds.* Doubleday: New York (2004).
Szalavitz, M. "The Love Hormone." *NewScientist* (May 10, 2008), 35-38.
Thimm, K. "Rivalen fuers Leben." *Der Spiegel* 2 (2006), 142-153.
Tyre, P. "Boy Brains, Girl Brains." *Newsweek* (September 19, 2005), 59.
Tyre, P. and J. Scelfo. "Why Girls Will Be Girls." *Newsweek* (July 21, 2006), 46-47.
Underhill, R. *Red Man's Religion.* University of Chicago Press: Chicago (1965).
Underwood, A. "How to Read a Face." *Newsweek* (October 23, 2006), 65.
Ungar, P (editor). *Evolution of the Human Diet.* Oxford University Press: Oxford (2007).
Vugt, M. "Follow Me." *NewScientist* (June 14, 2008), 42-45.
Wall Street Journal. "World-Wide." November 30 (2005), A1.
Wall Street Journal. "The Man Problem." April 24 (2007), D1,D3.
Walker, B. *The Crone.* Harper: New York (1988).

Walker, B. *Restoring the Goddess.* Prometheus: Amherst (2000).
West-Eberhard, M. *Developmental Plasticity and Evolution.* Oxford University Press: Oxford (2003).
White, J. *Pole Shift.* Doubleday: New York (1980).
Wilford, J. "Fossil Skeleton from Africa Predates Lucy." *New York Times* (October 2, 2009).
Wikipedia. "Cro-Magnon." en.wikipedia.com/wiki/Cro-Magnon.
Wilkerson, C. "Teenage Bullies." *NewScientist* (April 9, 2005), 25,
Wilson, C. "Teenager Special: Bully Boys." *NewScientist* (March 5, 2005), 49.
Wilson , C. "Glad to Be Gullible." *NewScientist* (January 28, 2006), 37-39.
Wrangham, R. "Beyond War: The Human Potential for Peace." *NewScientist* (March 10, 2007), 49.
Wrangham, R. and D. Peterson. *Demonic Males.* Mariner Press: Boston (1996).
Zahavi, A. and A. Zahavi. *The Handicap Principle.* Oxford University Press: Oxford (1997).

Part II
Scientific Answers to the Ultimate Questions about Everything and Nothing

With Part I having laid the historical groundwork for understanding the meaning of life, the universe, and everything, this Part II builds upon that foundation to expand the depths of knowledge beyond prior limits that were created by the preconceived biases of contemporary religions and ideologies. These insights even enable an understanding of the essence of everything, which is nothing. Humans have evolved to seek answers to the ultimate questions of purpose that have previously been largely incomprehensible (Urban, 2007), and this part provides the most likely answers based on available information.

This second Part, which is far longer than Part I, surveys many cross-disciplinary research findings to develop an integrated understanding of everything. A recognition of how everything evolved is essential to such a comprehension of life and the universe itself. While this book logically draws conclusions about the universe based on a diverse and extensive body of known relevant facts, the evolutionary history of everything is vital for a complete understanding of both the present and future possibilities.

Evolution is a chaotic process by which forms of life develop in conjunction with the environment of all else to continue to exist together (Bennett, 2010). The very state of all the quantum components of all evolves in a similar fashion (*NewScientist*, 2010aaq). Evolved living organisms, whose forms can vary greatly due to small differences in starting conditions or additional random inputs to the system, actually have fractal structures (Bennett, 2010) that may reflect the evolution of their quantum parts.

In spite of the overwhelming evidence indicating the laws of evolution are valid, an increasing majority of Americans that represents a larger percentage of the national population than all other countries except Turkey doesn't believe in evolution, largely due to widespread scientific ignorance in the USA and the influence of Christian religious fundamentalists (Hecht, 2006a). While Christianity is only one of today's major systems of religious beliefs that developed from the original patriarchal fairy tales invented by cruel barbarian conquerors long ago for the very purpose of manipulating and controlling people, the other major religions today aren't much better in terms of distorting perceptions of reality. The world's contemporary religions, however, are destined to "land somewhere near astrology, witchcraft and Greek mythology on the scrapheap" (Harris, 2010), and their zombie followers duped into believing in their fairy tales as opposed to the scientific processes of evolutionary reality will have a difficult time in obtaining any insights into the true meaning of life, the universe, and everything.

The actual facts about everything indicate that the universe, which self-created itself out of nothing, promotes ever more variety, coherence, and beautiful structure where everything possible will occur in the future eternally. Within that context, humanity itself has moved from a communal coherence with the universe to a consciously individualized incoherence, and that situation is evolving into a powerfully self-ware harmony that incorporates an individual self-reflection of the group consciousness of all human beings. This collective hypnosis permits unlimited creativity, resonating power, and communal ecstasy. This process provides human beings with the capacity for creation in the very universe in which they evolved, as well as the ability to reconstruct it in more wonderfully diverse forms.

While everyone has the freedom to choose any actions without any constraints on intent, humans have evolved to be happiest when they coherently strive to act as a unified group of mutually cooperating individuals. People have been programmed by evolution to make their own creatively different contribution to the common good of humanity, and failure to follow that genetic code results in unhappiness that derives from a distasteful, worthless feeling of emptiness. In addition, each human being has maximum power in exerting their free will if the intent is to serve humanity as a whole and thereby resonate with the diverse but unified human natures of all others.

Freedom of choice without power is slavery, and the extreme powerlessness associated with pure selfishness results in maximum restrictions on an individual's free will that, in conjunction with the emptiness it brings regardless, creates pure hell. In contrast, the freedom to choose one's own actions and even one's own purpose creates unlimited power that enables each individual to choose from a virtually unlimited set of desirable possibilities as long as the person's intentions are coherent with the infinite potential of the probability wave functions of all existing and future human wills. This heaven on earth is made possible through the mere exercise of humane will power that enables the paradise to continue in ever new forms forever.

These deductions are based on logical analysis of scientific evidence, although personal experience has been very useful in guiding that analysis. Although an effort is made to keep this research objective and ensure that the conclusions are well supported by scientific evidence, at least a few of the concepts, especially those in the Appendices, were partially arrived at via intuitive trains of thought and thinking based on personal experiences. As indicated by Tyson (2008), it is often far more efficient to reach conclusions by means of intuition or artistic methods that are more creative than pure rationality.

An integrated use of the brain's capacities that employs both logical and intuitive means may sometimes be the only feasible way to understand reality. In particular, since everything is connected, quantum physics implies that it really isn't possible to become totally detached even from observations of seemingly external forces (Capra, 1999). The very nature of life itself needs analysis of the subjective tendencies of the human mind that require personal experience and reflection. One leading scientist has indicated that it is useful to try and think like the object of an investigation, even if the object is just a corn cell (Turkle, 2007), and so it seems especially useful to incorporate subjective insights when evaluating life, the human mind, and consciousness (Warren, 2007). However, all information has also been evaluated from an objective point of view through multiple re-readings in a state of detached self-awareness.

This fully informed investigative process has even enabled unraveling the mysteries associated with the puzzle of human consciousness. Consciousness has already been determined to reflect "coordinated activity of the entire brain" that is characterized by greater synchronization of neural firings in different parts of this structure (*NewScientist*, 2009zg). This reflective self-awareness represents more of a process than an entity, insofar as external stimuli are perceived in the brain via patterns of chemical activity and electrical firings of the neural cells that are further processed by a method of Bayesian logic. The rational analysis itself that is innate to all organisms' brains but that is uniquely aware of itself in human beings takes place via coherent electrical activity of entire systems of brain cells.

The brain's cells create electrical activity in the form of voltage that varies depending on the ions flowing in and out of those neurons (Thomson, 2010a). The connections or synapses between the neurons act as "memristors" or "resistors with memory" which facilitate the powerful processing in computer chips (*NewScientist*, 2010aaq). They enable inputs to be translated into meaningful output. Indeed, the synchronized firings of many neurons at the same frequency create electric brain waves that provide simultaneous signals in the thinking portion of the brain to create a holistic perception of the environment across the five senses (Thomson, 2010a).

Each of the neurons involved in creating consciousness learns to fire electrically over time via Bayesian learning that rewards them with chemicals like dopamine for activity that meets the needs of the organism. Such Bayesian learning involves adjusting the probability of firing based on past rewards in a situation with a given set of other brain cell firings that themselves are affected by external stimuli.

The genetic ability of human brain cells to create the special electromagnetic process called consciousness evolved from forecasting of oneself as well as others and everything else. Conscious perceptions and predictions were learned to be communicated via the development of speech and language that are based on categorizations of self-aware perceptions and thus enable better organization of the reflective self that is consciousness. This predictive capacity likely developed out of seeing patterns like w€€omen's reproductive cycles (Shlain, 2003), and the ability to predict is now reflected in the structure of the human brain (Marshall, 2007). This evolved structure enables humans to imagine or "construct alternate realities which can be played out as simulations in the mind's eye" (Semeniuk, 2008). In fact, about 90% of the human brain is made up of glial cells that underlie imagination and dreams (Thomson, 2009b).

Actual play, which also enables trying out new things and situations with minimal risk (Motluk, 2009), serves a similar function that has the advantage of more interactive external inputs, as is especially useful for evaluating the response from different organisms. However, play, which also occurs even in organisms like ants (Douglas, 2009d), is constrained by the limitations of the existing environment, including the availability of willing playmates at any moment. In contrast, the empathetic imagination of human beings enables much broader application of learning and forecasting at any time in any environmental situation.

Consciousness itself is based on a system of simple logic, consisting of "if A, then B", that is applied to the problem of forecasting to enable better decision-making. Information from the senses, emotions, and memories are fed into that logical processor to create the sensation of reflective self-awareness that evaluates probabilities based on past learning that is updated continually for new experiences. This consciousness that is especially useful for self-forecasting for better predicting the future environments that each individual affects more strongly because of such predicting. The entire process is based on probability associations between stimuli and events learned from past experience or communicated knowledge. This reflective self-awareness, which adapts or adjusts itself in an individualized fashion that is both caused by and affects the system of neural firings involved, and which terminates at death, provides direction to the individual's subconscious, which typically operates most efficiently when there is no conscious interference or doubt with respect to accomplishing the tasks assigned by the conscious. The conscious self is not all powerful in determining direction, however, as it is impacted by innate human nature through its capacity for self-awareness, which is the source of its very being. Human nature evolved to motivate individuals to cooperate in accomplishing the task of enabling humanity to survive and thrive into the future in a manner that promotes further evolution of humanity to be more consistent with the environment in which it exists.

The findings of this research indicate that each individual has the complete freedom to choose, but humans have evolved to be happiest when they are striving for the good of not only themselves but all other humans, including future ones whose reverence and wills can be felt via quantum processes in the present. Most current religions don't reflect this reality and meaning but instead were imposed unnaturally on humanity for exploitation by past elite rulers who fabricated gendered deities, and belief in such religions represents a pathological condition of slavery to cruel monsters that long ago decayed into dust.

As opposed to being created by man-made deities, the universe is shown in the book to be a self-creating one that develops endless variety and beautiful structure via unlimited energy and other forces, such as gravity, that are interconnected but sum to nothing. The energy created derives from an attraction of the separate parts of the universe for each other, as those components of nothing are divided by time and space, which were created by the divisions of nothing.

Space and time themselves are curved by concentrated energy or matter (Childress, 1999), which pops spontaneously out of a vacuum all the time (*NewScientist*, 2008zp). Matter itself consists of energetic points in different fields (McAlpine, 2010d) that reflect tugs through space and time via the electromagnetic attraction between matter and antimatter, which are merely two opposite divided pieces of nothing that have an irresistible pull on each other through the temporal and spatial dimensions. Gravity, which appears to attract smaller objects toward larger ones in an interplay of mass, time, and space, may be caused by energy movement within a holographic system of information transfers that reflects the amount of energy concentrated in matter (Calmthout, 2010). The energy itself is created by unifications of opposite divided pieces of nothing that become light photons which can be absorbed by still divided pieces of nothing to enable their movement toward each other.

Quantum reality, which indicates that the four dimensions of time and space are mere parameters that help define the relationships between everything, implies that nothing has any absolute existence (Penrose, 2004). Reality is composed simply of probable relationships of split pieces of nothing with each other that are divided into separate points of space-time.

The forward direction of time itself that humans perceive reflects the natural phenomenon of entropy in the universe, whereby everything has a natural tendency to move toward a state of greater randomness (Childress, 1999). Entropy stems from the fact that the laws of physics make it more likely for particles to move into less ordered random structures, thereby causing the universe to move from its original state of low entropy (Callender, 2010). However, some phenomena such as living beings themselves actually reverse the process of entropy to become more ordered locally as time progresses from the past into the future (Childress, 1999).

Human beings make especially large contributions toward more order in the universe because they themselves have evolved to have greater control over the external environment. They are therefore able to create more structure within the constraints of the continuing variety resulting from existing diversity and continued random creations out of nothing. The infinite possibilities created by creation out of nothing by the mere division of nothing by itself (i.e., 0/0), as happens all the time in a vacuum in the form of pairs of positrons and electrons popping up out of nowhere, makes the existence of humans possible.

This research shows how human beings evolved from nothing in a façade of relationships between pieces of nothing that have no absolute meaning or context. The mere possibility of human beings makes it inevitable that they appear because everything possible does happen in some set of parallel universes connected by time. The inevitable evolution of human beings with will power actually results in them willing themselves into existence. They are thus deities that created themselves in the multiverses in which they exist, and their will power therein is only constrained by their evolved human nature which enabled their existence today with that power.

The human mind isn't naturally aware of the reality of this universe in its entirety, as humans' senses evolved merely to enable people to survive and procreate (Gazzaniga, 1998). For instance, all matter and even light itself consist of probability waves of different charge densities or vibration frequencies that are not directly observable (Barbour, 1999). These waves affect the probability waves in the cells of organisms' brains which have evolved to learn to categorize the various electromagnetic waves of reality into useful perceptions through sensory cells detecting patterns in the waves. These perceptions vary widely among different species. For instance, one type of snake sends sensory signals from hairs around its mouth to the area of its brain involved in processing information about the environment transmitted from the snake's eyes, resulting in that organism actually being able to "see" not only with its eyes but also its moustache (*NewScientist*, 2010c).

All particles like light photons and electrons move in packets or quanta even though the quantum uncertainty principle makes them often act like waves (Radin, 2006). Those waves merely reflect all the possible locations, speeds, and other characteristics of the particles relative to other quanta that also exist as points on other probability wave functions. This property of particles allows light to diffuse around corners, as each photon moves and spins in every feasible direction with a specific probability. Unless a probability wave function is collapsed on a particular point via physical processes that exclude other points, the light spreads out in a weighted average fashion like a wave. While physicists and philosophers continue to be baffled by what it means for something to be in more than one point in space at any moment in time, the mathematics of imaginary numbers in the probability amplitude makes it a reality (Brooks, 2010e).

Perhaps some imagination is required for understanding the processes involved. In particular, existence in more than one point in space for everything requires multiverses that represent many parallel universes which are merely imaginary when a particular point on a probability wave function is selected that excludes the other possibilities in the then imaginary universes. However, until the probabilities of the alternative possibilities in those multiverses being connected in time to any particular event in the present are reduced to zero, the particles and antiparticles continue to exist in all the possible situations in all multiverses not excluded from feasibility by quantum mathematical relationships.

There are other interpretations of quantum reality, with the most popular being many different string theories which specify everything is composed of vibrating strings that reflect the waves of mathematical possibility. One special string theory hypothesizes the existence of a "superspace" that includes dimensions that are square roots of the observable four, but such theories and their extra dimensions have not been verified (Ellis, 2010). The origin of the existence of the hypothetical strings and their dimensions has never been explained, whereas the hypotheses presented in this book on the creation of all out of nothing do explain the "imagination lands" of quantum physics.

The most basic probability wave function is the electron, which has an electromagnetic charge that reflects its attraction to a positron, which is merely an electron moving in the opposite direction in time. The electron has a magnetic moment that causes it to act like a magnet, but there is an infinity of possibilities for the strength and direction of the magnet's electromagnetic properties (Inman, 2006) that are defined by quantum probability wave functions (Penrose, 2004). All other probability wave functions represent some combination or system of potential relationships between such basic quanta.

All particles have a wave function of possible spatial positions relative to others that is determined by the particle's frequency. The frequency of a particle equals the amount of energy absorbed by it divided by the Planck constant, which is the smallest packet of energy that exists. A higher frequency for a quantum probability wave function results in a smaller wavelength of possible points in space for it as the light energy concentrates it in a narrow location in space-time.

Light itself is also a mere electromagnetic wave, albeit one with a neutral aggregate charge (Rosenblum and Kuttner, 2006), because the basic quantum of light is the photon which originates from the unification of an electron and positron. Light waves themselves refract or bend when they pass through a different medium (Courtland, 2010) because their speed is slowed by matter influencing the photons' electric and magnetic fields (Service and Cho, 2010). Their frequency can shift to create a different light color pattern (Childress, 1999), even when the speed of light is unaffected by the frequency (Courtland, 2010e), as the magnetic fields of spinning matter that curve space-time around itself diffract each passing light photon into a specific wavelength that is perceived as a particular color because the bending of the points in space-time through which the photon travels affects the photon's spin frequency per quantum of time.

On a macroscopic scale, the frequency of light and the color patterns it reflects can be affected by the force of gravity that is induced by matter with large mass (Childress, 1999). Mass derives from the existence of charged probability wave functions that have a concentrated location in the spatial dimensions and that thereby gravitationally curve space around themselves through their own spin which creates a magnetically drawing force. A gravitational tug can be so strong that it weakens the electromagnetic force at extremely high energies (Ananthaswamy, 2010d), and gravitational curving of space-time can even collapse the waves of potential light into various forms that appear to human beings as having different colors.

Without gravitational curving or other influences, light appears as an average of all potential frequencies that results in a white color, just as the movement of light across divided points of space-time assumes the characteristics of an averaged wave of possibilities. All waves of light and matter are actually just probability functions

(Barbour, 1999) that can be collapsed to a single point or particle form upon focused observation (Radin, 2006).

The probability waves of everything obey the laws of quantum physics, which describes how the individual microscopic packets of energy or quanta that make up everything in the universe interact (Radin, 2006). It isn't possible to measure both the location and momentum of those waves simultaneously because the very short wavelength of light energy involved in observing a particle like an electron on its probabilistic wave function changes the wave's overall momentum and makes it uncertain (Childress, 1999). While observation causes a collapsing of a probability wave function into an observed particle (Radin, 2006), other events can also transform a wave into a particle by merely interacting with the wave set of potential to exclude all but one possibility (Brooks, 2010b).

Each particle is merely a spinning ball of charge that generates a magnetic field (McAlpine, 2010b). Even an atom that has a net neutral charge creates a magnetic field that exerts a weak effect (McAlpine, 2010g) because of the circulation of the different charges within the atom that is caused by the movement of the electrons around the atom's nucleus. Any moving charge that creates a magnetic field has a spin which reflects the rotation of the spherical particle around its axis that exerts its magnetic effect in either an upward direction at the northern pole, or downward at the southern pole (Zandonella, 2010). A magnet flowing through a wire creates an electrical current (Highfield, 2010).

The magnetic fields have various different possible locations at any point in time, and they appear to be spread over all of them unless they are collapsed by interactions with other fields. Collapsing the wave of potential locations for a magnetic field on a single point in space makes it impossible to measure the particle's speed or momentum (Penrose, 2004).

Waves, which represent coherent energy, lose individuality when there is no energy, as a deficiency of energy causes operation of all as a singular force (Childress, 1999). Energy is typically transferred to matter via contact between the wave functions of electrons and light photons, as such overlapping waves results in the light energy being at least partially absorbed by the electrons in the probability averaged wave functions of velocity and position that characterize the behavior of particles uncollapsed on any particular point. While an individual light photon can only be absorbed by an electron when the photon's energy is sufficient to knock the electron into an excited state (Crow, 2010), the wave function of different possibilities for photons and electrons can make the interaction result in a combined wave function of mere partial absorption. On the other hand, such contact between light and matter can result in virtually complete absorption of the photons by some atoms or molecules called metallic conductors (Crow, 2010).

Although the strength of the interaction between photons and matter may actually vary in different parts of the universe, it is normally defined by an alpha "constant" (Brooks, 2010f). This constant magnetic moment for the electron affects the force of electromagnetism and thus virtually everything, including the strength with which the electrons bond to atomic nuclei and even the behavior of the particle-antiparticle pairs that are constantly popping up in a vacuum, besides also determining the photon emissions of electrons that can then lead to other pairs appearing out of nothing (Inman, 2006).

Sufficient variation in the states of atoms' orbiting electrons make absorption of a photon by at least one electron virtually certain in the conductors that therefore prevent light from passing through (Crow, 2010) or reflecting. It's also possible that many of the atoms may absorb the photon all at once (*NewScientist,* 2011c). When a particular interaction leads to a result that is 100% sure to occur, the wave functions are collapsed on single points of possibility that cause them to act like particles as opposed to waves of averaged possibilities. The energy spectra of electrons bounced back from atoms by high-energy beams of electrons sent by advanced microscopes actually enable deductions of the composition and structure of atoms (Campbell, 2010).

Any wave or particle coming into contact with another becomes entangled, meaning that each takes on the properties of the other (Aczil, 2002). The superposition of possible future states for each entangled wave function is thereby so linked as to be considered a "whole" (*NewScientist,* 2010ap), and this linkage can exist between dissimilar particles of varying size and complexity (*NewScientist*, 2011c). Only when the coherent energy of something else affects the unified wave function through contact can the relationship be broken.

The net coherent energy of anything equals its infinite self-energy of coherence minus the infinite incoherent energy that exists in a vacuum, and each wave of particle potential represents a form of net coherent energy (Childress, 1999). The fact that water boils in any area of low pressure like that of a vacuum (*NewScientist*, 2011d) provides an example of the effect of incoherent energy. While the energy that has been observed to pop spontaneously out of a vacuum all the time (*NewScientist*, 2007) is incoherent, it can lead to the creation of coherent energy in the form of fermion pairs. Since externally directed radiation can affect the creation of fermion pairs in a vacuum (Close, 2009), and since everything emits light, it can be deduced that even incoherent energy can be controlled by coherent energy, including by harmonious systems of resonating waves of coherent energy like human beings. In particular, the normal randomness of a vacuum's quantum fluctuations (*NewScientist*, 2010aag) is at least partially influenced by the very products of those vibrations or divisions of nothing, as will be discussed later.

The properties of any molecule can vary significantly at different temperatures (Cartlidge, 2010), which are affected by light energy. Under the normal conditions of positive temperatures, additional energy increases entropy or disorder, as when a solid melts into a liquid, but energy flows from higher temperatures to lower energy states (Shiga, 2010e). thereby causing the opposite effect where the temperature falls.

Some strange behavior exists at very low temperatures because of the singularities caused by such states where energy levels approach zero. When particles, including photons, are "at their lowest possible energy state", they "behave as a single entity" in what is referred to as a "Bose-Einstein condensate" (*NewScientist,* 2010aaaf). This coherence may result because particles at low temperatures don't have sufficient energy to engage in any frictional interactions such as change their spins. At temperatures below 2 kelvin, helium set in motion will move forever and even defy the laws of gravity (Brooks, 2010a). Low temperatures, which impact the amount of space between molecules (Cartlidge, 2010), causes the movement of electrons in atoms to be reduced to extremely low levels that enable measurement of the electrons' momentum, but their position becomes highly uncertain as a result of the laws of quantum physics, and empty space between slow-moving atoms can therefore even pass through the solid atom itself

(Brooks, 2010a). The latter movement in time is feasible because the low temperature collapses the probabilities of most potential velocities for the atoms to near zero, thereby resulting in quantum singularities in their probability wave functions that create discontinuities in their spatial positions over adjoining points in the temporal dimension.

Temperature also affects the percentage of ordered versus disordered molecular structures in groups of a particular type of molecule (Cartlidge, 2010). In fact, temperature is a measure of the degree of disorder (*NewScientist*, 2010aal). Electrically charged bonds between each of the molecule's individual atoms are continuously breaking and reforming in a process that is affected by the molecule's structural strength (Cartlidge, 2010), and lower temperatures result in greater order between systems of atoms and molecules (*NewScientist,* 2010aal). As the temperature approaches absolute zero, atoms and molecules start to act as a unified whole without any friction between the individual atoms (McAlpine, 2010f).

The behavior of electrons and molecules and how they tend to react and combine in groups is a probabilistic process (Battersby, 2009d). When their superposition of different possible states is collapsed on a particular point through measurement, their temperature and level of disorder falls (*NewScientist,* 2010aal). The narrowing of a probability wave function to a single location reduces the chance of interactions between them. However, some molecules naturally combine into arrangements that are fairly stable such as crystals (Battersby, 2009d) even when they remain in a superposition wave of different potential states and locations because the spread of different potential points spread out for their component electromagnetic fields of attractions. For instance, silicon crystals used as transistors in electronic devices are held to together by four of the fourteen electrons in each silicon atom existing as a wave everywhere in the entire crystallized group of such elements (Rosenblum and Kuttner, 2006).

Although all particles and even entire molecules are merely packets of vibrating waves (McTaggart, 2008), humans normally perceive many things as solid objects (Barbour, 1999) that reflect the rigidity (*NewScientist*, 2010aaae) in the combination of probability waves of electromagnetic fields in those "solids". Because all electromagnetic fields are subject to an uncertainty principle whereby their exact location in space and their speed can't be simultaneously measured (Penrose, 2004), the observation of them can only collapse either their wave of potential speeds or their probability wave function for location, not both.

The characteristics and behavior of all matter and light can be fully described as probability wave functions that interact with one another and that sometimes lead to interference patterns which cause them to exclude many possibilities on the wave function in any particular instant in time (Rosenblum and Kuttner, 2006). Both the spin, which creates a magnetic effect that is different from the electrical effect (*NewScientist,* 2010aaq), and the location in space at any moment are determined by those functions which define their relationship to the waves of other electromagnetic fields. The exact phase of their electromagnetic fields can't be determined, but relative relationships can be computed with complex geometry that includes imaginary numbers (Penrose, 2004).

Although the interactions between the wave functions can sometimes cause them to collapse on a single point on a wave (Rosenblum and Kuttner, 2006), they otherwise exist in a superposition state of many possibilities that appears as an average of the potential points on the waves. While individual molecules can act as waves, they tend to appear as particles because of other fluctuations in the environment that collapse their wave functions upon contact (Ananthaswamy, 2009g) if other points on the contacting waves aren't coherent with one another. Without such collapsing of the wave functions into a single point, each particle of matter and light appears as a probability-weighted average of the wave function of electromagnetic forces (Barbour, 1999). These electromagnetic fields are observed to always have two poles that merely reflect the circulation of the electrical charges in the fields (Penrose, 2004).

A light wave itself is merely an oscillating electromagnetic field that can point in any direction perpendicular to that of its movement (Rosenblum and Kuttner, 2006). The field of the light photon points alternatively at the north or south poles (Penrose, 2004). While most light consists of photons that have electromagnetic fields pointing in a random or unpolarized fashion, light can also be polarized in some particular direction, such as occurs when light reflects off water and roads to be polarized in a horizontal direction that can be blocked with sunglasses (Rosenblum and Kuttner, 2006). Polarized particles, such as light photons from lasers, spin non-randomly by pointing their intrinsic waves of electric fields systematically in some particular direction (Smith, 2003). Lasers typically operate via partially reflecting surfaces that focus light photons in the direction and frequency desired (Rosenblum and Kuttner, 2006).

Other particles such as isotopes, which have a different number of neutrons than the standard atomic elements, are also polarized whereas most atoms spin randomly (Smith, 2003). Different numbers of neutrons may collapse spin wave functions to a particular coherent direction or at least cause them to average out as if they had. In addition, a greater coupling of the electrons' spins in heavier atoms that have nuclei with greater positive charges can permit them to conduct electricity better because the free electrons flowing in electrical power aren't slowed by the frictions associated with a change in spin caused by contact with the electrons orbiting those nuclei (Zandollena, 2010). These atoms with more protons, neutrons, and electrons than others apparently also either have their spin wave functions collapsed by the structure of charges in them or their superposition state average out to create such a phenomenon.

While light photons have no electric charge (Rosenblum and Kuttner, 2006) and can only have perpendicular polarity, the nature of charged particles enables them to be polarized in any direction for their own special electromagnetic effect (Smith, 2003). The restricted directional polarity of photons derives from their instantaneous movement through space that results in them only being able to point forward or backwards in time. In contrast, charged particles moving at finite speeds can exhibit polarity in any direction in space because their movement is divided between the temporal and spatial dimensions. Particles with a charge therefore spin on their axis with an angular momentum which has a magnitude equal to half that of light photons. The angular momentum of photons themselves equals the Planck constant divided by two pie, which may reflect their timeless existence being divided by the circle of the time dimension in which they can travel in either direction.

Particle spins create electromagnetic fields that become entangled into a coherent oscillation upon contact with others. Electromagnetic fields cause anything in their deepest midst to take on both their polarity and frequency, as well as their velocity (Childress, 1999), because the particle spins that create their electromagnetic fields have their probability wave functions entangled. This phenomenon characterizes light photons that are absorbed by the charged fields of electrons and positrons and thus move with them.

Everything is merely an electromagnetic field that exerts oscillating physical forces on other electric fields, primarily via the electromagnetic force, which attracts fields of opposite charges and repels like-charged ones (Rosenblum and Kuttner, 2006). The spinning of electromagnetically charged particles draws space and time around themselves in the process of creating magnetic fields (Brooks, 2009d), which connect two or more points in space electromagnetically (McTaggart, 2008). While the electromagnetic tug can be cancelled when fields of the opposite positive and negative charges interact (Rosenblum and Kuttner, 2006), the magnetic curve of space-time can still exert a gravitational effect.

Matter and life are created by the electromagnetic interactions of such charged particles or fields in chemical and biological processes (Lipton, 2008). Electrons are negatively charged particles or fields that often revolve around atomic nuclei, which consist of positively charged protons and neutrally charged neutrons, to form the atoms of matter that themselves are merely probability wave functions (Rosenblum and Kuttner, 2006). Atoms can exist freely, but they can also bond together with other atoms to form molecules, as is spurred when the combined number of electrons around the nuclei fills up the space of an orbit. There is space for two electrons in the first orbit around the nucleus, but the number that can exist in the other orbits varies in a fashion that may reflect the geometry of their spins.

Many electrons also exist freely outside an atomic orbit, and such electrons can thereby create electrical current that reflects moving negatively-charged fields which emit various wavelength frequencies of light (Radin, 2006). Even within atoms, electrons emit light photons that reflect different vibration frequencies of the electrons that vary with the amount of energy the electrons absorb from the light and heat emitted by other electrons (Rosenblum and Kuttner, 2006). Light photons or radiation is also emitted upon chemical reactions that cause atoms to bond together in a process which results in photons being released from the electrons because the molecular bonding restricts their movement.

Light exists in discrete packets called photons, and different levels of energy absorbed by the electrons prior to emission of such quanta of light energy result in different frequencies for the electron probability wave functions (Calleman, 2009). Although the light photons absorbed by electrons, as well as their energy, exist in only discrete as opposed to continuous values (Childress, 1999), both the photons and the electrons appear as waves that have all feasible points on their probability wave functions occur at once unless their wave functions are collapsed on a single point resulting in a discrete particle. The electrons, which themselves are merely electromagnetic force fields, move, exert a charge, and spin with a specific probability in a superposition state of potential unless they are collapsed into points as particles.

Each atom of matter itself is actually just a vortex of spinning and vibrating energy, and those vibrations affect other atoms (Lipton, 2008). In particular, the "jiggling" of the charges of matter create electromagnetic waves that transfer the energy of the vibrations to other particles at the speed of light (Rosenblum and Kuttner, 2006). The amount of light energy radiated by any object is a simple mathematical function of the object's temperature, as is reflected in the frequency of the light vibrations that vary with the length of the light wave or wavelength (Seife, 2000). Different jiggling frequencies of matter cause varied light wavelengths that are reflected in different colors of light, such as infrared for lower frequencies, ultraviolet for higher frequencies, and white light, the latter of which encompasses the entire spectrum of light frequencies (Rosenblum and Kuttner, 2006). A different number of electrons revolving around the nucleus in orbits (Lipton, 2008) create varying vibrational frequencies for each type of atom, thereby resulting in each particular chemical element exhibiting its own unique color, with the structure of the orbits creating the different vibrational frequencies (Rosenblum and Kuttner, 2006).

Upon contact with matter, the probability wave functions of light photons appear to be collapsed on various larger wavelengths and lower frequencies that are reflected in different colors being perceived by human beings. In particular, "the information from the photons hitting a particular spot on the retina is restricted to their wavelength which" is consciously perceived as a "color" after processing through the ventral lower portion of the brain for categorization in the "master map of location" existing in the dorsal stream (Robson, 2010c). Visual perceptions of everything thus merely reflect the eyes picking up some wavelengths of light photons that reflect off some molecules because of their chemical properties at a given temperature, whereas objects appearing as black have properties that result in the entire light photons being absorbed by pure black objects. The varying colors of different things thus exist because their chemical compositions interact with light photons differently, with light being absorbed more and reflected less by some objects because of their system of component electrons and positrons.

The surface of a normal object like a metal is largely composed of a sea of electrons, and when light hits that sea, it causes a wave of vibrations in the electrons called plasmons that reflect the light waves which become trapped in the metal's surface (Mullins, 2010a). Light photons sometimes liberate charged electrons off of substances like metals and cause electrical current (Radin, 2006), but they can also be absorbed by electrons they hit, resulting later in the emission electromagnetic waves of various light frequencies in a process of radioactive decay that creates heat energy (Rosenblum and Kuttner, 2006). On the other hand, light passes easily through air because the thinness of the atoms and molecules in such space result in fewer light photons being absorbed by or reflected off the matter there.

Some of the unreality of the typical perception of reality through the normal five senses is indicated by the fact that a few people perceive that they can both see and hear a picture (*NewScientist,* 2008zs). Some humans without eyesight can even use vibrations from echoes emanating via clicking of their tongues to fairly accurately determine objects near to them (Kish, 2009). These seemingly unbelievable phenomena illustrate the effects that can occur on the macro level which is perceived by basic human senses. However, given that both light and sound are mere electromagnetic waves of different frequencies (Szego, 2002), and given that the plasticity of the brain enables it to develop a great assortment of different interconnections between brain cells that cause them to become electrically charged or fire in all sorts of combinations to create conscious perceptions (Bernstein and Nash, 2008), a capacity to perceive hearing pictures shouldn't be all that surprising.

Light reflecting off an object creates a visual perception of colors in most people, depending on the wavelengths of the light photons that are reflected, but unusual neural firing patterns in the brain cause some people to perceive a sight as a sound. In the latter case, there are atypical connections between the temporal lobe portion of the brain that is involved in hearing and the brain neurons that fire when light initially enters the eyes. Some powers of such sense-joining synaesthesia, like seeing numerical symbols as colors, may represent evolutionary advantages which enable better detection of some stimuli in a few situations and may have developed from the higher reasoning abilities of human beings to optimally adapt to environments that normal senses aren't as capable of discerning (*NewScientist,* 2009zzzz). Such an ability in just a few individuals would be beneficial in the human groups in which people normally lived in the prehistoric past.

While not directly perceivable by humans as such, sound itself is caused by energy from some source causing low frequency vibrations of the electromagnetic fields called particles that are in its path. Light is a higher frequency form of energy that has a different impact on particles in its path, and it moves through some systems of particles like air and other gases but reflects off or is completely absorbed by systems of particles called "solids". In contrast, sound is merely a force that moves through existing matter in via a transfer of energy that the ears have evolved to be able to detect.

It must be emphasized that, although objects are perceived by humans as being impermeable matter occupying empty space, "solid" objects actually consist of mere electromagnetic force fields that repel other fields and therefore only appear solid (Peat, 2005). These electromagnetic fields largely move in waves, but they act like particles wherever the fields are very concentrated in thin waves with high amplitudes at a single point (Capra, 1999). Particles, which have a minimum size defined as a plankton (Rosenblum and Kuttner, 2006) are only vibrations or processes that have characteristics of an object (Capra, 1999).

Light reflecting off electromagnetic fields of vibrations thus enables visual perceptions via electrical charges being induced in the occipital lobe of the brain by neurons firing to relay signals sent there by the eyes via the stringy axon and dendrite parts of brain cells that provide connective communication between those neurons (Bernstein and Nash, 2008). The brain and its neurons themselves, as well as the entire human body, are mere electromagnetic force fields that relay, attract, or repel other fields and information. Human bodies are therefore optimally thought of as fields of meaning and processes of information, instead of as machines (Peat, 2005). The "mental stuff" of human beings is "just the neurons carrying out mathematical calculations" in internal brain mechanics that process the information and create cognition (LeDoux, 2002), which itself is merely a categorization of information to develop knowledge, including about the various electromagnetic fields that can be visually detected and interpreted.

While magnetic fields are created by all spinning atoms, including ions (Reich, 2009a), the properties of individual atoms vary with the system of electric forces contained within them. Ions, which are atoms with nonzero electrical charges because of missing electrons (Mullins, 2009a), can cause atoms to have much different characteristics, including with water existing at a temperature above 2000 degrees centigrade under heavy pressure, such as that which may exist on giant planets like Neptune, that exhibits a yellow color (Shiga, 2010a).

Electrons in atoms travel around the atomic nucleus in orbits that are determined by the integral wavelengths of vibrations that fit in its atomic structure (Rosenblum and Kuttner, 2006).This structure gives each atom its particular chemical properties (Lipton, 2008). One of those orbits requires electrons to "dive in and out of the proton" of its atomic nucleus (*NewScientist*, 2010am), as the averaged probability wave functions of those nearby particles overlap in the same space. The mutual destruction of the component electrons and positrons is inhibited by the positions in the quarks of the protons being frozen in two-dimensional space, as will be explained later.

The basic building block of the atomic nuclei of all matter is assumed to be the quark, which has an electrical charge that is only a fraction of that of an electron (Smith, 2003). The quark is the basic component of the neutrons and protons that provides both with internal charges in different mixes that are bound up to form the nucleus of each atom (Close, 2009). Atomic nuclei themselves act as magnets with a north pole which can be "spatially quantized" into a superposition state via electromagnetic waves of the proper frequency that results in that nuclear pole pointing up and down simultaneously in a wave form of potential positions (Rosenblum and Kuttner, 2006). Such a superposition of different possible points on a probability wave function existing at the same time will decohere into a single collapsed point upon the occurrence of some event that results in a detection of which point has occurred (Brooks, 2010b).

Since knowledge from observations of the characteristics of matter can collapse the superposition states or waves of potential into a single position, mere observation can cause a wave to act like a particle (Rosenblum and Kuttner, 2006). Other phenomena unrelated to deliberate measurement, such as a stray photon that gives away the position of a wave, can also collapse a superposition state of different possibilities into a specific point on the probability wave function (Brooks, 2010b). Entire organisms exist in superposition states (*NewScientist*, 2009zzzo), thereby resulting in them acting like waves of potential instead of whole entities, although they can be collapsed into a particular state upon observation by others or via other phenomena, such as the occurrence of random events that interact with the organism (Brooks, 2010b). Anything that results in an event which causes behavior like a particle and is thus incoherent with the existence of an uncollapsed wave function destroys the superposition state at least momentarily by making other possibilities on their wave functions inconsistent with these interactions.

The electromagnetic force fields that largely make up matter and that are formed by various electric charges (Smith, 2003) do not have a boundary that is distinct from all else (Capra, 1999). Instead, even tiny electrons "smear out over large areas of space" (*NewScientist*, 2009zi), and the smallest particle is more like a wave that is generally never in only one particular place (Capra, 1999). For instance, at extremely low temperatures, each electron in about a ton of metal atoms can be "simultaneously everyplace" in that mass, thereby enabling creation of superconductor devices like a Magnetic Resonance Imaging (MRI) machine that needs no electric power beyond an initial push to maintain an electron current flow and magnetic field (Rosenblum and Kuttner, 2006).

The deepest essence of matter is a process or movement as opposed to an object (Peat, 2005). In fact, matter is merely a patterned manifestation of interrelated interactions between various fluid processes, as opposed to actual permanent substances, and these electromagnetic systems follow spontaneously but consistently their own tendencies, which themselves are determined jointly with the tendencies or properties of other aspects of the interconnected universe (Capra, 1999). The various electromagnetic fields themselves can combine to form new fields (Close, 2009), but all field patterns typically exist only with a certain probability despite a continued normal observance of a probability-weighted average state for any moment in time (Capra, 1999). However, greater interaction between the wave functions of the various fields restricts their potential variation and thus makes them more stable particles as they combine into larger systems such as living organisms. The hardness of a substance that is measured by the difficulty involved in sliding one layer over another (*NewScientist*, 2010aay) is determined by the interactions between the object's electromagnetic fields.

According to quantum physics, everything, including both waves and particles, are just relationships (Radin, 2006). Time itself may represent nothing more than an averaging of a statistical state of all cosmic molecules at a given temperature, which itself is merely an averaging of the speed of the molecules (Gefter, 2008a). Many multiverses of potential can therefore be disconnected in time when interactions at a particular point in space-time collapse wave function into forms that make some otherwise possible quantum mathematical relationships impossible in connected points in the time dimension.

Quantum physics, which is based on linear mathematics (McTaggart, 2008) that is extremely complicated because of the myriad of interrelationships between all the particles in the universe (Griggs, 2009a), indicates that all possible events and all possible paths that subatomic particles can take do indeed occur (Segre, 2002). Individuals normally experience only some averaging of the probability function of all such paths or quantum waves (Barbour, 1999) that determine the speed of molecules at different temperatures.

However, concentrated focus can collapse the wave functions on other points. In particular, every particle of matter is merely a wave function that measures the probability of observing it at every point in space at a given point in time (Seife, 2000). With each particle existing at any particular point in space only with some probability at any time, only the probabilities of a particle being at a particular point in space, as well as the average of those probabilities that is normally perceived, moves with time (Capra, 1999). If there is concentrated focus on one such state of events, only one path or point on the probability wave may be observed that is likely to be the most probable one unless there is sufficient will power to cause a different result, as is explained in more detail in Chapter 16.

For instance, when a single light photon is detected with a mechanical device to go through one of two slits of an otherwise solid screen before its wave function collapses on the averaging of the wave function by a second solid screen behind the first one, it creates a pattern on the second screen indicating it went through just one hole instead of resulting in the different pattern that normally exists when its wave function is not first collapsed to go through just one slit by observation (Radin, 2006). The photon thus acts like a particle once its wave function is collapsed on a single point by mere detection before it can regain its wave-like effects that would be inconsistent with its observation as a particle.

The power of concentrated human focus (DeLaurence, 1910) implies that humans can affect the point to be observed or experienced. The human conscious is optimally geared to focus on one matter at a time, and that focus likely collapses the observed wave functions into particle points that are effectively willed into occurrence. The interaction of the photon emissions caused by the emotions that can be consciously directed can actually result in other wave functions collapsing on points coherent with an individual's intentions in the power of positive thinking that is emotionally and confidently directed.

The conscious often uses connections between bits of information, including categorizations of them, in order to permit concentrated attention while also considering a great amount and variety of information. Delegation of much processing activity to the subconscious, which also categorizes information, allows the subconscious analysis and operation that enables the emotions to carry out the assigned tasks with maximum efficiency. The subconscious operates using simple but effective algorithms in an emotional process that has incredible power. The ability of some to remember over 20,000 different numbers in a sequence by drawing connections between them (Tammet, 2009) represents an example of the power of the brain despite the limited memory capacity of consciousness itself. While the conscious can specify some of the algorithms used by the subconscious and interfere with its processing in various ways, it generally operates best when the conscious sets the overall focused direction and then lets the subconscious handle the details of achieving the intended results in the process of exercising will power.

Quantum mechanics, which indicates that an infinite number of parallel universes exist, implies that there are an infinite number of histories which lead to the current point in time that itself is merely a particular configuration of the waves of energy or oscillating strings of electromagnetic fields closest to other configurations in the perceived past and future (Barbour, 1999). There are an infinite number of historical situations which can lead to any present situation, but situations which are inconsistent with the existing one are relegated to parallel universes that aren't connected to the contemporary one in space-time. Quanta of space-time that can't be connected according to the mathematical relationships defined by the laws of quantum physics can't be experienced by particles existing in the coherently connected set of universes. Quantum reality indicates that there is nothing absolute in these systems of parallel universes, as everything in these multiverses is created by interrelationships between divided pieces of nothing. This reality created out of nothing by aggregate human will power consists of an infinite number of possible parallel universes that all occur simultaneously in each moment of time unless a person's will power collapses these aggregated wave functions of all possibilities onto a particular feasible point of space-time selected to be experienced.

The very mind-boggling nature of such phenomena is actually understandable (Buchanan, 2007b), and such a comprehension can assist in comprehending the mysterious nature and purpose of life itself (Capra, 1999). For instance, the existence of an infinite number of parallel universes that merely consist of different mappings for everything in the 4-dimensional universe in which humans live (Barbour, 1999) implies that everything possible will happen and has happened.

However, there are some phenomena that aren't possible in quantum reality. For instance, it is impossible to precisely measure simultaneously two particular aspects of a particle such as speed and position in any particular universe because everything that is perceived is merely an electromagnetic wave, where the probability of measuring any point on the wave is determined by the wave's amplitude, and the wave's frequency or energy at a measured point in space and time may vary from the average frequency of the wave that is normally perceived or measured (Capra, 1999). The very collapsing of a particle's wave function on some point in the present eliminates some contemporary possibilities that also affect the past and future probability wave functions of that particle and others that interact with it.

A person's will power is enacted via this process, as an individual's focused intention collapses wave functions in the present to thereby influence the future at the same time that the probability of past histories connected to that future are also impacted. The neural firings that create an individual's thoughts and collapse particle wave functions through cellular microtubule openings via emotionally willed forecasts have their free choice restricted only by the quantum reality of mathematical relationships with everything else that includes the intentions of other human beings. Since harmony or coherence in the frequency of vibrations magnifies their effect, while incoherence nullifies them (Lipton, 2008), a willed reality appears only when there is coherence in separate vibrations of nothing.

Thus, while everyone has free will to do and experience everything that is possible, what is actually feasible can be restrained by the environment and others' free will. Human will power can be collectively harmonized via a hypnotic group consciousness that was the norm in prehistory and that can each individual's potency of intent via the resonance. In any event, via humane beliefs which are confidently felt, will power enables people to maximize the probability of experiencing what their human nature leads them to desire, including through psychic powers whose influence can persist even after death. Will power can even affect the past, thus implying that future humans may have an impact on the universe today, and this power may be overwhelming given they likely will have greater knowledge and ability to enact their will power.

All living things are comprised of just complex systems of electromagnetic fields that themselves merely represent the process of divided pieces of nothing seeking blissful reunion back into infinite nothing. Living cells can detect the photon and electromagnetic waves emitted by the cells of other beings and by inanimate objects, and it is very likely that they send signals, directly or indirectly via the gut and pineal gland, to the brain where the information about those waves can be subconsciously categorized in the form of useful emotional feelings, intuition, and hunches. From an evolutionary point of view, it would seem very improbable that the dominant species on earth, homo sapiens, didn't develop the capacity to interpret and affect the basic electromagnetic waves created by those fields with more than the five "normal" senses that alone aren't capable of perceiving such underlying quantum realities.

Although many believe in an ideology of survival of the fittest where human beings are merely biological robots battling each other (Lipton, 2008), the evolution of human nature has resulted in people and their will power being most potent when they are helping humanity to continue to exist into the future. Human beings also evolved to be happier and experience pleasant feelings of self-fulfillment when they aren't actually purely selfishly and instead direct their will power toward serving humanity as a whole.

However, the patriarchal societies that have largely ruled the world for the last 5000 years have brainwashed people into believing in an intrinsic selfishness of human beings. Within the resulting pathological environment of individuals exploiting each other, attempts to love and help all other individuals can actually be counterproductive to humanity. Nevertheless, human nature and will power can still be blissfully and potently directed toward serving all humanity, especially the powerful human beings of the future.

Chapter 10
The Beginning and Essence of Life

A living organism is typically defined as something that has processes which are self-perpetuating and permit self-replication, including in modified new versions of themselves (*NewScientist*, 2009zf), and that also have some means of self-directed motion (Lipton, 2009). Calleman (2009) has specified living organisms as systems of molecules that maintain an energetic imbalance with their environment and that employ coordinated biochemical reactions to external stimuli. However, given that even an inanimate particle like nickel sometimes "behaves like some live object" by moving and absorbing other particles (*NewScientist*, 2009zb), the definition of life may be a matter of interpretation.

Life functions, which include accessing nutrients for energy, waste disposal, and reproduction, are largely carried out by living cells cooperating together through various means (Lipton, 2008). Microtubules under the direction of the cells' centrosomes help coordinate those functions by using their long narrow tubes to position the various organelle components of the cells at different places in the cells' cytoplasm, and the relationship and content of those organelles to each other enables cells to specialize in different functions (Calleman, 2009).

According to Cornish-Bowden (2004), life is just chemistry out of equilibrium, where the various chemical reactions are largely driven by enzyme catalysts that in turn are directed by a living organism's genes. In particular, the chemical reactions needed for life are spurred by enzymes that appear to hold waves of energy vibrations in specific parts of those molecules via a crystalline structure (Buchanan, 2008). The enzymes, whose molecular cores are metals of certain types (*NewScientist*, 2010aan), only release the energy locked up in their structure upon binding with specific molecules that react to the particular enzyme that thereby serves as a catalyst (Buchanan, 2008a). One of the most important biological processes involved has each cell using oxygen and glucose to convert adenine diphosphate (ADP) and phosphoric acid into adenosine triphosphate (ATP), which then slow drips the energy needed for the various cellular functions (McTaggart, 2008).

Chemistry itself results from the actions of electrons and the force of electromagnetism (Smith, 2003). In particular, the laws of chemistry only reflect a deeper set of quantum processes, and the biology of life emanates from these interactions in cells, each of which is surrounded by a membrane wall (Lipton, 2008). Life probably originated in rocks with labyrinthine pores on the bottom of the ocean floor that formed when water mixed with olivine minerals of the earth, and these foaming stones likely formed according to natural chemical processes into membranes in which organic molecules naturally developed from the inorganic matter within them (Lane, 2009b).

Life today involves each cell using the binding of molecules in different combinations to form their electromagnetic charges and shapes into a balance that causes the cell's activities (Lipton, 2008). The energy for this activity derives from food sources, including light energy derived from the sun, that "pump" ions and protons through membranes to create different electromagnetic charges in processes called chemiosmiosis and the more complex glycolysis (Lane, 2009b). Behavior similar to that of living organisms themselves can be observed in a drop of oil that can similarly move in a seemingly organized fashion to even navigate a maze in a process that merely reflects changing chemical imbalances (Barras, 2010).

Since there are natural chemical tendencies to return to steady state without catalysts, proper control of enzymes is crucial for the activities necessary for life (Cornish-Bowden, 2004). Such control is largely exercised by cell membranes, which use protein-receptors on the outside of the cells to control entry, and which employ signals from those receptors to initiate actions via protein-effectors that can enable enzymes to transform various proteins into desired molecules (Lipton, 2008). Cell membranes essentially take inputs from the environment and process them like a computer to produce behavioral "outputs" by modifying the shape and chemical composition of proteins that are combined in different ways (Lawton, 2009e). In the eukaryotic cells that characterize multicellular life, the various tasks are directed through a centrosome, which operates on information flowing through tiny microtubule openings in the centrosome's centrioles that directly coordinate the various cellular processes (Calleman, 2009).

Cell membranes are also responsible for causing the DNA in the cells' nuclei to create new cells (Lipton, 2008). DNA represents the coding material for the cellular production of proteins and exists on 46 chromosomes in human beings (Hooper, 2006a). This code consists of 6 billion "letters" (A, G, T, and C) that exist in different combinations to make up the human genome, and the genes are separated by non-coding sequences in the over 300 billion base pairs of the DNA coding letters (Trivedi, LePage, and Aldhous, 2010). This DNA code determines the proteins that can be manufactured in the cell (Cummings, 2006), but the information in the membranes determines what is actually produced at any time (Lipton, 2008). In addition, the production of some proteins is inhibited because methylation is attached to some C base pairs during early life in an "epigenetic" process that seems to serve the purpose of adapting an individual optimally to the environment that is experienced then (Trivedi, LePage, and Aldhous, 2010). Epigenetic processes can also affect the genes of sex cells and thus be inherited by future generations that result from the subsequent sexual reproduction of that individual (Coghlan, 2010c). The environment apparently can thus affect the genes because chemicals like methyl groups can bind to the DNA and cause it to fold up more tightly around packaging proteins called histones, thereby making it more difficult for a cell's protein-making machinery to access the its code (Spinney, 2010b).

Each of an organism's genes contains code to produce as many as tens of thousands of different proteins after the gene is spliced into RNA copies (Trivedi, LePage, and Aldhous, 2010). RNA acts to produce proteins by exerting osmotic pressure inside the cells that places tension on the membrane, which grows by absorbing molecules from the environment (Szostak, 2010). Some of the proteins produced are called transcription factors and inhibit the production of specific types of RNA (Trivedi, LePage, and Aldhous, 2010). The DNA from which the RNA is derived thus affects the type and occurrence of proteins produced by the cells, but other parts of the cells, as well as the environment around it, have an important influence.

DNA itself represents a form of an optical D-isomer, which is a molecule that reflects light to the right, whereas the basic amino acids of organic matter reflect light to the left (Calleman, 2009). The latter protein molecules of life tend to be characterized by such left-handedness, probably because light radiation circularly polarized in the earth's atmosphere tends to destroy the right-handed molecules (*NewScientist,* 2009zzzj). In particular, while the right-handed DNA, which corkscrews around itself to form a thermodynamically stable double helix in cell nuclei (Calleman, 2009), may be fairly well shielded from the destructive or mutating influences of light photons, the amino acids may be more exposed to bombardment by light photons in the environment and thus need to reflect light to the left in order to maintain sufficient stability for the orderly processes of life. DNA remains susceptible to fairly minor mutations that can even enhance the adaptability of organisms (Cummings, 2006), but the cellular membrane protects DNA from the totally destructive changes that would inevitably occur without that shield.

Because the DNA, which is safely inside the cellular membrane, is negatively charged, cells are naturally out of balance with the external environment, thus enabling the processes of life (Calleman, 2009). These processes are driven by differences between the charges inside and outside the cell (Lipton, 2009). In particular, individual cells, including those of single-celled organisms as well as the neurons of the human brain, act like electronic circuits by electrically responding to charged current from chemicals in the environment to create a magnetic field which fluctuates in a fashion that is measured as voltage, and the electrical signals cause chemical reactions within the cell according to their frequency, strength, and location (Mullins, 2009c).

The membrane of each living cell, which forms the basic building block of life on earth and has evolved to cooperate with other cells in organisms like human beings, is so essential for the processing of the signals in the environment that it has been labeled as the brain for the basic unit of life (Lipton, 2008). Cellular life has developed an increasing awareness of external events in order to interact with that environment in a manner that permits ever more control of it for the benefit of the organism (Calleman, 2009). Evolution is actually causing the cells or groups of cells to develop an ever-higher level of awareness of their environment that makes life forms in the image of that environment (Lipton, 2008).

The "memristor" nature of individual cells enables them through basic electromagnetic processes to remember and learn from electrical patterns in the environment in a process whereby external current causes the cells' internally charged chemicals or ions to change back and forth from being semiconductors (i.e., partial resistors of inputs of the electrical current that create the flux in the current-induced magnetic fields or voltage) to full-blown conductors of the current (Mullins, 2009c). The functions of electronic transistors, which can both amplify and switch off electric current flowing into them (Rosenblum and Kuttner, 2006), are thereby carried out to enable cells to operate like miniature computers (Lipton, 2008). Alternating electric current through two nano-sized electrodes can even drive a tiny molecular motor when another microscopic electrode serves as a gate to create an oscillating electric field surrounding a compound which has negatively charged electrons systematically congregated on one side and a positive charge on the other like a magnet that cause it to rotate as the electrical current varies (*NewScientist*, 2010aav).

Cells apparently even react to the tiny current of light that all living cells emit, thereby enabling communication between cells that aren't directly connected (McTaggart, 2008). The special change in the polarity of light that can be detected when reflected off living tissue (Shiga, 2009b) is perhaps the means for such expanded communication and awareness between cells.

The chemistry of life is driven by quantum forces such as electromagnetism and quantum entanglement, and tiny nanoparticles, which can travel directly between cells of the same organism via microtubules as well as between the environment and the cells via receptors and synapses, may play an important part in these processes (Lipton, 2008). Quantum entanglement, which links the superposition of the quantum states or probability wave functions of two particles (*NewScientist*, 2010al), correlates the direction of the spin of all particles that have ever been in contact with one another and may be the key to life itself (McTaggart, 2008). When two particles collide with each other, their probability wave functions become integrated with one another, although subsequent separate collisions with other particles can swamp the effect of an earlier entanglement. The impact of subsequent entanglements is analogous to the effects of a hurricane on ocean waves that can make a pebble previously falling into the water unnoticeable.

Since it appears that life forms may be artificially creatable in the laboratory, perhaps in the next decade (*NewScientist,* 2009zf), there is nothing especially magical about life beyond the seemingly magical character of quantum mechanics. The quantum laws themselves will also be scientifically explained subsequently in this book without the need to invoke any particular mysterious "magic" or an external creator.

As previously explained, life on earth probably originated as molecules constantly combined in new forms and systems that became ever more complex and eventually developed into the first life forms (Hamilton, 2008a). Such a possibility is especially feasible within the context of some quantum crucible which permits many different possibilities at once (Merali, 2007zl) that can branch off into an infinite number of parallel universes, many of which never develop life. However, the process of organic life developing from inorganic matter seems to have been a very probable event given the composition of materials existing on earth after its formation (Lane, 2009b).

It's also possible that some of the original molecules involved in the development of life on earth may have actually originated from meteorites, as is evidenced from the contents of fragments of a meteorite that struck the earth in 2000 (*NewScientist*, 2006j). Meteors, which typically originate from the dust of comets whose ice is liberated when they come close enough to the heat of the sun and then burn up completely as "shooting stars" when they hit the earth's atmosphere (*NewScientist,* 2010aah), could carry organic globules formed long ago in temperatures of 10-20 degrees Kelvin in the outer reaches of the developing solar system (*NewScientist*, 2006j). Amino acids necessary for life have indeed been found in significant quantities in meteorites (*NewScientist,* 2008r). Some evidence has even been found of nanobacteria in meteors from Mars and other parts of the universe that have crashed into the earth (Calleman, 2009), and entire microbes from other places in the universe could theoretically have survived a crash landing (Anderson, 2008). Thus, the raw materials for life on earth, and perhaps life itself, may have originated from other parts of the universe, although such a possibility may not be the most likely source of life on this planet.

The specific molecules necessary for the creation of a basic building block of life, RNA, could also have come to earth via meteorites in unfinished form (*NewScientist*, 2008zf). However, it is unclear why the evolution of life, or even the origin of its basic organic compounds, might be more likely elsewhere in the universe than on earth itself

(Calleman, 2009). For instance, natural chemical reactions of materials already on the earth itself could also have formed RNA (Lane, 2009b).

RNA "can catalyze chemical reactions", "can sustain its own replication", and is thought to have preceeded its "sister molecule DNA" (*NewScientist*, 2009e). RNA, which is essential for existing life on earth (Cornish-Bowden, 2004), is the molecule that forms from the basic DNA code to enable the creation of the proteins used in necessary cellular processes (Cummings, 2006). Strands of RNA spontaneously nestle with amino acids in natural chemical reactions that form various proteins for which the RNA codes (Brahic, 2010). This RNA basic building block of life could also have originally evolved on the surface of the earth via ultraviolet light (UV) waves from the sun hitting molecules found on the earth in its earliest days (*NewScientist,* 2009zt). Alternatively, in environments like hydrothermal vents, amino acids and nucleotides would be formed by the natural laws of chemistry (Brahic, 2010), and replicating DNA could have formed and recombined in new combinations there (McAlpine, 2010c).

Experiments sending an object made out of the standard materials of meteorites, which generally consist of carbon, iron, and nickel, into water at a high speed imply that reactions on earth could have created the basic "organic molecules needed for life" (*NewScientist*, 2008zzzd). These amino acids tend to be left-handed possibly because of the effect of cosmic rays and water (*NewScientist*, 2009zk). It's therefore very feasible that the organic molecules that formed early life could have developed naturally from materials that were on earth itself (Hooper, 2006b) after that planet formed from rocky debris revolving around the sun (*NewScientist*, 2009i). Some of that material may have arisen early in the solar system from asteroid collisions which sprayed radioactive material into the forming earth that creates the planet's molten core (*NewScientist*, 2009zzw). The reactions in that center enable it to have a magnetic field that provides it with some protection against some of the energy released by the sun and thereby retain an atmosphere (*NewScientist*, 2010aaaa).

One possible source for the development of life might have been deep beneath the surface of the planet (Gold, 1999), or in sulfuric acid on the surface near mud boiling as a result of heat from below (Fox, 2009a). While the extremophile bacteria that continue to exist in such extreme environments may not be directly related to the cellular life that predominates in the less severe environments on earth today (Calleman, 2009), they certainly could have represented a beginning from which other life evolved. However, it has also been shown that organic life could have evolved naturally from more normal reactive molecules available on earth after its formation (Lane, 2009b).

While life may have begun in an environment on the early earth that doesn't exist today (Ravilious, 2008), studies indicate that RNA and the related DNA of more advanced life likely developed from natural chemical processes that have been found to exist in the beginning days of the planet (Lane, 2009b). The earliest life may have utilized the radioactive decay of the earth as energy, as some existing bacteria still do (*NewScientist,* 2008zzq), but energy for the origination of life could also have arisen naturally from more normal reactive molecules like methyl sulphide (Lane, 2009b).

Some of that material used by organisms on earth to form their key RNA and DNA elements, such as phosphates, may have originated from lightening strikes that stripped oxygen atoms from phosphate compounds (*NewScientist,* 2009zzx). Laboratory experiments have shown that combining materials that were on earth during its earliest days with lightening and steam to mimic likely volcanic effects results in the creation of 22 amino acids that are the building blocks of life on this planet (*NewScientist*, 2008zzv).

The energy for life is made possible by the force of electromagnetism that living cells harness by maintaining a charge inside their membranes that is different from the opposite one outside it, as in a battery, just as each human cell recharges its battery by excreting 3 positively charged sodium atoms for every 2 positively charged potassium atoms it admits (Lipton, 2008). Such reactions may have originated from the electrical current created in hot underwater vents that release electrons through the vents' walls into the seawater that could cause chemical reactions naturally provide such energy for early forms of life like bacteria (Evans, 2010). An increasing body of evidence does indicate the most likely origin for life was from materials on earth (*NewScientist*, 2009zy).

In any event, the first life forms on earth may have been viruses, which appear to have preceded the bacteria and other organisms, and which themselves helped both form and evolve animals and humans (Hamilton, 2008a). In particular, the enzymes produced by viruses and other parasites sometimes change the genetic code of the host cells, including that of sexual cells involved in reproduction, by retroposing the host cell's RNA which is sometimes inserted back into the DNA or genes (Page, 2008). A very large portion of the human genome derives from human ancestors having incorporated viral RNA into their heritable DNA (Zimmer, 2010). While a large portion of such DNA is related to diseases in humans (*NewScientist*, 2008zzn) or is totally useless, some of the DNA derived from viruses actually helps protect against diseases, and at least some is essential for mammalian life to exist (Zimmer, 2010).

Hazen (2006) and West-Eberhard (2003) provide substantial details on how life evolved from the earliest species, but the very important influence of viruses in that process has only recently been discovered (Hamilton, 2008a). Microscopic organisms like viruses and bacteria tend to spread genes to others through not only reproduction but also via other means (Buchanan, 2010). Over 40% of human genes, including a higher portion of disease-related ones than for other primates, actually come from bacteria (*NewScientist,* 2008zzn), and bacteria themselves may have their genes spread among different bacterial species via viruses (Doolittle, 2010). While such sharing of genes between existing organisms was most important in the early development of life, it has continued to have a significant effect (Buchanan, 2010).

Evolution is the long process that naturally selected life which is adaptable enough to survive and reproduce within the changing environment of the earth. For example, early bacteria species developed to form types that could photosynthesize light from the sun (Hecht, 2006b). These cyanobacteria that appeared 2.5 billion years ago were soon thereafter engulfed by more complex organisms called plants that transformed them into internal parts called chloroplasts ((MacKenzie and LePage, 2010). These plants had advantages in access to virtually unlimited energy through the photosynthetic process (Hecht, 2006b).

Photosynthesis involves light splitting water molecules into oxygen, hydrogen ions, and electrons (Shiga, 2010f). When light photons excite electrons in quantum dots or tiny crystals (*NewScientist*, 2009y), plants are able to "grab" electrons from proteins higher up in the chain reaction to manufacture the energy they need (*NewScientist*, 2009zzzh). One byproduct of photosynthesis is oxygen, which plants release into the atmosphere to give it a greater concentration of that element in the environment and which permitted the further development of non-plant life (Lovelock, 2009). In addition, changes in the content of the earth's crust may have resulted in variation in the emissions by early bacteria which fed on nickel that contributed to higher levels of oxygen in the environment and thereby enabled more complex life (*NewScientist,* 2009f).

While the first cellular forms of any type to develop may not have had reproductive abilities, a "quantum leap" forward occurred at some point that enabled reproduction through division (Calleman, 2009). Conditions do exist at the bottom of the earth's oceans that could enable the concentration of polymerase and nucleiotides to form replicating DNA which would, with the help of fatty acids creating membranes there, enable recombinations of the DNA (*NewScientist*, 2010c). Even after reproducing cells began to spread, the multiplicative divisions of single-celled species like some algae could still be inhibited by a particular gene that was activated during periods of extreme environmental stress and that later evolved via a mutation to suppress reproduction of only some cells in multiple-celled organisms (Kleiner, 2006).

Single-celled organisms like bacteria and archaea appeared 3.5 billion years ago (Ward, 2008) after bombardment of the earth by cosmic objects like asteroids became much less intense about 3.8 billion years ago (Mackenzie, 2009). However, the eukaryotic cells with a nucleus and other organelles characteristic of multi-cellular beings first developed 2.7 billion years ago (Ward, 2008). Eukaryotic cells evolved from bacteria living inside other bacteria that developed via mutations into the mitochondria portion of cells with a nucleus (Lane, 2009a). The mitochondria, which is the primary basic protoplasmic substance of most cells today (Calleman, 2009) and which produces almost all the energy used by such organisms, are the internalized bacteria that eventually became part of the cell as the mitochondria (Lane, 2009a).

Plants, which use sunlight to convert carbon dioxide into energy and an exhaled oxygen byproduct in the process of photosynthesis, evolved about 2.3 billion years ago (Ward, 2009). As a result of the exhalation of oxygen by plants, the earth's atmosphere became so "polluted" with that element about 2 billion years ago that it spurred the development and spreading of organisms, including multi-cellular ones, which utilized oxygen as an energy input and which actually fed on the plants (Lovelock, 2009). However, oxygen levels appeared to have fluctuated greatly for hundreds of millions of years thereafter, as organisms evolved that could adapt to the new atmosphere, which itself was being changed by the varying oxygen levels until an equilibrium between life and the environment was reached (Lane, 2010a). Oxygen-induced "burning" is ten times as efficient as alternative means of extracting energy from food and was therefore important in the development of more evolved forms of life (Lane, 2010b).

Multicellular life began about 2 billion years ago (*NewScientist*, 2010e). One of the impetuses for multi-cellular life was an increasing size of single-celled life-forms, which eventually grew too large to form many new receptors on their membranes for optimal monitoring and managing of their environment (Lipton, 2008). The merging of various single cellular organisms into multicellular ones (Lawton, 2009a) followed after they first started cooperating as independent single-celled organisms like amoebas still do today (Lipton, 2008). Species with multiple cells dependent on one another may have actually developed out of such cooperating but independent single-celled organisms that eventually became dependent on each other in unified new multi-cellular species.

Prior to such cooperation and combinations, reproduction by division likely evolved as individual single-celled organisms mutated into dividing ones that were better able to adapt to the environment on earth than both non-reproducing cells and others that may have included single-celled dividing entities which only evolved to be larger and potentially too bulky to survive. The multi-cellular forms of life evolved sexual methods of reproduction, instead of the cloning methods employed by single-celled bacteria, probably because of the heavy mutation rates that characterize their eukaryotic cells (Lane, 2009a). A mixing of such mutated genes through sexual merging apparently help complex species become adaptive to the environment more effectively than mere individual random combinations of mutations, as many sets of mutations that might be non-functional by themselves may contribute positively in the evolutionary battle for survival when merged via sexual selection into larger and more complex sets.

The sexual form of reproduction that mixes DNA or genetic capacity across two separate organisms into the offspring began about 1.2 billion years ago (Calleman, 2009). This form of reproduction involves the combining of a sperm cell with half of the father's genes in its DNA and an egg cell with half of the female's chromosomes into an embryo cell that grows into a fetus and an eventual new being (Cummings, 2006). The insemination of the egg cell, which lacks a directing cellular centriole, normally results in the activities of the egg's mitochondria being directed by a centriole emanating from the inseminating sperm cell, which lacks a mitochondria, although it has also been found to be possible to have the egg cell's mitochondria controlled by a mere centriole artificially inserted to enable a "virgin birth" (Calleman, 2009).

Multi-cellular organisms' cells, which "stick" to their neighbors, continue to reproduce their individual cells by mere division, using some of the same genes that bacteria do for cooperation in controlling the size of their colonies, although eukaryotic cells utilize special chemical signals to each other in order to enable greater coordination (Holmes, 2010). Some genes, like the one termed p53 found in single-celled placozoans that detects and deals with damaged DNA which could trigger cancers, have been passed down to multicellular organisms like humans over a billion years (*NewScientist*, 2009zzzzp). Thus, there is indeed continuity in the development of sexually procreating life-forms out of the earliest organisms that reproduced by division.

Organisms like sponges are the most likely mother of all multi-cellular life (Holmes, 2009). Such eukaryotic life-forms evolved after bacteria dominating life below the surface of the ocean had polluted the planet with their hydrogen sulfide waste which contributed to a series of super ice ages between 900 and 500 million years ago that created a "Snowball Earth" (Fox and LePage, 2009). The multi-cellular sponges themselves along with multi-cellular algae and other photosynthesizing plants may have been the major factor causing such cold waves as they absorbed carbon dioxide out of the atmosphere that warms the earth (Brasier, 2010). While the extreme conditions during that period may have made active life impossible in many places for most organisms, some species like the sponges evolved forms that enabled survival through those times (Fox and LePage, 2009). Today, there is even a type of bacteria that can hibernate for one hundred million years until a suitable environment exists to awaken them from their dormancy (*NewScientist,* 2010aae). While extreme environmental conditions result in the extinction of many life-forms, they pave the way for the emergence of new species of organisms, of which about 10 million different types exist today (Calleman, 2009).

Several hundred million years after sponges evolved to hibernate in cysts in order to survive the prior cold environment that was extremely inimical to most life-forms (Fox and LePage, 2009), abundant multi-cellular life began about 600 million years ago (Ward, 2008). Widespread animal life started between 542 and 520 million years ago (Fox and LePage, 2009). The catalyst may have been cold periods over a half billion years ago that led to glaciers stripping phosphate compounds in the ground onto the surface, and this phosphate provided a fertilizer for greater plant growth which then increased the production of emitted oxygen that enabled the development of more complex life at that time (*NewScientist,* 2010aax).

A great deal of diversity developed about 500 million years ago that could have been spurred by the greater usefulness of different types of evolved ingenuity as conflicts among organisms increased because of rising competition for scarce resources in an increasingly populated world that also experienced a spurt of more volcano eruptions and meteor showers at that time (O'Donoghue, 2008). The development of more complex life may have been made possible by the evolution a half billion years ago of genes which create special microRNA in vertebrates that engages in complex control of protein production in cells by binding to messenger RNA, which is made by other genes in DNA (Nicholls, 2008). MicroRNA inhibits Messenger RNA from acting as templates for the production of proteins (Trivedi, LePage, and Aldhous, 2010) and seems to function against pathogens that get inside the body (Nicholls, 2008), as may have been very important as larger and more complex species developed to exist outside water.

Organisms that evolved to live on land apparently developed around this time, having shells like crabs to enable living in the "new environments" of "the sandy title flats" that "fluctuated between the vast wet and dry zones" existing then (*NewScientist,* 2009zn). After the first animals appeared 565 million years ago (Ward, 2009), the first vertebrates with backbones evolved about a half billion years ago (Nicholls, 2008). The first live births, i.e., offspring emerging with umbilical cords instead of in eggs, occurred at least 380 million years ago, as evidenced by some shark-like fossils recently discovered (Young, 2008c). However, breast-feeding mammals remained insignificant creatures on earth for hundreds of millions of years after their development. Milk was actually used originally by the first mammals to wet eggs, and only later did live offspring evolve that were breast-fed (*NewScientist*, 2008ss). After the largest mass extinction of life in the history of the world about 250 million years ago that eliminated 95% of animal species, the first dinosaurs appeared that were initially about the size of a cat and were likely food for crocodiles early on (*NewScientist,* 2010aak). The dinosaurs grew to dominate the planet for over a hundred million years before they were also exterminated.

Since the initial appearance of animals, there have been five major and ten minor mass extinctions, some of which may have been caused by large extraterrestrial objects like comets impacting the earth, but most seem to have been related to global warming caused by wastes emitted by life-forms on earth itself (Ward, 2009). The evolution of life on earth from the simplest organisms occurred through a process of mutation, development, and perpetuation of the genetic combinations that were the most successful in surviving and reproducing in the changing environment there (West-Eberhard, 2003). Many genetic mutations may have initially provided no positive benefits to the organisms, but large genetic changes would have been encouraged in stable environments that harsher conditions would only later prune or encourage depending on the usefulness of the changes to enhance survival and reproduction of the genes (Kenneally, 2008c). Some changes in hospitable environments can even enable moving to a more adverse environment.

Given that viruses apparently can move genes from one organism to another (*NewScientist,* 2008zzw), it is likely that the evolution of complex species has been heavily affected by spreading viral infections, which can not only be destructive but also lead to rapid evolutionary changes. Genetic parasites can invade the genomes of organisms and cause gene duplication that leads to the development of many new genes, as appeared to have occurred on a widespread scale about 45 million years ago (Page, 2008).

Since not all of the large changes in life-forms and life processes can be explained by mutating individual genes (Calleman, 2009), it is likely that many of the largest developments in organisms, such as those that result in new species, are caused by viruses and bacteria. Viruses themselves were apparently useful in the evolution of many genes of many organisms including humans (Hamilton, 2008a), as the enzymes produced by the viruses sometimes change the genetic code of the host cells, including that of reproduction or sexual cells, by retroposing the host cell's RNA which is sometimes inserted back into the DNA or genes (Page, 2008).

Life forms have built in lots of room for errors and gene mutations without being fatal (Cornish-Bowden, 2004), and many gene mutations enhance the flexibility of organisms to change and be able to adapt to a varied environment (West-Eberhard, 2003). Many of the developments in the evolution of life forms are extremely efficient, such as bones that have extraordinary strength and flexibility through interconnections which have back-up connections (Groshong, 2007), and such as snail shells that have outer membranes which cause the impact of energy hitting one area to fan out across and be absorbed by other areas, thereby protecting an inner membrane from the full applied force (*NewScientist*, 2010f). Evolution also allows complexity and unnecessary abilities to de-evolve in many instances (Spinney, 2007a). For instance, many less efficient forms of life evolved like giant fungi actually had advantages over plants in earlier moister environments but later disappeared after conditions changed (O'Donoghue, 2007). Viruses, which are the most abundant life form on earth, appear to be important in speeding evolution by entering and affecting the DNA of host organisms in "an interconnected network of continuously circulated genes" that represents more a process of "give-and-take" than survival of the fittest (Hamilton, 2008a). Genes can also apparently be affected by other factors in the environment (*NewScientist*, 2010p), as explained in general by Gynn and Wright (2008).

Organisms can, upon the mutation of a just a few genes, become incompatible for procreating with unmutated forms of life, thereby creating separate species (*NewScientist*, 2006i). The mutants must be adaptable enough to survive and propagate among themselves, and since the likelihood of two organisms born with identical mutations would happen to meet and mate, it seems probable that most of the changes in the genetic code that caused the appearance of new species are generated by viruses and bacteria that might have infected many organisms at once with the mutations that differentiated species.

Even after separation, each separate species continues to need other species and their cooperation for that purpose. For instance, humans need a myriad of bacterial species in balanced amounts in order to digest food, and human cells are actually outnumbered by bacterial cells by 10,000 to 1 there (Spinney, 2007c). In addition, bacteria "may play a key role in making our weather, and even form part of the global feedback that evolved to keep the earth's bio-systems in balance in what is called the Gaia hypothesis" (*NewScientist,* 2008k). Some bacteria actually "evolved to summon water from the sky to help them survive" by excreting rain-inducing surfactants as well as through direct contact when they are swept up by the atmosphere into the clouds (*NewScientist*, 2009b). Some different species of bacteria even cooperate in an assembly line to draw their own form of energy from different elements in a sort of electrical connection that is partially enabled or wired by the hairs they grow (Brahic, 2010). Thus, despite separation into different species, life continues to require balanced cooperation.

The benefits of cooperation, as well as the costs of recognizing and differentiating between genetically related organisms, resulted in a genetic altruism in many cases (Gigerenzer, 2007). The cooperative tendencies may have been magnified as a result of some organisms like mammals evolving to develop brain chemicals such as oxytocin and dopamine to motivate caring for their young via rewards of experienced pleasure (Klein, 2003). While this process could be characterized by a theory of purely selfish genes, or combinations of genes, in some organisms that merely seek to survive and reproduce themselves, the usefulness of cooperation among both similar and dissimilar organisms caused many genes and organisms to evolve to naturally coordinate actions to maximize propagation of themselves (Fanelli, 2008). In particular, evolution not only created species that seek to perpetuate themselves or those closely related genetically but also motivated the development of cooperation among species. The massive amount of bacteria in people's stomachs that aid in human digestion (Spinney, 2007c) is a good example.

Today, life actually consists largely of extensive cooperation among organisms and individual cells that began very early in the evolution of living organisms (*NewScientist*, 2006f). For instance, a portion of the cells of complex life-forms were even once separate organisms themselves that took cooperation to its fullest extent by giving up independent life to become part of a larger unified whole as cell mitochondria (Rose, 2005). As previously mentioned, the mitochondria existing in the cells of most advanced organisms today was indeed originally independent bacteria living inside other bacteria (*NewScientist,* 2008zzza). Cooperation among cells is required for complex life to exist, and much of the necessary communication among cells for that cooperation apparently occurs through nanotubes made between cells that are so small and temporary they have only recently been discovered (Ananthaswamy, 2008e).

Brains and thinking especially require extensive cooperation and communication, as "mental information is stored not in single cells but in entire populations of cells and patterns of their activity" (Eagleman, 2008). At least part of the mind's operations is carried out through the stringy dendrites and axons of brain cells, which are called neurons, that connect to other neurons (Bernstein and Nash, 2008). The speed of the signals carried by the dendrites and axons, which are called "white matter", is affected by the quality of the myelin sheath between the neural processors themselves, which are called "grey matter" (Pearson, 2009). However, microtubules may provide more direct communication, which can be enhanced by the brain's production of melatonin during sleep (Gynn and Wright, 2008). Microtubules are small enough to enable collapsing of the probability wave functions of quantum chemicals in the brain and thereby enable psychic powers (Radin, 2006), as will be explained later in Chapter 16. Regardless of the means of coordination, brain cells in advanced organisms like animals actually die sooner if they don't communicate with other brain cells (*NewScientist,* 2006g), possibly because they don't receive needed chemical rewards when they don't fire in synch with other neurons in a manner that leads to evolutionarily adapting thoughts and actions.

It is interesting that all the many different types of cells in a complex organism form from the initial identical embryonic stem cell, which itself is constructed upon procreative insemination of the female egg cell, and which divides into differentiated cell types due to chemical signals causing different voltage levels across cell membranes (*NewScientist*, 2008zzzc). At least some signals for cell differentiation may come via more invisible means for differentiation, while all the various cells still have special self-receptors that enable immune cells to recognize an individual organism's own cells to inhibit attack by the immune system (Lipton, 2008). Both cell specialization and differentiation require communication between the cells.

Despite the differentiation that has occurred since the universe in which human beings live originated from a huge explosion called the Big Bang about 14 billion years ago (Magueijo, 2003), all native life forms here are made from the same essence. All particles of matter, regardless of how complex, remain merely manifestations of different processes or vibrations (Capra, 1999), and the quantum nature of the universe connects everything as well through interacting wave functions (Peat, 2005).

A. Life Within the Context of the Whole Cosmos

Consciousness, life, and the universe are organized according to patterns defined by quantum mechanics (Davies, 2004). This science indicates that the universe is made up of discrete quantum particles of infinitesimally small size (Seife, 2000) that are actually mere excitations of electromagnetic fields which follow specific rules (Barbour, 1999) which can be quantified mathematically (Rees, 2009). The consistent patterns followed by the different fields or processes result in their categorization as particles (Capra, 1999).

The most advanced computer models indicate, however, that the universe is composed of a single connected fabric as opposed to a collection of particles (Das, 2007). The continuous electromagnetic fields and emissions of everything that affect all else in waves of potential puts each individual field and emission directly and indirectly in contact with one another. In addition, the entanglement of particles that cause their actions to be correlated after contact (Capra, 1999) may reflect one aspect of the universal connectivity. The fact that very low temperatures cause objects to blur because their position is very uncertain (Brooks, 2010a) provide an illustration of the unified nature of the universe. In particular, the uncertain position of particles in their probability wave functions results in them having different interactions in different parallel universes that cause their fields to overlap in the uncollapsed averaging of those universes that is normally observed.

The laws of quantum mechanics indicates connections across matter that are most consistent with the ancient human religions like Taoism and I Ching, which also incorporate a belief that individual atoms have a vibrating interconnected consciousness (Sherrill and Chu, 1989).[1] The conscious mind itself can possibly be liberated into understanding this concept by viewing the universe as a labyrinth, with a center everywhere and a circumference nowhere (Cont, 2002). A mere concentration on breathing can also be enlightening as to the interconnectedness of all when one sees breathing out air as giving back energy to the universe that one has taken in (Chia and Wei, 2003). Laszlo (2006) summarizes some of the expert opinions and findings on the integration of mind, matter, and the broad environment of the cosmos.

The universe in which humans live was initiated in a Big Bang, which resulted from energy being emitted via a decay in an inflating field or particle (Chown, 2007b). The energy had been stored in a sort of magnetic or gravitational field that drove the expansion of space from a tiny area initially at a speed exceeding that of light itself (McAlpine, 2010d). Once energy and matter were released from this field into the expanding space, the variety in the universe developed through random quantum fluctuations and interactions (Chown, 2007b). This process essentially involved an initial quantum object following a multitude of alternative paths in a wave function to expand to the current situation (Merali, 2008e).

Quantum particles include basic substances like quarks, which are relatively permanent and are therefore called fermions, and entities like photons of light, which are forces that act on fermions, are called bosons (Close, 2009). These original particles of the universe later combined to produce the simplest elements like hydrogen and thereafter mixed further to create more complex elements and compounds (Chalmers, 2008). The first atoms formed 380,000 years after the Big Bang (Battersby, 2010b).

All atomic elements are characterized by electrons orbiting an atomic nucleus, which contains protons and neutrons that are called nucleons. The nucleons are composed of three quarks (Smith, 2003). These atoms form into molecules or compounds of atoms which bond together largely due to the magnetic attractions that are affected by their structures (*NewScientist,* 2010az). Atoms and fermions in general are held together and move because of bosons (Close, 2009), including light photons which can be absorbed by fermions. All atoms regularly emit microwave radiation whose frequency varies by atom and may be affected by gravity (Chalmers, 2009).

Many of the heavier elements may form when some of the neutrons entering the nuclei of existing atoms are later transformed into protons after collapsed stars merge (*NewScientist,* 2010ah). The neutrons for this process originate in atoms split by nuclear fission that shoot them into other atomic nuclei to cause the nuclear fusion involved in creating elements with mass greater than that of the simplest elements like hydrogen which has only one proton in its nucleus. When some of the neutrons merging in existing hydrogen atoms decay into protons over time, the more complex elements with more protons and neutrons are formed. Many such elements forged in a deep layer of a star are "carried to the surface by convection, and ejected as part of a powerful stellar wind" (Battersby, 2010d). After the enormous energy from the Big Bang about 15 billion years ago caused the universe to expand from a single point, large individual stars arose as the hydrogen and helium atoms from that initial explosion clumped together in large mass, but many of these stars subsequently exploded in supernovas that spread heavier elements throughout the estimated 125 billion galaxies in the universe that formed as newly formed star systems merged with each other (Calleman, 2009).

Some evidence exist that massive black holes with the mass of 10 billion suns concentrated in a very small area of space formed within a billion years after the Big Bang (Chown, 2010a). A black hole is a heavy concentration of mass in a small space that exerts tremendous gravitational pull on nearby space (Seife, 2000). Anything coming within 15 million kilometers of a black hole is sucked into it, and even stars approaching 100 million kilometers of such a gravitational field are ripped apart (Chown, 2010b).

Black holes can form when massive stars collapse on themselves via the gravitational pull of their concentrated inner mass on particles located in outer portions of the large objects (Seife, 2000). Not all stellar objects end up as black holes, however. For instance, brown dwarf stars, which don't have enough mass to sustain the nuclear fusion that creates light and heat in most stars and which may be common in the universe (*NewScientist,* 2010m), represent an alternative end for stellar objects that occurs when their mass becomes too small to sustain the pressure needed for them to burn with light in atomic fission and fusion, much less collapse from an overwhelming gravitational tug into a black hole.

A black hole typically evolves from a neutron star, which itself forms when the star's hydrogen atoms are condensed into such a small area that the electrons in those atoms smash into the protons to create neutrons, and which can later turn into a quark star that forms when gravity pressures the neutron star collapses into tiny enough to decompose neutrons into quarks which themselves eventually condense into the zero space of a black hole (Seife, 2000). Once formed, black holes spin (Battersby, 2009c) to create a magnetic field, and they suck in all matter within a certain distance into themselves, and the energy around that event horizon is limitless (Griggs, 2009b).

Spinning black holes are thought to be at the center of most galaxies (Battersby, 2009c). There is some evidence they create those galaxies by causing jets of radiation from attracted nearby objects to spray into interstellar dust clouds which coalesce into systems of stars that suck the black holes into their midst (Chown, 2010a). When there is enough gas to prompt significant amounts of star formation from all sources, the newly formed stars orbiting a black hole naturally align to create an elliptical disk (*NewScientist,* 2010r). The typical elliptical shape of galaxies (*NewScientist*, 2010r) is thereby explained.

Black holes can combine with other ones (Battersby, 2009c) and grow to a size of over 6 billion suns (*NewScientist,* 2009zzze). Black holes, some of which contracted such "massive gas clouds" into a tiny area that they have the mass of 20 billion suns, and which form the center of galaxies of star systems (*NewScientist,* 2009h), can also represent wormholes to different points in time and space if they have been compressed into zero space and thus have no event horizon (Seife, 2000). For instance, the enormous black hole believed to exist at the center of the earth's galaxy could send the matter and energy it collects to somewhere else in the universe (Shiga, 2009a). Wormholes that enable travel to different places in the universe in both space and time require matter with negative energy (*NewScientist*, 2010aj), but since black holes suck in more antimatter than matter (Shiga, 2010d), a sufficient amount of such anti-energy may exist there.

The gravitational field from which the Big Bang of this particular universe arose over 14 billion years ago may itself have been a black hole. Such a gravitational field would emit jets of hot plasma gas without light (*NewScientist*, 2010q) to keep the universe expanding even while the large mass of the black hole causes it to exert a strong gravitational pull on nearby space. The universe might then take the shape of a spinning spiral, just as galaxies formed from black holes created after the Big Bang do.

There is some evidence that the various galaxies in this universe revolve around a central axis, with those above its plane orbiting in a direction opposite that of those below it (Calleman, 2009). It is therefore quite feasible that all the matter in the universe's originating black hole didn't completely spew out all its essence in the initial moment. As a result, the existing galaxies and everything else in the universe seem to be attracted toward what remained of the central black hole in a spiraling orbit that takes the same elliptical shape of individual galaxies.

Within this context, the universe continues to expand outward because of both the initial energy from the Big Bang and subsequent emissions from that black hole at the center. Dark energy originating from particle pairs being created out of the expanding space and mutually annihilating themselves may also contribute to the expansion of the universe.

One theory called loop quantum cosmology that is consistent with both the theory of general relativity and quantum mechanics indicates each universe springs out of another one which eventually collapses on itself because of the tug of gravity until the mass concentrated in the area of a single proton equals a trillion suns (Ananthaswamy, 2008f), and it then inflates back after an initial period of "superinflation" (Ananthaswamy, 2010c). The energy from the gravitational pull reverses itself because it isn't possible to reach Planck density, and so at that point the gravitational compressing force causes space-time to bounce everything back out again in a Big Bang (Ananthaswamy, 2008f).

The original black hole may have actually been born from a wormhole in another universe. It is well recognized that this universe may actually co-exist with many other universes similarly formed but with potentially far different characteristics that are independent of this universe (Gefter, 2009a). The origin of the black hole in this universe could even have been a universe of antimatter. Such anti-universes would consist largely of antimatter, which is basically matter that travels in an opposite direction in the time dimension. Antimatter universes might thus consist of fermions and bosons that move backwards in time in a process that leads them to collapse on a black hole that forms the beginning of a matter universe when it bounces back on itself to go forward in time.

The existing universe consists mostly of matter that moves forward in time, but the particles can move toward or away from each other depending on the force exerted. A strong force exists to hold subatomic particles called quarks together to form the protons and neutrons that make up the nuclei of atoms (Close, 2009). Other hadrons exist such as mesons and baryons (Capra, 1999) that are also composed of quarks (McAlpine, 2010), but atomic nuclei consist of only the combinations of quarks that make up protons and neutrons.

Particles called gluons are theoretically thought to exist to confine quarks together in a "color charge" within nucleons that is stronger than electromagnetic binding, but they are never observed outside their quark confinement (Smith, 2003). Gluons and other virtual particles which can't be observed represent "negative energy states" (McTaggart, 2008) that are merely modeled by existing theories as particles.

The typical atomic nucleus is circled by the most common type of leptons called electrons (Capra, 1999), which revolve at 1% of the speed of light in different orbits, and those nuclei, which give atoms most of their mass, occupy only 1/100 of the space covered by the entire atomic electromagnetic fields (Close, 2009). The concentrated mass of atomic nuclei decays slowly over time in an interval that varies inversely with the speed of motion of the object (Childress, 1999). The decay process theoretically involves a boson acting in what is called the weak force that effectively transfers a negative electric charge from a neutron to convert it into a proton with a byproduct of an electron and an antineutrino (Close, 2009). Neutrinos emanating from the sun may be involved in the process that switch their spin and thereby transform themselves into antineutrinos (Mullins, 2009b). It is also possible for decay to work in the opposite direction, with electrons, protons, and antineutrinos coming together to form neutrons, but such decay is much rarer (Childress, 1999).

Neutrinos and antineutrinos, which effectively represent neutrally charged electrons (Close, 1999) and have little mass, were first determined to exist because there was a difference in energy between a neutron and the proton-electron byproduct of their decay (Kumar, 2010). Neutrinos are a type of lepton, which also includes charged electrons and muons (Capra, 1999), the latter of which are the standard negatively charged electrons but with very large mass (Close, 2009) that actually decay very quickly into the smaller electrons (McAlpine, 2010b). Electrons themselves don't decay as they represent the basic particle of the universe. Most of the remaining particles are ones that actually appear for only a moment while they travel at close to the speed of light in man-made experiments within large accelerators that utilize magnets and the electromagnetic force to induce high speeds in particles before they disappear (Close, 2009).

Three of the major four forces of the universe, including the strong and weak forces and the electromagnetic force, are assumed by existing theory to be carried by other particles (Smith, 2003). For instance, photons, which are one of over 100 different types of force-carrying bosons, carry electromagnetic forces between those various fermions in the form of X-rays, radiation's gamma rays, and light that includes visible as well as ultra-violet infra-red light, where the exact force and color is determined by the photon's frequency (Close, 2009). Since light photons travel at the speed of light and have no mass, it is possible that the other bosons that both travel at slower speeds and have mass are composed of the same basic energy, which is merely moving at a different velocity through space because each has a different concentration in any point in time.

The origin of these particles and forces is deduced in Section C of this chapter, but it will be summarized intuitive here for improved understanding of the other aspects of cosmology and creation.

Energy is merely unified nothing vibrating at different frequencies due to the electromagnetic tug of matter that curves the nothingness of space-time. The energy, which oscillates in alternative directions in time, causes movement of the divided pieces of nothing through the illusion of separate intervals of time and space.

Moving through space over time can't occur without the existence of the illusory time dimension that is created by the division of nothing into separate parts consisting of positive temporal energy pulling forward in time and negative energy tugging backward in time. The positive temporal energy takes the form of the electron, which exerts a negative charge in space, and the equivalent other half of nothing called a positron, which has the opposite electromagnetic charge and moves in the opposite past direction in the time dimension. The attraction of those divided electric fields moves toward unification when the past meets the future after a past and future of equal length, and this temporal pulls along with their attraction in the spatial dimension cause them to spin and thereby exert a magnetic field that curves space-time back on itself. The positive and negative energy force fields, which are created by divisions of nothing that inherently create the dimension of time which divides the fields, have mass in the space they occupy because of their separate existence in each moment in time until these divided pieces of nothing unify by occupying the same point in space-time.

The unification of the divided pieces of nothing that occurs when they merge creates unified temporal energy in the form of pairs of photons that can alternately oscillate between forward and backwards movement in the time dimension without experiencing any passage of time because they aren't temporally divided. This light energy, whose temporal oscillation gives it a spin, can't exhibit mass because its unified temporal energy that gives it its timeless nature isn't concentrated in any point of space-time. Without being divided by time, the photons' energy has to be exerted in other dimensions, and so where space doesn't exist, photons movement create it, pushing it in three dimensions because it remains integrated with the temporal dimension that curves back on itself. The finite speed of light across space through time is observed only by anything that is divided by time and therefore moves at less than the speed of light, whose entire unified energy is directed in the spatial dimensions. Each photon can be absorbed by still divided pieces of nothing to cause them to move in space across time toward their opposites because the temporally undivided light effectively exists only in the spatial dimensions and thus doesn't occupy any particular point in space-time.

Thus, upon unification, the positive spatial energy field of a positron and the negative spatial energy of an electron are transformed into two light photons of energy with a net zero spatial charge. Each photon is both positive and negative energy that is unified temporally and spatially. Being thus unified in temporal movement, those photons are timeless but exist in different points in space, thereby enabling their absorbtion by still divided pieces of nothing that cross their path at particular points of space-time. The diverse collection of possible combinations of the divided and unified pieces of nothing enable the creation of an infinite number of universes filled with more matter and anti-universes of greater parts of antimatter.

The different combinations of divided pieces of nothing create matter like nucleons and antimatter like anti-nucleons that have various component fields of spatial charges, which are frozen in two spatial dimensions by their particular structures in space that exist in the form of triplet combinations of positrons and electrons. Those triplets make up the nucleons' and anti-nucleons' quarks and anti-quarks, each of which has two parts of one form of temporal energy and one field of temporal energy tugging in the opposite direction in time but prevented from doing so because of their trapped state in a triangular formation of space. Systematic differences in the amount of such frozen temporal energy result in systematic differences in the positive or negative temporal energy of positrons and electrons existing in such universes that cause the appearance of a tug in a particular direction in time.

The net amount of positive and negative energy continues to sum to zero across all universes consisting of more matter and all anti-universes consisting of more antimatter, thus leading nothing to be created out of nothing despite the reality experienced within the temporal dimension that also sums to nothing across the infinite possible histories and futures for everything in time. Those past and future potential universes exist only as probabilities until they are collapsed by interactions between the various probability wave functions of the component interrelated divided pieces of nothing. The probabilities of all potential future parallel universes are affected by the history of past possible universes that could have led to them, but those future probabilities also impact the past in an integrated fashion according to the laws of quantum physics.

The divisions of nothing that create new pairs of particles and antiparticles all the time in the vacuum of space are themselves a probability wave function that can be collapsed by interactions with timeless light photons that vibrate nothing to create new fermion pairs which are divided by time and whose combined positive and negative energy sum to zero. The mere possibility of the creation of divided pieces of nothing out of nothing increases the likelihood of various future universes evolving that include a living species with will power as does exist in the form of humanity as explained in Chapter 16. As the likelihood of such future universes thereby grows, so does the probability of the past feasible universes that enable their existence. This interactive process across time occurs simultaneously to result in the probability wave functions of divided nothing to be of a form that maximizes the probability of human beings evolving. Thus, the evolution of the reality in which human beings exist became virtually inevitable just because it was possible for nothing to divide by itself to create divisions of nothing out of nothing that led to the infinite variety of their universe.

In macro terms, the subatomic particles that were emitted via the Big Bang that initiated the existing universe of space-time in which humans currently exist coalesced into electrically neutral atoms within 400,000 years of that genesis, but there may have been another billion years before ultraviolet light from the first stars and black holes created any significant amount of ions or electrically charged atoms (*NewScientist*, 2010aaag). The enormous stars that formed later spread matter into varied galaxies when they themselves exploded in supernovas (*NewScientist*, 2007zi). Supernovas generally occur when a star collects sufficient amounts of various types of chemicals to blow apart (*NewScientist*, 2009zzj). Exploding stars emit cosmic rays, which include both positrons and the antimatter form of protons called antiprotons, that collide with electrons and protons in the earth's atmosphere to create gamma rays (Close, 2009). Light, which travels at the speed of 186,000 miles per second (Capra, 1999), moves through space from these various objects and processes via photons to enable their observance in the past on earth (Close, 2009).

It may not be possible to see areas that expanded in a direction opposite that of any particular point in it because the rapid expansion of the universe after the Big Bang may have exceeded the observed finite speed of light (Capra, 1999). That incredibly quick expansion likely resulted from the gravitational pull of a black hole in an anti-universe consisting of more matter that pulled on itself so strongly in space that it froze time in a point concentrated with matter it sucked into itself. The concentration of more positive energy frozen in time within the black hole created an opening in the antimatter universe that spewed more matter into this new space which expanded forward in time via matter's tug in this direction in the temporal dimension. Since only one part more of matter than antimatter per billion was sufficient to create the existing universe (Battersby, 2010c), it seems quite feasible that Hawkins' theory of different amounts of matter and antimatter being drawn into a black hole as a result of the simultaneous creation of such divided pairs on the hole's event horizon of that attractive force (Shiga, 2010d) can explain matter universes being created from antimatter universes and vice-versus. In particular, the systematic tendency for antimatter to be drawn into the black holes of this universe could lead to the creation of anti-matter universes where it would be emitted, just as the black holes of the antimatter universes might tug more matter into them, one of which likely caused the Big Bang and further emissions from it in the creation of this universe.

The reversal of direction in time through the singularity of a black hole could very well lead to quantum discontinuities and infinities associated with speeds faster than light. For instance, divided pieces of nothing could initially travel faster in space than the normally finite speed of light observed today in this universe because there was no gravitational tug in the newly created space. However, more matter entering this expanding space from the black hole, as well as the creation of more energy and matter from new divisions of nothing in the increasing number of additional points in space has caused the speed of light to slow to today's limit of 186,000 miles per second.

The freezing of time within the black hole anti-universe provides the energy for this process. In particular, the frozen temporal energy of the black hole in the form of timeless light photons push space outward in the expanding universe because they can't be frozen in time and thus had to move in additional dimensions called space that they themselves create as their unified temporal energy expands the spatial dimensions to enable their timeless movement.

As space expands over time in the new universe, additional light photons move timelessly into this created space and can upon contact with divided pieces of nothing be absorbed by those fermions to permit their own movement in the spatial dimension. Some of that light energy still in the black hole having merged there with the divided pieces of nothing in the black hole of space gives the divided pieces of nothing like electrons the spatial energy to move into the expanding space. That space expands over time as the greater positive energy of the electrons not frozen in a triplet formations pushes time forward in the newly expanding universe. This universe's space-time remains curved by the tug of that positive energy toward unity with the negative energy of the past, and even the unified timeless energy of light photons can appear bent in the newly forming space-time to give it the appearance of a finite speed over incremental points in time that don't actually divide the light.

Based on assumptions about the speed of the inflation of space resulting from the Big Bang, objects only about 45 billion light years away can be seen at the edge of the cosmic horizon that is observed by human beings (Gefter, 2009a). Bosons, which include forces with mass like pions, and which emanate in radiation from the explosion of distant stars via cosmic rays (Close, 2009), may only be experienced closer than 45 billion light years away because they travel at slower speeds than light photons.

Bosons don't have the alternating ½ and -½ spins of fermions because they don't have their energy divided equally between the spatial and temporal dimensions. As opposed to light photons whose temporal tugs are fully unified and thus exert no concentration in any point in space-time, bosons with mass consist of various combinations of electrons and positrons trapped into frozen attraction to each other because of their structure. Some quark-like bosons like pions can exist in either double form with no charge or in triplet form with a net full positive or negative charge. The existence of a nonzero charge for some pions may reflect a point on a quark triplets' combined wave function frozen in a triangular formation in space that allows for the possibility of separate positron and electron components to pull forward or backwards in time with a charge in space, as explained with respect to quarks in general in Section C of this chapter, while the doublet form represents the resulting separated existence of an equal combination of positrons and electrons. Because these quark-like bosons are at least partially frozen by separate tugs in time in their triangular formation, they can occupy and move through the same space as fermions without annihilating each other because the fermions exist freely in all four dimensions of space and time.

All fermions, bosons, and combinations thereof like people are affected by the relativity of the speed of spatial movement across time. For instance, a person traveling at close to the speed of light, as is theoretically feasible, could explore the universe 15 billion light-years away, but 70 billion years would have elapsed by the time the individual returned to the area where the earth was (Courtland, 2009c).

Time itself is thus purely relative in this universe. Einstein's theory of relativity indicates that movement close to the speed of light of 186,000 miles per second results in time passing more slowly for faster-moving objects (Capra, 1999). This widely accepted theory, which has been confirmed in many tests, including with quantum particles in accelerators, creates the hypothetical possibility of people in an extremely fast rocket ship moving away from earth coming back to see their friends having aged much more rapidly (Childress, 1999).

Matter also shrinks in size while it is moving closer to the speed of light (Buchanan, 2008f). Speed in space across time, the energy that enables such movement, and the mass measured for anything existing in any point in space-time are therefore all related. In particular, all matter, antimatter, and energy in the universe can be converted into each other via the equality $e=mc^2$, where e is energy, m is mass, and c is the speed of light in a vacuum (Close, 2009).

With the application of enough force, matter can be destroyed or transformed back into energy, which itself can induce the creation of pairs of particles and antiparticles as well as photons of light (Capra, 1999). Matter and antimatter, both of which are fermions, provide the structure in this universe while the energy and bosons obviously provide the force to have the structures move, change, and interact with each other. All these particles are, at the basic level, mere energy or vibrations of nothing within the interconnected nothingness.

While the shrinkage of mass to zero at the speed of light might normally preclude movement in excess of that speed (Buchanan, 2008f), especially since the light energy that moves particles through space over time only moves at that speed, it might be possible to exert negative energy that could result in negative mass. Antimatter that does move backwards in time may even be the result of such applications of negative energy which could enable travel backwards in time.

The zero mass of light photons creates the singularity of light moving instantaneously through all moments in the temporal dimension that enables it to represent positive or negative energy. The negative energy for time travel thus does exist, as does the form for concentrations of that energy in points in space-time in the form of antimatter that merely represents negative mass, which is concentrated energy moving backwards in the temporal dimension in divided quanta of time..

Antimatter, which has the same mass as matter but the opposite charge, acts essentially like matter moving backward in time (Smith, 2003). Antimatter, which may sometimes enter this universe through black holes (*NewScientist*, 2008a) that themselves arise from stars collapsing on themselves into an ever smaller rotating ring or point in space and thereby create quantum singularities leading to the possibilities of infinity (Seife, 2000) may represent the form that matter takes when it moves backwards in time. When it does so, it would annihilate with its identical matter forms, thus negating any further existence for them in sets of past and future universes connected to them. Thus, while both antimatter and matter represent concentrations of energy (Close, 2009) merely moving in opposite directions in time, and while energy can move fermions in either direction in time, the unification of matter and antimatter upon contact into timeless light photons largely precludes their co-existence for very long in areas of space-time within the same universe.

In any event, it would take quantum singularities such as speeds at or exceeding the speed of light that might reverse the sign and thus direction of energy to engage in time travel that would allow at least some momentary existence in the past. Observance of all historical events might, however, be possible via the mere movement at the speed of light, since such movement would eliminate the passage of time and enable experiencing all events simultaneously without actually existing in any particular point in time. It is actually very feasible for sending information and intended impacts across space and time through photons and antimatter, as well through a mere collapsing of wave functions in the present that exclude past and future possibilities, that is enabled through the very will power that enables all creation as is explained later in Chapter 16.

As previously mentioned, actual time travel for a being might also be possible via a wormhole created by the singularity of black holes. Such wormholes could also represent gateways to and from antimatter universes. Since there may exist universes of antimatter (Battersby, 2008c) that can even create anti-life forms given the evidence of the creation of anti-atoms created on earth (Close, 2009), antiparticles entering this universe through black holes may actual come from such antimatter universes through black holes in those universes. The spin of the particles that make up the black holes, which themselves spin, does enable them to potentially represent portals to different universes (Ananthaswamy, 2010a), including antimatter ones moving backwards in time.

The singularities involved in black holes could also allow movement between large areas of this universe without any passage of time. The universe in which human beings exist is shaped in the form of a self-contained dodecahedron, which consists of 12 identical pentagons that implies an exit from one end of the universe to be the entry to the other side of that same universe (Merali, 2008a). The wormholes in this universe that may permit instantaneous movement to another area of the universe or some other universe (Childress, 1999) as well as time travel (Brooks, 2008a) only seem to involve some freezing or reversal of temporal energy as does in fact apparently happen within the structure and components of this universe.

Evidence of a non-gaussian distribution of particles in the universe implies quantum fluctuations that aren't completely random (Brooks, 2008c). Everything instead interacts with a consistency that reflects both an internal spontaneity and an interconnectivity of all (Capra, 1999). While the particles of the universe behave according to a probabilistic function that permits some randomness, they also follow the specific rules of quantum physics (Barbour, 1999) that create some order through characteristics like entanglement. Whereas pure randomness would be very unlikely to create the order existing in the universe (Calleman, 2009), the basic nature of the universe that is characterized by quantum mechanics makes the existing universe inevitable, as will be explained.

There is approximately one proton of matter and 10 billion quanta of radiation for every five cubic meters of space in the universe (Close, 2009). Much of this universe seems to move in a turbulent pattern that can best be characterized by what is called Lagrangian coherence (Mackenzie, 2007). Such structure created by the laws of quantum physics represents why the universe has so much beautiful variety as opposed to the bland uniformity that would exist with pure randomness, although the historical path to today's structure in no sense represents the only way an original Big Bang could have evolved into today's situation (Barbour, 1999). The existing patterns certainly allowed for life to evolve as it has, but the path it has taken to arrive at the present moment is far from necessarily the only path (Urban, 2007). It also may not be any sort of "optimal" path with respect to any particular criteria like balance or beauty, especially since it includes a significant amount of past incoherence, violence, and blandness. It results from creation from nothing and just is.

A symmetric creation of particle and antiparticle pairs occurs all the time spontaneously in a vacuum of space of this universe (*NewScientist*, 2007v), and such equal amounts of matter and antimatter implies equal amounts of positive and negative temporal energy. The sum is zero, just as it is for the respective spatial charges, thus resulting in nothing arising out of nothing. Thus, despite all the wonderful variety perceived in this universe, there is thus no creation out of nothing, merely its division that creates the reality that is observed.

A vacuum is constantly producing an uncertain amount of energy (Seife, 2000), which seems to reflect divisions or vibrations of nothing. These vibrations, which arise from divisions of nothing to create opposite fermion pairs, appear to induce all forces in the universe in a particular chaotic but non-random structure according to the specific laws of quantum physics. Those laws don't create something out of nothing, since they merely reflect the nature of divided nothing, which by definition sums to nothing. In particular, $0/2+0/2=0$, and a division or vibration of nothing requires no energy by scientific, mathematical definition, as there is no actual net mass in nothing to move, vibrate, or split nothing. While the laws of this universe may differ greatly from those of the infinite number of other possible universes that could arise spontaneously out of the vibration of nothing, all such universes might be constrained to sum to nothing.

Some of the forces in this universe such as gravity are not directly observable (Merali, 2008b). Gravity tugs large cosmic bodies towards each others and can even cause a star to draw an orbiting planet into itself if they are close enough, as the orbital energy is absorbed by the star in an increased spin rate (*NewScientist,* 2010aap). Some believe that unobserved particles called gravitons carry gravity forces among objects of very large mass even though the graviton type of boson remains a purely theoretical concept since it has never been detected (Close, 2009).

The force of gravity is actually just the warping of the three dimensions of space by concentrations of mass and energy (Brooks, 2008a) that slow the passage of time (Seife, 2000). In fact, different states of the time-space geometry merely reflect different gravitational effects that may be caused by the collapsing of the quantum probability wave functions created by the stochastic structure of nature (Rosenblum and Kuttner, 2006). At extremely low temperatures and momentum, objects can defy the laws of gravity (Brooks, 2010a), perhaps because of quantum singularities in the interactive wave functions at a minimum level of energy that disconnects the waves across time, but matter is normally attracted toward other matter by the force of gravity.

Galaxies may have formed after matter began to clump together from the gravity of dark matter that didn't interact to the massive light energy from the Big Bang (Chown, 2009b). Dark matter is an unobserved substance in the universe that is hypothesized to exist because the universe would expand more rapidly than it does if there weren't the gravity of such matter (Clark, 2009) or some other force of negative energy.

On the other hand, since black holes in general do belch out hot plasma gas (*NewScientist*, 2010q), some of the matter of galaxies could originate from the black holes at their centers. In addition, the radiation from the heat caused by the gravity of black holes attracting other objects toward them can result in radiation-spraying quasars that may have helped give birth to galaxies (Chown, 2010a). Some of the particles and antiparticles popping up all the time out of the vacuum of space could also be sucked by the black hole in to increase the amount of hot gas revolving around it. When there is enough gas to prompt significant amounts of star formation, the newly formed stars orbiting a black hole naturally align to create an elliptical disk (*NewScientist,* 2010r). Many of the galaxies rotate in a direction opposite the spin of the central black hole because of hot matter drawing near to the black holes in a direction opposite the spin of the black holes that twists the magnetic fields (Chown, 2010c).

The Milky Way galaxy or cluster of stars, of which the earth's sun is just one, began forming 10-12 billion years ago (Chown, 2009b). Today, the estimated 175 billion stars and 200 billion planets of the Milky Way (Calleman, 2009) form a disk 2000 light years thick that rotates with spiral arms (*NewScientist*, 2010z). While some stars in the galaxy have "backwards" orbits that may reflect a collision with other older galaxies (*NewScientist*, 2009zzzzd), the majority of the estimated 175 billion stars and 200 billion planets in the Milky Way rotate around a central axis for the galaxy in the same direction (Calleman, 2009). This directional movement of most stars in the Milky Way around a galactic nucleus in spiral arms (Clark, 2009) that give it a spiral shape (*NewScientist*, 2010r) appears counterclockwise if viewed standing on the north pole of the earth.

The earth's solar system, which consists of a star and revolving celestial bodies, rotates around the Milky Way's center about 2/3 of the distance between the galaxy's black hole middle and the galaxy's outer edges (Calleman, 2009). This sun and collection of revolving planets formed from an explosion of another stellar object over 4 billion years ago within the Milky Way (Segre, 2002). The sun, around which the earth and other orbiting planets formed about 100 million years after the sun's birth 4.5 billion years ago (*NewScientist*, 2010aad), will eventually burn hotter and scorch all life out of the world in a few hundred million years before another explosion of the sun destroys the earth completely (Segre, 2002). The entire Milky Way will collide with another galaxy

(Andromedia) in a few billion years (Brooks, 2010,c), and so the current environment for life on earth isn't permanent.

It takes about 250 million years for the earth and its solar system to make a complete rotation around the Milky Way, and this trip includes four trips through the galactic plane in a pattern which represents a sine wave over and below this fairly flat section of the 3-dimensional galaxy (Calleman, 2009). Within this galactic system, the earth itself seems to be traveling through the universe at about 200 kilometers per hour (Lemonick, 2003), while humming with vibrations that are synchronized with variations in the electromagnetic field of the sun (Brahic, 2008), at a distance of about 28,000 light years from the galactic center.

Further perspective on the earth's position and movement is provided by the fact that, on two days per year close to the vernal and autumnal equinoxes when the sun is shining directly at the equator of the earth that spins tilted in one direction, the spin of the earth offsets its rotation around the sun at 80 degrees north or south, resulting in zero acceleration for about a second, thereby enabling an egg to stand on its head despite the tilt of the earth (Merali, 2007b). Equinoxes were considered special events by ancient people who studied and knew much about the cosmos (Baring and Cashford, 1993). The earth may be more balanced between all the pulls of the universe on that day thereby potentially enabling greater resonance of brainwaves across humans and with all else to enable greater enlightenment as to the integrated nature of everything.

Planets like the earth tend to form from discs of material rotating around stars that coalesce and revolve around the star's equator (*NewScientist*, 2008zzh), generally in the same plane and with an orbital direction the same as the rotation of the central star (Chown, 2010c). Dust and gas first gather into objects up to a meter long that then form into asteroids, which later collect enough other objects to create the planets (*NewScientist*, 2009zzzb). The earth, which itself took shape in that process, continued to be impacted by bombardments of meteors for the first few hundred million years of its existence (*NewScientist,* 2008zh). Water on earth probably originated from the ice of comets and asteroids that hit the earth after they were formed as a result of the "gravitational resonance" in the orbits of planets forming farther from the sun that shook loose portions of them (Shiga, 2010b). The chemicals in comets impacting the earth are also the most likely cause of the creation of the earth's atmosphere (*NewScientist*, 2009zzzzq), but since some planets in other solar systems also seem to have atmospheres (*NewScientist*, 2009zzzzm), the earth may not be unique in that respect. Many of the earth's heavy elements arose from impacts by comets, but the heavy melted elements in the core of the planet likely originated in a large collision with a planet-size object very early in its history (*NewScientist*, 2010aaak).

The earth's own moon was created around then when there were "numerous large projectiles careening through the inner solar system" that "walloped" the current lunar object out of the "primordial Earth" (*NewScientist*, 2008zh). That moon helps stabilize the tilt of the earth's axis to inhibit greater extremes in temperature (Shiga, 2010c).

Within this cosmic environment, the surface of the earth itself was formed by the extremely slow movement of seven plates beneath it (Williams and Nield, 2007). The earth's plates originated about 3 billion years ago due to further collision of the planet with asteroids or comets (Reilly, 2007). Past movement in those plates have resulted in catastrophic environmental changes to life, including when they caused a merger of the continents into one 290 million years ago, and large changes in the plates' positions are likely to repeat themselves every few hundred million years, (Williams and Nield, 2007). While much of the earth's components and structure was caused by inanimate impacts, most of the thousands of minerals on earth have actually resulted from the actions of earth's life-forms (Chown, 2008b).

The earth's sun, like all stars, is composed largely of hydrogen atoms whose extremely high concentration results in their gravity pressuring those atoms into smaller areas so that they fuse together to form helium and large amounts of energy (Seife, 2000). The result of the pressure is so great that temperatures over 10 million degrees exist in the center of the sun that cause its atoms to break into protons and electrons, and light emanates from all the energy produced to eventually reach the earth (Close, 2009). However, it can take 100,000 years for the gamma rays resulting from the nuclear explosions in the core of the sun to make it to the solar surface as those rays are repeatedly absorbed by electrons that emit new energy ever closer to that perimeter (Close, 2009). The "solar wind", which emanates from the sun and includes some electrons (*NewScientist*, 2010aai), creates a bubble around the solar system when it comes in contact with the interstellar gas of the rest of the galaxy (*NewScientist*, 2008zzt), which apparently still ripples with waves from a collision with another galaxy 2 billion years ago (*NewScientist,* 2009u).

The variation in this solar energy reaching the surface results in observable sunspots that seem to follow a rough 11-year cycle (Clark, 2010) inside a 200-year pattern (Mullins, 2009b) that cause regional weather effects which impact crop yields (Clark, 2010). These temporal weather patterns originate from differences in the amount of unltraviolet radiation emitted by the sun that are absorbed by the earth's stratosphere and that peak during solar maximas (Marshall, 2010b). The sun's electromagnetic field reverses its magnetic poles approximately every 22 years (*NewScientist*, 2009zzu) in a process that is apparently creates the average 11-year cycle in solar flares which affect life on earth (Braden, 2009). For instance, these periodically recurring events on the sun over very short time intervals appear to even cause cycles in biochemical reaction rates in living organisms such as humans on earth (Stray, 2005).

Astrological electromagnetic forces, which are "generated from any circulating region of electrically conducting molten material" in the core of celestial bodies (Clark, 2008a), interact with other cosmic phenomena to also create effects visible by humans. For example, such forces cause an electrical field to form around the earth that deflects the solar wind of hot charged particles emanating from the sun toward the planet's poles for further passage beyond the earth, and when the planet's electrical field is stretched to extremes in solar storms, beautiful auroras appear at the earth's poles that are sometimes called the southern and northern lights (Battersby, 2009a). These lights form out of counter-spiraling particles sent along the earth's electrical field that collides with molecules in the ionosphere 400 kilometers above the surface of the earth (*NewScientist,* 2009zp).

While the earth's own magnetic field deflects many intense solar particles emanating from the sun, some electromagnetic forces do have a significant impact on the behavior of organisms on that planet that often exhibits itself in cycles because of the cyclical nature of the forces involved (McTaggart, 2008). For instance, one especially illustrative example of an unseen "power" about which few are aware is provided by the underground electromagnetic forces stemming from the earth's core that affect the human mind, that are spiral-shaped, and that are influenced by stellar bodies like the moon (Sjoo and Mor, 1987). The earth's electromagnetic fields, which are caused by the swiveling of charged material at the center of the earth (Close, 2009), are also apparently are sensed through birds' eyes and thus actually seen (*NewScientist,* 2009zzzz). Given that other animals like cows sense the earth's magnetic field and use it to help them navigate direction (Callaway, 2008), as do even some bacteria (*NewScientist,* 2009zzzc), it should not be surprising that those forces affect humans in some fashion. These waves play a symphony of frequencies ranging between 0.01 and 300 Hertz, and they both impact and are impacted by the electromagnetic fields created by the 6.5 billion people on that earth (Braden, 2009).

Electromagnetic forces could also possibly explain the crop circle effects, which are seemingly unexplainable markings in farm fields which occasionally appear, and which seem to be related to cosmic events (Silva, 2002). Events happening in far-off solar systems and galaxies result in microwaves from outer space hitting the earth that can indeed have effects on this planet (Stray, 2005). Some cosmic rays include radiation, which consists of electromagnetic waves that act like photons with energies different than those of normal light, and larger high-energy protons which scientists have not been able to create in the accelerator laboratories of the earth (Capra, 1999). Cosmic rays tend to have larger effects at night when they are not swamped by the effects of photons from the nearby sun around which the earth revolves (Ananthaswamy, 2009a).

Given the historic synchronization of women's reproductive cycles with that of the moon (Gooch, 2006) and given the significant effect that the moon cycle has on the stock market (Yuan, Zheng, and Zhu, 2006), it seems plausible that electromagnetic or other waves emanating from or relating to the moon also have an effect on both human reproduction and behavior. The moon reflects sunlight onto the earth that can affect events in this world. Its reflection of sunlight is definitely different, at least in terms of intensity and magnitude, from that of light photons coming directly from the sun. The times of moonlight alone have been found to have special effects on the biological functions of some organisms on earth (Sugama, Park , Park, Takeuchi, Kim, and Takemura, 2008).

While the moon doesn't exert any magnetic field itself macroscopically (*NewScientist*, 2009zzzg), it does affect the tilt of the earth's axis (Calleman, 2009). Moreover, the gravity of the moon affects ocean waves and rocks on the surface of the earth and has been found to have a significant impact on even tiny particles as small as electrons on the planet (Close, 2009). The effects of gravity are two-way, as the moon has been found to have periodic volcanic eruptions like the earth that may be caused by a combination of radioactive decay and the earth's gravity (Chown, 2008a).

Events outside the earth's solar system can also affect phenomena inside it, as with cosmic rays that emanate from other stellar objects in the Milky Way and beyond. In recent times, however, the earth's solar system has been traveling through a region of space called the "Local Bubble", which includes clouds of dust and gas that formed from a supernova explosion, and which protects the world somewhat from cosmic rays, (Anderson, 2007). Over the last 44,000-150,000, this solar system has been passing through an even more protective area of the Local Bubble, and it is scheduled to remain in that area for the next 10-20,000 years (Anderson, 2007). While the sun provides some protection against cosmic rays, a weakening of that star's activity is currently resulting in nearly 20% more cosmic rays reaching the earth (*NewScientist*, 2009zzzr). The current pressure from interstellar particles diverts the solar wind itself and causes it to dissipate from even being detectible about 30 light years away from the solar system (*NewScientist*, 2010aaai).

The cosmic rays that make it through the various protective shields affect events on earth, including the brain activity of human beings. For instance, cosmic rays, which reach the earth, affect human hormonal secretions such as melatonin that is released during sleep (Stray, 2005). Melatonin is especially important in brain cell coordination and learning as well as possibly perceptions of consciousness (Gynn and Wright, 2008), and so cosmic rays could certainly affect human thinking abilities. Cosmic rays have been found to create electrical currents in humans in outer space that are caused by electron displacement and that result in distortions in perceptions such as the sight of hallucinating white lights with halos (Clark, 2008b).

In any event, given that plasma is air with special ionized or electrical properties (Mullins, 2009a), the earth's movement through a particular plasma area of the Milky Way can certainly impact the earth and life on it. While the earth's movement through that plasma band may affect human thinking, the fact that the planet on course to be physically and perhaps harmoniously aligned with the galactic center in 2012 could also affect human psychic powers and consciousness itself (Stray, 2005). In particular, the human brain is also influenced by changes in the earth's magnetic field that could be significantly impacted by events in 2012 when the earth's solar system will be in a process of returning its direction in its elliptic orbit back toward the galactic center at the same time that it is at a point on the galactic "equator" that divides the northern and southern halves of its galaxy (Braden, 2009).

The fact that concentrated focus and will power can exert external physiological effects (Delaurence, 1910) implies that human consciousness could affect the entire process (Stray, 2005). The partially holographic nature of consciousness with respect to the thoughts of one person affecting others makes such a conjecture to be especially feasible given the potential coherence in thinking about all people that might result (Braden, 2007). The earth's electromagnetic field has already been shown to found to be affected by human thinking as well as vice versus (Braden, 2009). There may thus be an actual interactive effect whereby physical cosmic forces affect human psyches at the same time that human psychic powers affect events in the cosmos that can lead to a magnification of both processes if they are coherent. However, the result can be the destruction of the psychic waves if human beliefs are inconsistent with the innate characteristics of the masses of forces in the universe. In particular, such disharmony causes the psychic waves to decohere with little effect on the cosmos as a whole.

The quantum entanglement of all things ever in contact with each other does seem to enable the psychic powers that can affect others and even inanimate matter (Radin, 2006). It is therefore scientifically possible for there to be a large change in the physical environment or in human consciousness, or both, in 2012. The supposed importance of the year 2012 stems from the Mayan astrological calendar, which indicates a high degree of astronomical accuracy in predicting cyclical stellar events (despite Mayan superstitions which included a belief in a flat earth), and which could imply a possible worldwide catastrophe on December 21 of that year (Henderson, 1997). This particular day in 2012 may actually represent the cyclical reappearance of a potential cosmic disaster like a comet coming near to the earth every five thousand years or so (Stray, 2005), or possibly a catastrophic polar shift (Braden, 2009). However, it is conceivable that a cosmic catastrophe for the earth could actually be deflected by human thought waves (Stray, 2005), and the alignment of the earth with the galactic core in late 2012 could instead lead to a new age of human harmony and enlightenment (Braden, 2009).

Calleman (2009) has concluded that the Mayan calendar actually implies October 28, 2011 is the central point for a new age of enlightenment. Braden (2009) has suggested that such a phenomenon could occur as a result of increasing galactic energy reaching the solar system from the galactic core over the entire time interval between 1983 and 2016 when it is in the process of reversing its orbiting direction at a line which divides the galaxy into a northern and southern half.

The earth's special position in the galaxy in 2012 could certainly lead to a modification of consciousness that could open up or even revolutionize human thought (Braden, 2009). This impact might even be indirect, such as with respect to gravitational and other influences on the sun that impact the solar waves reaching the earth. The massive and varying amount of solar energy emitted by the sun alone does affect the earth's magnetosphere shield and significantly influences the behavior on earth of all living things (McTaggart, 2008).

The earth's electromagnetic field might also be affected by both solar and cosmic effects that might indirectly offset human psychology. Electromagnetic waves from the earth's core do significantly affect human brains, as evidenced, for example, by their increasing the incidence of epilepsy that allegedly plagued many past religious leaders and shamans (Stray, 2005). Such phenomena relate to an unusually low level of electrical frequency at the theta wavelength in part of the brain (McTaggart, 2008). Those brainwaves are related to hypnotic learning effects that could open up people's minds to their interconnectivity with all. The feeling of a greater reality which is experienced in near-death, out-of-body perceptions that include entering a being of light and a feeling of peace (McClenon, 2002) may reflect the fact that normal human perceptions are mere distortions of the underlying nature of the cosmos that causes the complex system of particle waves that interact to be categorized into a form that can be interpreted by macrocosmic organisms for their optimal reaction. Such distorted perceptions of the microcosmic reality are useful from an evolutionary perspective in the aggregated macro world of normally impenetrable force fields but don't reflect the true nature of the electromagnetic waves that make up the universe.

Perceptions merely categorize the underlying reality of everything in a fashion that enables the group of electromagnetic force fields that make up humans to survive in the environment with other groups of force fields in this enormous universe. This cosmic fact provides a perspective that makes many of the illusions of perceived experiences seem less serious, if not outright unimportant.

The electromagnetic effects of tornados that result from particles becoming electrically charged by rubbing against each other and then creating detectable low-wave radiation by spinning (*NewScientist*, 2009w) have even resulted in the occurrence of seemingly impossible physical events, which are nonetheless consistent with quantum theory, such as solid objects like boards flying through solid walls and pieces of straw becoming stuck in the middle of unbroken glass (Stray, 2005). Such bizarre phenomena can help one recognize that everything that is perceived to be a solid actually is largely composed of just electromagnetic force fields that can be relaxed in some circumstances.

While scientists have been able to artificially create cosmic rays on earth that lead to the same visual illusions of halos that astronauts have experienced in outer space (Clark, 2008b), similar lighting illusions are also caused by natural electromagnetic events on earth like tornadoes (Stray, 2005).

The universe seems to be fractal with respect to both the time and spatial dimensions (Calleman, 2009), indicating that its macrocosmic appearance repeats itself on ever finer scales but with some deviations (Buchanan, 2009b). In particular, an object characterized by fractal geometry has a similar but not necessarily identical shape no matter how much it is magnified, and some processes like the weather or the atmosphere over time appear to be multifractal systems (Matthews, 2009). There are complex forces that create such effects. For instance, the weather and atmospheric conditions are significantly impacted even by man-made dams that increase rainfall in the area around the constructed concentrations of water (*NewScientist*, 2009zzzzo).

Time seems to repeat itself in cycles that are similar to prior cycles, perhaps due to various cosmic forces, but each cycle can also differ from prior ones because the cosmic forces only create conditions that are similar to those in the previous patterns (Braden, 2009). Evidence from the calendars and religious systems of various ancient peoples such as the Mayan Indians indicates that periodic cycles exist in larger time cycles that are 20 times longer in duration than the shorter ones (Calleman, 2009).

The fractal characteristics of the universe in the spatial dimensions can be thought of as fractional dimensions which macroscopically reflect deviations from the straight lines of a whole dimension (Jamieson, 2009). A "triangle" is "the simplest unit that can be stuck together in different ways to make curvy universes", such as the one that humans observe, and the "'true' shape of the cosmos should be the sum of all these possibilities" (Battersby, 2009b). That geometrical framework can enable construction of mathematical models of the universe that reflect its fractal nature, including one to be hypothesized in a later section of this chapter that indicates a triplet of fermions trapped in a triangular structure forms most of the mass and gravitational pulls in the universe.

The effect of a balanced tug of gravity at the galaxy's equator along with a return toward movement coherent with the pull of the galactic center around 2012 might actually enhance understanding of these phenomena. The continuing weakening of the earth's magnetic poles that has significantly accelerated since 1860 (*NewScientist,* 2006c) could also contribute to people awakening from entrenched views to open up their minds for greater enlightenment (Braden, 2009).

On the other hand, an actual reversal in the earth's magnetic poles from north to south that can occur over a period of only a few years could cause widespread chaos (*NewScientist*, 2010aab). Given that the earth's poles reverse on average about every half million years, and given that the most recent switch was 780,000 years ago (Braden, 2009), such a reversal is actually overdue today, and there is some evidence of a rapid pole shift that has been measured to be as much as 6 degrees a day in some areas of the earth's core (*NewScientist*, 2010aab).

A further reduction in the planet's magnetic field could actually cause a switching of its poles (*NewScientist,* 2006c) that would have catastrophic effects (White, 1980), most evidence on the current rate of polar shifting indicates that such an earth-shattering event won't happen for another 2000 years at current rates of magnetic weakening (*NewScientist*, 2006c). However, especially strong solar flares that are supposed to reach a 40-year peak in 2012, a "superwave" of cosmic radiation that impacts the earth every 15,000 years or so, a large meteor collision, a nearby appearance of a comet or asteroid, an unexpectedly large weakening of the earth's magnetic fields, or a tsunami could contribute to a reversal of the planet's magnetic poles sooner (Braden, 2009). Moreover, a mere large change in position for the poles from their current location that may happen every 7500 years could have serious consequences (White, 1980). Pollution caused by man that is leading to global warming and the melting of the earth's ice sheets (Braden, 2009) also increases the chance of a pole shift (White, 1980). Thus a catastrophic pole shift or reversal might indeed occur in 2012 or some nearby future year, unless more is done to reverse mankind's effect on the environment. Destructive physical forces could therefore potentially dominate the year 2012 (Stray, 2005) that might otherwise be just a relatively uneventful year in a long cyclical process that appears to be trending toward more positive coherence between human consciousness and the quantum forces of the universe.

In any event, a mere recognition of the vastness of the universe is enough to make people realize the pettiness of much hatred and the senselessness of wars (Brooks, 2008b). Human beings are merely stardust, composed largely of nuclear waste from solar explosions that occurred over 4 billion years ago (Close, 2009).

B. Characteristics of the Cosmos

According to quantum theory, everything is ordered by a set of mathematical relationships (Tegmark, 2007) that appear to be grounded in symmetric geometrical figures (Sautoy, 2008). Peat (2005) has suggested that there may be an underlying mathematical meaning to life and the universe, as Native American cosmology has long hypothesized.

In one interpretation of the quantum world, which characterizes reality at the most fundamental level, everything is composed of strings that vibrate at different frequencies to create electromagnetic fields which are associated with individual particles of matter that also act like waves (Barbour, 1999). While the laws of quantum physics don't require such a string theory, at least one of the incredible numbers of possible string theories seems to be an adequate simplification of the complex quantum mathematics that hold in this universe (Griggs, 2009a). An M-theory with 11 dimensions unifies five widely accepted categories of the string to explain the empirical processes that occur in the universe, but an unlimited number of different mathematical possibilities are still feasible within that framework which allows "all possible universes" to "exist" (Ananthaswamy, 2009c). One string theory proposes that the three spatial dimensions that are normally observed represent a "brane" that exists within a higher dimensional world, into which gravity can leak out due to interactions between the observed 3-dimensional brane and the other unobserved dimensions (Ferreira, 2010).

Human perceptions are generally restricted to only a bubble of 4 dimensions, time and the three dimensions of space, in a universe that may have 10 or more spatial dimensions, some of which may be infinitely large according to one leading physicist (Adler, 2005). The extra dimensions could actually cause time in our observable 4-dimensional universe to appear to slow down (Merali, 2007m), as does concentrated mass via gravity (Gefter, 2008b). One of the extra dimensions may even be other mutiverses existing simultaneously with the one actually being experienced (Rees, 2009), whereby such multiple universes represent the realization of different quantum possibilities (Barbour, 1999).

One of the additional unperceived dimensions may represent different mappings of the 4-dimensional space/time universe that may also be entered through black holes from which our own universe originated (Wesson, 2006). There may actually be an infinite number of such universes (Merali, 2006a), each of which may last about one trillion years before collapsing back into the same tiny quantum state from which it originated (Merali, 2006b). Some universes, including the one in which humans live, however, may also continue forever (Magueijo, 2003).

Some of the infinite parallel and other universes that exist, including those with humanity in them, may have many hidden dimensions that human senses don't detect because there is no useful purpose in an evolutionary sense to notice them. There is limited evidence of a fifth dimension may already be appearing in quark stars of this universe (Merali, 2007e), but it far from conclusive.

A scientific theory has even been developed which permits an infinite number of dimensions but which mathematically collapses to the 4-dimensional universe observed by humans (Gefter, 2008c). That universe may float around in a 10-dimensional hyperspace along with many other universes of different dimensions (Battersby, 2009b). The number of dimensions in existence merely represents the number of different directions in which something can move (Webb, 2009a), and since the possibility of patches of space-time being disconnected from each other in the quantum singularity of black holes allows them to expand or contract independently (Courtland, 2010e), there could be an infinite number of universes with different numbers of dimensions.

It is true that the existence of other dimensions could help explain the creation of the universe in which humans currently live, since energy can actually be created from a vacuum in an open universe with unobserved dimensions (Magueijo, 2003). It is known that a vacuum itself is teething with "countless subatomic particles and their antiparticles that pop fleetingly in and out of existence" (*NewScientist,* 2007v), and this creation out of nothing can result in energy that, when concentrated, is sufficient to make all matter in the universe (Gefter, 2008d). Such energy creation actually yields too much for the existing universe (Lemonick, 2003), and the surplus may escape into unseen dimensions.

It has even been asserted that other dimensions and a deeper reality may be perceived by humans naturally via the chemical DMT that is triggered for release into the brain by the human body itself in religious experiences (Strassman, 2001), and lucid dreaming (Stray, 2005). DMT, which can be promoted by merely eating more fruit, is even produced by laughter and feelings of happiness (Gynn and Wright, 2008) and has been documented to lead to extraordinary perceptions that are seemingly hallucinogenic (Strassman, 2001). Mere electromagnetic stimulation of the brain's temporal lobes can lead to similar spiritual and hallucinogenic experiences, including alien abductions (McTaggart, 2008).

DMT relaxes inhibitions on the activation of thought processes that don't contribute positively to an organism's survival and reproduction in normal times but that enhance adaptation to unusual circumstances. However, the firing of connected systems of brain cells enabled by DMT or electrical stimulation of the brain can also lead to fictitious observations and false deductions. While DMT can enhance creativity as well as mental health (Gynn and Wright, 2008) and may sometimes help the brain better perceive the underlying reality of the interconnected nature of everything, it can also reflect mere randomly fabricated illusions, and so the truth about the cosmos and other dimensions can't be determined from mere DMT trips.

In any event, even meaningful symbolic hallucinations may require logical rationality and scientific facts to understand. For instance, there is a frequent perception of being tested during alien abductions imagined under the influence of DMT that cause the hallucinating person to think he isn't supposed to be aware of the tests which are often perceived to be benign (Strassman, 2001). This hallucination could reflect a symbolic message from the powerful "alien" subconscious which generally communicates through to conscious thought via metaphorical imagery that can enable otherwise incomprehensible concepts and be generalized. The mental experience of an alien abduction under the influence of DMT could therefore simply be an attempt to indicate the complex forces in the brain acting on itself for evaluating and repairing neural connections that typically happens during sleep while under the influence of internally produced DMT and that might normally be experienced only in dreams which aren't remembered.

Instead of evaporating into dimensions that could be merely invented figments of a symbolic imagination, any imbalance in energy in the universe may go elsewhere, such as through black holes that could open up new universes, including antimatter ones that move backward in time. In addition, since such particles that pop into and out of existence arise due to the laws of quantum physics that makes the amount of energy anywhere uncertain (Brooks, 2010b), since many particles in the universe represent negative energy (McTaggart, 2008), since the light photons that contain energy are their own antiparticles (Capra, 1999), and since dark matter may represent concentrations of such energy in a vacuum that may represent dark energy when it is in less concentrated form, it seems quite feasible that the energy sums to zero within the reality of a universe that doesn't have more than four dimensions.

In any event, the creation of equal amounts of matter and antimatter out of nothing occurs all the time (Magueijo, 2003) and results in the mutual annihilation of the appearing fermions when they collide with each other, thereby producing energy (Close, 2009). In particular, the collision of particles with their antiparticle counterparts results in energy in an amount equaling the combined mass of the fermions that is displayed in the form of photons darting in opposite directions (Gefter, 2009b). Collisions of matter and antimatter maintain the energy within them, even when they are of different types, such as when quarks and antiquarks of different flavors collide to produce kaons or mesons that can eventually decay into electrons or positrons, and the matter of this universe is composed of concentrations of such energy that also create the force of gravity (Close, 2009).

The dark energy hypothesized to maintain the expansion of the universe despite the pull of gravity (Krauss, 2008) might be explained by the simultaneous creation of matter and antimatter out of nothing. While hidden extra dimensions could potentially cause the simultaneous creation of matter and antimatter out of nothing (Magueijo, 2003), such unobservable dimensions, however, are not understood, defined, or perceivable. There isn't even agreement on whether they are too large or too small to be seen

(McAlpine, 2010e). The hypothetical existence of unobservable extra dimensions could be characterized by a vast number of different feasible string theories that are difficult to test empirically (Gefter, 2007), and the origin of such hypothesized extra dimensions remains totally unexplained and unknown regardless.

Thus, postulating the existence of extra dimensions is something akin to invoking "magic" or a deity to explain things. Because the origin of the 10 or more dimensions of string theories, which is the "most preferred" by physicists (Anasthaswamy, 2009), isn't explained by such theories that themselves have never been verified (Penrose, 2004), these models may only represent an attempt to mathematically characterize the traits of simple nothingness. In that sense, the spatial energy created by the internal splitting of nothing that forms into particles and antiparticles spontaneously in a vacuum could be mathematically characterized by a counterbalancing anti-energy in extra spatial dimensions of nothing when it doesn't represent different pulls on nothing in the time dimension.

The discrepancies that are found between reality and quantum theory are typically resolved by adjusting the theory or assuming the existence of new particles (*NewScientist*, 2010an). For instance, the bizarre behavior of microscopic particles isn't always explained by existing quantum theory, especially in extreme situations, and a new "quantum effect" must be created for new anomalies (*NewScientist*, 2010ad). It might instead be possible to model these phenomena with the reality of nothing and its divisions being all there is.

Since the universe is fairly symmetric in many respects but has some violations of symmetry (*NewScientist*, 2008zzk), one string theory of supersymmetry has been developed that explains physical anomalies of non-symmetry by assuming the existence of many more dimensions than the perceived four that include time and the three dimensions of space (Barbour, 1999). This theory of supersymmetry implies the existence of some super partners for existing particles that existed only shortly after the Big Bang, with the super-particle partners eliminated by the expanding universe, thereby resulting in new vacuum bubbles that also created the observed force of gravity via gravitational waves (*NewScientist*, 2009zc). Such assumptions that may not be verifiable may not be necessary within the reality of everything being spawned by divisions of nothing.

While there are different theories as to what actually initiated the Big Bang, one hypothesis is that two giant cosmic membranes spanning the universe had pushed everything together as tightly as possible, and the eventual collision of those "branes" led to the explosion that initiated the existing universe (Battersby, 2008a). As reported by Merali (2007d), it may be possible to see some evidence of actual collisions with other universes that don't destroy ours in the process. A few of the multiple universes in existence may be more directly connected, and some evidence of such influences has been found that is consistent with string theory, which suggests that many different universes arise out of a quantum vacuum (Chown, 2007d). Whether valid or not, such theories ignore the question of creation and where everything in these universes arose.

While string theories and their extra dimensions may have some mathematical validity, and could even reflect one or more past potential histories that are feasibly connected to the present, they don't seem to provide any predictive power for the future or insights into existence. However, there are some known scientific facts about history that can. For instance, the universe in which human beings live arose out of a quantum vacuum (Chown, 2007d), and so the quantum nature of the current universe seems quite natural, thus leading to meaningful connections between the past and future that shed insights on the nature of existence and reality in the universe.

Quantum mechanics implies that everything at the most microscopic level exists in discrete packets, and while these quanta create phenomenon that seem continuous at the macroscopic level, some macroscopic objects appear discrete as well (Childress, 1999). All discrete quanta of any size are actually electromagnetic fields as opposed to solids (Rosenblum and Kuttner, 2006), and they arise spontaneously out of nothing all the time (*NewScientist*, 2007v). Despite being discrete, the quanta become at least partially entangled with everything with which they come into contact (Aczil, 2002).

Quantum theory actually implies that there can be infinite energy in a vacuum or zero-point field at any moment in time (Seife, 2000). Within this context, the universe is essentially a self-creating one that thereby makes the universe itself the closest thing there is to a "god" in the conceptual sense (Barbour, 1999). However, there doesn't appear to be any purpose, external consciousness, or favored ethical system for this "deity". It just is.

The concept of a self-creating universe is incredibly consistent with the symbolism of some belief systems such as that of many Native Americans who believe everything was created by mathematics, insofar as a void (0) represents a single whole (1) vacuum, and the two numbers (0 and 1) implied by nothing and its wholeness are then continually added to the prior sum in order create all numbers and the universe itself (Peat, 2005). In particular, adding one whole nothing, out of which pops a particle-antiparticle pair, to another continuously can, with some probability, create this universe and maybe others.

The golden ratio of 1.618, upon which the proportion of the sum to the prior number collapses in the infinitely additive process that creates the set of Fibonacci numbers of 0, 1, 1, 2, 3, 5, … (Braden, 2009), may reflect this additive property of the universe's creation. The golden proportion, which is based on an intrinsic duality, seems to represent the foundation of too many things in the universe to be merely coincidental (Calleman, 2009). This ratio, which has long been thought to have some deep significance (Braden, 2009), is incorporated into various DNA and cellular proportions of primates, as well as the distance of the electrons' discrete orbits around the nucleus, and even human mathematical systems such as the different ancient and contemporary cultures that revolve around the integers 2, 10, or 20, directly result from the existence of the counting of human limbs, fingers and toes in a fashion reflective of adding basic Fibonacci numbers (Calleman, 2009). These facts are consistent with the hypothesis that the universe created itself out of the splitting of nothing into matter and antimatter like electrons and positrons, respectively.

Others people, including ancient German Druids, had similar beliefs about the creation of all things being initiated by the wholeness of nothing (Davidson, 1990). Ancient Indian religion indicates that the universe came into being via the existence of nothing that exploded into space, which is the "stuff between the nothing", and it did so merely by existing (Braden, 2007). Other religions or philosophies like Buddhism and Taoism promote ideas such as the meaningfulness of nothing and such as the essence of the universe being a void (Capra, 1999).

The duality and the sacred mathematics that the material universe reflects in fractal form (Calleman, 2009) may stem from the simple division of that nothing. Mathematics at its foundation is merely a generalized comparison of sets of different sizes based on the number of items in them, and this fact has been hypothesized to be the key to understanding the complex knots of quantum physics and life (Elwes, 2008). The facts that the net energy in the universe is zero (Barbour, 1999), that pairs of particles and antiparticles continue to be constantly created in vacuums (*NewScientist,* 2007v), and that all observed matter represents additive wave functions (Barbour, 1999) which are entangled upon contact (Calleman, 2009), may comprehended intuitively by the mathematics of dividing by zero to get infinity and connecting the dots by adding the infinities.

The Fibonacci number pattern itself seems to mirror the infinitely additive nature of the universe resulting from the division of nothing (0) from the whole nothingness (1) to create the infinitely increasing reality of nothing. Recognizing such realities is something that evolution has allowed and encouraged most humans to ignore in order to survive in their limited fractal macrocosm of the infinite reality.

Also contributing to the inability to face the mathematical facts of life are the patriarchal Middle Eastern religions of Christianity, Judaism, and Islam that have been widely adopted in Western societies. These systems of indoctrination into beliefs in fictional phenomena have successfully focused people's attention on the cultural necessity in a patriarchy of serving an elite or god as opposed to living peacefully and coherently with others in interconnected, equal bliss.

Even some more "enlightened" Western societies of the past such as the "logical" ancient Greeks, who revered mathematics, deliberately prohibited the existence of the number "0" as it led to conclusions that defied comprehension within their finite hierarchical system of beliefs, such as 1/0=infinity (Seife, 2000). Systems that are based on sets of assumptions grounded in patriarchal lies naturally inhibit understanding reality and the truth of 1/0, as the patriarchs wanted their subjects to see service to an elite as the meaning and purpose of the universe.

As previously discussed, the Big Bang creation of the universe in which humans live may have arisen from a black hole in another universe that itself is just one of an infinite number of universes in existence (Dai, 2007). Black holes, which are hypothesized to consist of extremely dense matter, whose gravity keeps anything nearby from escaping, and which may have originated from collapsed stars (Lemonick, 2007), represent a quantum singularity that results in division by zero and thus makes an infinity of possibilities feasible. Like zero which sucks all that is multiplied by it into itself, black holes engulf all falling prey to its gravitational tug into itself (Seife, 2000). Black holes have been scientifically interpreted as phantom energy that disconnects from this universe to create other universes (*NewScientist,* 2008t) in a potentially eternal fashion.

Shiga (2007) indicates that, in addition to an infinite number of universes that have always existed, an infinite number of additional ones continue to be created via black holes. Such newly created universes have been shown mathematically to lose most memory of prior universes, leading to no universe being quite the same (Dai, 2007). Nevertheless, some hypothesize it may be possible to determine something about the nature of this universe's parent from the cosmic microwave background radiation at the time of the Big Bang (Battersby, 2008b), which originated the existing universe

(Magueijo, 2003). The Big Bang likely occurred due to a natural bounce from the contraction of a prior universe into a very tiny area like a black hole, and space-time consists of the subsequent "constant rearranging of interconnections" between everything from that origin (Battersby, 2008b).

The universes have always been, and time is merely a fourth dimension (Magueijo, 2003). Within that context, the moment persists eternally, and time itself may not even exist at all in all universes (Barbour, 1999). However, since there is no beginning or end (Laszlo, 2006), as scientific theories and facts seem to indicate (Barbour, 1999), the time dimension seems to exist for an eternity into both the past and future for the entire set of universes.

According to Magueijo (2003), creation imposes some limitations such as on the horizon of a universe, but there can be an infinite system of universes, and nothing exists outside the infinite system. Barbour (1999) has suggested that the entire universe is merely a single whole entity following a common principle that results in an infinite number of different configurations and particles within that entity. However, this configuration seems to include many unconnected parallel universes or multiverses, and so his concept of a closed system seems to reflect incorporating all possible universes into the infinite set of feasible universes.

The mystery of creation is answered by the splitting of nothing into pairs of particles and antiparticles, as happens all the time everywhere in space. The various forces in the universe, such as electro-magnetism and those that bind the particles of atomic nuclei together, act as if they arise from subatomic virtual particles, are constantly popping in and out of existence (Smith, 2003). Those forces create electromagnetic fields of opposite charges that effectively act as if they are moving in opposite directions in a temporal dimension (Close, 2009). A universe of four dimensions follows logically from these facts.

It is possible that the three dimensions of space perceived by humans may be merely holograms of the 2-dimensional quantum world, which characterizes the activity of matter in its smallest size (1 Planck length) that isn't perceivable by the human eye (Chown, 2009a). The hypothesis of 4 dimensions being perceived in a universe that is characterized by 2-dimensional quantum laws consisting of time and one dimension of space (Courtland, 2010e) may also be consistent with the cosmic vibrations or energy underlying the quantum world being 1-dimensional and with the nothingness at the depths of it all having a whole of 0 dimensions. The scientific laws of quantum physics cause all particles to act as zero-dimensional objects (Seife, 2000) in a 2-dimensional world (Gefter, 2008c), which, according to string theory, exists and operates due to the vibrations of 1-dimensional strings (Ananthaswamy, 2009c). The quantum fluctuations of electromagnetic fields or particles disconnect patches of space-time from one another, resulting in mathematical infinities for the electromagnetic forces at extremely small scales that remain linked to appear as if smoothly finite when observed at greater distances between the fields (Courtland, 2010e). Space-time thus seems to be integrated with the potential of an infinitely powerful attraction between electrons and positrons.

When electrons are trapped in a 2-dimensional spatial world at a very low temperature, they break down into partial particles called anyons each of which has a fraction of the electron's charge (Brooks, 2009f). Without a third dimension, the curvature of space caused by time bending back on itself may cause the temporal bending effect to be spread out into more than one point in 2-dimensional space. Since a particle is really nothing more than an electromagnetic field that is attracted to fields moving in the opposite direction in time, a separation of a particle into separate forces simply reflects a breaking up of the charge into separate fields. A full electrical charge may require the full three spatial dimensions, as is consistent with the hypothesis developed later in Section C of this chapter that the fractional charges of quarks results from triplets of electrons and positrons being trapped in a singularity of equidistant attractions between the fermion triplets in 2-dimensional space. Partial charges or fields that are products of the motion of other things can operate separately, such as in the 2-dimensional crystals of anyons created out of electrons (*NewScientist*, 2010aaq), and atoms, which are largely made up of quarks (Smith, 2003), themselves exist in two-dimensions of space (Geim, 2010).

An electron trapped in one dimension of a straight line in a very cold space "can be made to act as it were two particles, one with the electron's charge and one with its spin" (Brooks, 2009e). Restricted to moving in just one direction seems to require it to follow the path of either its electrical charge or its spin, as the inability to travel outside that single dimension restricts a particle from moving according to both of those forces simultaneously, thus resulting in their apparent separation. Since theories of quantum gravity indicate that electromagnetic fields and particles behave as if space is one-dimensional at very small scales (Courtland, 2010e), the other two spatial dimensions observable at the macrocosmic level may very well stem from the basic single-dimensional nature of the basic attractive relationships between those fields that is curved by opposite directional movements in time into forming the other two dimensions of space that are observed.

The fact that the location of electrons becomes increasingly uncertain at low temperatures (Brooks, 2010a) may also contribute to the creation of seemingly fractional electrons that may merely reflect the uncertain position of those fermions. At extremely high temperatures, thermodynamic forces behave as if there are only two dimensions (Courtland, 2010e). It is possible that the rapid movement in space caused by such high levels of heat energy blurs the effect of the effect of time curving space back on itself. In any event, even electrons trapped in 0 dimensions still emit exactly one photon when they are excited with energy input into that dimensionless point (Webb, 2009b), and the minimum spatial size for any matter remains 1 Planck length (Chown, 2009a).

Since quantum gravitational pulls become significant only at tiny scales of one Planck length (Courtland, 2010e), it is quite possible that the force of macrocosmic gravity largely stems from particles like quarks that are trapped in less than 3 dimensions of space. With a quark being just the triplet of the attracting fields of three fermions existing in a triangular form with its sides having the minimum spatial length, they are frozen in time in two spatial dimensions. The gravitational tug observed at macrocosmic scales is a function of concentrations of particles with greater mass that characterizes the nucleons which are composed of quarks, and so that force may actually stem in large part from quantum gravity trapped in the fermion triangles of quarks via the freezing of fermions' opposite directional movements in time. The quantum gravity is then smoothed into being an observable gravitational force only at very large scales. The quantum nature of the universe that merely reflects processes and interactions between forces (Peat, 2005) can thus help make such phenomenon such as gravity more comprehensible.

As will be explained later, spatial dimensions are necessary for opposite charges to exist in the same time period without the mutual annihilation that is caused by the merging of opposite fermions at the same point in space. The time dimension itself may therefore be associated with the attractive force between opposite divided pieces of nothing in one dimension in space, and the opposite temporal movements along with one-dimensional spatial attraction could create the observed spin of particles that curves space-time into two more dimensions, as will be discussed in more detail later.

Although the reality of four dimensions can be represented with the complex mathematics of string theory by a superposition of some smooth and curved structures of space-time within the framework of a single spatial dimension integrated with the observed temporal dimension (Courtland, 2010e), the theory of creation out of nothing enables explaining both the creation and nature of dimensions and everything in them. Divisions of nothing by a temporal dimension that cause the opposite halves to move in opposite directions in time while being attracted to each other in 1-dimensional space cause a spin, thereby necessitating a second spatial dimension to enable the magnetic spin or revolution to enable them to switch between the dual energies, while the curving of time back on itself to sum to zero net ensures that space does so also, thereby making the third spatial dimension of time necessary to enable such curvature. The fact that nothing can be created out of nothing makes it necessary for all possibilities to sum to zero on probability wave functions of potential. While only an averaging wave of each set of relationships between all the divided pieces of nothing might normally be observed, the probability wave function of one process can actually be collapsed on a single point by interacting with another. Such collapses of probability wave functions cause the existence of multiverses for the excluded possibilities that are disconnected from the collapsed point. The result of such a collapse is a quantum singularity which may be the microcosmic basis for the development of consciousness and will power that enable a diverse reality not characterized by averaged nothingness, as will be explained in Chapter 16.

In any event, the normal perception of a unidirectional nature for the time dimension that creates the appearance of cause and effect is set by the finite speed of light, but unidirectional cause and effect can be violated by an object inside the cone of light itself, which diverges from a point in space-time, as movement at that speed causes no existence at a single point in space-time (*NewScientist,* 2009zzzi). In particular, existence at multiple points in space-time for light makes it impossible to determine the cause of temporal position.

Matter and people, who exist at different points in a space-time cone, could conceivably escape that cone by moving faster than the speed of light, but the existing universe normally prohibits such speeds (Chown, 2009c). However, as previously indicated, singularities in quantum math such as wormholes and black holes could enable escaping points within the bubble of this universe. Discontinuities can and do exist in the quantum reality, which doesn't even operate in just one direction in time (Penrose, 2004). Discontinuities in the form of discreet building blocks existing for space-time at the microscopic level can even explain the existence of gravity (*NewScientist*, 2010k).

There is evidence that the time dimension exists in discrete intervals that equal the minimum spatial length (1 Planck) divided by the speed of light (Chown, 2009a), and that relationship may provide some clues as to the maximum distance travelled by anything over any time interval. In particular, a photon, whose existence in space isn't divided into increments of time because of that particle's timelessness, could move a total of one Planck of space multiplied by the number of quanta in the temporal dimension if it moved at its maximum speed of light. Since the distance across spaced travelled by a photon that moves timelessly is unbounded, the maximum size of the universe must also equal that particular product. That particular equality is consistent with the hypothesis that light photons create the spatial dimensions and their size by their unified temporal energy pushing outward into those other dimensions. Since anything that absorbed a photon couldn't travel faster than the photon(s) that propel(s) their movement, their distance travelled and hence space itself are similarly bounded. Nevertheless, if time is eternal, however, space and travel within it can be infinite. Even if the speed of light observed per quanta of time may have actually varied over time (Magueijo, 2003), such as due to quantum singularities relating to the Big Bang from a black hole origin, the same equalities would hold, although the distances involved would equal the sum of the products over the different time periods and speeds, thereby still making space infinite if the average speed of light exceeds zero per temporal quanta in a future that lasts an eternity.

Moreover, since space and time are not independent (Chown, 2009c), there may be a deeper connection between light and the dimensions of space and time than just the ratio of the maximum speed of the former to the minimum interval of the latter. In particular, light serves as an imaging device in the process of creating observable reality via the entanglement of its photons with those particles and waves with which it comes into contact (Radin, 2006). That process, like light photons themselves, transcends time.

A second time dimension may also exist (Chown, 2009d), and that dimension could very well include all the parallel universes of possibilities that don't occur in the universe that is experienced by an individual. In any event, all possible universes do actually exist (Ananthaswamy, 2009c). Many universes may have all sorts of different life forms (Inman, 2007), with an estimated 25% having some form of life that may, however, not resemble the organisms on earth (Brooks, 2008d).

Virtually everything is possible in the infinite universes, and all that is feasible according to quantum physics will happen, although the probabilities of various events vary widely. For instance, since a specific scientific theory of how human time travel can occur through wormholes has been developed (Merali, 2007j), visiting the past is quite feasible. Such backwards temporal travel wouldn't affect past universes that already occurred or exist, as the travel from the future would only affect parallel universes that are contemporary to the ones unaffected by the travel into the past. Thus, in any particular parallel universe, there can in fact be causative effects from travel from the future.

In fact, the nature of antiparticles, which act like particles traveling backwards in time, seems to make it almost certain that the present is affected by the future via the antiparticles that move back toward the present from there. There are an infinite number of possibilities that are experienced by humans as if in a sequence, but that experience may be merely due to the way human minds categorize things (Barbour, 1999).

Already seemingly impossible events, including teleportation, invisibility, and artificial creation of life, look increasingly possible even in the universe in which humans live (Kaku, 2007). Exact replication of matter in different places is already now possible via quantum entanglement (*NewScientist,* 2009k). Quantum physics indicates movement into the past as well as into the future is feasible, implying the possibility of time travel such as via a quantum computer (Brooks, 2007a). Some form of such a computer is likely to be developed given that functional quantum communication systems already have been created (Parsons, 2009).

The singularities and infinities of quantum mathematical reality make such seemingly impossible phenomena quite possible. Even the creation of particle pairs that occurs all the time results in singularities and infinities that are especially difficult to evaluate and interpret, although some have proposed assuming that a vacuum isn't really composed of nothing, albeit with no evidence supporting that very speculative conjecture (Penrose, 2004). Since researchers haven't been able to establish a set of plausible principles from which the quantum laws that characterize the relationships between observable fermions can be derived, the nature of the universe would remain based on "unnatural foundations" (Webb, 2010) even if it were found that an observed vacuum isn't really a vacuum.

Instead of trying to explain the reality of creation out of nothing separately from the context of everything, that creation can be used to explain everything. Quantum theory doesn't explain why its laws are followed (Brooks, 2010b) that are perceived by current scientific theory as existing without cause (Battersby, 2010b), nor does it explain creation, but divisions of nothing that sum to nothing do both.

C. A Scientific Explanation for Creation and the Laws of Quantum Theory

Instead of invoking magic or a deity, the creation of matter and antimatter can be explained by divisions of nothing by itself. Such divisions are symbolized by the mathematical singularities of dividing zero by zero, which mathematically equal zero, one, or infinity, resulting in 0/0 often being labeled as undefined. Many divisions of nothing, when united in a whole that is symbolized by the mathematical figure of 1, add up the 1's to create all the integers between 0 and infinity. Putting each individual piece of divided nothing above the whole, as symbolized mathematically by 0/1, equals 0, whereas putting the whole above the individual piece of divided nothing equals infinity.

While the sum of the parts still equals 0/1+0/1=2x0=0, the divided existence has significance because of the variety it creates when combined to form whole systems of interrelated particles between the parts, such as 0/0 and 1/0.. Such possibilities create the infinite wonder of the universe that includes light photon pairs, which consist of two opposite pieces of nothing that have combined to create a special united structure of nothing. The various relationships of pieces of nothing whose attractive energy is divided by space and time are thus exerted separately in those dimensions to create infinite diversity.

In the quantum reality of the microcosmic form the divisions of nothing take, pairs of particles and antiparticles are created in a vacuum that then combine in different forms to create everything in an additive wave of infinite possibilities. The potential is infinitely greater when each divided part of nothing puts the whole above the individual (i.e., 1/0) than when each puts its own individual attraction to other pieces of divided nothing above the whole (i.e., 0/1). Quantum reality makes those systems of particles that interact holistically with all others have far greater power through resonance to affect reality than a myopic attraction that dissolves into nothingness upon successful contact with the object of its attraction.

Since particles and antiparticles, which annihilate each upon contact, are composed of equal amounts of positive and negative energy, respectively (Penrose, 2004), nothing net is effectively created by divisions of nothing, and everything returns back into the whole nothing when reunified. However, all relationships, divisions, and merging are probability processes, many of which may only occur in separate parallel or imaginary multiverses. The complex mathematics created by zero and division by zero create require a complicated geometry that include both the computations with respect to dividing by zero and adding infinities (Seife, 2004), and quantum physics also incorporates imaginary numbers (Penrose, 2004).

Such an interpretation of quantum theory can also help draw the connection of reality to many ancient beliefs about creation and the nature of the universe, as previously explained. However, such old belief systems, whether Native American, Druid, or Taoist, essentially assumed everything originated from nothing, and they represented more intuitive views of the universe that weren't grounded in scientific evidence and theory. The fact that combination of particles and antiparticles appearing out of nothing return to nothing upon uniting does scientifically explain everything from nothing. This section provides more exact details, some of which are mere hypotheses that may need adjustment as more information becomes available.

The particles and antiparticles which appear out of a vacuum all the time do, upon contact that results in their mutual annihilation, create light photons or energy (*NewScientist,* 2008zp) that appear like the creation of something out of nothing. However, because light photons are themselves their own antiparticles (Capra, 1999), they also sum to nothing in pairs of photons. Photons concentrated into a small enough space, such as via a laser, can create pairs of electrons and positrons (*NewScientist,* 2010aw) that also add to nothing because such antiparticles are the exact opposites of particles.

Particles consist of positive energy while antiparticles are composed of negative energy (Penrose, 2004), and the energy of photon pairs also sum to nothing. Since photons moving at the speed of light don't experience time, light photon pairs created by the union of divided pieces of nothing are, from their perspective, never spatially separated by units of time after their creation. Without separation, their existence remains integrated across space for all time periods even if they appear to move a finite speed from the point of view of pieces of nothing that are divided into separate existences at different points in space-time.

For the divided pieces of nothing that are appearing constantly out of nothing and that don't immediately unify into light photons, space-time does exist because their division creates separate bits of positive and negative energy in the form of particles and antiparticles, respectively. Each of these divided pieces of nothing that haven't unified into light photons have their energy reflected in electromagnetic force fields that tug on each other toward unification in space, but the positive energy of particles moves forward in the time dimension into the future while the negative energy of the antiparticles moves backwards into the past.

If the time dimension is curved to have a finite or infinite future meet a past of equal temporal length, time also sums to nothing. Given the mathematics of nothing or 0 and its inversion, it is likely that the temporal dimension represents an infinite loop, with the past and future meeting, probably only after an eternity that results from dividing the whole of time by its net sum. Regardless, within this system of circular time, the spatial dimensions which are integrated with time and don't exist outside that dimension also sum to nothing. The four dimensions thus sum to nothing.

The four dimensions arose from nothing out of the process of the division of nothing by itself and had to sum to nothing, because nothing is only created out of nothing. The opposite but equal parts of nothing that resulted from the splitting or vibration of nothing were divided only by movement in two different directions in a time dimension. Created out of nothing as part of that process, time had to sum to nothing by one piece of nothing moving eternally backwards toward infinity and the other forward toward infinity where they meet and reunite after both an infinite positive and negative infinite number of units of time that add up to nothing. All this movement in two dimensions in time must occur in at least one point in space, and thus a spatial dimension is thereby created as a byproduct that also collapses to nothing with the time dimension with which it is integrated after a positive and negative infinity that unite everything back into nothingness.

The divided pieces of nothing consist of a positron and an electron, which are mere force fields with identical characteristics but moving in opposite directions in time. Any simultaneous creation of those two fermions in the same point in space would result in their mutual annihilation to form two light photons that exist in all points in time instantaneously. Once such light photons already existed, newly formed pairs of fermions resulting from additional divisions of nothing would exist in the same point in space with the photons. Since each divided piece of nothing separated by time couldn't merge with a photon that wasn't so divided to also move instantaneously through all points in the time dimension, the excess speed of the light photon in time had to be absorbed by the fermion as a movement in another set of dimensions. With only one dimension of space, fermions combined with photon energy had to create additional spatial dimensions to enable this movement in space over time that harmonized the differences in their speed through time. A second dimension was thereby created thereby, but the curvature of time back on itself also curved the linear second dimension that existed within temporal dimension, thereby creating a third dimension. In fact, the curvature of time back on itself in a circle required either a third temporal dimension, as opposed to just a linear forward and backwards in time, or the three spatial dimensions that are indeed integrated with time.

Once the time dimension and the three spatial dimensions had been created, existing and newly formed electrons tugged on their positron opposites with their observed negative electromagnetic charges, while the electrons' pull forward in time represents positive temporal energy. In contrast, a positron has an opposite positive electrical charge but negative energy in time. As unified pieces of nothing, each photon represents both positive and negative temporal energy that is everywhere simultaneously in time at each point in space it occupies, as is reflected in the energy for movement through space that a photon provides when it is absorbed in the same point in space with a fermion.

Within the context of the time dimension, all energy merely causes actions in space over time for divided pieces of nothing. Each divided piece of nothing has the same amount of energy as a light photon, but each such separated form of energy or anti-energy creates an electromagnetic force field that tugs on its opposite. That energy of a particle or antiparticle is divided into a pull through both space and time. The tug exhibits itself in the electromagnetic force in space, but it is reflected in a pull toward each other through the infinite future and past converging into unified nothing after an eternity in both directions in time.

Space-time itself is thus created out of nothing because of the division of nothing into energy that is divided into a tug on its opposite in space and a pull in time back on itself. The irresistible attraction of divided pieces of nothing for their opposites that is separated by time causes the curvature of the time dimension back on itself after an infinite amount of time while those particles and antiparticles tug on each either in the spatial dimensions in the intervening time periods that sum to zero.

Time itself has been deduced to be an illusion (Barbour, 1999), and so movement or actions themselves are also an illusion within the context of the summed nothingness. The energy or light photons that cause such illusory movements are therefore also illusions created by the temporal dimension that sums to nothing. Since everything is merely a wave of infinite possibilities of relationships in the time dimension, which itself is merely a parameter in that potential (Penrose, 2004), everything is an illusion created from nothing and summing to nothing that reflects a different reality in the four dimensions of space-time.

Each point of space-time represents a different combination of pulls and tugs of pieces of nothing that are divided by the illusory temporal and spatial dimensions. That illusory reality created by dividing time into infinite sets of separate points is what is experienced by separate pieces of nothing that are divided by the temporal dimension because they are opposite pieces of nothing that have their energy split among temporal and spatial tugs on one another. From the perspective of those divided pieces of nothing, timeless light photons are perceived to have a finite speed through the spatial dimensions because the perceptions of each particle and antiparticle are divided by points in time. All divided pieces of nothing may eventually reunite back into timeless unified nothing, but an experience of time is required for them to experience their temporary divisibility and attraction.

The curvature of space-time caused by concentrations of divided pieces of nothing pulling on each other across space and time that is reflected in gravity can slow the passage of time and can therefore even cause a perception of light photons traveling at speeds less than that of photons moving through a vacuum. In the extreme concentrations of divided pieces of nothing that are labeled as black holes, time is slowed into freezing, drawing everything traveling near them in space into their frozen state in time. However, in the significant number of cases of spontaneously created fermion pairs where the particles actually move away from their antiparticle counterparts instead of mutually annihilating each other (*NewScientist*, 2008zp), new divided pieces of nothing are created that enable an eternal new supply of fermions that continue to experience time, thus enabling time to continue to exist for those particles to tug on each other separately through space and time.

Fermions' movement across space over time represents a force in the spatial dimension that seems to be separate from fermions' movement through time at a single point in space. Fermions, whose presence at a particular point in space-time gives them their mass, can move through space by absorbing timeless light energy which enables movement of their electromagnetic force fields to new spatial locations, but fermions' movement in time seems to be self-directed. The temporal movement causes their force fields to spin in space as those fields alternate between attractions to their opposites in different directions in time at different points in space-time.

These separate spatial and temporal movements of fermions may actually be two sides of the same electromagnetic sphere that spin between them. The electromagnetic force draws divided pieces of nothing back toward unity across space over time, while the reverse directions in time for fermions of opposite charges reflects their attraction toward each other through the time dimension from the same point in space. This attraction in both cases is between force fields, which themselves are nothing but those forces of attraction that have different probabilities of existence at different points in the space-time that separates them. Those probabilities can change upon contact with the probability wave functions of divided pieces of nothing, including new ones popping up in pairs out of nothing and existing ones that have already been unified.

Each probability in a wave function reflects the chance of it doing so on a particular point that leads to a particular situation that excludes the other points on the wave functions. Once a particular point has occurred, those other points then take place in separate sets of parallel universes which are totally unconnected with and can't be affected by the set of parallel universes that is determined by the collapsing of the wave function on a particular point that excludes them. Such collapses of the waves into particles can be caused by interaction with other wave functions, including those of human conscious choice, as will be explained in Chapter 16.

On the other hand, if there is no collapsing of the wave functions into a particle, then a weighted average set of parallel universes occurs that keeps open the possibility that all potential events allowed on the wave function happen. A force field maintains its wavelike characteristics in terms of its behavior in space-time until it is collapsed by a past, present, or future interaction that excludes other possibilities for that moment in time.

While the opposite directional movements in time for particles and antiparticles seem to separate the divided pieces of nothing at the same time that the electromagnetic force appears to draw them together, the electromagnetic force can also repel fermions when they are of like charges. Similarly, just as the temporal movement can draw opposites together, such as when the past and future meet where the time dimension curves back on itself, opposites of divided pieces of nothing mutually annihilate upon spatial movement into the same point in space-time. When those opposite attracting forces that result from the divisions of nothing meet at the same point in time, they become timeless and spatial separation no longer exists. The force of attraction drives the various fields toward that end of reunion into timeless nothingness which creates spatial energy for movement over time for other split pieces of nothing that are still separate and thus appear as particles and antiparticle.

When particles and antiparticles absorb light photon energy to cause movement in space that can be pulled by electromagnetic attraction to their opposites, the light photons they absorb are slowed to their speed. The merging of light photons and fermions thus causes a reduced and increased speed over time for the former and latter, respectively. If a division of nothing is driven to the speed of light by the energy of light, those fermions experience no passage of time as with light photons. Time from their perspective of something moving at an infinite speed results in instantaneous presence in very moment in space, but such presence everywhere simultaneously may not be possible for divided pieces of nothing because their very division from their opposites involves opposite movements in time that aren't feasible if all moments in time exist instantaneously as for a unified light photon.

In cases where the separated divisions of nothing remerge, as in the collision of particles and antiparticles, they return to nothing, but that reunification or return to nothing transforms them into energy and anti-energy in the form of photons and anti-photons, which are identical unified pieces of nothing. The photons of light energy resulting from the union of particles and antiparticles thus represent a reunited form of divided nothing that moves in both directions in time instantaneously. The photons then holistically reflect their entire history timelessly, and this historical information may differentiate them from a pure vacuum of nothing that has never divided and unified, nor crossed space timelessly.

A fermion travelling through space at the speed of light would inevitably result in contact with an opposite at some point in space-time that would cause its transformation from a divided piece of nothing into unified nothingness. Such a change in form at a different point in time would be experienced instantaneously everywhere in space-time. Travel at the speed of light would therefore have to cause a fermion to transform itself into a massless photon.

Speeds close to the speed of light do reduce fermions' presence in any point in space and thus decrease their mass that measures their concentration in a particular point in space-time. Speeds in excess of the speed of light might also be possible that would change the direction of their movement in time and thereafter result in a speed less than the speed of light but with the opposite temporal direction and electric charge. Such reversals in time would require an enormous amount of energy, perhaps an infinite amount, as might exist within the quantum singularity of a black hole that freezes a massive amount of temporal energy in small space. A black hole can thereby result in the speeds greater than that of light that are thought to have occurred instantaneously at the Big Bang, which likely originated for a black hole in an antimatter universe.

On the other hand, without much absorbed energy, the movement of a fermion slows, as is reflected at lower temperatures. The lower movement in space over time allows greater knowledge of its speed, but the probabilities of a fermion's location at any moment in time become more uncertain. At extremely low temperatures, a fermion like a helium atom can even defy the curvature of space-time caused by gravity's pull (Brooks, 2010a). In the latter case, both the pull through time and the electromagnetic tug in space still exist, but the gravitational tug of larger more distant objects isn't sufficient to attract it if it doesn't have sufficient energy to move toward them.

Only when two fermions of opposite charges annihilate each other upon meeting does the energy of their charged fields that tugs on their opposites unify into a timelessness that experiences all space as one without any pull on anything else. Something travelling at the speed of light thus experiences the time dimension as an illusion, as it actually is. However, for the divided pieces of nothing that result from divisions of nothing, time exists because the divided form of those fermions consists of their movement in opposite directions in time in incremental temporal quanta. As divided pieces of nothing, the electrons and positrons remain nothing, but they exhibit an attractive charge in space for their opposite pieces of nothing that moves in an opposite direction in time. They are thus just relative processes within an illusory time dimension.

A division of nothing in nothing thus had to create something like a temporal and spatial dimension in order to permit the divided pieces or processes of nothing to exist at separate points in nothing. Thus was born the illusory time dimension with its separated units of time through which positrons and electrons travel in opposite directions. Within each of those temporal units, which consist of an infinite past and an infinite future that sum to zero because they represent the negative and positive temporal energy, respectively, of positrons and electrons, additional divisions of nothing occur that create unified pieces of nothing in the form of light photons that reflect all knowledge of the illusion of time. These photons move instantaneously through all moments in time, and so all the temporal energy of the unified pairs of positrons and electrons must be applied in some other dimensions. The dimensions of space are thus born.

When photons merge with charged fermions still moving in a single direction in time at a finite speed, the photons allow the absorbing particles to move space. Absorbed photons can also cause fermions to spin more rapidly, or oscillate between being directed in time or space. In any event, movement in space over time sums to nothing, however, as time runs in both directions, with matter travelling forward in time and antimatter moving backwards, thereby causing the points in space-time to sum to zero, as with everything else in nothing.

There is no original division of nothing that started this process, and there is no external cause. In fact, the lack of a temporal dimension in pure nothing results in the existence of no time constraints, thus permitting the passage of an infinite amount of time within the 0 existing points of a temporal dimension of nothing. With an infinite amount of nonexistent time within the 0 points of space that also exist in nothing, everything is possible, including a division of nothing that creates an illusory time dimension divided into separate points into the past and future that add up to nothing. The temporal dimension, which is sparked upon the division of nothing into separate halves of nothing that move in opposite directions in time, then prompts the creation of spatial dimensions integrated with the temporal points that sum to nothing. There is thus no first cause for the illusions that create reality. They were inevitable in nothing and had to happen.

Once the net nothingness of the temporal dimension was divided into separate units of time, constraints existed for any such divided point in time that are set by the laws of quantum mechanics which is the basic nature of everything in the illusory space-time. There remains no constraints over the whole of all temporal points because they sum to 0, and the mathematics of quantum mechanics allows for infinite possibilities within singularities in space-time, such as exist as the frozen time of black holes and the timeless energy of unified pieces of nothing that sum to 0 in the form of light photons.

Divisions of nothing create charged energy in the form of matter and antimatter that consists of only force fields that affect other force fields of divided pieces of nothing moving through space and time. They tug on their opposites in both the spatial and temporal dimensions, thus causing the energy of their attraction to be divided across those dimensions. The resulting spatial and temporal movements occur in discrete quanta.

The energy resulting from the collision of matter and antimatter in places is released in the form of light photons that can move existing divided pieces of nothing through space. Since all matter is made up of charged energy that is affected by the energy of light photons (Smith, 2003), this overall cosmology explains everything in terms of the creation of pairs of particle-antiparticle and energy out of nothing in a manner consistent with observed reality.

Systems of divided pieces of nothing, or particles and antiparticles, which are in reality merely electromagnetic fields that tug on each other, have even evolved into fractal macrocosmic forms of those component parts that are called human beings, who perceive an alternative macro form of reality consisting of solid objects where there are actually only electromagnetic force fields that inhibit penetration. Those fields of divided nothing that create the reality that humans perceive and experience return to nothing when they are reunified.

Thus, everything was created out of nothing by nothing, and everything exists within that nothingness. Nothing has no purpose for itself and everything that it creates by dividing by itself. It has no morals, nor does it seek anything from its divisions. Nothing and everything just are. However, the evolved systems of divided pieces of nothing called human beings have evolved from nothing to be able to control nothing by dividing the individual pieces of nothing into the whole, as will be explained in Chapter 16, and they therefore have infinite power to determine their own purpose and reality via the mathematics of 1/0. Even the illusions created by nothing thus become reality. Nothing and its divisions that created everything have become mere tools of its own creations in that reality, albeit subject to the nature of nothing and its divisions that enable the infinity of 1/0. In any event, the nature of the universe in which human beings exist, the laws of quantum physics that the universe follows, and all the events and scientific findings become clear within the context of everything originating from a meaningless nothing.

The divided pieces of nothing called matter move the universe forward in time in a type of positive energy, while the other pieces of divided nothing called antimatter moves the universe backwards in time in a form of negative energy, thereby summing to zero. Since there is more matter than antimatter in the universe, this perspective could help explain the expansion of the universe of going forward in time, given that the net mass of matter would theoretically move ahead in time, whereas universes with net antimatter would move in an opposite direction in time.

Matter is sometimes characterized as concentrated positive energy, but that energy merely represents the trait of one side of a divided piece of nothing to move forward in time while the other moves backwards in their tug on each other that curves time back on itself to merge the infinite past with the infinite future. The other trait of matter is its tug on its opposite in the spatial dimensions. Thus, the energy of matter exerts itself through both time and space, with part of the energy of matter causing it to pull in one direction temporally and the other rest effectively tugging on the spatial dimensions to curve space around itself. The attracting opposite tug of antimatter in both time and space may explain the component energy of both particles and antiparticles, which are pieces of nothing divided by the time dimension that separates them into separated existence in discrete temporal points until they merge and form the united energy that creates the spatial dimensions and enables movement through space as well as time. The attracting energy of both particles and antiparticles merely represents their tug on their opposites that curves time back on itself for merging after an eternity and that also warps space around itself for spatial merging. The spatial attraction of the divided pieces of nothing that is reflected in an electric charge and the temporal attraction that moves the divided pieces of nothing temporally in quantum increments thus curve both time and space around itself to create the force of gravity.

The attraction between matter and antimatter that is reflected in their opposite electric charges, which create the electromagnetic force or pull (Close, 2009) and also the division of time into separate units, thus provides the basic insight into the intrinsic nature of the universe. In particular, those characteristics of divided pieces of nothing cause matter and antimatter to be attracted to each other in space but move away from each other in the time dimension. Gravity, which causes concentrations of energy in the form of mass to pull other concentrations of energy toward each other, and which slows the passage of time by warping both the spatial and temporal dimensions (Brooks, 2008g), is related to the electromagnetic pull of matter and antimatter toward each other that curves both space and time back on themselves. This gravitational pull of divided pieces of nothing is in the opposite temporal and spatial directions for matter and antimatter, and so the gravity of matter curves or slows the forward passage of space-time to enable its curving back toward meeting with the eternal past while the gravity of antimatter curves or slows the backwards passage of space-time to enable its warping for merging with the eternal future. This pull of divided nothing separated by time back toward itself thereby slows the expansion of space that otherwise would expand at an infinite speed to create a relatively empty infinite space immediately.

Because gravity reflects the relation between energy and matter discovered by Einstein (Calmthout, 2010) that is a function of the speed of light, i.e., $e=mc^2$, it appears to reflect the relationship between light, energy, mass, space, and time. As such, it is not be a force itself but instead a reflection of the underlying basic structure of the universe caused by the mutual attraction of positrons and electrons for each other that are created out of nothing, just as everything is.

The fact that objects can defy the standard “laws” of gravity at very low temperatures and extremely low levels of momentum (Brooks, 2010a) is consistent with this hypothesis, as gravitational “forces” merely characterize the relationship between light, energy, mass, space, and time when they aren’t blurred together by a complete lack of movement at a zero vibrational frequency of nothing. In other words, each particle and antiparticle is merely a divided piece of nothing that consists strictly in an attraction for its opposite that is reflected in an internal energy that curves space and time around itself. When that energy of attraction is reduced to zero, it doesn’t exist and has been reduced to complete nothingness. As it nears that point of extinction, quantum singularities arise that cause its attractive energy to be less than the energy created spontaneously in a vacuum induced by the divisions of nothing there. Such singularities even permit the possibility of a fermion’s remaining attractive energy existing in more disconnected points in space-time, as completely non-existent energy simultaneously negates both its own reality and the reality of its environment that give it its reality through spatial and temporal interrelationships with other divided pieces of nothing.

While it is possible to actually eliminate the reality of space-time after an eternity when the future and past meet in a temporal singularity that merges space or when matter is concentrated into a point of spatial singularity that freezes time, the reality of a divided piece of nothing can only be eliminated by its merging with its opposite. That unification of divided pieces of nothing eliminates the reality of its separated existence in incremental quanta of time and thereby makes the time dimension a complete illusion for its unified form as it experiences everything instantaneously.

Given that pairs of positrons and electrons in a vacuum can be created by focusing radiation energy at it (Close, 2009), it seems that pieces of nothing which have made the time dimension illusory for themselves by uniting in the form of light photons can actually cause the excitations of nothing to produce all. In particular, such radiating energy provokes the vibration of nothing to enable its splitting into divided pieces of nothing that can also re-merge to create light photons on contact. All such photons can spur more divisions of nothing eternally, as is consistent with the uncertainty of the amount of energy existing in a vacuum at a single moment in time.

Because the very light photons that prompt divisions of nothing can move forward or backwards in time, there is no causality with respect to this eternally creative process of creation out of nothing. There is indeed no one-way temporal causality in quantum physics, and so future photons create the vibrations of nothing that divided it and led to its reunification which created the future photons that provided the energy for the creative division and reunification.

Since each photon that both sparks the divisions and is created out of it represents both positive and negative energy, there is no net energy created. Zero energy is required to divide nothing by itself within a time dimension separated into temporally discrete quanta, which are instantaneously experienced by the energy of united parts of nothing, but which impact the pieces of nothing that are still divided by time into force fields of attraction for their opposites by causing these fields existing in single moments of space-time to move through space or oscillate between such spatial travel and opposite directional movements in time.

Since time only exists for divided pieces of nothing that tug in opposite directions in the time dimension and thus move at less than the speed of light, there is no first division of nothing that led to everything. Instead, divisions of nothing instantaneously created the time dimension that enabled the separate existence of those divided pieces of nothing at separate temporal points in which further divisions of nothing can occur when they become empty, especially when such vacuums are excited by the net zero energy of light photons which are anti-photons as well and thus both positive and negative energy.

Because the time dimension curves back on itself when the tugs of one side of the divided pieces of nothing into the past meet with the opposites' pull into the future, the time dimension also sums to nothing. The entire process is a component of nothingness that is inevitable given the infinite possibilities that exist with no time constraints on time.

Such infinite potential continue to exist for united pieces of nothing in the form of light photons. Light photons carry holographic information about the history of all objects encountered (Laszlo, 2006) because time doesn't pass for such united pieces of nothing that move at the speed of light. In particular, since time doesn't pass for a photon, the moment persists forever inside the neutral electromagnetic field of light, and there is instantaneous retention of information about the past, present, and future movement of the photon through space that oscillates in both directions of time simultaneously to create the appearance of a particular spin in any moment of time that it experiences instantaneously with all others.

Although individual particles of matter itself are often characterized as concentrated energy (Magueijo, 2003) that has slowed down (Strassman, 2001) or decayed (Lemonick, 2003), they are really divided pieces of nothing that can turn into energy moving timelessly upon unification with their opposites. Particles of matter create electromagnetic fields that appear to arise from exchanging photons of light among themselves (Merali, 2008b), but these photons observed nearby are really nothing more than virtual particles that arise from divisions of nothing that instantaneously merge close to the particles. Those divided and unified pieces of nothing appearing out of nearby empty space exert alternating forces of attraction and repulsion, as well as positive and negative energy flow, that all sum to nothing even though they cause oscillating effects in charges, spins, and electromagnetic fields of the more permanent particles they momentarily impact.

All matter is sometimes described as vibrating strings or waves (Aczel, 2002) that display mass via a scalar field (Ananthaswamy, 2008b). The scalar field merely represents the degree of concentration in a point in space-time that the electromagnetic field of attraction has and that reflects differential spatial and temporal tugs which cause them to spin and curve or warp nearby space-time around them. The concentration in space-time is indeed related to the speed of a particle, as faster speeds lower mass and even eliminate mass when the speed reaches that of light, which is pure energy that isn't concentrated in any moment of space-time because it is instantaneously everywhere in the path that it takes through space.

Thus, it is the existence of pieces of nothing divided by time that enable them to move in space in directions that are affected by the electromagnetic tugs of other fermions with whom they are interrelated both electromagnetically and gravitationally. Their very mass stems from their charged force fields occupying separate spatial points at different moments in time, and those electromagnetic tugs that draw fermions with opposite charges toward each other are therefore integrated with the gravitational force which warps space-time to attract objects with smaller mass toward larger macrocosmic objects.

Some evidence that matter can be characterized as vibrating strings is provided by Chown (2007a). One can't see the quantum stuff because it's blurred (Merali, 2007a) in the form of a wave function of possibilities, although the processes involved in that potential can be detected with advanced scientific instruments. For instance, an electron particle with lower energy is characterized by a wave with a lower resolution and a higher wavelength (Smith, 2003), as is consistent with greater certainty of its speed but less clarity as to the location of that vibrating entity.

Evidence of strings consistent with the mathematics of quantum physics indicate that many of the individual particles of our universe enter through higher dimensional black holes, as summarized by Merali (2007g) and Gefter (2007). The gravity of black holes in antimatter universes freezes the movement of space-time into the past, thereby enabling the divided pieces of nothing pulled inside them to reverse direction into going forward in time by being emitted in matter universes, such as via Big Bangs. The gravity of black holes in this matter universes have the opposite effect of warping the forward movement of time to enable the emission of anything pulled inside them into backwards movement in time in antimatter universes.

Modern physics has found energy or vibrating strings to be the more building blocks of the universe, as opposed to the larger atomic particles, with evidence of the omni-present energy being found even in vacuums (Greene, 1999). Within these vibrating strings or particle fields, information of all experiences moves at speeds in excess of 1 billion times the speed of light in waves that can't be eliminated, just as ocean waves carry information on ship wakes that can be swamped by bigger factors, such as bigger ship wakes and hurricane winds, but can't be destroyed (Laszlo, 2006). Energy itself, however, can include information-charged relationships and connections that resemble particles in a massive quantum computer (Castelvecchi, 2006). These phenomena are all quite comprehensible in the context of all energy being merely the spatial and temporal attraction of separate pieces of nothing divided by space-time that can merge into a unified timeless spatial energy to push still divided pieces of nothing toward each other faster.

Energy is interconnected throughout the universe (Chaisson, 2006), and observable matter can be so interconnected through chaotic systems that "the flap of a butterfly's wings in Brazil could set off a tornado in Texas" (Graham-Rowe, 2008). Some scientists have hypothesized that "space-time" is merely "a constantly rearranging fabric of interconnections" (Battersby, 2008b). The tendency of gravity to pull objects together unless they are the same size (Laszlo, 2006) is just one macro aspect of this connection.

Everything, including human beings, is tied more deeply together by energy in a universal communal consciousness (North, 1997) in which each individual is merely a tiny wave in a gigantic ocean of relationships that actually define the individual wave (de Quincey, 2005). While everything at its core is composed of the same energy, magnetic fields, waves, and empty space (Peat, 2005), there is, on a more macro organizational level, also a direct connection between life on earth and the rest of the universe, insofar as the organic matter that later composed the building blocks of life on earth may have originated from comets that appear to contain such matter even today (*NewScientist*, 2006h). On the microscopic level, however, the electromagnetic fields of everything result from the vibration or division of nothing into those fields.

Given that matter and antimatter are constantly popping up in a total vacuum empty of anything (*NewScientist,* 2007v), the creation of electron-positron pairs out of such empty space might also be constantly created in spatial points filled with electromagnetic fields in the form of observed matter but not be observed because the antimatter portion is destroyed almost immediately upon creation. Such a hypothesis could explain the radioactive decay of all things that occurs periodically in what is called the weak force. In particular, the spontaneously created fermions might interact with existing matter, and contact between the created positrons and existing matter could cause the observed weak force. Since positrons are merely electrons traveling backward in time, these fermions might actually be created in the future after the decay occurs.

With positrons arising out of a vacuum traveling backward in time from the future, humans with coherent enough brains could have the power to will their psychic waves into the past through some effect the waves of energy created by their wills have on the antimatter that moves backward in time. Humans can indeed will different patterns of energy emissions (McTaggart, 2008) that could have such an effect. As previously mentioned, since there are many different histories that could have led to the present (Barbour, 1999), it certainly seems feasible for a person to will events into the past. Such pasts could also affect the experienced present and the future through the exercise of such "psychic" powers, and empirical evidence on this issue will be discussed in Chapter 16.

Regardless, since the natural recombination of matter and antimatter out of a vacuum results in mutual annihilation (Smith, 2003), there is a mathematical symmetry in joint creation of opposites out of nothing that return to nothing when they recombine, with the resulting light photons representing equivalently one photon with positive energy and one anti-photon with negative energy. Within this context, antimatter universes are likely to exist that go backward in time from a Big Crunch, toward which some hypothesize the matter universes move toward as gravity collapses it all into a huge black hole, from which an antimatter Big Bang would exist in a sort of backward symmetry.

The antimatter universe would actually be moving backwards in time to that Big Bang from a gravity-induced collapse that began its move backwards in time as well as well as started the forward movement time for the gravity universe, which may end in the same black hole in which the antimatter universe ends its backwards movement in time. The matter and antimatter universes would thus both begin at the same point in space-time and after some lapse of time for the antimatter universe, it might collapse back into the same black hole point into which the matter universe traveled forward to meet at the same point in space-time. The past and the future thus meet after some amount of time that might even be infinite for eternally expanding universes.

In any event, the antiparticles existing in this matter universe are moving backward in time toward the original Big Bang that started this universe, and so their future is indeed the past for the matter that originated the Big Bang. They would indeed meet at that same point in space-time if they don't collide with their opposites into unity prior to that point.

The "magnetic fields that thread through galaxies" remain seemingly "inexplicable", as does the limited existence of antimatter in the universe, unless some consistent theory of antimatter is developed (Merali, 2008c). In a vacuum, electrons display "super-fast trembling motion," possibly as a result of interactions with the subatomic particles and their antiparticles that appear simultaneously, meet, and immediately destroy each other there (*NewScientist*, 2007v). The action in vacuums actually results in nearby objects exerting attractive forces on one another as matter and antimatter that pop out of the empty space exert pressure on the outer edges of things, with which they come into contact, prior to annihilating each other (Das, 2008). That pull probably comes as a result of the pairs of positrons and electrons that tug on the electrons and positrons incorporated in the quarks of the nearby matter, as will be explained later.

As previously mentioned, antimatter has properties identical to matter except that it acts as if it is going backwards in time (Smith, 2003), and contact between a particle and its antiparticle normally results in their mutual annihilation (*NewScientist*, 2007v). Since each particle of matter is merely a perceived form of a particular electromagnetic force field, antimatter would seem to be that force field moving in the opposite direction through time.

The simultaneous creation of pairs of particles and antiparticles in a vacuum implies a sort of splitting of nothing that results in the two created force fields going in different directions in time. These force fields of split nothing thereby create the time dimension, as their vanguards move farther into the past and future. As they move farther away from each other, they would increase their distance in time, thus continuing to increase the size of the time dimension they themselves created, but they might after an infinite past meets an infinite future, as previously explained.

Although particles and antiparticles sometimes bounce off of each other, the majority collide and mutually annihilate (*NewScientist*, 2008zp), and the only remains of such collisions are two photons of light carrying the pairs' energy in opposite directions (Gefter, 2008d). Essentially, the two identical force fields moving in opposite directions in time meeting at a particular moment in space and time return them to a state of unified nothingness that has the separate movements in time neutralized by their collision. The pair consisting of a light photon and equivalent anti-photon has the energy of the neutralized force of time transferred into pure energy or movement through space that has to occur since they can't occupy the same point in space. That photon energy may expand space when it is at the edge of the universe, which is thereby moved outward by the spatial energy the photon reaching the limits of existing space, with the expansion occurring in all three spatial dimensions when different photon pairs pull in different directions.

Such a hypothesis not only explains the mysterious continuing expansion of the universe but also the initial creation of the three dimensions of space. In particular, when the matter and antimatter created out of the division of nothing attract each other to the same point in time and space instead of splitting off into separate directions in time, their mutual annihilation results in photons moving in opposite spatial directions that pushes existing matter farther into space, which they thereby create if there is none further out in that direction. That space is bent back on itself by the time dimension that curves space-time back into nothingness, thereby requiring a third dimension to allow for the bending of two-dimensional space.

The mutual annihilation that creates spatial energy might alternatively be thought of as a productive union of the electron and positron opposites, which results in a conservation of the number of particles and antiparticles at two per union at the same time that it unifies the two fermions into chargeless, timeless photons that can't tug on other fermions in space or time because they have already unified with their partners. When the photons collide with divided pieces of nothing, they either merge with them to provide them with spatial energy that harmonizes their movement in space over time, as in the case of contact with electrons, or they bounce off in the case of thick force fields like atomic nuclei. Unified pieces of nothing that are observed to meet at the same point in space-time that is divided into separate increments of time become entangled as their directional movement in the temporal dimension becomes unified in a correlated spin that reflects their timeless unity. The photons themselves move instantaneously through all points in time simultaneously into both the past and future, but dividing that infinite speed into increments results in them sometimes moving forward and sometimes backwards in time in the form of their spin that reflects the temporal tug in a point in space.

The fact that light photons are created from the reunification of electrons and positrons at particular points in time and the fact that light photons can be absorbed by such divided pieces of nothing implies that they have a beginning and an end in time in their pure form of united nothingness. Their own timeless existence as a unified group of just two opposite fermions begins when the two divided pieces of nothing unite and ends when they are absorbed by other divided pieces of nothing.. Even in the united form, the time-dependent wave function of the photons' spin and polarization, as well as their bending in space that is caused by divided pieces of nothing tugging on them via the force of gravity, may be impacted by the divided pieces of nothing existing at separate moments in time at different points in space. Although the light photons themselves may not experience any passage of time, the fact that the spatial and temporal dimensions are integrated causes divided pieces of nothing to have an effect on space through gravity and other forces differently over time to affect the photons' spatial experiences that occurs everywhere instantaneously.

Since light moves at faster speeds through some gases than others (Battersby, 2010a), since light actually bounces off solids, and since light can be absorbed temporarily by matter, it appears that different combinations of divided nothing can impact unified nothing in different ways. In particular, while those unified pieces of nothing experience called photons may experience everything simultaneously, at least when moving through a vacuum, the fact that light is impacted by the divided pieces of nothing provides evidence that those divided pieces of nothing do control nothing, at least to some degree. Since those very controllable unified pieces of nothing can affect nothing by promoting its division, the divided pieces of nothing can influence nothing in all its forms, including in its purest undivided form that is a vacuum, in its unified form that is a photon, and in its divided form that is a charged electromagnetic field of attraction moving in opposite directions in time.

In any event, the three dimensions of space develop in a manner somewhat similar to, but not identical to, the process of the creation of the time dimension. In particular, the creation of both time and space result from the spontaneous creation of particle-antiparticle pairs in a vacuum of nothing. That phenomenon essentially represents the splitting of nothing into identically opposite electromagnetic fields that pull in opposite directions in time but towards each other in space. The pull of the temporal dimension therefore sums to nothing just as the basic electromagnetic force sums to zero because of the separate charges of electrons and positrons, so that nothing is created from nothing. The pull of the positron to create the past represents negative time that offsets the tug of the electron that produces the positive time of the future to maintain a net of zero time and preserve the nothingness of divided nothing. In simple mathematical terms, negative infinity plus positive infinity equals zero or nothing.

As previously mentioned, the time dimension likely curves back on itself to unite at the end of some period of time that might equal an eternity. Whether there is an infinite or finite past and future, the electrons moving into the future might be drawn toward the positrons traveling into the past, which is a future for an antimatter universe, and the result is a curvature of time that has them reunite in an infinite circle.

If one of the dimensions had to occur first in time, it would have had to have been the creation of the temporal dimension, as, before that dimension existed, there was no "time" to be first, and there couldn't have been any movement in space after a division of nothing into its component parts to enable them to avoid immediate reunification. A separate movement in time for the divisions of nothing is necessary in order for them to do more than merely reunify back on themselves immediately. If there were no temporal or spatial dimensions, there would only be that same nothing that had to exist on top of itself, i.e., $0/0=1$, which mathematically represents the whole nothing. Without any dimension in which to exist separately from nothing, the whole nothing had to be above nothing, i.e., 1/0, which mathematically represents the infinity of the time dimension.

Once time existed, additional vibrations of nothing at any point in the temporal dimension would result in the mutual annihilation of the electron-positron pairs created by the resulting division of nothing to create two photons that were unified and thus moved instantaneously through all moments of time. The photons thus collided with the still divided pieces of nothing that existed at separate moments in time, and the resulting excess temporal energy had to be transformed into movement in another set of dimensions that became space. As the photons interacted with positrons and electrons in each moment in time, spatial dimensions were created for each moment in time, and that additional space created the opportunity for further splitting of nothing to produce electron-positron pairs that could move through the expanse of the created space, either in unified timeless form as two photons of light moving in different directions through that space in the present or in time-separated form traveling into the future as electrons and traveling into the past as positrons.

Whereas the majority of particles and antiparticles created out of nothing continue to annihilate each other to form light photons that travel in opposite directions in space, Movements in opposite directions in time apparently continue to occur in a vacuum for some of the pairs of electrons and positrons that emanate from nothing, Their very movement outside the same space of nothing where they are created may be related to the energy of light photons external to the point of space-time entering the point and thereby creating the force that enables their separating travel. Electromagnetic and gravitational attractions from other charged particles outside that space could also promote their divided movement into separate existence in other points in space-time.

Each division of nothing is a distinct event creating pairs of separate fermions that represent the minimum size of anything. Those fermions are merely electromagnetic fields that occupy a minimum amount of space that has been measured as the Planck length. The evidence that points in space-time come in indivisible chunks with a minimum size of one Planck length (Ananthaswamy, 2009a) seems to be consistent with the hypothesis of space and time being created by the same divisions of nothing that create fermions which have the same minimum size and whose attraction for each other in time and space determines the observed speed of light that is observed.

The divided pieces of nothing exhibit their attraction for their opposites via electromagnetic fields of space and time that take the form and characteristics of electrons and positrons, with their charged energy tugging on each other across all dimensions to create spatial and temporal movement as well as a spin that reflects separate temporal pulls in opposite directions in time. Antimatter essentially pulls space through time in a temporal direction opposite that of matter. However, to the extent that the past space already exists, antimatter travels backward in time just as matter moves forward in future time periods already created by the existence of prior electrons.

The theory of relativity that time passes relatively more slowly when objects travel faster through space and doesn't pass at all at the speed of light (Magueijo, 2003) is thus integrated with all cosmological creation in a fashion consistent with the reality of quantum physics. In particular, a slower passage of time caused by faster travel through space may reflect a relatively lower amount of antimatter existing to tug on it backward in time. Although light photons aren't affected by the passage of time and actually travel at an infinite speed through space that circumvents the time dimension, even they can be trapped by the gravitational singularity of black hole into be redirected instantaneously into the opposite direction in time for timeless travel in another universe. Divided electron-positron pairs don't experience the singularities of instantaneous existence everywhere although they can incorporate a finite amount of light energy and information by absorbing photons in some of the particular points of space-time that divide them from their opposites.

Time never passes at all for travel at the speed of light because there is no tug of antimatter whatsoever on it, as matter and antimatter have already been unified therein in the form of a light photon. Light photons are therefore their own antiparticles that can't pull in a direction opposite to itself through time, and so there's no tug backward or forward in time for light photons that would enable time to pass in increments for light photons. Anything traveling at the speed of light therefore experiences everything instantaneously always, although a division of time into separate temporal points that represent illusions for the matter and antimatter contained in photons results in a perception of light moving at the maximum speed of 186,000 miles per second. This speed, which reflects the relationship between the minimum size for a unit of space and time that is created by divisions of nothing, is reality only for still divided pieces of nothing travelling at lower speeds. The light photons that come into contact with such divided pieces of nothing that exist in separate points in space therefore carry only their information between matter in different points in space that they have contacted in a particular temporal direction, such as between the electromagnetic fields of a star and a human's eyes on earth that is delayed by the finite speed of light observed by the human beings who are composed of divided pieces of nothing and thus experience time in discrete quanta.

As a result of their instantaneous existence everywhere and as a result of the transfer of information between photons that come into contact with each other, light energy transferring between universes through black holes may incorporate all their information about their instantaneous existence in all universes at all times in any universe. Regardless, both the history and future of all universes is contained in the aggregate in all light photons, although only information about the past may be ascertained by pure matter that continues to be divided by time into experiencing the future in discrete quanta of time. However, when particles merge with antiparticles, which can only ascertain information about the future, the information is aggregated into a whole that enables recognizing everything.

For any force fields that incorporate both matter and antimatter, such as quarks as will be explained later, such holistic information is incorporated into a mutual attraction that is represented in the fractal macrocosmic form of love, which can enable the coherent resonance in microcosmic interactions involved in consciousness, as will be discussed in subsequent chapters. Since quantum reality doesn't have a temporal cause and effect, as is consistent with the instantaneous temporal movement of holistic light photons, it can even be concluded that the timeless love between human beings caused the creation and evolution of everything in all universes. That love enables an ecstatic experience in separate quanta of space-time that may provide insights on everything as well as motivate the creation of everything, including reflective self-aware systems of electromagnetic waves caused by neural firings that enable conscious perception of love and everything else both timelessly and in temporal increments.

Unified pieces of nothing in the form of light thus transfer information and enlightenment, as is consistent with many religious views that are therefore consistent with reality. However, there is friction involved in such transmittal of information. For instance, light photons are only partially absorbed by some matter and are totally reflected by many solid objects. There is even some evidence that the speed of light photons through the vast regions of space varies slightly because of their interactions with that "foamy froth of particles … that spark up out of nothing" (Ananthaswamy, 2009a).

As previously indicated, light photons can be thought of as a unified group of an electron and positron whose tug on time nets out to zero to enable their instantaneous existence everywhere. The pull of electrons and positrons toward each other for such unification in a single space is the driving force of both electromagnetism and the passage of time, and existing unified pairs in the form of light photons provide energy for the process of moving through space. The virtual light photons that appear to be exchanged in these electromagnetic processes (Penrose, 2004) may actually be divided pieces of nothing in the form of an electron and a positron that immediately unite to create two light photons. Those photons then provide energy for fermions to move in space in the direction of their opposites as the fermions absorb those photons. For fermions of like charge that appear to be repelled by each other (Rosenblum and Kuttner, 2006), the light energy appearing out of the vacuum between them may be the force that pushes them away from each other in a probability wave function that is determined by the stochastic creation of pairs of electrons and positrons in space that results in such energy. Since electrons actually have an infinite charge that is only moderated by such virtual particles appearing in space, and since the electromagnetic force exhibits a wave of possible strengths and locations (Penrose, 2004), the virtual light photons appearing between fermions of opposite charge seem to be the cause of the varying force of the attractive form of electromagnetism.

The uncertainty principle that makes everything a wave of potentials could be caused by the stochastic creation of positron-electron pairs out of nothing. Collapsing those probability wave functions on single points may therefore only be accomplished by controlling the creation out of nothing, as has been shown to be possible by light energy focused into a vacuum resulting in the creation of fermion pairs. This deduction has interesting implications for consciousness and psychic powers, which may control light photon emissions to affect the creative power of nothing, as will be discussed in subsequent chapters.

Mass represents the more basic form of the 4-dimensional universe, as is measured by $m=e/c^2$, insofar as electrons and positrons are the more basic substances of the observable universe. However, it is the energy from the unification of electrons and positrons in the form of light that moves those substances in space over time, and this phenomenon is a mere consequence of divisions of nothing that appear as vibrations in the spatial and temporal dimensions which result from those divisions. The effect of energy on objects with mass merely pushes electrons and positrons toward their unity of opposites that is possible through movement in space as opposed to waiting an eternity for reunification with the positrons that attract them at the end of time.

Extensive amounts of light energy absorbed by a particle can cause it to travel at nearly the speed of light and experience a slower passage of time because the light provides it with relatively more pure spatial energy that reduces its concentration in a point in space-time. In particular, as a particle is loaded with so much energy to travel at near the speed of light, it approaches the infinite temporal energy of a light photon that is reflected in a slower passage of time and less mass that exists, as is consistent with the theory of relativity.

The spin of electrons, positrons, and other fermions might be explained by this theory of the tug of electrons and positrons on each other through both time and space. In particular, the combination of a separate pull in time and an attraction in space upon the creation of an electron-positron pair causes a spinning action when they split. The fermions' unidirectional movement in separate directions in time represent one possibility in their probability wave function while their electromagnetic force field attracting its opposite to a particular point in space-time is another possibility on that wave. A fermion's spin may reflect the stochastic fluctuations around those possibilities in the form of the smoothed average of each characteristic that is normally observed in an uncollapsed wave as the fermion alternates between attracting its opposite in space via its electromagnetic field and also through the temporal dimension where they meet when the past meets the future.

The direction of particle spin could thus reflect the different pull of positrons in the past and future on electrons. In particular, the spinning of particles with the north pole facing upward could be caused by the tug of the future being slowed by pull of electrical opposites in space while the spin of particles with the north pole facing downward might reflect the pull of the past. The single force of attraction for a divided piece of nothing is merely directed in different forms when the very division of nothing into separate moments in time creates space.

Fermions typically have only some of their energy being applied to reflect the pull in the time dimension and the rest in their electromagnetic field existing in a point in space, thereby resulting in both a finite speed with mass through space. However, since fermions trapped in less than 3 dimensions sometimes divide into partial particles with the spin separated from the electromagnetic field, that trapping may actually be collapsing their wave functions into separate possibilities that can then exist as separate units.

In any event, the spin of two charged wave functions intersecting in the same space, such as when they come into contact through orbits around a nucleus or through collisions, must have opposite spins (Penrose, 2004). Those charged spins create a magnetic field connecting more than one point in space because of the uncertain location of a particle within a probability wave function that allows it to be in more than one place at once. That magnetic field is largely neutralized in most macrocosmic objects like wood because the spins in different direction offset each other, but some metallic materials can have their composite electrons systematically spinning in the same direction, thereby creating a magnetic effect observable by the naked eye. Such magnets repel objects with magnetic fields spinning in the same direction but attract those with fields spinning in the opposite direction, just as electrons moving in different directions in time are attracted to each other.

The fact that charged fermions have spins that are opposite of one another when their whose wave functions overlap in the same space, including in the entanglement effect associated with two colliding electrons having spins that are inversely tied together, could be related to an intrinsic characteristic of each divided pieces of nothing being attracted in any moment to only one other opposite split piece of nothing. Upon occupying the same point in space-time, each charged fermion must be attracted to different opposites in both time and space, and the result is a different tug of time from that differing attraction. A charged fermion can spin in any direction in space because the fermion with the opposite charge to which it is attracted in a different time period can be located in different directions in space.

In contrast to charged fermions, light photons have spins that are positively related upon contact, as they are already unified with respect to charge, and because their timelessness also reflects a unity in time or spin as well. The fact that photons appear to have different spins when observed by systems of charged fermions like human beings reflects the splitting of time in those fermions that creates the experience of observing the photons to be drawn together in different directions in the temporal dimension at any moment in time when in actual fact they are continually unified by their timelessness at both ends of an eternal future and past without any difference in the passage of time toward the merging of the time dimension after an eternity.

In any event, the various spins of fermions seem to sum to zero in the cosmos, just as do the charges, as is consistent with everything adding up to nothing. The geometry of complex numbers that leads to spinning spheres, which can be simply made to invert their poles to reflect infinity and nothing (Seife, 2000), may provide the foundation for mathematically understanding these processes. For instance, since particle spins reflect the circulation of electrical charges in their fields of potential that are determined by such mathematics (Penrose, 2004), the nature of light photons that aren't affected by the time dimension and have no charge but spin and spatial energy might be explained by the theory of them consisting of a unified positron-electron pair. In particular, the charges of the positron and electron might exist alternatively at the poles of the photon, thus resulting in their averaged probability wave functions having a neutral charge but a spin that reflects the continuing pull of the unified electron and positron in different directions in time that the unified whole photon experiences holistically without any passage of time. providing the observed different spins at different points in time for the photon particle that itself experiences the different spins simultaneously as a unified whole because of its timelessness. Regardless, quantum math when also applied to measure the magnetic field and the electromagnetic potential may actually lead to insights as to the relative force of the temporal pull on fermions compared to electromagnetism.

The electromagnetic field of attraction moving through time that is the essence of each fermion or divided piece of nothing can exist simultaneously in more than one point in space according to its probability wave function of potential movement in the spatial dimensions, and a field existing at more than one location at once connects points in space. Those connected points in space in the field with which it is integrated draw the points in space around itself when that field spins due to the separate tugs in the spatial and temporal dimensions. This curving of space both creates the third spatial dimension which is integrated with time but also results in the force of gravity that merely reflects this curvature. Time, space, and electromagnetic attraction are thus integrated with one another, albeit in a purely relativistic manner, as quantum reality indicates the probability wave functions of all fermions are mere processes which indicate the relative position and movement of everything that are defined with dimensional parameters.

There is therefore no absolute time or space, as everything is merely a relationship between divided pieces of nothing that integrate across everything to a whole nothing. The effect of such summations is represented in the simplest form by a light photon that is everywhere at once and isn't any different from pure nothingness if it unifies with its anti-photon opposite. When the two opposite parts of a division of nothing that take the form of an electron and positron merge, they create two light photons, each of which is a unified piece of nothing but in divided form insofar as the two divided halves of nothing create two divided unified wholes of nothing with no net electrical charge or mass because they exist in all points in space at once and therefore aren't concentrated anywhere. While the photons don't move in a time dimension through space, they do have a spin that reflects the separate temporal tugs of the electron and positron halves at their north and south poles that are unified after an eternal past and future. While light photons experience that eternal past and future and all points in the spatial dimensions at once, they appear to fermions with mass to be moving at a finite speed with an actual spin affected by temporal pulls because such fermions are concentrated in points in space at particular moments in time that keeps them from moving at an infinite speed to experience all points in space at once like the light photons.

For both fermions and photons, the direction of particle spin changes over time according to the particles' probability wave functions that are affected by other probability wave functions. In particular, the probability wave functions of spins become integrated upon contact and therefore have a partial correlation in spin direction with all waves they have contacted. For instance, electrons orbiting the nucleus spin differently than their nearby ones due to their entanglement (Calleman, 2009). The nature of that entanglement could be caused by electrons naturally being attracted to positrons in different directions in time than the ones with which they have most recently come into contact because of an inherent cooperative, as opposed to competitive, nature for the most basic particles. The "mysterious" connection involved in quantum entanglement (Penrose, 2004) can also be explained by the particles having their probability wave functions collapsed on a single point at the moment of contact that thereby results in them having the same joint probability wave functions for their future spins. The joint collapsing of the superposition of possibilities that exist in any moment into only one actual occurrence causes the particles so entangled to have future probability wave functions with the same spin relationship as at the time of contact, unless either the wave function of either of the particles is impacted by another wave function that partially or fully collapses the wave function into an alternate set of possibilities.

In any event, the deep "mystery" of what vibrates in electrons (Rosenblum and Kuttner, 2006) could possibly be clarified by the hypothesis of spin being caused by separate pulls between electrons and positrons in time and space. In particular, particles may rotate the separate pulls on each other in the different dimensions of time and space to create vibration. The scientifically proven quantum nature of the universe that merely reflects interactive processes between forces, as opposed to being objects themselves (Peat, 2005), is also consistent with this theory of the attraction of positrons and electrons being the basic drivers of everything. The facts that electromagnetic forces reflecting both magnetic spin and electrical charge represent a unified force in the presence of four dimensions (Rosenblum and Kuttner, 2006), and that these forces separate when constrained to one dimension (Brooks, 2009e) are consistent with a hypothesis of each particle being merely intrinsic characteristics of a process that unite when there are enough dimensions in which they can exert themselves in a coherent whole.

The dual tug of electrons and positrons in time and space may even explain the continuous creation of matter and antimatter out of nothing. In particular, the vibrating attraction of existing charged fermions of energy could cause a vacuum to vibrate into a division of nothing, just as the energy of the unified form of divided nothing does. Within this context, the "original" vibration was perhaps caused by the tug of the fermions created in the future on the nothingness of the past that created those fermions into an infinity that sums to nothing. Nothing itself may also self-create vibrations out of its past, present, or future attraction of itself for itself, thereby also leading to a division of nothing.

In any event, a spin naturally results from the attractive pull via the time dimension that is separate from the attractive tug through space, and friction is created by these identical attractive tugs on each other because they are exerted in different dimensional directions. Such friction doesn't exist for light photons, which represent a unity of the attractive forces that comprise positrons and electrons, and which are therefore observed to travel at a speed faster than anything else. In particular, light photons contain the same amount of energy as electrons and positrons but in uncharged form (Close, 2009) that results in all its energy being concentrated in instantaneous movement in time through space. The photon form of energy reflects the quantum singularity of a simple union of an electron and positron that effectively has infinite energy in the temporal dimension.

Despite not having a net electrical charge, light photons do have a spin (Rose, 2004) that appears to reflect an alternating instantaneous movement into the infinite past and infinite future that is observed in the divided units of the inherently unified nothingness of time-space as existing at different points in the four dimensions of perceived reality. Light photons do emit an electromagnetic field that results from their oscillation (Rosenblum and Kuttner, 2006). The mathematics of $1/x$ oscillating to all possible values as x approaches 0 (Seife, 2000) could possibly explain the spin of light photons, which may rotate back and forth between their equal positive and negative charges or pulls toward an infinite past and future as they approach total unity at zero.

This phenomenon may be similar to the overall creative process which goes on continuously in the universe and which itself may be analogous to dividing zero to get more zeros. In particular, each created division of nothing occurs at different points in time and space that themselves are created by dividing nothing to create matter and antimatter, leading to the creation of light, which is unified matter-antimatter or nothing with zero charge or pull across space and time. The combining of two pieces of such divided zeroes at the same point in space and time results in them multiplying into two zeros that move in opposite spatial directions and cause yet more divisions of nothing, as is evidence by light photons prompting an increase in the creation of fermion pairs in a vacuum.

Within this context, future divided pieces of nothing could conceivably create themselves by sucking on light photons from the past via their future absorption of the photon energy to allow their own future movement in space. Those light photons, which are created upon the union of two other divided pieces of nothing that originate upon a division of nothing farther in the past, are thus caused in the past by the future divided pieces of nothing. Those divided pieces of nothing further in the past that are thus drawn into being the light photons that is absorbed by the future divided pieces of nothing may have originated from a prior division of nothing that is also thereby caused by the future fermions drawing the light energy. Since such a process can persist into the infinite past, it allows for the possibility of future fermions, or macrocosmic systems of them in the form of future human beings with will power, creating everything. Since the quantum nature of the universe allows for cause and effect to happen in both directions in time, it is also possible for any fermions existing in any moment in time to be involved in the process of creation, thereby making individual human beings with consciousness creative deities by mere enlightened will power that cause everything else in the past and future that merely reacts.

After an eternity in the past and future that then merge for a unified whole nothingness, light from that unity which meet forwards and backwards in time can manufacture everything before and moves everything after it. The meeting of the infinite past and future thereby produce the framework for the creation of everything out of nothing, including its own divisions that created the time dimension by such dividing of itself.

Since vibrations of nothing imply movement in time, they can only occur with a time dimension, which can't be created out of nothing unless it also is nothing. The vibrations of nothing that split nothing into two halves of nothing can also only be separated by a dimension that is nothing. The time dimension thus arises that must circle back on itself with the past and future meeting after an eternity of nothing in both directions. Thus, the time dimension is created simultaneously with the vibrations of nothing and its component halves. Further splitting of nothing requires the creation of the space across which the divided pieces of nothing are irresistibly drawn to reunite into nothing, with the spatial dimensions also collapsing to nothing as fermions do indeed draw space around themselves by their very existence. The temporal and spatial dimensions thus cancel to nothing in the aggregate like everything else. Nothing comes from nothing.

Despite the eternal creation of infinite zeros thereby, all the divided zeros remain attracted to each other through the basic force that pulls on opposites in time and space. Quantum entanglement also follows directly from this model of cosmology and creation insofar as all the zeros remain connected, as it's not really possible to divide nothing into separate pieces. Even unified light photons have their spins made coherent with one another upon collision that is determined by an alternating tug into the infinite past and into the eternal future at the moment of contact.

The holistic nature and fractal characteristics of the universe follow from the fact that each of the zeros is made of the same nothingness. Each piece of that united whole exhibits the same vibrations as the others, but they are expanded in complex combinations to form more macro versions of the same nothingness. The macro versions of reality therefore reflect the intrinsic changing and creative characteristics of the micro world. While some have inferred a holistic nature of all things from the facts about quantum theory (Braden, 2007), the effects of quantum entanglement that seem mystical to the ignorant are not completely holistic but instead are restricted to the actual contacts between the different wave functions that are both separate but partially connected at both the micro and macro level of reality.

At any rate, this unified theory implies that the electromagnetic attraction between electrons and positrons in space is tempered by their tug on one another in the time dimension through a possible union of past and future at the end of eternity. That temporal pull results in time curving back on itself in a loop that connects the past and the present. In a similar fashion, the repulsion of like charges in space itself possibly represents an attraction to the extent that space curves back on itself in space-time. Since the three dimensions of space are integrated with the time dimension, the space in the negative time of the past is fully offset by the space of the future to maintain the sum of nothing that meet at infinity to unify into nothing.

While time and space may be mere illusions that sum to nothing, they are the reality of the universe. However, as the famous physicist David Bohm hypothesized, that "reality" may actually be only the projection of something more real at a deeper level (Braden, 2007). Bohm hypothesized deeper reality to have an implicate order in contrast to the explicate order of ordinary observations (Radin, 2006). Within the context of this definitional cosmological framework, the "implicate" or deeper order may essentially represent nothing and its creative characteristics which are caused by vibrations in nothing that divides by itself to create divided pieces of nothing, as well as the unity of such in the form of the infinity of incremental units of unified nothing and their aggregated whole.

Divisions of nothing into the simplest forms of electromagnetic fields of attraction moving in opposite directions in time in the form of positrons and electrons may also explain the construction of more complex combinations of them that are currently assumed by science to be basic particles such as quarks. Quarks, which are the fundamental substances of atomic nuclei, include up and down quarks that have charges of 2/3 and –1/3, respectively (Close, 2009). They combine in threes to form protons, which have two up and one down quarks, and neutrons, which have two down and one up quarks (Smith, 2003).

The spontaneous creation of positrons and electrons out of nothing might also explain the origination and nature of quarks. In particular, quarks could form when two electron-positron pairs simultaneously pop up close enough to each other for the mutual electromagnetic pull of an electron in one pair to draw the positron in the other pair toward it, thereby possibly freeing the electron of the pulled-in positron from the attraction of its twin of opposite charge. The remaining electron could then conceivably be trapped in an equidistant pull between both of the positrons.

So trapped in a 2-dimensional relationship, the particle might break up into partial particles with fractional charges. Just as partial particles with partial charges form when electrons are trapped in 2-dimensional space (Brooks, 2009f), each quark may represent a fractional particle that is created when an electron and two positrons are frozen into a 2-dimensional triangular relationship. The summed charges of those partial particles would still sum to the one net positive charge of $-1+1+1=1$ of the particle trio of one electron and two positrons, but the partial particles would have the partial charges of 2/3 and –1/3. These charges sum to one in the quark trio that make up a proton, i.e., $2/3+2/3-1/3=1$ (Smith, 2003).

Antiquarks, which have opposite charges and appear like they are moving in different directions in time from quarks (Close, 2009), may be structured in a manner similar to that of quarks except that two of the trapped triplets are electrons. In particular, antiquarks might form from a positron being trapped in a 2-dimensional space between two electrons.

Since the charges of quarks (Smith, 2003) are opposite that of the partial charges of –2/3 and 1/3 for antiquarks, they mutually annihilate with quarks upon contact (Close, 2009). However, quarks have never been observed outside a bonded form that often lasts only for brief moment for pairs of quarks and antiquarks that are essentially virtual particles. Whether bonded together rather permanently in nucleons and antinucleons or momentarily in virtual particles, quarks and antiquarks exhibit their partial charges only via the observed strength of their tug that varies in different places in their bonded form, and those bonded structures always have only full charges or are neutral in the aggregate.

While the annihilation of some quarks via contact with antiquarks in the same exact point in space creates energy, neutrons have a zero net charge because they consist of particles with partial charges consisting of 2/3-1/3-1/3=0 in divided forms that are trapped into position by the separate tugs across time and space of their component electrons and positrons. It's alternatively possible that an extra electron is trapped between the 3 quarks of a proton that thereby transforms one of its 2/3 quarks into -1/3 quark and thereby turns the proton into a nucleon. As will be explained later, one of the –1/3 quarks of a neutron may actually include a positron that is attached to an electron in the form of an antineutron, thereby transforming a proton's 2/3 quark into the neutron's quark that gives it its neutral charge when a proton so decays..

Although protons and neutrons are far heavier than a trio of free-moving electrons and positrons, greater mass may be created by the trapping of those smaller particles in the triangular 2-dimensional relationship. In that relationship, two electrons would be moving backwards in time in the form of the positrons, while only one electron would be moving forward in time. Trapped by the opposite pulls in time, the probability wave function of the particle trio would be more heavily weighted in the present, thereby magnifying the mass.

Since the mass of a proton has been found to be higher in some places, as in gas clouds in the earth's atmosphere (Reich, 2010), it's possible that the varying levels of tugs on time are affected by the attraction of particle pairs popping up out of nothing that may occur at higher rates farther from centers of mass like the earth's core. The resulting different pulls in time could result in the change in mass. Alternatively, the probability wave functions of everything may simply have greater potential when they are farther from larger gravitational pulls such as emanate from planet centers, as they are there in a superposition state of virtually unlimited possibilities, whereas they are at least partially collapsed by contact with other wave functions that thereby restrict or reduce the chances of some states.

In any event, the constraint on the tug of the fermions in the time dimension caused by their binding in quarks may indeed cause neutrons and protons to have much greater mass than that existing in their component trapped fermions, and the difference between the measured mass of protons or neutrons may be related to the differing tugs of time in their component fermions. In particular, the foregoing hypothesis indicated that protons would have a 5-fermion pull forward and a pull of 4 backward. Because an antiproton would have the opposite net charge of -1 and 1/9 pull backwards in time, it seems likely that the neutron is merely the proton that has had an additional electron frozen in the position in some fashion, as through a neutrino or antineutrino. The neutron would then have a tug of 5 fermions forward and 5 backwards, therefore having greater mass than that of a proton.

The temporal energy trapped in nucleons is released in the spatial dimensions when an atom is split. Quantum mathematical analysis of this phenomenon might lead to important insights on the exact relationship between temporal and spatial energy. The relative strength of the electromagnetic force compared to the strong force that reflects the temporal bonding in atomic nuclei may provide some idea of that relationship.

The fact that one electron in this stable grouping is pulling forward in time while the other two in the form of positrons are pulling backward in time might also be what fragments the charges into thirds. In particular, a probability wave function might arise whereby 2/3 of the time the combined system of fermions has a 2/3 charge going backward in time, and 1/3 of the time it has a charge of –1/3 going forward. After the creation of this quark wave, only the average of this probability function would normally appear in existence that would consist of two full positive charges which just came back from the future and a full negative charge which just came from the past, thus collapsing in the averaged form of two 2/3 quarks and a –1/3 quark.

While the simultaneous appearance of two pairs of electrons and positrons in the exact positions necessary for such an origin for quarks is very improbable within the vacuum of space, it is feasible. The great expanse of space could have made it possible for a large number of such formations to occur over the 14 billion years since the Big Bang, especially in the vast areas of space with low temperatures that would result in little movement of new electrons and positrons popping out of nothing. The fact that particles with partial charges can be detected at extremely low temperatures (Brooks, 2009f) makes it more feasible for quarks to be so created in the massive cold vacuums that comprise much of space in the universe. Extremes of cold imply little movement (Seigo, 2002), thereby making it easier for pairs of positrons and electrons popping into existence to be trapped into a 3-way electrical pull with just one member of another pair. Pairs of quarks and anti-quarks can indeed pop up out of a vacuum (Penrose, 2004) , and thereby providing some evidence that fermion triplets can be trapped into quark form in empty space, as when three pairs of electrons-positrons appear through divisions of nothing in nearby points in space-time to create a quark-antiquark pair.

Other possibilities also exist for the creation of such fermion triplets. For instance, electrons traveling back from the future (i.e., anti-electrons) and electrons moving forward in time might be similarly drawn into such a fixed triple relationship that would form quarks and antiquarks in places of low temperature and such little movement. In particular, given that the location of particles becomes so uncertain that they tend to blur into adjacent spaces (Brooks, 2010a), a trio of electrons and anti-electrons might easily blur into appearing as single quark particle. Because gravity can concentrate the virtual particles popping out of the emptiness of space into such densely packed forms that their quantum field causes a star to explode (*NewScientist*, 2010u), it seems quite feasible that gravitational curving of space-time and other forces might contribute to the trapping of the fermion trio to form quarks within the context of 3-dimensional space. Although modern scientists aren't very successful in accurately modeling and predicting the reactions between a positron, an electron, and other particles in a normal environment because of the "competition" in attractive forces (*NewScientist,* 2010aaab), it might one day be possible to use infinity mathematics to define the quantum singularities involved in a freezing of two positrons and one electron in a triangular formation with sufficient precision to know exactly how and when quarks might be formed.

Since a trapping of an electron equally between two positrons may represent a singularity, quarks could have also formed in some fashion within the environment of singularity around the time of the Big Bang. Such trappings of fermion triplets might have occurred at or immediately after the Big Bang due to the singularities the incredible energy involved in that event may have created.

Quarks may also have originated prior to the Big Bang in a black hole from which this universe likely originated. In particular, because a black hole restricts movement (Seife, 2000), because particles with partial charges tend to form when there are restrictions on movement in spatial directions (Brooks, 2009f), and because the characteristics of black holes appear to be very similar to those of exotic quantum particles (Zaanan, 2010), a black hole origin for quarks formed as per this hypothesis seems especially feasible. Given that this matter universe appears to have originated in a tiny space where there are singularities that cause gravity to be infinite (Anathaswamy, 2008f), such an attraction in a small space could no doubt transform electron-positron trios into quarks. Many scientists do believe that the big bang exploded a hot plasma of quarks, as well as other fermions like gluons that consisted of slightly more matter than antimatter (Courtland, 2010a).

Quarks could have also been formed and arisen in the space of other universes, or anti-universes. There is actually some evidence that black holes form from quark stars that result from extreme gravity of extremely concentrated neutron stars breaking the bonds of their component neutrons. The enormous energy resulting from the busting of the neutron's bonds that could reflect a collapsing of the quark probability wave functions in a chain reaction of powerful singularities would also be trapped into the black hole and thus provide the energy for the explosion of the quarks and photons into another universe. Regardless, whether the quarks are formed prior to entry into the black hole or inside it, they were blasted out during the Big Bang that created this particular observed universe.

Quarks could have formed according to a mix of the different methods, although they may also be feasibly constructed by only one such means. Antiquarks may also therefore have a similar origin in the vast expanse of space or in black holes. Since there are many different paths that lead to the present, all such possibilities may be valid.

In any event, three electrons moving in different directions in time existing equidistantly in the same moment in a triangular formation may give rise to a sort of quantum black hole of singularity that creates the quarks. In particular, that triangle can also be perceived as three fermions of different electromagnetic charges that might normally be subject to the uncertainty principle, whereby their probability wave functions might create the possibility of one particle with a particular electromagnetic charge moving closer to another with a different charge than to the second with that different charge, thereby resulting in the mutual annihilation of two of the fermions. An equal probability of mutual annihilation by one fermion or the other might create an average probability wave function that would lead to a freezing of the triangular relationship that makes up a quark. Only some event like the observation of which fermions annihilated each other would actually cause a decomposition of the triangular relationship into two light photons, resulting from the mutual annihilation of the two fermions, and one fermion that would, upon collapsing of the wave function, be an electron moving backward in time for a quark, or an electron going forward in time for an antiquark.

A fermion equidistant in space away from two opposite charged fermions in the same time period creates the singularity of a black hole freezing time because the trapped fermion is locked into moving in two different spatial directions at once that isn't possible over time due to the required movement toward one or the other. The inability to choose a single closer opposite exists because the quarks outside a cosmic black hole normally exist only in the form of a nucleon that is composed of a triplet of quarks that also restricts the movement of the other fermions into a locked position.

It might be possible to collapse the wave function of a quark in a fashion similar to existing scientific procedures that enable collapsing the normal wave function of a photon into a particle by observing through which of two holes in a wall it goes in a setup that allows the photon an equal probability of going through each. Such an experiment should allow seeing two electrons moving backward in time and one electron moving forward in time before their transformation into a positron and two light photons that would exist after the electron moving forward in time annihilated one of the ones moving backwards in time. An enormous amount of additional energy might be released as the space which had been curved by the triangular quantum black hole that freezes the movement of the electrons in different temporal directions would be eliminated, as would be consistent with the extent by which the mass of a quark exceeds that of three separate electrons according to the equation $e=mc^2$.

The mass difference between a quark and its component three fermions may not be as large as between a nucleon and three separate quarks because the three quarks bound in a nucleon are trapped more tightly. In particular, isolated outside a nucleon, quarks might split up quickly into three separate components, such as 3 fermions, or 1 fermion and 2 photons upon mutual annihilation of 2 of the fermions. On the other hand, there exists a far smaller probability that the component electron and positron components of each quark will move outside the triplet bonding of the fermion triplets in a proton or neutron, and so the probability wave functions of the quarks average to a much larger concentration in the present where the movement of the positrons and electrons through the time dimension is frozen. Various other combinations of quarks in bonded form, as with quark-antiquark pairs, have different degrees and probabilities of their component electrons and positrons remaining frozen in position, as is also reflected in their differential masses.

The temporal energy that is frozen in quarks and nucleons causes that energy to exist more in the spatial dimensions, but the electromagnetic trapping of the three fermions in a single point in space causes the trapped temporal energy to exist in the form of mass that exerts its extra energy through the force of gravity while still bonded and that explodes the energy in atomic power when the fermions are released from those electromagnetic bonds. Thus, the fermions' separate energy in a point in space-time that is especially concentrated when it is strongly frozen in formations of quark triplets creates the mass that curves space-time through their freezing of time in the space around them. This curvature exerts the force of gravity on other fermions to affect their movement through space-time as well.

In particular, the combinations of particles and antiparticles in the form of quarks bend space-time for a gravitational effect because their opposite temporal tugs inhibit the movement forward in time for the space occupied by the quarks. The slower passage of time in areas that have such concentrations of mass compared to spatial points with less or no such temporal constraints has to draw points in space-time around those with mass. The curvature in space-time around objects with mass that is reflected in the force of gravity follows directly from this fact.

Despite the pulls in the opposite directions in time by the fermions trapped in the quarks' triangular formation of balanced spatial charges, the points occupied by quarks remain pulled forward in the temporal dimension because of the existence of more electrons tugging other points in space into the future. The separate points in space-time that are moving forward temporally, combined with those points space that are frozen or even tugging backwards in time, result in time moving forward in a non-linear and non-uniform fashion that curves space into three dimensions.

This triplet theory of quarks might also shed some insights on some of the interrelationships between particle spins. For instance, since larger atoms with more protons in their nucleus tend to be more susceptible to a coupling of their component electron spins (Zandellona, 2010), it seems likely that the coupling of electron spins in different directions at opposite poles of the magnet is somehow influenced by the existence of more quarks in such atoms with more nucleons. In particular, more quarks create the opportunity for systematic positions of the component fermions with different charges and spins that then result in spins of the orbiting electrons being more coherent on the surface on each side of the magnet.

The mutual attraction of opposite spins and charges that are associated with the integrated electromagnetic force could itself be related in some fashion to the opposite spins and charges of electrons and positrons. Magnets might therefore create or reflect an effect that is similar to opposite movements in time. Such phenomena could even be related somehow to the different probabilities of atomic nuclei moving backwards or forward in time due to the positrons existing in their own quarks. The fact that atomic nuclei themselves create a smaller magnetic effect than the electrons may also relate to the existence of the nucleons with electrons and positrons in their core as their opposite spins could partially offset, especially within the context of different directional movements in time.

The implications of this new theory of quarks would seem to allow empirical testing of it, but collapsing the wave functions of the fermions inside the quantum black hole of a quark might prove to be infeasible. However, nuclear fission, which results in the trapped temporal energy of an atom's quarks being released as an atom is split into several divided components, may represent a form of significantly changing the probabilities on the quark wave functions. Such partial wave function collapses represent discontinuities in their probability wave functions and result in the untrapped temporal energy in the form of mass being released in space according to the formula $e=mc^2$. Focused application of this energy or force in chain reactions therefore turns into powerful nuclear bombs or atomic power that provide an indication of the effect of discontinuous changes in the probability wave functions of fermions with respect to the division of their energy between points in space and travel in time.

Energy can be reflected in velocity v, which equals the force or energy necessary to stop a particle over some distance in space, and momentum p, which equals the energy necessary to stop a particle over some "distance" in time, according to the equation $e=c^2p/v$. In particular, the force of attraction between divided pieces of nothing is normally split between the temporal and spatial dimensions according to a particular ratio, whether it be a electron moving backward (i.e., a positron) or forward in time, and that force exhibits itself in energy. Because the energy of a positron or electron is exerted at a different point in time in each point in space, it exhibits mass or concentrated energy there. However, travel at the speed of light, as with a merged form of divided nothing in the form of a photon, exhibits no movement in the time dimension and exerts all its energy in the spatial dimension. Any fermion moving at the speed of light like a photon experiences every point in space simultaneously and therefore has no mass since it exhibits no concentrated energy in any single point in space in any particular time period.

On the other hand, a greater concentration of a fermion's unidirectional energy in any point in space in any time period due to restrictions in travel in the temporal dimension results in greater mass being exhibited there. The trapping of the temporal energy of three basic fermions into a single point in space in a quark results in a larger focused presence in its trapped space-time point and thus much greater mass than for three fermions moving freely through space and time. The discontinuity of the fermions being so trapped instead of moving backwards or forwards in time causes them to exhibit more energy in that point in space, with the degree of that concentration and thus mass being a function of the probability of maintaining such a frozen state of energy. For quark triplets trapped inside a nucleon, the probability of being so frozen is very large, and so nucleons exhibit very large mass compared to their quark components. Quarks existing in an even more frozen state, as with combinations of nucleons in atomic nuclei, can have an even larger probability of being trapped in their triplet formations and thus greater mass. The large amount of atomic energy that is released when an atomic nucleus is suddenly split into components with smaller probabilities of the quarks being trapped reflects the large discontinuous change in that probability.

The freezing of the temporal energy of fermions in a single point in space with a greater probability that causes greater concentrations of gravitational energy or mass thus traps the passage of time. This bending of space-time results in the force of gravity that even tugs on other fermions through that bending of space-time. A larger probability of a freezing of temporal energy thus results in a greater mass that bends time more as exhibited in a greater force of gravity. Black holes that completely trap fermions' temporal energy in a single point of space in a singularity of 100% probability exert the greatest gravitational pulls.

The speed of light itself is reduced by the existence of fermions more frozen or concentrated in particular areas of space, as the force of gravity slows photons from their observed movement of 186,000 miles per second in a vacuum. The speed of light is decreased even further by nearness to more concentrated masses of protons and neutrons, and it is even slowed to the zero speed when it nears extreme concentrations as with a black hole. The energy of unified nothing is thus controlled by the substances that it creates through nothing's own divisions.

Existing theories of physics are unable to provide any reason for the differential particle masses without fabricating particles that have never been detected (Ananthaswamy, 2009f). Inventing particles that may not exist isn't necessary with the theory of quarks, as it implies the separate pulls in opposite directions of time for the fermion triplets in a stable relationship increase mass in the present as they are caught between going forward and backwards in time. The fact that the mass of quarks is so much larger than for electrons (Close, 2009) is also explained by this theory that indicates the fermions trapped in a quark formation persist longer in the present in a manner that increases their mass to a level many times that of the electrons and positrons that form them. Since standard particle physics doesn't have a verified explanation for the mass of objects, having had to resort to invent particles like unobservable Higgs bosons to clarify the mass attributes of fermions (McAlpine, 2010b), the foregoing hypothesis about quarks seems quite plausible.

Even the greater mass of neutrons compared to protons (Close, 2009) is explained by this theory of quarks, as neutrons could conceivably have a balanced pull of 5 fermions forward and 5 backwards, whereas protons would have a 5-fermion pull forward and a pull of 4 backward. Protons are thus pulled net forward in time, while neutrons have no net pull and are thus fully concentrated in the present moment. Neutrons' mass reflects this greater presence in space over time. While the averaged wave function of neutrons wouldn't travel in time if it were totally frozen, there are chances of it collapsing in one direction other, as may occur when other forces moving through space and time interact with it, as in the process of decay.

The fact that a 3-sided geometric shape, the triangle, can explain the curved nature of the universe (Battersby, 2009b) seems to be consistent with this "grand" theory of quarks and antiquarks. In particular, since quarks that are the basic components of atomic matter and antimatter consist of 3 fermions trapped in a triangular relationship, the macrocosmic universe is naturally characterized by closed 3-sided figures, as merely reflects the fractal nature of the microcosmic quarks. The over one hundred fermions that have been observed (Smith, 2003) may also merely represent different combinations of electrons and positrons and their triangular quark combination. The apparent existence of those other seemingly basic substances may only reflect some mix of electrons moving forward and backward in time and their electromagnetic interactions in space with the new fermion pairs that are constantly popping up out of a vacuum.

The fact that there are exactly three dimensions of space may even have something to do with the fragmentation of the fermion triplets trapped inside a quark combination. In particular, because the probability wave functions of the component fermions' tug on one another are collapsed in a quark formation to have only a net 1/3 tug in the time dimension, that system of collapsed wave functions also causes them to have a 1/3 pull in spatial directions that must equal three in order for those partial tugs in space to sum to one.

Since the discrete nature of the laws of quantum physics that characterize this universe is caused by the compactness of space into 3 dimensions (Penrose, 2004), some of the parameters in that quantum reality may be determined by the trapping of three divided pieces of nothing into a frozen quark relationship. The freezing together of other numbers of particles could therefore themselves lead to much different universes, including ones with dimensions that are continuous instead of discrete.

The strong force, which binds quarks together into nucleons and nucleons into atomic nuclei (Smith, 2003), seems to be fully explained by the frozen combination of pulls in opposite directions in time via the travel of particles and antiparticles into the future and past, respectively, and toward each other in space via the electromagnetic forces that exist within each quark. The strong force is simply a mix of the frozen attractions of positrons and electrons for each other across time and space. Even the fact that the strong force becomes weaker as quarks near each other due to the confinement principle, in contrast to the quantum laws which cause other forces like electromagnetism and gravity to be stronger with proximity (Penrose, 2004), is explained by the 3-fermion theory of quarks. In particular, as quarks near each other, their component fermions with the same charge become too close and repel each other, thereby preventing them from completely merging even though they do maintain stable relationships in quark triplets inside nucleons.

The inability to observe quarks alone (Smith, 2003) can possibly be explained by the 2-dimensional nature of quarks that lead to partial charges that inhibit interactions with electromagnetic fields existing in three spatial dimensions. Only when the quarks are attracted into bonded groups of three inside a nucleon do they take a 3-dimensionsal shape. These triplets of fermion triplets are held together by the attraction of their individual component fermions in their trapped 2-dimensional triangular form.

The nucleons themselves are often bonded by the strong force inside atomic nuclei, often with an equal number of protons and neutrons. Atoms in existence that are characterized by an equal number of neutrons and protons tend to be more stable, whereas an imbalance in neutrons and protons is associated with greater instability (Smith, 2003). The relative stability may have something to do with the relative number of 2/3 and -1/3 quarks involved, with each neutron-proton pair having three 2/3 quarks and three -1/3 quarks that combined create a triangular integer force of 2 positrons and 1 electron. The interrelated electromagnetic tugs of the fermions involved are potentially stronger than for atoms that don't have their triangular form reinforced at a higher fractal structure.

Because the positrons of positrons in the nuclear quarks are frozen in structures with nearby electrons, atomic nuclei with net positive charges can attract and hold electrons in nearby orbits around them without any merging or mutual annihilation because of the repelling effect of the electron components of the quarks. The alternating opposite spins of electrons in the same atomic orbit may reflect the changing tugs of positrons in the nuclear quarks that vary over time and area of the nucleus that cause them to switch between spatial and temporal attractions.

The fact that the positrons and electrons trapped in a nucleon aren't fully unified into light photons results in them having different sections of their electromagnetic fields vary in electrical charge, thereby causing them to be attracted to other charged particles at different points in their total electromagnetic field. The position of the various quarks in the nucleons that tend to smear the two identical quarks on the outside around the loner quark in the middle (Smith, 2003) may reflect this phenomenon that creates the strong force.

The weak force, or the radioactive decay of atomic nuclei, may represent some form of interaction between different mixes of positrons and electrons trapped in 2-dimensional forms. According to existing theory, neutrinos, which are neutrally-charged electrons whose negative charge is effectively offset by an assumedly absorbed positive charge, can spark beta decay by their positive charge collapsing –1/3 quarks into an electron and a 2/3 quark, thereby resulting in a neutron turning into a proton and an electron, while the neutrino is transformed by the interaction into the anti-neutrino component of beta decay. The replacement of a neutrino with an anti-neutrino in this process essentially reverses the spin of the neutrino to form the otherwise equivalent antineutrino. This process of transforming a neutron into a proton, an electron, and an anti-neutrino may actually involve the collision of a neutrino and an antineutrino with a neutron that can effectively transfer the temporal direction of a –1/3 quark to produce a 2/3 quark because of the particular charged components of the neutrinos and quarks.

Neutrinos, which move at or near the speed of light and have no electric charge but have a fixed spin, may represent some form of an equal number of electrons and positrons somehow trapped in a superposition state of different types of mutual annihilation that would normally yield two photons with entangled but oscillating spins but that instead is collapsed into two particles similar to photons except with fixed spins, with one being an antineutrino that is identical to a neutrino except for an opposite fixed spin.

Neutrinos, which are not well understood or easily detected, could form from some trapping of 2 electrons and 2 positrons in a two-dimensional square formation that results in 2 neutrinos and 2 anti-neutrinos because their equidistant positions results in an equal tug in two directions with equal probability, with each such possibility separately forming as a neutrino and antineutrino that reflects the opposite spins of the two attracting opposites entering near the same space. This square structure of 4 fermions might arise when the movement of electrons and positrons through space-time results in a trapping of them in positions that equalize the energy of their spatial attraction and the temporal energy of their opposite directional travel in time, as might be mostly likely to arise in places like black holes. The separation of the byproducts of this process that consist of a neutrino with a fixed ½ spin and an antineutrino with a -½ spin could potentially be caused by the spin of both one positron and one electron in the square being trapped by the separate equal attractions and repulsions of the formation into a wave that averages their two possible spin states at a ½-½=0 spin, thereby resulting in the other electron and positron attracting its effectively non-spinning opposite into travelling with it through time and space without being able to merge because of non-like spins for the fermion pair that effectively make up neutrinos and antineutrinos. In any event, the spins of the resulting neutrinos and antineutrinos would reflect a moving forward and backwards in time, respectively.

This hypothesis on neutrinos may help explain how protons decay into neutrons. When a neutrino and an antineutrino together, or a square of 2 positrons and 2 electrons in the form of a neutrino and an antineutrino still combined, collide with a proton, one of the proton's positrons may swap its position with the antineutrino. In particular, if an antineutrino collides with one of the positrons in a proton, the positron half of the antineutrino can dislodge the positron and take its trapped position in the quark triplet, but the electron half of the antineutrino continues to be electrically connected to the replacing positron component of the antineutrino in a two-dimensional space separate from that of the quark except on the point of the connecting antineutrino positron. This antineutrino thereby retains its positive charge within the connected point of the quark's

2-dimensional space, but the electron half of the antineutrino remains attached to the quark's two dimensions through the positron half that binds together the system of two sets of different two dimensional structures at the point where the positron half of the antineutrino exists. The existence of an antineutrino in its set of two dimensions that is connected at the single point in space-time where the positron half of the antineutrino is located prohibits any merging of the various quarks with the antineutrino because the quarks and the antineutrino exist in separate spatial dimensions except at the connecting point of the neutrino's positron half.

This process could explain how at least some neutrinos become separated from antineutrinos. In particular, when one of the positrons in a squared structure consisting of 2 electrons and 2 positrons frozen in position collides with one of the positrons of a proton's quarks, the quark's positron can be replaced but only a particle with the same temporal movement as a positron. As a result, only the portion of the square's wave function that has it move backwards in time like a positron becomes part of the quark in the form of an antineutrino, while the other half of the square's wave function is divided out into a neutrino.

An isolated antineutrino can't normally cause a proton to decay into a neutron because it exists in a separate set of spatial dimensions from the quarks that are frozen in their own 2-dimensional form of a proton. However, an antineutrino can move into the same point in space-time as one of the positrons in one of the proton's 2/3 quarks if sufficient energy is applied that causes the positron half of the antineutrino to replace that quark's positron in the quark, thereby transforming the 2/3 quark into a –1/3 quark while releasing the displaced positron that becomes "free" to move in 3-dimensional space. Since such reactions aren't normally observed without a neutrino also appearing as a byproduct, proton decay may only be possible outside the singularity of a black hole when there is both sufficient energy to cause the collision at a single point and when the antineutrino is also paired with a neutrino in a squared formation that has each of the 2 electrons and 2 positrons of the combined neutrino-antineutrino structure attracted to their opposites with equal probability. The separate pulls of the 2 electrons and 2 positrons on each other that freezes them in their 2-dimensional space with identical chance of being attracted to their opposites to create the wave form of the squared structure enables one positron of that square to take the place of a positron in a 2/3 quark that is then held in place in the quark's separate two dimensions through its attraction to the quark's electron. The squared formation of 2 electrons and 2 positrons thereby has an antineutrino made part of the quark while one half of that square in the form of a neutrino is separated out to become "free" to travel in its unconnected two dimensions at the same time that the positron dislodged from the 2/3 quark is "freed" to travel in three dimensions since it is no longer trapped by any offsetting charged attractions.

Alternatively, within the tightly bound structure of a stable nucleus consisting of protons and neutrons, a collision of an electron and an antineutrino in the same point as one of the positrons of the proton's 2/3 quarks may cause the 2/3 quark's positron to be drawn into a trapped position with the electron, perhaps in a temporary square formation consisting of the quark's positron, the input electron, and the antineutrino. Within the equal attractions in both the quark and the square consisting of two electrons and two positrons that form a combination of a neutrino and an antineutrino, the antineutrino must immediately take the place of the quark's positron, and a neutrino byproduct is created that effectively consists of the quark's old positron and the input electron. The specific structure of the charges within a stable nucleus of at least one neutron and one proton may so strongly bind the proton's positrons that it takes an electron to dislodge it and replace it with an antineutrino's positron half that is effectively bound by both the quark's other electron and positron and by its attraction to the electron.

The origin of neutrons may actually stem from such decay of protons. That decay might be especially likely in black holes or similar situations of quantum singularity that allows particles to be trapped in a frozen attraction in two-dimensional space that has a probability wave function for the combined whole in three spatial dimensions.

Neutrons themselves may also decay into protons in a similar fashion but in reverse where the antineutrino is replaced by a positron to transform one of the neutron's –1/3 quarks into a 2/3 quark, thereby creating a proton out of a neutron. Here the electron half of the antineutrino portion of the –1/3 quark is drawn out by an attraction to the positron portion of a neutrino traveling freely in space. It must move close enough to pull the antineutrino's electron with a 0 spin away from the antineutrino's positron portion that is held in place by the electrons in one of the neutron's quarks. Once the electron is dislodged, it is no longer trapped in 2-dimensional space and therefore can't merge with the antineutrino's positron still trapped in 2-dimensional space. As a result, it travels as the "free" electron that is a byproduct of such decay while the positron portion of the antineutrino stays in the quark and thereby changes the quark's charge from -1/3 to 2/3.

The antineutrino that is also observed after this transformation of a neutron into a proton is merely the reaction-sparking neutrino which has had its spin reversed because of the attraction that also slows it down into being detectable. In particular, the positron component of the neutrino is freed from its averaged 0 spin by the attraction to the antineutrino's electron that is escaping from its attachment to its positron half in the quark, thereby causing the positron to assume its spin of -½ while simultaneously freezing the electron half of the neutrino into an averaged spin of 0 that transforms the neutrino into an antineutrino with its fixed -½ spin. The neutrino's electron is transformed into that superposition state from its frozen ½ spin in the same instant that the positron half of the neutrino is pulling out the antineutrino's electron and changing its own spin to -½, thereby resulting in the freezing of the neutrino's two fermions in that state of spins that forms the antineutrino in the next Planck instant, as the electron from the quark's antineutrino moves on because it has become three dimensional and thus unable to merge with the neutrino's positron

As previously mentioned, neutrinos and antineutrinos are typically difficult to observe because they normally don't react with other fermions, as they have no net charge and may move at the speed of light when they aren't involved in some process such as decay that enables their detection. In cases where neutrinos and antineutrinos are attracted to fermions trapped by their own fermions' equidistant charged fields in two other dimensions of space, the speed of the neutrinos and antineutrinos is slowed by the resulting quantum singularity. That slowed travel may be only temporary unless they themselves are trapped in the separate two dimensions of space of fermions like quarks through their own frozen position in a point in overlapping space-time that binds them to the fermions of the quarks there.

As partial particles trapped in 2-dimensional space, neutrinos and antineutrinos may normally not directly react with fermions existing in three spatial dimensions, just as quarks can't, even though the net electromagnetic force of their combined whole does exhibit itself in a probability wave function in three spatial dimensions. However, such 2-dimensional combinations of fermions can directly interact with other 2-dimensional structures like quarks when they meet in the same point in space-time if a quantum singularity occurs there due to equidistant spatial charges that result.

"Free" fermions traveling through all three dimensions of space can interact with groups of fermions trapped into 2-dimensional formations if they have a net electrical charge, as quarks do. For instance, electrons are attracted to protons even though they can't merge and annihilate with the composite fermions of the proton's quarks. The attraction causes electrons to orbit around the protons in atomic nuclei, but they don't annihilate even when the paths of those orbits intersect with the nuclei. Such a phenomenon may be what enables orbiting electrons to actually pass through the nucleus to which they are attracted without effect. In particular, the attraction of the opposite charges of the orbiting electrons and the protons in the nucleus keeps them bound together in the standard atomic formation, but they can't merge because they exist in separate sets of dimensions.

This theory of 2-dimensional partial particles being attracted to but restricted from colliding with 3-dimensional ones may also help explain why neutrinos generally pass by other particles without any effect, since they also don't have an electrical charge. As mere separated directional tugs in time of the component electrons and positrons that have a neutral charge and thus no concentration in any point in space-time, neutrinos and antineutrinos may have zero mass and the speed of light, with the neutrinos moving forward in time and the antineutrinos traveling backwards in the temporal direction. While their neutral charge causes them to have negligible spatial energy in the form of an electric field in a single point in space-time that is reflected in their relative timelessness and mass, they are different from light photons because they are 2-dimensional objects that only react with other 2-dimensional forms like quarks.

In particular, neutrinos and antineutrinos have no electrical charge because they consist of unified forms of electrons and positrons, but they aren't fully unified 3-dimensional particles like light photons. Their 2-dimensional form causes them to exhibit different properties, reacting only with other basic particles trapped in two dimensional space like quarks. The fixed spin of neutrinos and antineutrinos is also different from the varying spin of light photons, which alternate between forward and backwards movement in time, as neutrinos and antineutrinos, as partial particles, only move forward and backwards timelessly, respectively.

Neutrinos and antineutrinos could potentially also react with the positron-electron pairs that pop up out of a vacuum in a 2-dimensional form in some fashion. Since neutrinos and antineutrinos aren't fully unified pairs of positrons and electrons, their still separate charged components frozen in a tiny space could be attracted to positrons and electrons that suddenly appeared near enough to them. The attraction would tug on them in that point in space-time, potentially giving them an electromagnetic field for at least a moment that might slow their speed to below that of light and give them a tiny mass. In the process, the electromagnetic pull of one fermion of the recently created pair may draw it away from its opposite sufficiently to prevent their immediate mutual annihilation, thereby being at least one potential cause for such divided pieces of nothing remaining split as opposed to unifying into light photons.

Such electromagnetic effects are somewhat different from the direct interactions between 2-dimensional objects that involve them merging or bonding with each other in some fashion. Such direct interactions occur in the weak force when neutrinos and antineutrinos enter into the space of the quarks of nucleons. The net neutral charge of the neutrons can allow such collisions without additional energy, whereas the net positive charge of protons creates a field that repels the positron portion of neutrinos or antineutrinos and thus requires energy for penetration by those 2-dimensional objects.

At least part of beta decay seems to be related to neutrinos emitted by the sun in its nuclear fusion reactions, with these neutrinos later colliding with atomic nuclei on earth in amounts that vary seasonally because the effect is diffused when the earth's slightly elliptical orbit causes it to be further away from the source of those neutrinos (Mullins, 2009b). Neutrinos, which are also created by interactions in the earth's own atmosphere, as well as by cosmic events such as "convulsions of supermassive black holes" in the center of far-away galaxies that produce high-energy gamma rays (Ananthaswamy, 2009a), are difficult to detect but are assumed to be involved in beta decay (Mullins, 2009b). The neutrinos are created in the process of nuclear fusion involving a transformation of protons into neutrons, and while they might mutually annihilate upon contact with any antineutrinos in their path, many reach the earth where they pass through most "solids" without detection because they are uncharged partial particles trapped into 2-dimensional space.

Neutrinos are also a byproduct of nuclear fusion, as might be partially made possible by a trapping of 2 positrons and 2 electrons in square formation in 2-dimensional space with intense heat and pressure. In addition, antineutrinos are a byproduct of the weak force and are therefore created by normal matter on earth, including human beings. The form of the weak force that results in the transformation of a proton and an antineutron into a neutron and a positron may occur much less frequently partially because there are more byproducts of neutrinos from nuclear fusion from stellar objects than antineutrinos arising from neutron decay. Mutual annihilation of neutrinos and antineutrinos upon contact might eliminate most antineutrinos while still having a net surplus of neutrinos that cause the more frequent decay of neutrons into protons. The fact that neutrinos and neutrons are separated, neutrally-charged partial particles might also help explain the rarity of the limited decay of protons, which represent sets of such partial particles and thus like electrons normally don't interact with neutrinos and antineutrinos because of both their charged electromagnetic field and their wholeness with respect to composition of components from which a set of actual whole particles could form.

The frozen positions of positrons and electrons that make up the neutrinos and antineutrinos could form in areas similar to the triangular trappings of fermions like quarks. For instance, neutrino-antineutrino pairs could be created in black holes where the singularities there might indeed allow them to be trapped in 2-dimensional squares. They could alternatively be created in other ways, just as with quarks, such as by a positron-electron pair popping up out of nothing just as another already existing positron and electron are approaching each other that results in the square formation that creates neutrinos and antineutrinos.

In any event, the combined wave functions of the fermions that compose quarks could actually form into baryons and hadrons that represent the collapse of that particular system of wave functions in similar conditions of high levels of energy and temperature into systems of separated quark components. Baryons, hadrons, and neutrinos do indeed arise out of similar extreme conditions that exist within solar objects, just as neutrinos do.

A careful evaluation of the quantum mathematics involved here may lead to important conclusions on some of the parameters involved in the relationships that exist in the universe, especially with respect to the relative energy involved in the tug of divided pieces of nothing across time as opposed to across space. For instance, the incredible energy released when an atom is split, resulting in powerful atomic explosions and power, may reflect a releasing of the separate, fragmented tugs across time and space that are bound in the quark components of the neutrons and protons in atomic nuclei. That energy may be magnified manifold because of an immediate change in their probability wave functions that was collapsed into fragmented pulls across the four dimensions into whole tugs of triple the power. Because the fermions that were frozen in time are suddenly freed to move in their separate directions, enormous energy is released in the spatial dimensions that reflects the size of the force that had trapped them in a timeless state. In particular, the change in the fermions' wave function at the point of the splitting an atom is discontinuous in time, thereby leading to a discontinuity in the amount of energy actually existing in the atom's component fermions. The tripling of their spatial tug on each other by itself doesn't create the enormous power that exists from splitting of an atom, but the near singularities caused by the discontinuities in their probability wave functions does. The parameters of the resulting nuclear power may provide useful insights on the relationship between spatial and temporal energy at discontinuous points of singularity.

The far greater energy released upon collisions of atoms and anti-atoms than exists in collisions of their unbounded electrons may also provide important mathematical insights on the force involved in freezing fermions in triangular formations. For instance, although 18 free electrons and positrons merging with their opposites transform into pure spatial energy equal to only 18 light photons that experience the time dimension as a unified whole, the 9 fermions trapped inside a proton release far more spatial energy when they merge with their opposite structure in the form of an anti-proton. That difference reflects the near-singularity of their wave functions going from being almost completely frozen in time with a large probability to being completely freed, resulting in an enormous release of spatial energy in the present that exceeds that of 18 light photons. Quantum math should reveal the exact effect of this discontinuity.

Since most atomic nuclei have a positive charge caused by a surplus of positrons over electrons trapped inside, it seems likely that their positrons actually cause them to move backwards in time. It may therefore be their electrons frozen in electromagnetic attractions that have unreleased temporal energy that gives atoms their extra mass that is released when they are at least partially freed from those bonds. For instance, when a neutron is pushed out of the nucleus of an atom, its electrons are no longer driven in a direction in time opposite that of their innate temporal energy, and so a great deal of spatial energy is released as a result. Free neutrons would be truly frozen in time in a vacuum, but they can be tugged in one temporal direction or the other by external electromagnetic fields attracting their component electrons or positrons, and they do decay soon into protons that move backwards in time.

Given that protons tend to move backwards in time, the decay of any nucleon may actually work in reverse order to the way it is normally observed. The forward decay of a neutron could involve a proton moving backwards in time to create a neutron when it collides with an antineutrino, which pushes out and replaces an electron in a quark's triangular formation, thereby enabling the electron to go forward in time. Proton decay may result from a positron travelling backwards in time to impact with a neutron and replace the neutrino trapped inside one of the neutron's quark that can then forward in time.

Most atoms have no net charge, but their electrons aren't frozen in positions and thus move forward in time even while their nuclei are moving backwards temporally. These orbiting electrons along with free electrons outside atoms contribute to maintaining the appearance of the spatial dimensions of the universe moving forward in time that limits the universe's observed speed limit of 300,000 kilometers per second. The natural state of entropy in the universe, whereby things intrinsically move toward a greater state of disorder (Childress, 1999), may at least partially reflect the effects of those free electrons whose attraction to positrons causes them to tug space forward in time as well as pull in space on the positrons encased in the quarks of ordered matter. These free electrons thus lead to the disordering process that characterizes entropy.

Even the additional electron-positron pairs that are created out of the vacuum of space in the future and that obviously also contribute to entropy may also be caused by electrons' pull forward in time that makes room for such phenomena in space-time. The pull of electrons through time and space toward positrons thus contribute to the vibration of nothing that leads to the creation of fermion pairs in a vacuum.

As has been separately observed, there is less antimatter than matter in the universe, possibly because antiparticles are more likely to be sucked into black holes than matter (Merali, 2008d), and so the positrons that are co-created with electrons whenever nothing divides don't offset this flow forward in time. The facts that time travel requires negative energy (*NewScientist*, 2010aj) and that black holes can be wormholes allowing time travel (Shiga, 2009a) are consistent with black holes being composed of mostly antimatter, which itself is composed of negative energy and moves backwards in time. Since light photons are their own antiparticles (Capra, 1999), thus having either positive or negative energy in space, depending on whether it is absorbed by matter (i.e., positive concentrations of temporal energy in points in space) or antimatter (negative temporal energy concentrations), the negative energy of antimatter that allows it to flow backwards in time is reflected in any light photon, but all antimatter seems to originate from divisions of nothing that creates positron-electron pairs.

The paucity of antimatter compared to matter in this universe that keeps it moving forward in time could stem also partially from the existence of more matter than antimatter in the universe prior to the Big Bang. Such an imbalance between matter and antimatter may be due to more matter being sucked into a black hole in largely antimatter universes, whose own black holes led to the Big Bang of this largely matter universe that largely consists of matter. This phenomenon seems to lead to a circular process that would create an infinite number of matter and antimatter universes.

In addition, some transformation of antimatter into matter has been found to occur under certain circumstances (Quinn and Nir, 2008), such as when a tiny particle called the meson transforms itself into its antiparticle and back again 3 trillion times per second before decaying into other particles (Jamieson, 20008). The latter event implies that a fermion can actually reverse direction in time. A meson represents a combination of a quark and an antiquark of different "flavors" that are determined by their spin and "color", which is like the attracting force of electromagnetism except that there are three "color" varieties ("whimsically" labeled "red", "green", and "blue") and two spins of ½ or –½ (Smith, 2003). As such, the colors may reflect different temporal directions that exist for quarks and antiquarks (and their traveling meson pairs) that include the three "colors" of time: the past, the present, and the future. Alternatively, the three colors could represent the dividing of their charges into thirds in an averaging of their probability wave functions, with each of the three possible wave functions representing a color. While it is unclear what would cause a particle to switch direction in time, it is quite possible that the electromagnetic forces arising out of continued divisions of nothing into pairs of opposite fields could very well be responsible, as it is for so many events in the universe.

The strong force itself may partially involve in some fashion virtual particles that can each separately attract their opposites inside the quarks. However, because the quarks are frozen in a 2-dimensional state, they can't actually merge with the newly created fermions, and as they near them, they compress them in a smaller space and eventually cause the pair to collide there and annihilate each other. The light photon energy that results then pushes the quarks back out, thereby explaining why the strong force actually grows weaker as quarks near each other.

The weak force may also at least partially stem from the spontaneous creation of particle-antiparticle pairs, as the antimatter that pops into existence in places of could form into neutrinos and antineutrinos that spark nuclear decay. In addition, such spontaneous creation could potentially contribute to other neutrinos and antineutrinos interacting with the quarks to cause neutrons to transform into protons and vice versus. The bosons with mass but without electrical charge that seem to carry the weak and strong forces (Close, 2009) may in fact be pairs of electrons and positrons that naturally arose spontaneously out of nothing and traveled together temporarily without mutually annihilating each other because of the electromagnetic tug of other positrons and electrons within the surrounding quarks.

Both the pion particles, which consist of the temporary combination of a quark and antiquark that seem to bind the protons and neutrons of an atomic nucleus together in the nuclear force, and the impermanent virtual gluon particles, which appear to hold quarks together to form the atomic nucleons (Smith, 2003), may thus be related to the spontaneous creation of positrons and electrons in between quarks. In particular, the positrons of quarks would be subject to the electromagnetic pull on the newly created electrons, while the newly created positrons would tug on quarks' electrons to bind them together. The positrons and electrons would then mutually annihilate each other upon their eventual collision that is forced by the push of the attracted quarks. The negative energy of the strong force (McTaggart, 2008) is balanced with energy from mutually annihilating particles that is then transferred to the electrons orbiting the nucleons, thereby helping keeping them in perpetual orbit there. A similar scenario might unfold when pairs of positrons and electrons popped out of nothing between neutrons and protons to form the quark and antiquark that make up a pion, with the nature of the quark that combines particles already partially moving in opposite directions in time with opposite charges enabling its traveling with its antiquark opposite, perhaps in a smeared orbit of probability waves.

Pions appear in cosmic rays that emanate from exploding stars (Close, 2009), as might be expected given the foregoing hypothesis on the composition of atomic nuclei that would emit pions in those explosions. In any event, the three different "color charges" of quarks that bind them together in the strongest force of the universe are discrete as opposed to continuous (Smith, 2003), and these discrete "colors" may reflect not only their different electromagnetic pull of positive and negative but also their direction in time. Both the strength and discreteness of the colors may depend on only the sign of the electromagnetic charge and the direction in time. The negative energy thereby created that holds the quarks together via the virtual particles of the strong force (McTaggart, 2008) may thus be created by that complex interaction that draws space and time together but that is more than offset by the continued creation of other new positrons and electrons in the vast vacuum of space that keep the universe expanding or inflating.

The empirical observations of particles and antiparticles sometimes not mutually annihilating themselves when colliding in the same space (*NewScientist*, 2008zp) that can lead to the formation of electrons and positrons can also be partially explained by this hypothesis. In particular, in the case of the hypothesized formation of quarks via three trapped particles, a free positron emerges from two positron-electron pairs, whereas a free electron develops out of the formation of antiquarks. As previously mentioned, there might be a similar explanation for the creation of quarks and antiquarks themselves. In particular, in cases where there is simultaneous creation of electron-positron pairs close enough for two particles in the separate pairs to pull on each other but not in a precisely linear parallel position, the result might be the freeing of one electron and one positron and the mutual destruction of the other two fermions. If a new particle-antiparticle pair is created near one of a separated electron or positron, the identical fermion to the separated one could team up with its identical fermion in the newly created pair to trap the other newly created fermion of the pair of opposite charge, such as in a separated electron trapping a positron of a created pair with the help of the opposite pull of the electron of the pair. Thus, might both quarks and antiquarks very well be created.

Given that existing established theories force seem inconsistent with the characteristics of the universe, especially with respect to the cosmological constant, it has recently been hypothesized that the strong force, which binds quarks together and gives nucleons mass, may be characterized by something different than gluons or other virtual hypothetical particles (Courtland, 2010c). The theory presented here of quarks formed from fermion triplets trapped by their electromagnetic tugs on each other from traveling in time or space seems to explain the mass and characteristics of nucleons without having to resort to assuming the existence of particles that have never been observed alone, and it might also help explain other aspects of the cosmos as well.

In any event, this general theoretical framework for the origin and fundamental nature of quarks could explain the electromagnetic attraction of electrons to protons that apparently isn't strong enough to cause them to merge. In particular, the electrons frozen in the quarks that make up protons keep the orbiting electrons from actual contact despite the movement of the nearest ones' probability wave functions through atomic nuclei.

The actual spin of light photons may reflect a gravitational pull of matter on the combined positrons and electrons that make up the photons, thereby leading to an observed finite speed for light because of matter's pull on them in space. Matter, which is created by various combinations of charged energy that emanates from separated electrons and positrons to create mass, effectively bends time and space upon themselves in a process that results in movement slower than the speed of light and that is reflected in the laws of gravity. Gravity may actually emanate from the force of time trapped in the fermion trio in quarks that slows an infinite speed of time to enable the present moment.

This theory might therefore help explain gravity's relationship to the strong force at the more microscopic fractal level. In particular, concentrated mass, which is largely composed of quarks, is what bends space and time in gravity, and the separate pulls between space and time in the quark components of matter may therefore be the underlying cause of gravity, just as those tugs are the cause of the strong force at the microscopic level. In particular, the opposing pull in time of the electrons and positrons that make up quarks and matter curve time back on itself in space in the force of gravity that may also reflect the separate electromagnetic pulls of electrons and positrons within quarks on the positrons and electrons within other quarks. Given that quantum physics indicates the particles like gluons that make up the strong force represent negative energy (McTaggart, 2008), this hypothesis on the nature of gravity seems consistent with the tug of time drawing energy from particles in space.

This hypothesis is also consistent with the overall theory of the basic force of the cosmos that creates and causes everything being the pull on each other of pieces of divided nothing that are separated by the time dimension in which they travel in opposite directions toward a unified end. These electromagnetic fields of opposite charges are attracted to their opposites in both space and time, with movement in space over time facilitated by unified pairs of divided nothing appearing in the form of light photons, and this attraction of electrons and positrons creates a natural curvature of space-time. The basic force of electromagnetism draws the fermions together in space that is curved as positrons moving farther into the past seek unification with electrons moving into the future that may unite them after some period of time. Gravity merely reflects that curvature and attraction across both space and time, but it is especially strong in positron-electron triplets trapped in quark relationships, which restrict the passage of time in either direction, thereby permitting a present moment to exist that is separate from the past and future.

Gravity itself may be a mere macroscopic form of the strong force that itself derives from the trapping of positrons and electrons into triplet formations that enables the creation of partial particles called quarks whose frozen form creates mass. As such, the force of gravity then emanates largely from particles which are composed of a combination of electrons and positrons. In particular, only matter like particles that consist of both positrons and electrons, such as protons, neutrons, atoms, and molecules composed of constituent quarks, have their tug on time restricted by fermions with an opposite pull. That attractive force is more powerful in the spatial dimension in the form of the electromagnetic force that causes the strong force to hold many such particles together. The resulting increased pull in space combined with a reduced net tug on time leads to greater mass. Gravity, which is proportional to mass and bends the space-time geometry (Childress, 1999), is caused by such concentrated forces in points in space that slow the pull of time.

The spinning of any particle draws space and time around itself (Brooks, 2009d), thereby causing an effect on a microscopic scale similar to gravity. As a result, the curvature of time and space reflect more generally in fractal form the attraction of any particle, including electrons, for their opposites through both the spatial and temporal dimensions. The transformation of the tug of time into greater pulls through space increases mass and therefore gravity, insofar as the pull through space slows the movement of their electrical fields through the time dimension even while their attractive force in the spatial dimension continues to be warped by the tug of time to bend space-time. The attractive pull in space is stronger for fermions like quarks that are composed of combinations of both positrons and electrons because their simultaneous tug in both directions in time leads to the intrinsic attractive force being concentrated more in the spatial dimension, and so the atoms and molecules which contain the quarks have greater mass and a stronger gravitational pull, thereby bending space-time more than mere electrons.

The force of gravity can concentrate matter in space further to form black holes, which reunify everything back toward nothingness where time and space do not exist. However, the energy of photons traveling at the speed of light in opposite directions after the continual annihilation of matter and antimatter that continues to occur subsequent to their creation in divided pairs out of nothing provides the countervailing force which keeps the universe expanding, albeit at a speed that doesn't rip everything apart. All matter would collapse on itself were it not for the energy which exists in between the particles and atoms (McTaggart, 2008), and which derives from the matter and antimatter being created everywhere and mutually annihilating everywhere to create positive spatial energy.

The fact that the electromagnetic force is stronger than the force of gravity by a factor that equals the atomic age of the universe implies that gravity is becoming weaker over time, although any such weakening hasn't been verified empirically (Childress, 1999). Gravity weakening over time might indicate the electrons and positrons creating the time dimension not only pull a potentially infinite future toward a possibly infinite past for a cyclical reunion back into pure unified nothingness again but also draw space itself in the past and future back toward itself in the present moment. The latter tug weakens as time goes on, thus curving space less, at least until such tugs are more than offset by the pull of time both backwards and forwards toward itself in the distant future and past. In any event, the relationship between the electromagnetic force and the force of gravity could provide some clues as to the amount of time toward a union of the past and future, insofar as that proportion might be a function of the average number of Planck units of space separating electrons and positrons in the present and the average number of Planck units of time to a unification of those fermions at a different temporal point, either from a complete unification of all back into nothing or from a past or future unification with individual opposites. The ratio of self-destructing to surviving fermion pairs popping out of nothing may actually affect this function and thus whether time continues infinitely into the past and future without ever collapsing back on itself. On the other hand, the process of continued creation out of nothing that definitely occurs all the time may be integrated with that function.

Regardless, gravity or negative energy seems to be more than offset by the vast amount of new energy or dark energy which is created by the continuous creation of pairs of particles and antiparticles in the huge expanse of space that keeps the universe expanding. In particular, net increases in energy to the universe result in expansion of space as opposed to increased energy within existing space that may be required by the inherent nature of divided nothingness to remain a net zero.

Dark energy, which provides the most likely explanation for the rate of expansion in the universe (Krauss, 2008), may actually exist in much of what appears to be empty space in this universe, consisting of light photons that are continuously popping in and out of existence (Merali, 2007c). Light photons have an inherent energy to move (Clark, 2007), but they are also their own antiparticles and can exist indefinitely because they have no mass and therefore require an indefinitely small amount of energy (Capra, 1999). The rate of expansion of the universe itself may depend on whether the light photons arising out of the vacuum of space provide net positive energy in the aggregate, as could easily occur in a system of zero energy if antiparticles are being sucked into black holes that could originate antimatter universes, as indicated in the prior paragraph.[2]

Dark energy is indeed recognized to possibly arise from simple "quantum fluctuations in the vacuum of space" (Courtland, 2009a) that continuously create energy through the spontaneous creation of matter and antimatter which occurs there. In particular, the collision of matter and antimatter creates photons of gamma rays (Close, 2009) that could explain dark energy which might even vary stochastically with a chaotic splitting of such nothing into self-destructing fermion pairs. Since the mutual annihilation of antimatter and matter upon contact can create energy that is 1000 times greater than that of an atomic explosion and 1 billion times greater than a conventional bomb of equal size (Close, 2009), even tiny particles appearing in the huge expanse of space could produce a very large amount of energy that has a very large effect. The widely accepted hypothesis that the cosmological constant, which is the mathematical representation of the mysterious force that seems to stabilize the expanding universe and that appears to be fairly stable, emanates from dark energy (Courtland, 2009a) is consistent with a seemingly "random" creation of matter and antimatter in a vacuum following a function that tends toward some average constant rate of creation over large amounts of space and time.

Some of the negative energy of gravity itself could result from that creation out of nothing, insofar as many positrons and electrons split apart after their formation in the vacuum of space and survive. The surviving positrons appearing out of creation in a vacuum between two bodies would attract the negatively charged objects in those bodies while the electrons would attract the positively charged ones. Such new creation of matter that creates the attractive negative energy of gravity provides some offset to the dark and other positive energy that drives everything outward.

Such matter could explain the additional gravitational pull that is observed in the universe but isn't explained by observed matter. Dark matter, which hasn't actually been observed empirically but which is deduced to exist because of such observed cosmological behavior (Garfinkle, 2008), may thus have its observed effects explained by the creation of particles and antiparticles in the vacuum of space. Dark matter is thought to surround galaxies in massive "halos" (Clark, 2009) and may be the source of observable matter funneled into spiraling galaxies (Reich, 2009b) that spin at speeds too fast to enable them to hold their stars in place without the existence of forces like unobservable dark matter (Chown, 2009b). Since about 1/3 of all fermion pairs split apart without annihilating each other after popping up in the vacuum of space, it is possible that particles from such spontaneous creation are the source of the matter that exists on the periphery of the clusters of stars that form galaxies. On the other hand, the approximately 2/3 of the fermion pairs that mutually annihilate upon popping up in a vacuum may be the source of the universe's dark energy.

Dark matter, whose actual composition, existence, and origin are unclear, but which does seem to explain the "cosmological constant, or lambda", exerts "a repulsive form of energy" to keep the universe together despite its continued expansion (Ferreira, 2009). Dark matter, which has been deduced to be distributed in a spherical fashion in the Milky Way from the speed of the movement of stars in that galaxy (Calleman, 2009), pulls space around itself just as other concentrated matter does via gravity, and this attractive force keeps energy, including the repelling force of dark energy that probably emanates from the mutual destruction of the fermion pairs appearing spontaneously in the vacuum of space, from tearing the universe and galaxies apart. Dark matter, which has been estimated to make up less than a fourth of the universe at 22.7% versus 72.8% for dark energy and 4.5% for ordinary matter (Cho, 2010), may actually arise from divisions of nothing arising out of a vacuum that don't result in annihilations and that thus exert a gravitational pull that bends space-time.

Since the spinning of particles does indeed draw space and time around itself (McTaggart, 2008), perhaps due to the separate tugs of electrons and positrons through space and time, the tug of gravity may merely reflect this bending of space-time via such spinning. The spin of particles with greater mass like quarks have an especially large impact on the force of gravity because they are more trapped in the moment in time. A black hole is an example of the extreme effect of that spinning pull that may completely unify particles whose tug on time also warps or attracts nearby areas of space toward itself. Unlike unconcentrated mergings of individual electrons and positrons that result in the transfer of energy to photons which can push matter outward and offset the gravitational pull of that matter that curves space-time back on itself, concentrated matter in the form of black holes may actually have its energy released into the creation of anti-universes that are similar to this universe except that they move backwards in time. The freezing of all the spatial and temporal energy in a small space in space-time creates a singularity that allows infinite possibilities for its release. For instance, black holes, which trap energy moving backwards in time in a phenomenon called white holes (*NewScientist,* 2010av), could form in anti-universes, and the energy in one of those white holes might have given rise to the universe in which human beings exist.

Once formed, light in this universe contains heat energy that causes contacted particles or electrical fields like electrons to either move through space or vibrate faster if they are trapped by atomic charges. Upon the transfer of one quantum of energy per photon of light, the energy-carrying photons disappear, and if the transferred energy causes electrons to oscillate fast enough, the electrons will release light of varying wavelengths to exhibit colors over time (Rosenblum and Kuttner, 2006). The colors reflect different intensities of the emitted light energy, which is transferred through photons that increase the spin of electrons. The change in the spin may reflect a rise in the opposite tug of time created by the light photon which consists of a unified whole of a positron and an electron pulling in separate directions in historical time. On the other hand, light and energy can instead cause movement in space if there isn't any force that inhibits such movement in space. Even electrons circling atomic nuclei because of their attraction to the positively charged protons there can be dislodged from their orbits when sufficient energy is applied to them.

The process of electrons absorbing energy involves a charged electromagnetic field absorbing the uncharged magnetic field of light photons. They are thereby transformed into more rapidly moving or oscillating particles. If mere uncollapsed probability wave functions of an electron come into contact with the probability wave function of a light photon, the electron has only a greater chance of moving or oscillating more rapidly.

The concept of probability wave functions can help clarify the "mystery" of quantum entanglement that causes the most recent contact between two like particles to result in their spins being completely correlated at any moment in time across any distance in space. In particular, the contact between two particles integrates their two electromagnetic fields to have the same future probability wave functions with respect to the separate tugs across space and time that create their spins. However, unlike the absorption of a light photon by an electron, the two integrated probability wave functions of two entangled electrons, positrons, or photons exist in separate points in space-time.

Entangled particles can't occupy the same point in space at the same time, and so they move in different directions when they meet. Electrons coming into contact in the same point in space are entangled to have opposite spins that reflect their natural attraction to unity with different positrons, with one electron attracted to a positron in the past and the other drawn to a positron in the future. The fact that electrons in the same space, whether in orbit around a nucleus or in free collisions, are restricted from having identical spins (Penrose, 2004) may help explain why they are entangled to have opposite spins after contact. In particular, their probability wave function is collapsed on opposite spins at the moment of contact, while, being identical particles, the probability wave function of future spin states will also be identically opposite. In contrast, light photons are entangled to have like characteristics because they have already achieved their positron-electron unity and don't seek unification with other separate particles or antiparticles.

A set of quantum particles in the brain that become entangled with those in another's brain can remain entangled with those other particles despite a change in their state caused by a new thought because the particles entangled in the other brain switch states automatically to be coherent with the new thought in the initiating brain. This process thereby allows the exercise of psychic powers that result in a similar thought in the other's mind, as will be explained in Chapter 16.

The spins of two entangled particles will be less than perfectly synchronized when those waves intersect with differing probability wave functions of other particles. That contact results in a nonzero possibility of entanglement with those other particles. One or both of the previously entangled particles can become fully entangled with other particles upon a collapse of their probability wave functions at intersected points on their waves, as can occur through observation of such new contact. A particle then becomes completely unentangled with the other particle with which it was entangled. Nevertheless, despite a termination of the prior entanglement for the collapsed wave function of one particle, the other uncollapsed wave function may remain entangled with it.

This phenomenon can help explain why psychic powers are stronger in those with greater focus, as that concentration collapses wave functions to be consistent with those of the focused thought, and others will therefore be strongly affected by it. However, if the other people have their brain particles entangled with other thoughts, such as those emanating from themselves by their own independent focus, or as directed by others, that partially or completely entangle the particles involved in the initial entanglement, the original entanglement correlation can be weakened or eliminated. Such decohering thoughts can certainly be heavily affected by the environment, and so a person's psychic powers would also be affected by the individual's attempt to make the environmental situation consistent with any directed thought.

The possibility of a concentrated point on wave functions creates the infinite uncertainty that characterizes the universe. Nevertheless, any focused collapsing of wave functions on particular points must be consistent with the wave functions of other quantum particles. This process results in some wave functions collapsing on a single point while others collapse on more than one possibility. While it might somehow be feasible in the future to move between parallel universes, such as through worm holes, any collapsing of the wave function of quantum uncertainty in an existing universe would always remain unchanged in the particular universe where the collapse occurred.

In practical terms, this facet of quantum physics implies that an unobserved tree in a forest will remain in a state of quantum uncertainty as to whether it has fallen or not, due to the quantum uncertainty of the fractal wave functions of the particles that make up the tree, until its observation affects the wave function of some other entity. Even after its observation by as a person, animal, or robotic camera, etc., it will remain in a state of quantum uncertainty for all who do not observe it and don't have their own wave functions affected by any future certain encounter with an entity whose wave function is so relevantly impacted, consciously, subconsciously, or otherwise, by the resolution of the tree's quantum uncertainty. Interaction with the wave function(s) of non-observing entities causes the collapse of the wave function of the tree in the past for the so-affected entity, albeit only partially with just a higher probability than otherwise, if the contact between the wave functions permits other possibilities according to the laws of quantum physics. In the macroscopic form of the latter phenomenon, only a partial collapse of a wave function occurs if there is some chance that a non-observing entity told of an event like a tree's condition is being told a lie. Given the extensive amount of deceit in today's world, such a probability might normally be perceived to be nontrivial, making a complete collapse of wave functions often only possible upon direct observation. In any event, there are many possible past realities that can lead to the present reality for an entity's probability wave function (Barbour, 1999), but contact with other wave functions affects the set of past multiverses that are consistent with the current situation.

Quantum uncertainty (Radin, 2006) might also at least partially derive from the random creation of particle-antiparticle pairs at different points in time and space that may be subject to some ordered probability process. The uncertainty caused by newly created fermions leads to some uncertainty in the particles that comprise quarks, which only exist with a probability, and which themselves are created by the random quantum entanglement of the fermions of which they are composed. However, it would be surprising if the trapping of matter and antimatter into a cohesive whole that pulls backward and forward in time didn't have something to do with the unusual aspect of quantum uncertainty relating to the collapsing of the wave function upon focused observation.

Thus, the modeling framework developed here may provide a deeper understanding of many of the characteristics of quantum theory. In a nutshell, divided nothing pulls time in either direction. Once the time dimension exists, new divisions of nothing can result in one part of the dividend nothing (antimatter) going backward in time, and the other part (matter) going forward. However, the initial pull of each part on the other in space (electromagnetism) causes the parts to spin harmoniously, thereby creating an electromagnetic force field that represents a wave of possibilities for the particles' positions on the probability function. As previously mentioned, since quarks are a mix of three fermions that are trapped in a small space in the present by their electromagnetic tug on each other and that have separate pulls in the time dimension, they move backwards and forward in time with specific probabilities that are collapsed on an average of their probability wave function's temporal tug, resulting in far more mass because of that trapping of the temporal energy in the present, whereas an electron that isn't so trapped moves rapidly in space unidirectionally in time.

If electromagnetic fields moving in opposite directions in time meet at the same point in time and space due to their attraction to each other in the form of opposite electrical charges, a collision occurs that annihilates both particles. Since the temporal energy to move forward or backward in time is thereby neutralized, it is transformed into spatial energy. If empty space already exists around the point of impact, then the spatial energy from the meeting of matter and antimatter causes two light photons to be created that travel in opposite directions through that space, holistically carrying information about the collision and everything else with them. Time doesn't pass for the light photons because they are created by the union of two particles with equal pull in both directions of time, although the photons' movement in space is observed to be restricted to 300,000 kilometers per second because the tug of particles with mass bends perceived light in space, possibly drawing both information and energy from it in the process.

The timelessness of light may actually be involved in the process of quantum entanglement that correlates the probability wave functions of particles that have come in contact with each other. In particular, given that quantum mechanics permits events moving backwards in time, the information necessary for quantum entanglement to be instantaneous may actually be drawn holistically from any light photon that ever reaches, or has reached, the particles, including photons that have been absorbed by the entangled particles. On the other hand, particles with mass can also directly exchange information upon contact to entangle their characteristics so that their spin states are instantaneously correlated. That exchange of information can affect their probability wave functions in the same way regardless of any subsequent distance between them.

It is interesting to speculate on the "experience" of electrons which combine with positrons into light photons. Since photons moving at the speed of light are characterized by no passage of time during their time as pure energy in space, the fermions therefore "experience" instantaneously every point in space they traverse. During the time when those electrons and positrons were separate, both before and after they were light photons, their movement in space is slowed by the tug in the temporal dimension, thus enabling them to "experience" each point in space in a separate moment. That phenomenon would also occur for anything moving alternatively at the speed of light and at slower speeds at other times.

The overall hypothesis of everything in the universe arising from divisions of nothing can also explain the variable structure in the universe. In particular, the fact that particles and antiparticles mutually annihilating each other in only 50-70% of the cases where they collide in the same space (*NewScientist*, 2008zp) implies 30-50% of divisions of nothing result in surviving particles and antiparticles that could cause macrocosmic patterns reflecting fractal formations based on those microcosmic probabilities.

For instance, the "golden proportion" of 1.618 that characterizes many patterns of matter in space and that is widely considered to be the most beautiful proportion for things (Braden, 2009) could result from such probabilities. The entire beautiful structure of the universe (Barbour, 1999) may be related to the golden ratio, which itself is the rational number toward which the "irrational" ratio of one integer approaches as the sequence of Fibonacci numbers expands, where Fibonacci values, starting with the integers 0 and 1 equal the sum of the prior two, as in 0,1,1,2,3,5,8,13,21,... (Braden, 2009). Such structure might arise from an average 1/1.618=0.618047 or 61.8047% probability of a division of nothing leading to the mutual annihilation of the resulting particle-antiparticle pair that causes the creation of two light photons, instead of divided matter, and that is then reflected in the macrospheres of the universe.

Alternatively, the number of electron-positron pairs popping up out of nothing and mutually annihilating in any Planck unit of time might average the golden proportion times the number of pairs of electrons and positrons that are simultaneously created and separate. In a fashion similar to a Fibonacci series, the latter number itself equals the number of mutually destroying pairs created in the prior instant via an ever growing process that reflects the constant expansion of the cosmos from the original pair(s) created out of nothing. While creation out of nothing might have a random component, the golden ratio might provide some sort of average periodicity that results in both variation and structure in the universe.

Other relationships in the macrocosm might collapse on the golden proportion as the universe gets ever larger by the laws of probability. For example, the quantum entanglement of all matter arising from the original contact just prior to the Big Bang (Braden, 2007) doesn't result in perfect correlation between the characteristics of all particles because of contact with new particles and antiparticles that pop out of the vacuum into which they move. If the golden ratio reflects the probabilities of such creations from further divisions of nothing, the level of entanglement between particles would be affected by the golden proportion because of contact with those new particle pair, thereby influencing entanglement as well as other macroscopic spatial and temporal relationships.

The uncertainty principle that causes everything, including energy in the vacuum of space and the magnetic field of each particle, to act as a wave of potentials (Penrose, 2004) may actually be caused by the stochastic creation of fermion pairs out of nothing. The energy created by their annihilation on as they are attracted to each other would therefore also be stochastic, or subject to a wave function of infinite possibilities, each of which will happen in a set of infinite parallel universe. The exact stochastic function followed by the divisions of nothing that in turn may affect whether the resulting fermion pairs annihilate each other or remain separate particles with mass could itself have an infinite number of forms, of which that existing in the universe in which human beings exist is related to the golden ratio, which in turn may be affecting the cosmological constant.

In any event, the universe has a mathematical foundation (Peat, 2005) that may be based on simple relationships between numbers in a set (Elwes, 2008) and that has macrocosmic forms reflecting simple additions of their microcosmic structure in both space and time (Braden, 2009). For instance, as previously mentioned, the basic microcosmic phenomenon such as the probability of pairs of particles and antiparticles mutually annihilating each other might naturally be reflected in the structure of the macrocosm in the form of a golden ratio. Because quantum uncertainty creates fluctuations around the probability of light energy or particles arising from divisions of nothing, the macrocosmic forms naturally also include randomness and variation within the environment of the beautiful average. There is also some mathematical and empirical evidence of rings of three particles in bound relationships in the universe (Brooks, 2011) that might reflect the 3 fermions frozen into the structure of each quark.

Even without any inherent structure, and even without the existence of divisions of nothing and the resulting creation of the dimensions of time and space in this universe, there were always the infinite possibilities in nothing. Those possibilities include the chance of such divisions that could lead to the evolution of human consciousness, whose very evolution enabled the collapsing of that possibility into its occurrence. The "magic" of the creation of a universe that enabled the evolution of human beings was therefore inevitable within the context of the infinite possibilities in nothing that include the division of nothing into separate parts of nothing and that can lead to the evolution of consciousness.

This explanation for events observed in the universe is thus simplified to be based in one basic fundamental law that reflects the electromagnetic attraction of divided pieces of nothing for each other. The electromagnetic force involved in the behavior of such fermions across space and time and their travel in time may actually represent two sides of the same coin. In particular, the fact that the behavior of a positron can be described as an electron traveling backwards in time that reflects in its opposite charge (Close, 2009) implies that the different natures of positrons and electrons can be explained by either electromagnetism or the time dimension. The electromagnetic attraction results in a curvature of space back on itself to enable unification with nearby opposites, whereas their own intrinsic energy drives them toward unity at the end of an eternity when the past and future meet. The effect of opposite electromagnetic charges is therefore at least analogous to opposite movements in time. Since their electromagnetic fields can actually move in space toward their opposites when they absorb the timeless energy of light photons, it appears that the two forces can be made essentially identical by fermions' absorption of unified temporal energy in the form of photons which arise out of the merging of two opposite pieces of nothing. With energy in time being thus transferred to spatial movement that directs the divided pieces of nothing toward unity, electromagnetism and the time dimension are reduced to the same currency.

As previously mentioned, the fact that there are only three observed spatial dimensions in this universe may reflect the nature of triplet fermions like quarks and antiquarks whose trapping in time reduces their charges into thirds that requires three dimensions of space for them to sum to a whole charge. In particular, their fractional charges imply fractional dimensions that are inconsistent with untrapped fermions and that may therefore cause additional vibrations of nothing to create two additional dimensions in space, thereby also permitting the movement across time of those otherwise frozen triplets. The divisions of nothing that enable such particles thus cause the creation of three dimensions.

Thus is developed an internally consistent theory of the creation of the universe from nothing explaining the laws of quantum physics that it follows. While intuitively reflecting all the scientific facts of which I am currently aware, each of these various hypotheses on cosmology, creation, origin of the quantum traits of all things, and a unified theory of the forces of nature would need to be appropriately examined by others with more knowledge of quantum theory, mathematics, and observations. In addition, each valid hypothesis would have to be tested empirically using scientific procedures. Although the hypothesis provides a unified theory of physics, it certainly requires analysis and empirical testing for verification and possible modification (or refutation) by those with greater knowledge of the extremely complex mathematics of theoretical physics. In addition, the model could certainly be developed further by explaining other special laws and characteristics of quantum theory. At the very least, the merging of quantum theory with the theory of relativity that seems as problematic as it did in Einstein's days over 50 years ago might be made possible by the intuitive conjectures provided here.

In any event, given that vibrations appear to be the root essence of the universe (Barbour, 1999), it is likely that it is some vibration of nothing that causes the splitting of the electric charges and thus results in the eternal creation in those four dimensions. Vibrations of nothing require no energy, and so they indeed are scientifically and even happen all the time in a vacuum. Since quantum physics and various meditative philosophies or religions like Taoism, Buddhism, and Hinduism, as well as those of the Native Americans, imply that all matter is a perceived manifestation of a movement or process like an electric charge, force field, or energy (Capra, 1999), this theory of creation seems consistent with the evidence.

In essence, then, the cosmos, including all matter and energy perceived through the senses, is created by divisions of nothing by itself that have evolved. Those divisions in nothingness create the electrons and their positron opposites that often collide to create light energy, which in turn had to create the time and spatial dimensions in order for the energy to exist in separate places in nothing, and those dimensions allow fermion pairs to separate and move. Those fermions, which are really only electromagnetic fields that tug on each other in both the temporal and spatial directions, are inevitably trapped sometimes by their own electromagnetic tug on each other in space to form bonded triplets called quarks and antiquarks, which require exactly three dimensions of space for the freezing of their temporal energy of three fermions to exist in the single area of space in which they are trapped. Everything evolves from that reality.

The various vibrations of nothing are real; everything else is a illusory or fabricated categorization of the divisions of nothing that create electromagnetic forces which make things look solid to the brains of humans to enable them to adapt to their particular environment of vibrations or energy by categorizing observations of such phenomena.[2] As science advances, the true reality becomes ever more important to understand, as with nanotechnology, which focuses on tiny particles where the patterns of quantum physics apply, and which is posed to radically affect almost all products (Uldrich and Newberry, 2003). A fascinating example of the importance of vibrations and quantum theory is provided by the fact that crops actually grow better when music is played to them (*NewScientist,* 2007zc). The latter fact makes some seemingly superstitious Native American rituals like singing and dancing in the fields and beliefs that unseen spirits exist in all things (Underhill, 1965) seem more consistent with true reality than Western religious practices and beliefs (Peat, 2005). The magical effect that music and dancing have on human beings is also inconsistent with the traditional Western beliefs and norms (Knight, 1999), and the scientific facts of quantum physics in general are also much more consistent with many Asian religions (Capra, 1999).

D. Insights from Ancient Religions

Ancient religions may have also provided greater insights on the nature of the universe than "modern" religions. For instance, the widespread but mysterious labyrinth constructions by prehistoric humans symbolized all things and represented the foundation of human mentality in ancient times (Conty, 2002). Although much about the mysterious labyrinths and the rituals relating to them are unclear (Sjoo and Mor, 1987), they were associated with astrological symbols and may have symbolized the cyclical nature of the universe (Conty, 2002). The swirling dance to the center of the labyrinths that occurred (Sjoo and Mor, 1987) allegedly represented the reality of everything and may have facilitated a return to Paradise from the endless paths of imprisoning mazes created by directionless logic (Conty, 2002). The ancient oriental religions like Hinduism assert the universe is merely a dance of processes, and understanding that phenomenon, as opposed to focusing too much on perceived reality, can lead to a higher consciousness or nirvana (Capra, 1999). This illuminating belief might have descended from the knowledge of earlier prehistoric humans who created labyrinth rites and cultures.

Given that some of the earliest non-human representations found of the original matriarchal deity over 10,000 years ago were of snakes whose coiled forms (Gimbutas, 1989) often resembled labyrinths, it seems likely that there was an original symbolic connection to snakes. Since coiled snake forms are reputed to have been worn as headdresses of many males and females from legendary advanced cultures like Atlantis (Joseph, 2004), it is possible that the entire labyrinth concept originates from such balanced societies. Snakes symbolize a cyclical regenerating power with respect both to their annual shedding and growing new skin and to their hibernation and spring awakening (Gimbutas, 1989). Labyrinths were often in the past reconstructed into knots (Conty, 2002) that also resemble some forms of the coiled snake representation found in prehistoric art.

Since the labyrinths also symbolized the female reproductive tract (Sjoo and Mor, 1987), it is possible that there is a reason for the word "labyrinth" to be similar to the word "labia", which is the outer part of the women's genitals that have similarities to a labyrinth in the interior. Such a connection would be consistent with a further hypothesis that the sanctified labyrinths of antiquity were entered and circled around to get to the vulva or Heaven in the center, after which one left only to return later (Conty, 2002). Given that loving sex can promote a deeper understanding of everything (Deida, 2005), such entries into the "Heaven" of sexual pleasures may have represented a pleasurable means to deeper meaning. The rituals might very well have resulted in virtual religious experiences as a love for, belief in, and service to partners of the opposite sex enabled the entry into Heaven of complete spiritual and physical unity with another human being that resulted in a group consciousness. Regardless, such "trips" into a labyrinth would be more productive and mutually satisfying than narcotics, as well as far healthier. A further speculative hypothesis on the link between labyrinths and the meaning of the universe is provided in Appendix B.

The word "labyrinth" itself stems from the word "labrys" that refers to the double axes that were shaped like butterflies and created by matriarchal cultures worshipping a goddess in antiquity (Gimbutas, 1989). These double axes, which were artistically crafted by humans for over a million years (Miller, 2001), were widely associated with labyrinths and were typically hung above them on the outside (Conty, 2002). Humans made and decorated these double axes in extensive numbers with little evidence of actual use (Miller, 2001), and their purpose is widely considered by many knowledgeable anthropologists to be unclear (Gamble, 1993).

Gimbutas (1989) provides evidence that the prehistoric double axes symbolized the internal reproductive organs of women, while Sjoo and Mor (1987) hypothesize that the double axe, which was a holy object used for ritualistic purposes only by women, was a symbol for the waxing and waning of the moon. Such diverse symbolism may have reflected the correlation of women's reproductive cycles with the lunar phases.

More recent Lappish shamans used double-axe symbols, and the Nordic god Donar/Thor had a double-headed hammer of similar form that made a swastika symbol of fire when two of them are crossed (Davidson, 1990). It is possible that any initial weapons wielded by women to overthrow the alpha male societies of our ape ancestors, as explained in Part I, may have actually been double edged, with that particular shape being potentially more useful for those ancestors swinging wildly due to inexperience in weapon use. Although this hypothesis is especially speculative and difficult to prove or disprove, it would be consistent with some patriarchs later seizing this symbol of power for their own gods after they had conquered prehistoric societies that had employed this icon. The identification of such symbols with fire in the ancient Nordic patriarchies may provide some evidence of the association of double-axes with the harnessing of fire that could have originated from the rubbing of two stones, possibly initially having been discovered to create fire while the rocks were being sharpened for use as tools or weapons. The Nordic continuation of that tradition may have been motivated by their more benevolent treatment of women than existed in many other patriarchies, as well as by an identification of the double-axe symbol with fiery sex with women.

The labyrinths and related double axes held special meaning for humans for thousands of years at least (Baring and Cashford, 1993). A two-faced male Roman god, Janus, was also later associated with a labyrinth, and so labyrinths as well as labrys may have developed many meanings, including the two-sided nature of the universe that can be split along an axis to permit observation of the individual components and enable an understanding of the cosmic whole Conty, 2002).

Knots were onsidered sacred and magical in ancient times (Sjoo and Mor, 1987), and one particularly important knot is related to labyrinths (Conty, 2002). In particular, a mathematical knot with eight dimensions (the Calabi-Yau) that integrates via string theory the perceived forces of classical physics with the underlying reality of quantum physics (Merali, 2007k) can be untied to construct a labyrinth (Conty, 2002). This particular knot can also be transformed into a shape that represents the building block structure of the universe in which humans live (Ananthaswamy, 2008a).

The labyrinth and related knots, which include even the modern necktie, may have enabled ancient people to understand with their sixth sense the ultimate reality that is not perceived by the five unaided senses (Conty, 2002). Recent findings have indicated that one key to such an understanding may have to do with a recognition of the fact that math is merely a generalization of a comparison of size for different sets generically categorized by the number of objects in them (Elwes, 2008). Such a basic insight may assist at least an intuitive understanding of the incredibly complex mathematical relationships involved. A full comprehension of reality may also require a loss of ego (Sjoo and Mor, 1987), as the separated self inhibits recognition that everything is merely a set of interrelationships and can even restrict the flow of information external to the individual. Children, who typically don't have a well-developed sense of a separate self (Bernstein and Nash, 2008), and who are also able to make mathematical judgments without any training at a very early age (Elwes, 2008), may therefore actually be able to develop a better understanding of the universe than most adults.

The unifying themes of the labyrinths and ancient religions may be their biggest contribution to an understanding of everything. A belief in the essential unity of all things was widely held by the prehistoric worshippers of a Goddess across most of the globe (Gadon, 1989). In addition, the ancient Egyptian cultures, which came after the demise of the Goddess religion, believed in an interconnection between everything for thousands of years (Budge, 1987). The essential wholeness of all was also a core belief of many indigenous peoples (Jones, 1999) such as the Incas (Polich, 2001) and other Native American Indians (Taylor, 1994), as well as many Buddhists currently in Asia and elsewhere (French, 1999).

Many people have long believed in a spiritual, religious, philosophical, and physical unity with the entire universe (Freke and Gandy, 1999). While some cultures such as Asian ones have tended to promote more holistic views of everything, even people in societies such as the USA that promote fragmented viewing of things can learn to see the complete picture (Yong, 2009). Such insights and knowledge don't require any particular ritual or superstitious process like eastern religions (Sherrill and Chu, 1989) or like a labyrinth-centered goddess cult. In fact, the continuing scientific investigation of matter and the universe provides increasing evidence of the unified but diverse nature of all. In particular, even fragmented research is revealing a fractal reality that repeats itself on ever finer scales (Buchanan, 2009b), insofar as the characteristics of fractal systems are commonly found in nature, with the most basic microscopic forms of reality repeating themselves in similar but not identical forms at larger scales using simple additive mathematics (Lipton, 2008).

Macroscopic phenomena based on such microscopic particles may be a factor that even contributes to the fractal structure of the macrocosm of stars and solar systems. That structure involves revolving macroscopic bodies just as the microscopic atoms have electrons revolving around them, and the fractal nature of that structure could even cause more macrocosmic formations such as a galactic brain that has two brain-like hemispheres with two different yin-yang polarities (Calleman, 2009). The universe itself can therefore represent a macrocosm of life on earth, whereby the spinning of celestial bodies and their attraction to each other in the form of gravity may be mere fractal forms of the spin and attraction of the basic fermions that could, at least one day, lead to a cosmological consciousness or harmony in wavelength. The fractal nature of the cosmos seems to reflect the inherent properties of divided nothing, which exhibits specific characteristics but which is affected by the random creation of additional pairs of electrons and positrons that arise out of new divisions of nothing.

Fractal systems seem to explain not only fractal systems within the spatial dimensions but also within the time dimension (Braden, 2009). Shorter time cycles do seem to nest within longer ones that grow in duration by a factor of 20 for each jump in the normal time scale, including at cycles of 18 days, 360 days, 19.7 years, 394.2 years, 7900 years, 158,000 years, 3.2 million years, 63.1 million years, and 1.2 billion years that each continue 13 times until all of them end on October 28, 2011, as is consistent with ancient Mayan calendars (Calleman, 2009).

Some time cycles on earth can easily be explained by celestial cycles, such as the annual seasonal patterns relating to the earth's 365.25-day orbit around the sun and the much longer 26,000-year cycle of in the celestial zodiac (Braden, 2009). Temporal patterns may also have some root cause in other cycles in the universe, such as caused by systematic cycles in sunspot activity that average about 22 years and such as result from the rotation of the solar system around the center of the galaxy that takes 250 million years. However, in their essence, these patterns may merely reflect the quantum nature of the universe that is based on the addition of probability waves of microcosmic phenomena to create larger phenomena that reflect the intrinsic microstructure in similar form. The micro and macro worlds are identical because of the randomness in quantum mechanics that includes a stochastic appearance of particles and antiparticles out of nothing.

Calleman (2009) has hypothesized that various sacred numbers like the spatial golden ratio and the time fractal ratio of 20, both of which are related to the electromagnetic force and the wavelength of light, reflect the space-time organization of the universe. Any systematic tendency of the time fractals to increase by a factor of 20 at higher scales may in some way reflect the periodicity of pairs of particles and antiparticles being created out of nothing in the time dimension that may occur every 20 units of time quanta. However, such a hypothesis is based on mere speculative logic based on an analogy to the repetitive fractals in the spatial dimensions like the golden ratio, which itself may originate from the likelihood of pairs of fermions being annihilated as opposed to separating. Such hypotheses merit and need empirical testing.

The universe has 40 known basic particles and forces (Merali, 2007k) that are categorized into fermions, which are particles of matter and antimatter that create the structure of the cosmos, and bosons, which cause changes and movement in that structure by moving the fermions (Close, 2009). Those particles and forces are more basically composed of divided pieces of nothing that take the form of electrons, which move in the opposite direction temporarily to their other halves but are spatially attracted to them. Since everything else is formed by an infinite number of combinations of those electromagnetic fields, it makes sense for the macro universe to reflect the microcosmic reality.

Morrell (2005) has suggested that the microcosm of the universe at its core consists of a universal superconsciousness which is aware of everything and which humans can tap via a unification of the subconscious, conscious, and a superconscious.

This superconscious could reflect light photons, which timelessly contain information about everything with which they ever come into contact, and which form from unions of opposite pieces of divided nothing.

Since people under the influence of hypnosis experience enhanced brain fluidity, more emotional openness, and greater merging with the environment, as well as a feeling of one with all that is more real than the reality normally perceived (McClenon, 2002), hypnosis may facilitate a unity between the subconscious and conscious that can enable accessing the informational light of a superconscious. In particular, hypnosis can eliminate the filtering of subconscious feelings and thoughts into the conscious, thereby enabling not only the union of those parts of the mind but also facilitating information received subconsciously from others through psychic waves in the form of emitted light photons. People under hypnosis have been found to have less signals pass between the rational left-side of the brain and the emotional right-side of the brain (Geddes, 2010a) possibly as a result of less rational filtering of subconscious perceptions and thoughts experienced through emotions that thereby enable a union between the subconscious, conscious, and superconscious.

Jung's hypothesized collective unconscious that is thought to cause or affect dreams (Shamdasani, 2009) may be reflective of the psychic waves of a superconscious. Many believe that a superconsciousness which transcends all polarities to create a "pure mind" that absorbs all streams of consciousness is made possible through love and sex that cleanses perceptions of separateness (Knight, 1999). Given the hypnotic nature of human love, a special group consciousness may very well derive from mutual experiences of that spiritual emotion. However, a unification of the subconscious, conscious, and superconscious can possibly also be achieved with self-hypnosis, as is often attained by meditation rituals. Since hypnosis in general is a form of submission to others, self-hypnosis, whether achieved through meditation or otherwise, is a form of submission to oneself and to one's entire subconscious, conscious, and superconscious faculties.

Meditation, which results in the pineal gland producing mind-healing melatonin just as in normal sleep (Gynn and Wright, 2008), and which is labeled by ancient Taoism as "No Mind" that is just feeling (Chia and Wei, 1994), activates the right-side of the brain by leaving the inhibiting conscious self behind (Gynn and Wright, 2008). It can result in DMT-type "spiritually" creative experiences (Strassman, 2001), which may be related to a unification of the subconscious, conscious, and superconscious. Whether created internally by the brain, such as by the pineal gland during sleep, imbibed from external sources, or allowed through the brain barrier during extremely stressful events, DMT apparently allows subconscious thoughts and perceptions (including any psychic ones) to enter into consciousness, where they are translated for better understanding by the conscious self, generally via visual impressions and stories, especially for psychic feelings (Gynn and Wright, 2008). Many chemicals, including those released by various regulatory cells that can actually reduce stress and enhance learning, do indeed penetrate the brain barrier and affect thought and feelings (Geddes, 2011).

In any event, hypnosis enables the brain to follow suggestions without conscious questioning or inhibitions. It works by shutting down any inhibitions to carrying out the suggestions of the hypnotizing agent(s) and feels relaxing because it enables the thalamus filter to rest, thereby enabling directions to pass unfiltered to the subconscious. The result can be a complete surrender of the subconscious to hypnotic suggestions without any doubts or questioning, thereby enabling a strong will or determination to do something that people can be hypnotized, or self-hypnotize themselves, to follow.

Hypnosis is achieved via intense focus. Hypnotizable people do have an area of the brain involved in focusing attention (called the rostrum) that is 30% larger than others (Jones, 2009b). The focus involved in hypnosis may very well turn off the brain perceptions that are useful in normal daily life but that inhibit seeing the true cosmic reality of electro-magnetism that unifies all things. The extreme focus involved in passionate love and sex has a similar hypnotic effect that enables maximum coordination of the subconscious, conscious, and superconscious for the specific purpose of unifying two people emotionally, physically, and spiritually.

People normally experience an averaging of all possible states of the world, with the average determined by the weighting of each possible state by its probability. However, a person can experience one particular state of the world via hypnotic focus. Upon focus, the experienced state can be collapsed on an alternate point on the system of interconnected probability wave functions that can be willed into existence if it is something consistent with one's innate human nature and thereby permits the most efficient functioning of the subconscious to resonate everything else with that intent.

For instance, if a person focuses on another person, the person experiences the other individual in one particular state as opposed to an average of states. This process can result in the focus of the other individual on the subject individual, and a sort of mutual hypnosis can result. The person initiating the focus can lead the direction of the concentrated thought so that the other follows that person's suggestions. However, the earliest mutual focus likely led to a group consciousness of people following each others' suggestions for the good of the group without any exploitation for the benefit of one or the other. Group consciousness, which was the earliest form of conscious thought (de Quincey, 2005), likely began as the hypnotic communal consciousness of romantic love, as is explained in subsequent chapters.

Despite the unity of all, everything is always in flux, and there is no static equilibrium (Peat, 2005). The reality of quantum mechanic indicates that the infinite number of possible paths taken by everything and everyone does occur (Barbour, 1999). This factual theory therefore seems to imply that there is an infinite number of parallel universes that reflect the occurrence of all possible events. Within the context of an infinite number of these future parallel universes, everything possible will happen. Everything possible will not only happen but occur an infinite number of times. In addition, everything that could have happened in the past did happen in a parallel universe, although it may not be possible to know where those past parallel universes that one didn't experience would have led. Thus, any holistic understanding of the universe should be tempered by a recognition of the infinite and fluctuating variety which exists.

Chapter 11
The Development of the Mind and Happiness

Within the context of the existing cosmos, life evolved to create minds to coordinate the activities of each organism to optimally co-exist within the environment of the universe. Brains themselves developed from early organisms' primitive nervous systems, which evolved into more centralized control processes that initially merely reacted subconsciously to events (Rose, 2005).

In the process of evolution, a set of 13 chromosomes determining species' physical shapes mutated into 4x13=52, and after several superfluous chromosomes disappeared, other new ones evolved to create the 46 current chromosomes that are the raw material for human beings. Some of the new ones include a set which controls brain stem development and another which is responsible for the portion "of the brain that controls facial expression" (Hooper, 2006a). Although the human brain remains tied to a primitive reptilian brain (Begley, 2007c), people's minds have evolved to have numerous distinguishing characteristics. One is an innate moral sense that acts in the best interests of the community (Semeniuk, 2007). Others include complex problem-solving abilities and consciousness.

Problem-solving starts with trial and error learning that has developed into forecasting based on relationships observed between perceived causes and effects (Lisbon, 2007). Learning from experience can occur both consciously and subconsciously (Bernstein and Nash, 2008), and "making predictions and re-evaluating them seems to be a universal feature of the brain" (Huang, 2008). Learning from experiences is apparently even possible in the cells of plants (McTaggart, 2008), but learning in humans can be much more complex because their brains draw connections between pieces of information for recall, including through conscious thinking of how new information is related to other information already stored in the brain (Gynn and Wright, 2008).

While the subconscious processes information very quickly using very simple but effective decision-making rules (Gigerenzer, 2007), the conscious developed for more complex decision-making processes that require recall and analysis of information in a working memory which can help to make more optimal decisions, especially with respect to forecasting possible future events. Since humans' brain recall methods make the most sense if conscious memory evolved in order to forecast better the future (Marshall, 2007), it seems likely that predicting the future for decision-making purposes prompted the special development of human consciousness. The self-awareness involved in consciousness enables better forecasting of oneself, as well as of others through empathetically feeling what others would feel and predicting what oneself might do in the situations others face. The increasing effect that each person and other human beings have on the environment due to their forecasting abilities made the further development of self-forecasting and empathetically predicting others even more important. This conscious forecasting process enables even the manipulation of other organisms living strictly by their simple subconscious rules or tendencies.

A. The Minds of Others

Perspective on the evolution of the human brain's characteristics can be obtained by examining the advanced minds of other organisms. These minds are in many respects very similar to those of humans, as people are largely just a different type of evolved living being that is not particularly unique except in terms of relative intelligence, which already enabled them 1.8 million years ago to harness fire instead of fear it as other organisms do (Kenneally, 2008b). Just as 98% of human genes had already appeared 400 million years ago in the fish ancestors of humans (*NewScientist,* 2008zzn), including 2 genes for smell that mutated into the 400 that exist in humans today (Page, 2008), and including genes for controlling human fingers that evolved from the ability to control individual fin parts in a primitive form of fish (Hecht, 2007), human brains evolved from those of earlier organisms.

Extensive evidence of similarities between human minds and the minds of other organisms remains. For instance, there is significant evidence that human brains aren't even especially large compared to those of other primates, and even some organisms like bees with incredibly small brains exhibit extraordinary mental abilities (Motluk, 2010).

The brain systems of all organisms adapt based on environmental experiences that lead to reinforcing useful neural connections and firings, which themselves can change in response to new stimuli. For instance, the brain cells of a salamander can be ground up and then returned in scrambled form to the fish, and it will still be able to quickly learn to function normally (Douglas, 2009d). There is also a great deal of plasticity for some areas of the brain to take over new functions over time, such as the visual center taking on new tasks in blinded people, in a process of new neural connections replacing ones not used (Klein, 2003). Such basic fundamentals of brains and minds exist in humans as well as other organisms.

Humans' mental abilities to communicate between each other also don't make their minds unique. Whales have brains that apparently permit communication between themselves via noises (Coghlan, 2006), as do dolphins which actually combine sounds into meaningful groups of phrases (*NewScientist*, 2007zm). Mice also emit noises, which have at least 10 different variations but which have too high a frequency to be heard by humans, and those sounds are used by mice to communicate with each other (*NewScientist,* 2008zze). Although such noises are different than human verbal language, mouse sounds can actually be made similar to baby cries with the insertion of human versions of a pair of particular genes common to both that seem to affect muscles associated with speech (Callaway, 2009a). The FOXP2 gene, which produces a particular set of amino acids or chemicals that regulate the production of other proteins by other genes to allow the full pattern of human speech, is only slightly different from that of other animals like rats, which split evolutionarily from primates like humans 120 million years ago (Yong, 2008).

Bats make different ultrasonic calls that cause other bats to react in specific ways, although the reactions are stronger when the bats are in the same colony (*NewScientist*, 2010ab), thus indicating a clear ability to differentiate noises. Some birds use the complexity of their varied songs to attract mates (*NewScientist*, 2008zzf), and prairie dogs employ different sounds to announce the appearance of different colors and sizes (*NewScientist,* 2009zr). Monkeys can even detect differences in the voices of different monkeys for easy identification (*NewScientist*, 2008h) and can vocalize novel cries or sounds to signal or attempt to communicate something new (Spinney, 2008b). Baboons appear to have developed a fairly advanced form of oral communication and voice recognition (de Waal, 2007), and orangutans create their own customized sign language between themselves (*NewScientist*, 2007x) as well as even construct musical instruments (*NewScientist*, 2009zzzzr). Since bonobos kiss, even this seemingly special mode of human communication, which doesn't actually exist in all human cultures (Kirschenbaum, 2009), and which was first recorded in human history only 1800 years ago (Blum 2011), is shared by some nonhumans.

Communication between organisms thus represents a development that preceded human beings. All organisms, including fleas and algae, engage in 2-way communication via light photons, which they are constantly emitting, and to which they often react, including sometimes across species (McTaggart, 2008). Even plants communicate with each other (Dicks, 2007), and communication even exists across species (McTaggart, 2008).

Some animals, including rats, have memories (Williams, 2007), indicating that such a capacity is also nothing unique to humans. An early gene mutation for KLK8 that is responsible for memory and learning in both humans and mice may actually differentiate humans from other apes (*NewScientist*, 2007j). Although mouse learning and memories does seem to be similar to that of humans in terms of being dependent on a mutatable gene that interacts with a brain enzyme to make a particular protein (*NewScientist*, 2008zg), mice appear to have developed their learning and memory systems separately from humans, as many species have. The human form of the memory gene permits the production of neuropsin II, which has 45 additional amino acids not existing in other mammal neuropsin (*NewScientist*, 2007j), thus making humans somewhat unique in this respect Nevertheless, since apes have been found to trust the implications of their memory less when more time has passed (*NewScientist*, 2010o), the human ability to recognize the imperfections of memory, or at least the uncertainty in the universe that can cause unobservable changes, isn't unique.

The cortex, which is involved in integrating large amounts of information in more advanced brains that have memory and can learn, and which forms the basic processing area for human thinking, apparently originated 600 million years ago in a type of worm that may have enabled the rapid development of complex and diverse organisms shortly thereafter (Marshall, 2010a). The brain frontal cortex that enables an organism to understand the consequences of actions and the ventromedial prefrontal cortex that gives an individual self-awareness (Hamelzou, 2010c) may have advanced thinking greatly over the last few hundred million years, but the basic processes are widespread among many species.

Tool usage, decision-making, and even creativity are not unique to humans (Kenneally, 2008b). For instance, dolphins use sponges "to dislodge prey from sand", and even octopuses often carry around half shells for protection against predators (*NewScientist*, 2009zzzzr). A few animals like chimps actually create weapons like spears (Hooper, 2007a) that they can throw (McClenon, 2002). Chimps also engage in multiple-staged tasks that require the use of several tools, as when they access honey from beehives (*NewScientist*, 2009zza). Orangutans can even create and apply herbal medicine to themselves (*NewScientist*, 2008zo), providing an indication of the ability to produce health care "drugs".

Complex tool utilization that includes learning to adapt new implements to new tasks occurs among some birds as well as primates (Kenneally, 2008b). For instance, crows have learned to crack open desired nuts by letting cars run over them and even utilize pedestrian walk signals to navigate through traffic (Blum, 2009b). These birds, which can even recognize and learn to react differently to the faces of different people (*NewScientist*, 2010i), appear able to optimally select tools in multi-step processes, although there is no evidence that they actually engage in planning and forecasting (*NewScientist*, 2009zzza).

Monkeys, which diverged from the evolutionary path of humans 25 million years ago, can even learn to utilize out-of-sight tools via observing a television screen that shows a live video of them (Spinney, 2008b). Monkeys also engage in at least implicit "trading" among themselves, offering grooming for food, and even have the capacity to naturally adjust the social "price" of food in terms of the amount of grooming in different situations that reflect the relative scarcity and competitive position of the sellers (*NewScientist*, 2009zzs). According to Begley (2007d), rats can actually think and make decisions based on probabilities of the success of their own actions. Even fruit flies may actually make rational choices (*NewScientist*, 2007m). Decision-making of course requires an ability to discern variation in the external environment and to react differently to changed situations, but even the simplest bacteria can do that (*NewScientist*, 2008b). Most animals seem to even have some ability to distinguish between different numbers of objects and make decisions based on that evaluation (Callaway, 2009b).

Some animals display behavior that is even consistent with conscious analysis as opposed to just instinctive reactions or copying (Barras, 2008). For instance, some primates have a limited working memory (Coghlan, 2008a) that may represent a form of consciousness (Kenneally, 2008b). However, since memory can be activated subconsciously for some problem-solving tasks, reasonably complex decision-making such as working crossword puzzles can occur without any conscious processing or recall (Bernstein and Nash, 2008), and so the existence of a working memory or decision-making capacity doesn't necessarily imply consciousness. For instance, birds can recognize themselves in a mirror even though they don't have the neocortexes in their brains (*NewScientist*, 2008zb) that appear to be involved in human consciousness. Thus, the evidence on the existence of consciousness outside of humans remains mixed (Kenneally, 2008b).

Animals do experience some unusual mental states that are similar to humans. For instance, birds and mammals experience all the dream states of people (*NewScientist*, 2008zi). The animal play-dead reaction to danger that apparently evolved to enhance survival in the face of predators seeking live moving bait are similar to the hypnotic state of mind that exists in humans, and animal hypnosis can be similarly prompted through repetitive behavior, such as through the repeated stroking involved in primate grooming (McClenon, 2002). Some primates, the tree shrews, which consume a fermented nectar found in nature, apparently get drunk, although they seem to have evolved a better tolerance for their alcohol than humans (*NewScientist*, 2008zq). Dogs also appear to have psychic powers, and there are widespread animal mating rituals that have hypnotically relaxing effects to reduce anxiety (McClenon, 2002).

In addition, animals exhibit the physical signs of many of the same emotions as humans, although they do not necessarily consciously feel them (Klein, 2003). The widespread evidence that many animals experience empathy, joy, grief, gratitude, love, and awe (Bekoff, 2007) is provided by "strong similarities in physiological responses" across organisms (Eagleman, 2008). Many of these emotions apparently existed in reptilian brains hundreds of millions of years ago, although some of the newer feelings such as love developed only with the need to care for offspring, as is promoted by the brain chemical oxytocin that also motivates security-enhancing group cooperation, which is further rewarded chemically with endorphins (Klein, 2003). Dogs, which appear to have been first domesticated in the Middle East (Aldhous, 2008b), and which evolved from wolves after the initial domestication 10,000 years ago to live with humans, appear to have developed some of the most similar human emotions in order to be able to adapt to their environment that consists of co-habitation with people (Douglas, 2008).

Even the empathy required in deception is possible among animals (Kenneally, 2008b). For example, squirrels seem to recognize when they are being watched and therefore use varying levels of deception when burying their nuts (*NewScientist*, 2008d). Male chickens make sounds indicating the existence of food even when there is none in order to attract female chickens (Blum, 2008). The fact that most birds exhibit some of the same human tendencies for monogamous sexual relationships which are often combined with occasional polygamous sexual intercourse in processes that appear to be motivated by the complexity of human love (Fisher, Aron, Mashek, Li, and Brown, 2002) implies that those organisms may even experience the complex emotions of love.

Moreover, orangutans appear to have virtually all the human emotions, even smiling and crying, as well as make complex decisions (Schwartz, 2005). The emotions may actually be essential in advanced decision making (Klein, 2003). Even human reason itself generally doesn't operate independently of emotions (DeLaurence, 1902), and it has even been proven that a human being can't make decisions without emotions (*NewScientist*, 2007i). Thus, it isn't possible to assert that humans are different from animals because they can live "rationally" without emotions or instincts.

However, not all animals possess the full range and complexity of emotions that humans have. For instance, although selfless helping of others or altruism appears to be programmed by evolution through the rewarding of such behavior with brain pleasure chemicals in some species, including some primates like chimps and bonobos (Schultz, 2007a), there are differences. For example, chimpanzees won't help other chimps unless asked to do so, suggesting that chimps have a lack of understanding of others' desires (*NewScientist,* 2009zzzu). In addition, chimps don't display empathy to non-relatives and won't help unrelated chimps irregardless of whether such help to non-relatives is costless and effortless even when they know each other and are raised together (Nettle, 2007). Some primates like baboons show no empathy whatsoever (De Waal, 2005).

Evidence has been found of ants recognizing the fact that their bodies are nearer to death and risking or sacrificing themselves for the good of their community (Holmes, 2007). However, such insects, which are motivated by the same endorphin pleasure chemicals as humans who also internally produce them in response to biologically useful endeavors (Klein, 2003), hardly have the developed brains and emotions of humans. They are more like slaves or programmed robots in comparison to people who can consciously choose to disregard pleasure incentives whereas other organisms are totally controlled by such stimuli-response systems.

Humans also seem to be unique with respect to their sense of humor, which reduces stress, creates dopamine pleasure chemicals in the brain, and acts as a "social glue" (Douglas and Robson, 2010). While some primates have developed a sense of humor based on slapstick comedy, it apparently doesn't approach the level of complexity of the human sense of humor (Kenneally, 2008b). Given the extremely contagious nature of human laughter, especially in stressful situations, and given that laughter enhances social bonding (Douglas and Robson, 2010), the peculiar aspects of human humor may have originally developed out of a need for humans to bond together to deal with stressful situations. By relieving stress, humor enables people to calm down a "fight or flight" reflex in the face of danger, thereby enabling rational evaluation of solutions to the situation. Humor also rewards with a pleasant feeling the recognition of errors in decision-making that enables learning via a conscious judging of such errors. The strength of humor in a group setting implies that it originally evolved in the context of a group consciousness, which itself reduces danger because of the enhanced power of a bonded group compared to numerous individuals acting alone, and humor promotes this functional state in an especially productive fashion.

Humor thus enables improvements in humans' adaptation to and control of the environment. While judgmental errors, including those based on both conscious and subconscious thought processes, create stressful negative judgments, humor offsets that negative effect and even provides a pleasure reward for consciously learning from one's own mistakes as well as those of others. Such a capacity for self-aware judging and learning from errors is beyond the abilities of other organisms that act more like manipulatable robots.

Non-human organisms on earth are as manipulatable as robots because they appear to operate on a subconscious system of robot-like programming. The subconscious rules that other organisms employ to govern their activity include extremely efficient ones that enable changing the decision-making procedures to adjust to the environment. However, such organisms don't appear to have the ability to consciously adapt their own programming, and so they largely just react to their situation without analyzing future consequences. They therefore are also largely incapable of consciously predicting others or themselves.

One of the primary differences between humans and other live organisms may lie in people's ability to consciously forecast that enables more complex thinking. Although some animals do engage in behavior that indicates preparation for the future, such as burying food for future finding and eating, there is no conclusive evidence that apes, birds, or other animals can think beyond the present moment (Nicholls, 2008). Non-human organisms react currently to stimuli, including internal ones relating to memory and genes, as well as external ones, just as human beings do (Bernstein and Nash, 2008), but they can't consciously forecast and plan for the future. The fact that even tools employed for current use by apes are typically discarded instead of held somewhere for future re-use (Nicholls, 2008) provides strong support for this hypothesis of human conscious forecasting abilities making them different from other organisms.

Although one 30-year old ape in a zoo has, after over a decade of throwing rocks at some zoo visitors, been observed to stockpile rocks in apparent preparation for future throwing, even that unusual case doesn't necessarily indicate conscious planning for the future (Geddes, 2009c). The latter behavior may have been initiated through random or playful activity by the ape to rearrange or put stones in a particular place, and the location was later recalled for easier throwing them at bothersome observers. The process involves an automated reinforcement via pleasure chemicals for both the memory of the location and for the actual act of piling up the stones, and it doesn't in any way indicate any conscious planning. Crows' ability to select an optimal tool to extract another tool for ultimate use to obtain a reinforced stimulus (*NewScientist,* 2009zzza) may reflect a similar unconscious process that enables actions for future benefits that aren't perceived in any consciousness.

The seemingly complex behavior indicative of the ability of some animals' brains to engage in tasks that unconsciously prepare them for the future without any actual conscious decision-making or realization of why the task is performed probably represents a step in the process through which human ancestors long ago passed. Human forecasting abilities probably developed originally from that unconscious process which effectively enables subconscious planning for the future.

Human self-forecasting evolved from such predicting of the future as humans and the environment became ever more integrated, as was possibly prompted initially by some observed repetitive relationship between humans and external events like women's reproductive periods that are tied to the phase of the moon. For optimal predictions of the self, it was necessary to develop an integrated sense of self that coordinated all the relevant information from memories and the environment in order to predict how one would behave and what effect that would have on the environment. Predictions of one's own behavior and the effects on the environment enabled improved subconscious decision-making that caused human beings to affect the environment ever more. As a result, the predictions of oneself that affected the future environment would also impact one's behavior farther into the future, and so the self-forecasting had to develop into an analytical process of predicting one's own decision-making. Consciousness thus integrated self-awareness, forecasting, and decision-making into one coherent brain wave that included inputs from the subconscious via feelings and intuition.

For maximum power, the conscious was given control over decision-making that was based on simple if-then logic utilizing a Bayesian learning process. Consciousness was even provided with some capacity to affect internal brain functions, such as with respect to filtering subconscious inputs to the conscious through the thalamus. However, human beings at this prehistoric time remained weaklings subject to being eaten into extermination if their cooperative nature were consciously willed away, and so the conscious was provided with strong incentives to always act in the best interests of humanity as a whole. In particular, positive feelings of happiness and self-fulfillment resulted from decisions that were made for the benefit of humanity.

Even when decisions were made that turned out to be suboptimal, the pleasure from humor and learning from those mistakes enabled overcoming any feelings of any unhappiness and feeling of empty worthlessness that resulted from failures as long as the intent of the decisions was for the good of humanity. As a result, as long as one was trying to contribute positively to humanity, simple pleasures relating to satisfaction of human needs and desires also continue to be rewarded with brain pleasure chemicals as well as feelings of happy self-fulfillment. In contrast, mere selfish hedonism fails to reward the brain cells involved in the integrated self-awareness process of consciousness with feelings of happy self-fulfillment. A feeling of emptiness therefore results from acting hedonistically that is so consciously negative that it more than offsets the restricted chemical reinforcement for pure physical pleasures, which are required in ever increasing amounts to be satisfying enough to offset the empty feeling in consciousness from acting purely selfishly.

Human beings are thus unique in being motivated by a feeling of happy self-fulfillment instead of just temporary hedonistic satisfaction of needs and desires. In fact, human beings are even motivated to sacrifice their own lives for the good of humanity if they think doing so is optimal for humanity, as with martyrs. They can also suppress the sex drive if they believe that is optimal, and if they have a feeling of self-fulfillment from trying as best as they can to serve humanity as a whole.

Humans' ability to consciously make and recall memories in the process of complex decision-making and forecasting for the benefit of humanity is also unique. Human memory is in fact a useful "tool" in making successful predictions for the future (Eagleman, 2008) that are facilitated by conscious recall (Bernstein and Nash, 2008). Human memory consists not only of an episodic form based on prior experiences but also a semantic type that derives from what an individual has consciously figured out, learned, and stored in memory (Douglas, 2009e). Without a conscious self-awareness, other organisms can't obviously be aware of their memories.

Possibly also related to the unique human characteristic of being self-aware of their own memories is the tendency and capacity of humans to teach other humans, especially their children. In particular, unlike simple behaviors that can be learned by mere copying, the complexities of forecasting and planning for the future by learning from complicated past events cannot be learned by mere mimicking behavior. It must instead be taught in some communication process that eventually evolved into complex languages to explain the concepts. Although apes can learn by copying, they don't teach their offspring or others (Brooks, 2008h), and their communication abilities remain primitive (Burling, 2004). The unique human abilities to teach complex ideas associated with planning and the related necessary predictions further increased the need for a consciousness to predict humans as well, and the empathetic cooperative nature of human beings resulted in that self-awareness originally being a sort of group consciousness.

As humans expanded the powers of their consciousness and the communication that facilitated their decision-making capacities, the size of the human brain grew over millions of years by the force of evolution. The greater success in controlling their environment that was caused by the resulting improvement in humans' decisions derived from self-forecasting increased the evolutionary incentive to enhance such predicting. A larger brain was needed for the conscious self-awareness that enhanced decision-making abilities in a cycle that fed on itself. Humans' continued cooperative nature, which greatly enhanced the power of the species, also required increasing thinking capacity to enable forecasting other humans who were also becoming increasingly complex and required ever more complicated language to communicate the complex ideas to each other. This process further promotes more increases in brain size via the force of evolution that continued to enable expanding forecasting and conscious analytical capacity.

One hypothesis for the origin of human forecasting and expanded intelligence is the tie-in between women's menstrual cycles and the phase of the moon (Shlain, 2003). Such a recognition allowed human ancestors to prepare for the monthly period when scavengers and predators would be attracted by women's blood, as explained in Part I. The preparations and reactions to that cycle were more complicated than mere dopamine reinforcement of a simple behavior like gathering stones for throwing. The ambush of animals smelling women's blood would have required cooperation between human beings that relied on large parts of the brain for effectiveness. There was likely a chemical reinforcement for the integrated neural firings that were necessary for planning such actions that represented the beginnings of consciousness. The mere recognition of a relationship between the stage of a menstrual cycle and the phase of the moon necessitated an integration of brain cell activity with respect to both the internal situation and the environment. A recognition of such patterns and reacting to their future cyclical effects would have naturally developed into a reflective self-awareness that reinforced the neural firings involved because it led to positive stimuli in the future, such as with respect to decisions to hide in trees when a specific phase of the moon neared that was associated with menstrual cycles and the related attraction of predators. The rewarding of the firing of the brain cells involved in creating the electromagnetic waves of the consciousness expanded planning, likely eventually resulting in ambushes for the predators. The predators likely developed a distaste for the human scent as a result by much slower evolutionary forces associated with the inheritance of genes of those predators that preferred prey less dangerous than humans and were thus more likely to survive.

Originally, the recognition of predators coming near during menstrual bleeding might have been observed to be related to the phase of the moon. As the moon phase approached that correlated with increased presence of predators, both male and female human ancestors may have learned to be in or near trees for safety purposes. However, at some point, someone may have learned that the animals were only attracted if there were females bleeding during that particular moon phase, and the cause of that attraction might have been figured out to be the menstrual bleeding. Such a relationship between an individual's or group's internal situation and the environment would certainly have motivated a self-awareness that would have rewarded recognition and communication of such information with higher social esteem and greater mating opportunities. The larger brains in the new species of human ancestors called Homo Erectus, which were able to create more advanced tools like symmetrical hand axes, which may have developed the earliest forms of protolanguage, and which appeared almost 2 million years ago (Fitch, 2010), might have both enabled and been spurred by such developments.

This initial step in consciousness may have occurred in Homo Habilus. This predecessor of the Erectuses, which themselves originated about 1.8 million years ago, had already been developing tools that could be used to scare off predators and kill scavengers that competed with Habilus for scavenged food. After group planning for menstrual cycles began, the first group hunting activities may have easily followed. Group cooperation would have been facilitated by communication between the members of the group, and the development of a group consciousness may very well have occurred simultaneously through communicative chanting that helped maintain the courage of all to stand together to kill both predators and scavengers. Such cooperative behavior led to greater nourishment and safety for all, while greater mating opportunities were likely available with those most successful in forecasting and innovative actions, and so the group consciousness was concurrently and evolutionarily rewarded.

The self-awareness involved in self-forecasting was reinforced not only because human nutrition and security were directly enhanced by perceiving new patterns and relationships between the environment and human behavior, but also because it facilitated predicting the environment that was affected by current and future forecasted human activity that would result. Increased consciousness therefore spurred learning that prompted further evolutionary development of the brain. The first specially revered "magicians" may have arisen out of such innovation and been rewarded with sexual benefits to motivate the development of such skills as well as pass more of the genes associated with them to the offspring of successful "soothsayers".

As this process developed, there was the incentive for humans to develop ever better weapons, including fire. Humans' control of fire that likely originated about 1.8 million years ago and contributed greatly to humans' development, as explained in Part I, may have coincided with the evolution of consciousness. Harnessing and use of fire for various purposes required rather complex thought processes, and it might have originally even been used as menstrually bleeding women gathered near it to protect against predators. Eventually they learned to harness fire more deliberately, by either figuring out how to start a fire or keeping an existing fire going during the days of menstrual bleeding that could be predicted by the phase of the moon. At any rate, the forecasting and complex thought processes needed to harness fire 1.8 million years ago seems to indicate an origin for consciousness around that time or before.

The harnessing of fire allowed human beings to leave the forest permanently, as they could utilize the fire as a weapon that could be kept permanently available for campfires. Such campfires also enabled moving to cooler areas of the world, as well as consumption of a greater number of calories from meat because digestion of it in human stomachs was facilitated by cooking it. Such cooking may have initially been spurred by humans spitefully eating predators after fighting them off with fire and learning that heated meat was easier to chew and digest. While all these developments required expanded consciousness, the meat in the diet itself provided the calories and fat necessary for increasing the size of human brains to enable that expansion of consciousness.

Given that vitamin B12 is vital for human brain development even in the early months of gestation (Williams, 2009), as well as later, and given that this vitamin essential for many other biological processes in humans can only be obtained naturally from eating animals or fish (Sizer and Whitney, 2003), it would appear that the special nature of human brains and consciousness is actually dependent on them eating flesh. This fact is consistent with a hypothesis that a diet including flesh actually helped prompt the evolution of the special characteristics of human nature, or at least coincided with its development. Strict vegetarianism is therefore both unnatural and likely very unhealthy for human beings, although a meatless diet can be artificially supplemented with vitamins to enable life to continue.

The hypotheses listed in Part I of this book with respect to humans figuring out how to successfully fight their predators millions of years ago with weapons and other means supply an explanation for such co-development with the human trait of frequent meat eating. Gynn and Wright (2008) provide a feedback loop that would have caused initial meat eating to reinforce itself, especially if it weren't balanced by sufficient fruit eating, through enabling more internal testosterone released in the human body that motivates more aggression, which leads to more hunting, which leads again to more testosterone, aggression, and hunting, etc. Since adding meat to the diet tends to increase the amount of iron individuals have (Shlain, 2003), and since more iron in the human body enables greater psychic abilities (DeLauren, 1910), meat eating may have contributed to human evolution by both increasing human aggressiveness and psychic abilities, as well as expanding human brain power in general.

In any event, the travel to new environments outside African jungles about 1.8 million years ago motivated further increases in human intelligence and creativity. In particular, humans' increasing impact on their new environments motivated better self-forecasting for adapting to the changes in the environment that they themselves were causing.

The conscious self-awareness and decision making that expanded the adaptability of human beings therefore isn't anything magical. The actual process merely involves more coordinated electrical firings of brain cells across larger sections of the brain that enables an integrated understanding of the internal situation for the individual as well as the external environment. This situation is experienced as an overall feeling that can be labeled with words in a language. The integrated system of neural firings that cause conscious thoughts utilizes the same Bayesian learning process to make decisions as individual brain cells do, whereby probability estimates are formed for firings being rewarded with chemicals like dopamine in response to different stimuli and combinations of stimuli, such as from various other brain cells firings. By following that Bayesian learning process, which enables adjusting those probability estimates based on new experiences, individual brain cells learn to fire in synchronized fashion with other neurons across the brain in order to be rewarded when the reflective self-aware experience leads to decision-making that results in an increased ability of human beings to survive and thrive.

The actual decision-making process that utilizes this self-awareness is based on the same logical procedure followed by individual neurons in determining when they fire. In particular, when the probability estimate of a certain neural firing leading to a dopamine reinforcement exceeds a particular level, the neuron fires, and the same goes for the entire system of coordinated neural firings that cause conscious thought and deductive decisions based on that thought.

For optimal efficiency, the logical decision-making capacity of consciousness evolved to be concentrated in the left side of the human brain, while the emotional subconscious is more concentrated on the right side. The two sides remain connected and integrated, as the "rational" processor on the left receives emotional signals from the right. The integration of the two sides of the brain halves enables conscious experience of feelings that can permit emotions to influence decision-making, while also allowing conscious thought to affect subconscious processes.

Although many other organisms, including reptiles as well as apes, have specialized tasks for each side of their brain, the splitting is most pronounced in humans, whose conscious experience of self is concentrated in the logical left side where an awareness of the past and future exists (Gynn and Wright, 2008). Other organisms may only process decisions in the present there. Many other species similarly use inputs from the emotions to enhance decision-making, as well as some type of memory of connected past events, just as the human conscious does but without the added insights of logical forecasting and conscious self-awareness.

Another significant difference between humans and other organisms that could have contributed to the large difference in intelligence capacity consists of people's unique combination of traits which include both spiteful and altruistic tendencies that facilitate humans' individualistic yet cooperative natures. While empathy and chemical reinforcement in the brain may promote altruism, human spite collectively helps enforce cooperation through a non-authoritarian system of individualistic punishments of selfishness. This complex combination may have spurred humans' logical thinking capacities, as the humans' own cooperative nature made other human beings a vital aspect of their environment. In particular, their own self-forecasting of actions would be heavily affected by whether they decided to let feelings of spite or cooperative tendencies dictate actions toward other human beings. Greater self-awareness was thus necessary for purposes of making such predictions that would strongly influence the environment. Such self-awareness also expanded the capacity for empathetic forecasts of others' reactions to the person's own present and future actions that had to incorporate a prediction of what the individual thinks others believe the person will think and do. This interactive complexity and uncertainty merely increased as humans developed greater intelligence and ability to empathetically forecast others, albeit in conjunction with subconscious processing that enabled efficient evaluation of massive amounts of consciously and unconsciously perceived information via rapid algorithms to deal with complex and fluid interactions.

Empathy motivates somewhat different behavior than spite, but the high level of human empathy may have at least partially developed out of the usefulness of feeling what others feel in predicting the spite of others in various situations. People interacting in groups with other individuals displaying varying actions and reactions to their own environmental stimuli needed to develop a method of forecasting both themselves and others in order to optimally interact in those groups. A conscious self-awareness and empathetic decision-making capacity was an ideal tool for that purpose. Since other organisms on earth don't have to deal with non-adaptive social problems, they never evolved a consciousness or ability to predict the future.

As previously mentioned, some organisms have altruistic tendencies similar to those of humans (Fanelli, 2008), but most aren't as complex. For example, in some species such as chimps which provide care for others only if they are genetically related, the altruism is more limited than for human beings, who, even as children, naturally exhibit empathy for all other humans, (Gigerenzer, 2007). However, rats sometimes demonstrate more general altruism that is similar to humans, such as helping other unrelated rats to get food with only a possible, unreinforced hope that the kindness will be repaid even though it never has been (*NewScientist,* 2007t). Rats also sometimes refuse to accept a benefit if it results in another rat being penalized (Blum, 2009a).

Bees also display altruism, including across species when they pollinate flowers, but the cause seems to be their enslavement to their sexual drive that is manipulated via chemical pheromones which the plants emit that create scents similar to those of female bees (Rathaus, Nevid, and Fichner-Rathaus, 2008). In a similar vein, ants are fooled into feeding the larvae of a particular type of caterpillar by the larvae having evolved to make sounds which are similar to those of the queen ant that the ants obediently serve in an unquestioning relationship analogous to worship (*NewScientist,* 2009q). Within another insect species, normally loner locust insects herd into swarms when levels of serotonin in their brains spike, and the insects remain herded even when serotonin levels drop (*NewScientist,* 2009n). Thus, various simple chemical processes, which may be similar to the human body's production of serotonin from foods like bananas and avocados that is then converted into melatonin in the pineal gland, thereby stimulating the production of oxytocin to motivate social bonding (Gynn and Wright, 2008), induce seemingly altruistic behavior in non-humans. However, it is strictly an automated evolved response as opposed to being based on an emotion-influenced conscious choice that the individual can freely decide to vary.

Human spite may also be based on brain chemicals and neural firings that exist in species other than humans. For instance, wasps instinctively deliver extra punishment to other wasps if they are deceived about their true strength via visual perceptions (*NewScientist*, 2010aaa). While the combination of human altruism and spite may have had similar beginnings to animals and insects based on basic electromagnetic forces, human nature has evolved to encourage cooperation on a much more complex level, including via conscious human empathy and forecasting of others.

Much of the empathy involved in animal behavior seems to be related to the same brain neurons firing when another animal performs an act as fire when the animal itself performs the act, thereby enhancing performance learning, but human empathy appears to be focused in the areas of the medial frontal and temporal cortices involved in planning, controlling, abstract thinking, and memory (*NewScientist,* 2010n). This difference provides some evidence that the thinking portion of the human brain may have developed at least partially to evaluate empathetic feelings. Those feelings would not only help human ancestors forecast the behavior of others but also enable weighing the relative benefits of acting with empathetic care or vindictive spite for perceived injustices against oneself, others, or the group as a whole.

Spite itself may at least partially derive from empathy, insofar as any perceptions of exploitation may cause an individual to empathetically feel the exploiting entity has contempt for the exploited. Contemptuous exploitation of any human being naturally creates negative feelings for the exploited that motivates treating the exploiter like an inhumane predator which must be taught via vindictiveness not to prey on people again. Exploitation of a person may also be resented as an injustice to be spitefully redressed because it violates innate human nature of egalitarian cooperation, insofar as it makes the exploiter feel superior to others while the exploited feels like an inferior idiot and not much different than ignorant animal prey. The reputation of all is therefore at stake, as allowing an exploitative act to occur without spiteful reactions may be perceived as acceptance of a hierarchy against which humans naturally revolt.

Feelings of spite apparently cause people to resist simple chemical rewards for cooperative behavior when there is a perceived injustice. Human brain production of chemicals like dopamine upon helping others does make altruistic behavior pleasurable (Begley, 2007c), and increased levels of serotonin, which are inhibited for offspring raised in isolation (Gynn and Wright, 2008), also contribute to greater cooperation among humans with no change in mood despite exploitation by others (Clark et al., 2008). However, humans appear to be complex enough to overcome such incentives to cooperate. The impact of chemicals promoting cooperation can indeed be overridden by feelings of spite that arise anytime there is a violation of the natural human egalitarianism which existed for millions of years.

Thus, although brain chemicals and empathetic feeling what one perceives others to feel contribute to herding behavior in humans, spite prevents it from being a slave-like obedience. Spite actually facilitates cooperation since it inhibits exploitation of the natural urge to want to cooperate by supplementing the chemical and empathetic incentives for cooperation with risks of punishment for failure to do so. Cooperative behavior is thus encouraged further in a manner that allows for some independence in actions and thus permits individualistic and creative thoughts.

The feeling of spite actually lights up the striatum area of the brain associated with rewarding experiences (Whitfield, 2009). It is assumedly therefore reinforced chemically. Humans may be the only species that exhibits spite (Barras, 2007a), at least with respect to taking punitive action against those who act selfishly and inconsistently with the welfare of the group or species as a whole.

While apes sometimes demonstrate a primitive sense of justice by effectively going on strike when an alpha male provides no reward for their obedience (de Lange, 2009), these actions seem to reflect a simple subconscious Pavlovian response. Such apparent evidence of spite in non-human primates may actually stem simply from the fact that a particular behavior isn't rewarded and therefore isn't repeated. Chimps exhibit some evidence of reciprocal treatment, such as sharing food only with others who have shared with them (Klein, 2003), but this behavior also seems to result from the same Pavlovian cause. In particular, when the sharing isn't rewarded, it isn't repeated.

Although some animals like monkeys with a hierarchical group culture impose punishments for failure to share food (Angier, 2009), spiteful revenge or actual direct punishment of another for an act that is perceived unjust to oneself or others that actually harms the punisher seems to be unique to humans (Whitfield, 2009). For instance, monkeys have demonstrated feelings of frustration and inaction after receiving worse treatment than others receive for the same actions, implying that they have developed some sense of fairness (Buchanan, 2005). However, such behavior may truly stem from empathy motivating expectations of a reward similar to that of others for the same deed, so that when the actual act isn't reinforced as expected, the incentive to have repeat the action dissipates.

Pavlovian responses to adverse stimuli may have formed the foundation for human spite, but humans' sense of justice that creates the unique spite of human beings is far more complex than a simple subconscious stimulus-response reflex. For instance, humans' sense of justice seems to be heavily based on intent, insofar as exploitation of oneself or others is considered unjust, but inadvertently hurting oneself or others is typically judged as an innocent mistake. An evaluation of which case holds requires more than a simple subconscious response and may therefore have contributed to the development of conscious analytical abilities that relied on empathetic self-awareness and forecasting what oneself would justly do. Consciousness also facilitated optimal decision making with respect to whether an individual should act out of empathetic love or spite for the good of the group as a whole. That very system of rewards and punishments for others helped avoid selfish exploitation of individuals or the group, thus promoting a group consciousness where everyone empathetically sought to help each other unless there was evidence of someone acting selfishly or malevolently.

Human spite appears to have developed out of a combination of the more primitive sense of justice that other primates experience and their tendency for reciprocal benevolence in treatment. Combining those two tendencies with a forecasting of others' reactions to one's own actions likely led to the development of spite, as humans learned that punishing another for unfair treatment at some point would result in others being more likely to act fairly to avoid such punishment for themselves in the future. At the same time, behavior that was coherent with others' needs and wishes was rewarded. The forecasting capacity of humans for feeling empathy eventually led them to evolve genes that promote cooperation with other cooperating humans to create the unique combination of spite and altruism that has proven to be so successful in the human species.

In particular, people most often feel spite in situations of perceived injustice, such as when others don't display reciprocal empathy. Human empathy leads to altruism unless others don't display similar empathy and hence altruism, as spite results when there is a perception of unjust exploitation. Spite might even partially stem from a forecast of cooperative behavior that isn't rewarded and therefore causes adverse feelings of disappointment that leads to anger against the cause of the problem. The spiteful revenge inhibits recurrence of injustices.

Spite itself represents a form of altruism called altruistic punishment which is essentially suboptimal behavior from a self-centered point of view that penalizes non-cooperative behavior for the benefit a group of people (Whitfield, 2009). Empathy and spite may merely be two sides of the same altruistic coin that is human nature. The combination of empathy and spite may have enhanced the social group cohesion that was necessary in order to survive a dryer environment that resulted in Africa several millions years ago from weather changes and forced humans to come down from the trees into the savannah that contained faster and stronger predators.

The combination of spite and empathy may very well have been a very important factor in causing human intelligence capacities to evolve to be so strong. In particular, the resulting complications associated with determining how to optimally balance the two feelings, along with the need to try to predict others' decisions on this choice, would create great evolutionary pressures for expanded analytical capacity to deal with these issues. Since feelings of spite and empathy sometimes conflict and therefore lead to inconsistent behavioral choices, a logical processor like the conscious that could "rationally" weigh each along with forecasted future repercussions was no doubt spurred by evolution. While greater social interaction among animals tends to promote an increase in the size of the brain and the thinking cortex in particular (Adovasio, 2002), the extreme complexities of human interactions that optimally involve an unpredictable mix of cooperation and spite (Gigerenzer, 2007) may have created the impetus for the evolution of humans' superior thinking capacity. An "ability to see the big picture" is required for recognizing injustice and determining when to replace cooperation with revenge (Whitfield, 2009). This capacity differs from the simple stimulus-response systems of primitive animal cultures which reflexively punish transgressors against their genetically ingrained norms. However, human spite may have originated from an early simple reflex which evolved to promote cohesion in animals operating in groups, but it later evolved with human intelligence into the complex form of spite that characterizes human beings.

The mix of empathy and spite that characterizes human interactions results in "game" dynamics that make application of simple rules, such as those of the subconscious, easily manipulatable by others who can consciously predict them (Gigerenzer, 2007). Empathy by itself is certainly important in forecasting the behavior of others (Bernstein and Nash, 2008), but it was the complex games between empathetic and spiteful humans that likely resulted in the evolution of ever greater human intelligences (Gigerenzer, 2007). While it has been shown that more intelligent individuals of other species die out sooner than less intelligent ones of the same species (*NewScientist*, 2008zd), humanity's mix of empathetic altruism and spite helped ensure that the more intelligent humans would survive and procreate, leading to ever greater human intelligence.

A combination of empathy and reciprocity not only necessitated the development of "floodlight intelligence" that "combines many cognitive inputs over time" (Whitfield, 2009), it also seems to have promoted an analytical ability of humans to forecast, including with respect to predicting punishments for purely selfish, non-cooperative, and unjust behavior. Such forecasting was enhanced by empathy, which enabled each person to predict others to act as the individual would, and such empathy-aided forecasting motivated more analysis of oneself in order to predict others. That self-evaluation resulted in ever greater self-awareness or consciousness, and self-forecasting may have co-developed with this type of analysis. Forecasting the complex interactions between people over time can be especially complex with respect to predicting one's own spiteful response to others' possible spiteful response, which itself may stem from one's original spiteful response to some selfish behavior of another person. As previously explained, the growth in human intelligence and forecasting abilities co-developed further with self-predicting. This self-forecasting evolved into and with the human conscious self-awareness that is truly unique in its complexity.

The natural human tendency to try to discern a cause behind all events, including phenomena that may have seemed explainable only by a hypothesized spirit or assumed soul in all things (McClenon, 2002), would certainly have been a useful component in the development of the forecasting abilities that provided humans with an evolutionary advantage. Learning from perceived causes and effects of past events requires logically figuring out hypotheses on reasons for the actions of all things.

Empathy, which enables development of hypotheses on the future actions of others by empathetic simulation of what others feel, is very important for forecasting the behavior of an entire species of organisms that have, like humans, evolved to such an extent that they significantly affect their own environment. When such learning utilized to forecast and make decisions reaches a certain point where individual or groups of organisms begin to exhibit a large degree of control over the environment instead of merely reacting to it, self-forecasting becomes evolutionary vital for predicting and decision-making.

Human empathy therefore likely developed in conjunction with human predictive capacities, as it would have helped considerably in trying to predict the actions of others and other things. Human empathy begins at an early age and is felt for all things and not just fellow human beings (Bernstein and Nash, 2008zi). Attempts to predict the behavior of other creatures or even other things has to be founded on some analytical framework, and, lacking other information, forecasts based on an assumption that other organisms feel and therefore would act as one's own self may be optimal. Therein may lie the evolutionary foundation of the universality of human empathy.

Thus, the far superior intelligence and decision-making capacities of humans may be attributed to humanity's unique combination of empathy, which promotes cooperation and learning, and feeling of spiteful justice, which inhibits exploitation of human beings. The development of these abilities may have stemmed from the very weaknesses of humanity in terms of strength and speed in comparison to predators that forced humans to creatively cooperate in groups in a manner which appears to have been optimized by evolution into their unique combination of altruistic empathy and spite. Social animals like human beings naturally evolve value systems that enable them to function in groups, and the analytical neocortex portion of the brain responsible for predicting others (Church, 2010) appears to have advanced far more in humans than in other animals because their relative physical weakness required more adaptive solutions that mandated both greater cooperation and innovation.

As human thinking abilities grew, so did the evolved size of the brain. The increase in the size of the human brain over time seems to have promoted a growth in burrowing of the fetus inside the mother's uterus (Coghlan, 2008b). Such need for additional burrowing for the fetus to access more blood to supply the increased thinking capacities of humans may have been one reason for the development of menstrual bleeding. In particular, women's menstrual bleeding represents a discharge of a portion of the walls of the uterus that grow in thickness over the menstrual cycle but partially disintegrates when pregnancy doesn't occur (Rathaus, Nevid, and Fichner-Rathaus, 2008). Although it has been hypothesized that this bleeding cycle and its tie-in to the moon cycle actually prompted humans' capacity to forecast for themselves (Shlain, 2003), the special characteristics of the human menstrual cycle may have actually been caused by the human brain growth needed for creative cooperation outside the trees when they were more vulnerable and only later contributed to the human forecasting.

In any event, it's possible that the very electromagnetic nature of the cosmos magically caused the moon and the menstrual cycle of humans to be intertwined, it isn't necessary to invoke magic to explain the evolution of human intelligence. If humans hadn't evolved from apes, or if apes hadn't evolved at all from other primates, and if other things hadn't happened, it's still quite likely that evolution would have eventually developed a life-form that could forecast for itself, with or without any relationship between the moon and reproductive cycles. In fact, mutations in humans that caused a correlation between the moon and human menstruation periods, whether actually caused by the light or electromagnetic waves from the moon or not, would have been naturally selected by evolution if it resulted in increased adaptability of humans through a facilitation of their ability to forecast that led to humans' unique consciousness and language abilities.

Although some organisms like chimps may have some limited consciousness as indicated by the apparent existence of a working memory, chimps' working memory is less than half that of humans and is therefore challenged by even simple tasks (Coghlan, 2008a). Birds may also have some sort of consciousness that incorporates empathy and forecasting the reactions of others, albeit in a different part of the brain from the neocortex where human consciousness and higher-thinking seems to exist (*NewScientist*, 2008zi). Even if some animals have a form of consciousness, there is little evidence that any organisms except humans consciously forecast their own selves.

As a result of their capacity for conscious thinking and forecasting based on a self-aware logic, humans tend to utilize their subconscious capacities and senses less than other animals. For instance, unless their cultural and environmental situation mandates it, people tend to have a sense of direction that is inferior to that of other organisms, as humans learn to rely more on conscious understanding of the relationships in their environment than on its geometric realities, thereby reducing their connection to their present physical location but enabling them to imagine themselves to be in different places so that they can change their "reality" (Bezirk, 2009). Similarly, human cognitive abilities also motivate them to rely more on their logical thinking than on subconscious reactions to stimuli received via the five physical senses, and people therefore generally don't develop those senses as well as other animals.

As previously indicated, perceptions of reality provided by the physical senses represent distortions of the true electromagnetic nature of things. Human consciousness enables a comprehension of that actual microcosmic reality underlying normal physical perceptions. The potentially fractal nature of the universe (Buchanan, 2009b) may have made it inevitable that some macro life forms evolve to be aware of the basic nature of divided nothing, and humanity may represent this evolved macro life form on the planet earth.

B. Evolutionary Differentiation of Humans

Stringer and Andrews (2005) indicate that evolution has no purpose and is affected by random events, whose complete impact is unknown. For instance, it is not yet totally clear what effect a catastrophic comet hitting North America 12,800 years ago had on the Native Americans there (Pringle, 2007).[3] The effect of reversals in the earth's poles, which have occurred at 0.8, 0.9, 1.6, and 1.8 million years ago (Stringer and Andrews, 2005), could also have had unknown creative effects on human development via catastrophes. Native American philosophy perceives the variability and cycles in Nature as often being without a rational preconceived purpose and frequently pointless (Peat, 2005). Ancient German religions also incorporated a "trickster" deity into their cosmology (Davidson, 1990) that is somewhat random and can have positive or negative effects on humanity.

There is a cycle in the shape of the earth's orbit every 95,000 years, in the earth's tilt every 42,000 years, and in the time of the year when the earth is closest to the sun every 21,000 years (Stringer and Andrews, 2005). All three cycles are caused by the pull of other bodies in the solar system (Muir, 2008) and affect the earth's temperature (Stringer and Andrews, 2005). The longer cyclical period seems to be related to a 100,000-year solar cycle that can cause a reversal of the earth's magnetic poles (Stray, 2005). However, the biggest effect seems to occur when they all hit in the same direction at once as last happened over 700,000 years ago (Stringer and Andrews, 2005).

The effect of such cycles on human evolution might be expected to promote greater adaptability in the species in order to survive such environmental changes. For instance, one of the impetuses for the evolution of the human race may have been a cold spell between 6.5 and 5 million years ago, which may have reduced the amount of rain forest available for arboreal living and therefore resulted in genetic changes in human ancestors that enabled upright walking, which allowed terrestrial movement with less energy, greater capacity to see the four-legged predators and prey in the savannah, and an enhanced ability to carry babies and other objects (Adovosio, 2002). Human cooperation would have been essential for such life exposed to predators, and so many of the special cooperative traits of humanity such as spite and empathy may have begun then.

Humans have changed more genetically than chimps since human ancestors split from chimpanzies about 6 million years ago (*NewScientist*, 2007g). The enhanced ability of upright hominids to carry objects such as tools may have been one impetus or reason for the earliest increases in the size of their brains (Jurmain, Kilgore, Trevathan, and Nelson, 2004). However, some of the human mental retardation diseases like phenyketomuria, which results from genetic mutations of normal human genes, are relics of what are normal genes in more distant apes like macaques or monkeys, from which human ancestors split 25 million years ago (*NewScientist*, 2007f). Findings such as the latter ones indicate humans' continued relationship to other beings with a very distant shared past.

Today's apes that evolved more recently together with human ancestors or hominids include chimpanzees and orangutans, which all seem to have originated from a prehistoric ape that was widespread in the world about 13 million years ago (Schwartz, 2005), including in Europe (*NewScientist*, 2009zze). While orangutans and hominids may have split into separate species around that time (*NewScientist*, 2007za), there is a gap in the fossil record between 8 and 12 million years ago (Stringer and Andrews, 2005), making it currently impossible to draw definitive conclusions about whether humans evolved from orangutans or chimps (Schwartz, 2005). The formation of two huge mountain ranges forming vertically across the plateau of eastern Africa that caused large variation in climate and rainfall about 15 million years ago promoted greater genetic adaptability of the more primitive ape species that eventually resulted in hominids and other primates which could migrate and survive in the less permanently lush environment that resulted (Ananthaswamy, 2010b).

Both orangutans and early hominids may have originally had some ability to walk on tree limbs (Jones, 2008), and they share many other traits as well that chimps don't have (Morgan, 2009a). For instance, unlike chimps, orangutans and humans both have long hair, thick tooth enamel, an ability to construct shelters and beds, and a preference for face-to-face matings (Lawton, 2009).

Although chimps and humans have a larger number of genes and general characteristics in common (Schwartz, 2005), many genes, including in humans, originated via spreading across species via viruses (Lawton, 2009a). The genetic similarity between humans and chimps may therefore result from viruses spreading between unique genes among them in their shared African environment of origin. In fact, humans and orangutans could have evolved from the same prehistoric ape, but humans diverged more genetically from orangutans through viral or other merging of chimp genes with hominids in the African environment that was geographically separate from the Asian habitat of the orangutans. Even without interbreeding and viral spreading of genes, the fact that chimps and hominids lived on the same continent of Africa would have also contributed to them naturally evolving similar characteristics to adapt to that particular environment.

Within the context of orangutans evolving in a separate environment, it would be difficult to explain the unique traits that humans have in common with them unless they evolved from a common ancestry that is different from that of chimps. For instance, without such a shared past, the fact that orangutans and humans display the rather unique trait of not being restricted to sexual intercourse only at particular times, unlike chimps which show outward signs of estrus and therefore don't engage in sex year round, would be especially perplexing (Morgan, 2009a). Orangutans are many times stronger than human beings, and they are also more solitary (Schwartz, 2005), but those different traits likely evolved as a result of their separate environments that encouraged selection of group cooperation over brute force for the hominid descendants who lived in Africa.

Early human types may have actually interbred with the ancestors of chimps for millions of years (Holmes, 2006). Some of the alpha male genes that exist in today's chimpanzees (Jurmain, Kilgore, Trevathan, and Nelson, 2004) may still persist in some humans because of such sexual reproduction as well as due to viral spreading of genes via diseases. Since female chimpanzees sometimes kill the offspring of other females (*NewScientist*, 2007m), an alpha-male culture of violence can affect both sexes in a species. Given that the earliest human species several million years ago exhibited large differences in size between the males and females (Stringer and Andrews, 2005), these early human ancestors may very well have lived in alpha male societies before they evolved into bodies and minds that were more consistent with greater equality and cooperation.

However, the genes of bonobos match even more closely to humans than those of chimps (Graham-Rowe, 2005), and so it seems more likely that humans' innate genetic characteristics are more like those of the peaceful, matriarchal, and "free-loving" bonobos (Murphy, 2008). Bonobos, which also evolved from the prehistoric apes, are similar to chimps except that they are smaller, don't outwardly exhibit estrus (and therefore have sex all the time like humans), and didn't retain the alpha-male culture of the more ancient guerrillas that separated even earlier from hominids (Wranghan and Peterson, 1996). The lack of outwardly visible signs of estrus in bonobos may have developed by some hominids spreading the gene to some smaller chimps that then allowed them to establish their own groups with cultures of reproduction not determined by dominating alpha male control. Without gene selection based on sex with the biggest and strongest chimp, their size would have easily declined from there under the influence of a more cooperative matriarchal or egalitarian culture.

As was explained in Part I, violence between humans likely first began about 6000 years ago in spiteful revenge against matriarchal hierarchies, and, as explained later in this book, the first matriarchies likely appeared about 35,000 years ago. Egalitarian cultures were the norm for humanity through most of its evolution, possibly because the creative cooperation involved in human nature requires equality between the sexes, as was long ago necessary for human survival. Bonobos were able to survive their mild form of matriarchy because they lived in a less hostile environment.

Bonobos like humans and other apes such as monkeys cooperate to violently kill prey, and so it can't be concluded that human violence in hunting is related to violence between humans (*NewScientist,* 2008zzo). The hunting capacity of the normally non-violent bonobos is also an indication that patriarchal or dominating tendencies are not a prerequisite for hunting. There is no evidence that human ancestors ever had a violent gene or nature, as any organisms with the weakness of human ancestors would have perished without offspring long ago.

It is possible that orangutans, bonobos, chimps, and humans share a common ancestor from over 10 million years ago, but orangutans remained in Asia, while the ancestors of humans, bonobos, and chimps evolved in Africa. Fossils from 4.4 million years ago indicate that human ancestors had already greatly deviated from chimps by that time, enabling some upright walking (*NewScientist,* 2009zzzs). The separation between chimps and humans may have actually been much earlier that 4-5 million years ago.

There is evidence that these human predecessors were already less aggressive and patriarchal than chimps then (Lovejoy, 2009), thus indicating human ancestors were more like bonobos than chimps over 4 million years ago.

Just as male orangutans tend to be less violently dominating than chimps (Schwartz, 2005), bonobos are very peaceful creatures, and are today more similar to human beings than any other species today even though they resemble a sort of smaller version of chimps (Wrangham and Peterson, 1996). Although human ancestors, which stood at 4-foot and 50-kilograms 4.4 million years ago (*NewScientist,* 2009zzzs), were significantly different from bonobos at that time, the greater similarity between bonobos and humans today may have been caused by the ancestors of bonobos mating more with human ancestors in Africa than chimps did. Bonobos, which, like humans and orangutans but unlike chimps, are not characterized by sexual activity being restricted to the special periods of estrus (Schwartz, 2005), could have alternatively developed out of a merging of human and chimp genes.

Viral spreading of genes in the shared African environment may also have contributed to the similarity in the genes of those humans and bonobos. Since orangutans tend to be very solitary animals (Schwartz, 2005), bonobos and humans apparently evolved to co-exist in groups, perhaps because of an environment with more concentrated food resources and more dangerous predators. This African environment also resulted in the evolution of the separate species of chimps, which also live in groups but are characterized by hierarchies dominated by violent males. While bonobos and chimps probably share a common ancestry separate from that of humans, the similarities between bonobos and humans could possibly be explained not only by their sexual encounters but also by a prehistoric life in closer proximity. Physical nearness and possibly even some form of co-habitation would have facilitated viral spreading of the same genes among them. Such closeness might have been facilitated by a greater compatibility between their cultures than for the more patriarchal chimps, and that compatibility would have grown with the similarity of their genes.

This hypothesis is consistent with bonobos surviving only in a secluded area of Africa devoid of violent chimps, and with humans being forced by dominating chimps into less habitable areas like swamps and eventually the savannah, where they had to either evolve to cooperate more closely in a powerful group consciousness or die out. The bonobos may have actually developed from smaller ancestral chimps, who were motivated to favor more egalitarian, matriarchal cultures and therefore split off from larger dominating patriarchal ones to create separate groups with females in secluded areas to escape the dominating alpha males. Greater contact between matriarchal bonobos and human ancestors might have then contributed to making both less patriarchal.

The much larger size of males compared to females in both bonobos and chimps (Wrangham and Peterson, 1996) as well as in orangutans (Schwartz, 2005) is similar to that for humans over 3 million years ago. The relative lack of such dimorphism amongst human beings today began about 1.8 million years ago. If humans already had equality between the sexes then, the increased size of females may have stemmed from other developments, such as the harnessing of forces like fire that enabled humans to fight predators but only if they had some minimum size.

Much of the adaptability of humans may stem from their having to learn to survive in many different environments, or at least mating across hominoids that had adapted to live in varied conditions of life. There have been at least 7 "natural" catastrophes in the last few million years, and the 95,000-year cycles in the earth's orbit have continued to cause ice ages in recent times, with the most recent one being 20,000 years ago that resulted in sea level rises of 100 meters and worldwide flooding as it ended (Lovelock, 2009). These situations certainly required some adaptation. However, most of the last 500,000 years have been colder than now, and some of the greatest progress in human development and adaptability has occurred in the last few hundred thousand years when the climate has been unusually stable and mild (Stringer and Andrews, 2005). Since the most recent catastrophe associated with ice melting about 10,000 years ago may have been at least partially induced by increasing human populations raising carbon dioxide levels in the atmosphere through activities like the clearing of forests for settlement (Lovelock, 2009), humans' ability to adapt recently seems to be more greatly challenged by changes in the environment that they themselves cause.

In any event, human ancestors developed through several categories of hominids, starting over two million years ago with homo habilus in Africa (Stringer and Andrews, 2005). These habilus creatures were unique insofar as they not only used tools but also made them from more basic raw materials (Calleman, 2009). While the climate between 3.5 million and 1.8 million years ago actually "was generally warmer than now" (Jones, 2009a), it has been hypothesized that an ice age about 2.5 million years ago enabled the growth of the larger brains in human ancestors that have such extensive cooling needs they could not have otherwise evolved in such warm climates (*NewScientist*, 2009zzz). Although speech didn't begin until 1.8 million years ago (Fisher, 2004), there is evidence that human mathematical abilities may have begun before speech (*NewScientist*, 2008zza), possibly during that ice age 2.5 million years ago that may have facilitated the geometric shaping of tools and weapons.

Homo habilus was followed about 1.8 million years ago by homo erectus, who emerged from Africa a few hundred thousand years later and subsequently evolved in China, Europe, and Africa into homo heidelbergenis (Stringer and Andrews, 2005). The evolution of this species to walk erect in parts of Africa may have been caused by the less lush and more dangerous environment there that provided less tree cover than Asia, where one of the most similar apes, the orangutan, never was forced by a tree-less environment to walk and think in such an advanced fashion (Schwartz, 2005). It was around then that human compassion (Cruttenden, 2006) and human harnessing of fire (Ungar, 2007) developed. The former may have been related to the need for cooperation outside the protected jungle environment, while the latter provides an indication of the unique forces that humans may have needed to harness in order to survive. People today tend to feel more compassion when they are on relatively "higher ground" (Fleming, 2011), as makes evolutionary sense given that the prehistoric usefulness of being with fellow humans was magnified in such positions that enabled better observation and defense.

There is evidence that at least some human ancestors may have evolved to live in more diverse environments like flooded areas near rivers (Gynn and Wright, 2008). Such a past might help clarify the virtual absence of human body hair that contrasts so sharply with other primates not living in such habitats (Morgan, 2009b). The ability of even human babies today to naturally swim (Lipton, 2008) is difficult to explain without a past existence in such environments for humanity. Such an environment would have also favored bipedality to enable standing in deeper water, as well as greater intelligence and cooperation to enable survival in such an environment. Given the relative weakness of human ancestors, a migration to swamps or lake areas 5-10 million years ago might have been the only means for humans to survive their prior forest homes that were populated by increasing numbers of alpha male chimps and other dangerous animals, thereby also explaining the development of human characteristics that are different from their orangutan ancestors. The evidence that aquatic animals like turtles, crocodiles, and catfish were the first meats eaten by human beings, about 1.9 years ago (*NewScientist*, 2010y) provides some indication that human life near water continued into more recent times.

Evidence exists that humans in the last few million years were more likely to live in uneven areas that included caves as well as lakes that provided areas for humans to hide for hunting and other purposes (Marshall, 2010c). There they could escape from land predators which might have been foiled by extensive time spent in the water, possibly even with mud, stone, or wood pilings formed in the water that would inhibit attack by crocodiles. Within this context, the movement of human ancestors out of Africa in more recent times might have been motivated by the deadly crocodiles, snakes, and mosquitoes that populated wetlands in Africa.

The brains of human ancestors appear to have grown significantly over the last 1.8 million years to permit ever more sophisticated communication, associations, and thought (Jurmain, Kilgore, Trevathan, and Nelson, 2004), and today human babies are actually able to think of numbers as abstract concepts only a few days after birth (*NewScientist*, 2009zzi). Nevertheless, vowel sounds and complex language appear to have been restricted for the earliest human species that included the Neanderthals (*NewScientist,* 2008z), who began to appear several hundred thousand years ago as short stalky forms of homo sapiens (Gooch, 2006).

The Neanderthals started evolving in the colder climate of Europe from homo heidelbergenis 500,000 years ago (Stringer and Andrews, 2005). The sizes of some human brains 500,00 years ago were already within the range of today's humans, with the enhanced size permitting greater plasticity of neural connections for learning new associations between things, acts, and concepts, thereby permitting behaviors like more complex speech (Calvin, 2008). Some Neanderthal remains have also been found in the Middle East and central Russia (Stringer and Andrews, 2005), as well as in Siberia (*NewScientist*, 2007zg), indicating that they were adaptable enough to spread their population widely.

A second homo sapiens species called the Cro-Magnons evolved in the warmer climates of Africa (Gooch, 2006). The Cro-Magnons developed from homo heidelbergensis in Africa and then went out to settle the world (Stringer and Andrews, 2005). The earliest evidence of these humans has been found in Morocco that indicates they evolved about 150,000 years ago (*NewScientist*, 2007d). Although various hominids had spread throughout Eurasia by that time (Adovasio, 2002), the Cro-Magnon species didn't emerge from Africa until about 75,000 years ago when a stabilization of the weather may have enabled their population to increase, thereby promoting outward expansion (Jones, 2007b). A large volcanic eruption 74,000 years ago that could have severely affected the environment may have also promoted the migration (Ravilious, 2010). While the first such modern humans left Africa at least 93,000 years ago, those humans didn't settle very far from their origin until about 50,000 years ago during a wet period that may have enabled crossing the Sahara desert to arrive in Europe (*NewScientist*, 2009zzzzf).

The period about 75,000 years ago corresponds to about the time that genetic evidence indicates the use of continuous clothing by the ancestors of today's human beings (Holmes, 2008a). Clothes would have been very useful in allowing the newly developed homo sapiens to travel out of Africa into the colder climates to the north. Shoes followed within the next 30,000 years (*NewScientist*, 2008f). Those emigrating out of Africa eventually evolved whiter skin because it blocks less of the vitamin D from sun rays that are needed by human beings and because it makes people less susceptible to frostbite than for people whose black skin blocks more of the dangerous ultraviolet waves from the sun that are stronger nearer to the equator (Ananthaswamy, 2009b).

The actual most direct ancestors of today's humans evolved 89,000 years ago in Africa, spreading out from that continent 66,000 years ago in a group of an estimated 4000 people (Begley, 2007b). It has been estimated that the survival rate of an isolated group is very low when there aren't thousands of that type (*NewScientist,* 2009zzzv), and so it is very feasible that other smaller groups also left Africa but didn't survive.

The large group that left Africa 66,000 years ago had a particular type of brain chemical (FOXP), which developed 50,000-200,000 years ago, and which may have sparked human culture and art (Begley, 2007b). The earliest evidence of art exists in the form of a serpent carved in a rock 70,000 years ago in southern Africa that could even reflect at least a subconscious recognition of the wave nature of the cosmos (Calleman, 2009). Interestingly enough, a FOXP2 gene, which is responsible for providing better coordination of the neural noise and messages in the brain that enable speech (Yong, 2008), was also present in Neanderthals (Adler, 2010).

The development of the bow and arrow, which may have occurred as much as 100,000 years ago, and which facilitated hunting (*NewScientist*, 2009zzk), could have been a related development, insofar as the human version of the FOXP gene may have enabled the creativity to develop the bow and arrow. In addition, such weapons enabled more food with less expenditure of energy, which then could be directed more toward leisurely endeavors like art and culture. Such creative activities may have been further spurred by sexual selection that might have resulted in creative people being honored by the culture and more likely to be chosen by the opposite sex. The reduced importance of brute strength with projectiles as opposed to hand-held blades increased the usefulness of creative intelligence in developing such new weapons and other tools, and so creativity naturally was spurred by evolution, but sexual selection hastened the process, as it not only helped pass along the genes of those able to engage in more creative and insightful thinking but also encouraged their applied use. The improved weaponry also facilitated movement out of Africa into areas with less plant nutrition available during the winter because it enhanced the ability to hunt for more food.

It is also interesting to note that, although the evolution of humans occurred during a period of frequent droughts, the exodus out of Africa occurred during a period of good stable weather (*NewScientist,* 2007zd) that apparently prompted both Cro-Magnons and the smaller group of modern human ancestors to emerge from the African continent. The reason for the emigration may have been to relievc the stresses related to overcrowding. Some of those who stayed in Africa were already growing in size to over 7 feet tall 10,000 years ago (*NewScientist*, 2008zzd), and shorter, crafty ones may have chosen to settle to the north as opposed to try to compete with giants.

While Neanderthals were taller than the Cro-Magnons and had bigger brains (Jurmain, Kilgore, Trevathan, and Nelson, 2004), a larger portion of their brains (the cerebellum) was involved in psychic activities as opposed to conscious thinking (Gooch, 2007), and they largely disappeared about 35,000 years ago after the Cro-Magnons arrived in Europe (Gooch, 2006). The cause of the Neanderthals' disappearance may very well have been due to a crowding out by the more recently developed humans who were able to adapt even better to the European environment. Shorter species tend to evolve or survive when the danger of predators is removed (Adovasio, 2002), as would have occurred once human intelligence had developed sufficiently to create weapons and tactics to dominate wild animals.

The Neanderthals didn't display as much forward planning capacity as more modern humans (Adler, 2010), as may have been caused by those beings having a more primitive forecasting ability, which would have put them at a relative disadvantage. In addition, although human language already existed a half million years ago (Callaway, 2009e), a full range of sounds necessary for a complete language didn't appear in the more modern humans until about 50,000 years ago, and it isn't clear if the Neanderthals were able to communicate ideas as easily as the humans who replaced them (Adler, 2010). As will be explained later, the development of language and forecasting may be intricately interconnected, and deficiencies in either or both could explain a replacement of Neanderthals with modern humans. The last vestiges of the Neanderthals, who seemed to have less ability to construct the symbols needed for language that more modern humans have (McClenon, 2002), were found in a desert environment of Spain that the more modern humans may not have been interested in settling (*NewScientist*, 2007n).

Although the earlier homo sapiens such as the Neanderthals may have mated with the later versions of humanity such as the Cro-Magnons and left some of their genes in today's humans (Jones, 2007a), some findings comparing some Neanderthal and homo sapiens genomes indicate no such interbreeding (*NewScientist*, 2009r). The earlier hominids could certainly have co-habitated with homo sapiens in the last hundred thousand years, just as the earlier versions of humanity may have shared quarters with apes three million years ago and just as wolves and dogs have for over thousands of years without mating (Holmes, 2008a). While recent evidence implies the possibility of interbreeding between Neanderthals and modern humans in the Middle East about 100,000 years ago, the fact that some Europeans and Asians having 1-4% of their genes in common with Neanderthals (Gibbons, 2010) could alternatively stem from viral infections resulting from mere nearby co-habitation or settlements.

On the other hand, given that some species such as camels and llamas can cross-breed to create new reproducing species (Gynn and Wright, 2008), it's conceivable that some matings between the various types of hominids, and perhaps even apes, created new species, of which some of the earliest human ancestors might have been a product. The fact that most of the unfertilized egg cells produced by women have chromosomal damage (*NewScientist,* 2009zq) might very well result from such somewhat incompatible reproductive mixes, which could make chromosomes more unstable. Genetic instability relating to interspecies breeding might even have contributed to some of the mutations that led to the unique nature of modern human beings. For instance, genetic mixing with chimps that have natures sometimes characterized by dominating violence could have contributed to humans being more violent than bonobos.

In any event, remains of some very early hominids have been found in Indonesia that appear to be only about 20,000 years old (Stringer and Andrews, 2005) and that may have survived as a group of "hobbits" living until just 1400 years ago (*NewScientist*, 2008n). They may represent the final line of the homo habilus or erectus (*NewScientist,* 2007ze) that emigrated out of Africa about two millions of years ago (Stringer and Andrews, 2005) and that evolved elsewhere into modern human beings.

Many details about human origins and the evolution of their minds is unknown or mere educated guess. However, it is interesting to point out that Neanderthals exhibited greater differences in size between males and females than other homo sapiens (Stringer and Andrews, 2005), but they had matriarchal cultures (Gooch, 2006) just like the more matriarchal bonobos. The Neanderthals apparently produced Venus figurines tens of thousands of years ago (Stringer and Andrews, 2005), just as did more modern humans for tens of thousands of years until about 6000 years ago when matriarchal society was replaced with a patriarchy (Baring and Cashford, 2003). As will be explained later, since cultures among non-Neanderthal humans were probably first characterized by matriarchies about 35,000 years ago, there may be a non-coincidental connection between the timing of the demise of the Neanderthals and the imposition of matriarchies among more modern humans. For instance, the first matriarchies among the Cro-Magnons may have developed from an exiling of older Cro-Magnon women, who then settled with the Neanderthals and learned some of their psychic or other "magical" abilities like hypnosis which, when combined with the technological and thinking skills of Cro-Magnons, could have enabled them to appear as deities to both types of human beings.

Besides the possibility of occasional inter-mating, there are many hypotheses that can be developed for evolved human traits, and the relationship between Neanderthals, Cro-Magnons, and modern humans. For instance, long hair on the head that isn't genetically coded to stop at some point evolved in human ancestors about 240,000 years ago, and this development, which precedes the existence of Cro-Magnons by less than 100,000 years, may have resulted from its usefulness as a sexual attractant since it doesn't seem to provide a "fitness" advantage (Barnett, 2006). Hair style may have initially served as a signal for a more modern brain, such as in helping distinguish desirable homo heidelbergenis mates from homo erectuses with smaller brains. Because longer hair enabled more elaborate styling, sexual selection might have caused the evolution of hair that didn't stop growing. Since Neanderthals, which descended from home erectus, had smaller portions of the brain designed for higher thinking than the Cro-Magnons even though they eventually exhibited similar cultural traits such as clothing (Stringer and Andrews, 2005), hair patterns could have also helped avoid interspecies mating between those two groups as well.

Today's human genome split off from the genes of the Neanderthals an estimated 536,000 years ago (*NewScientist*, 2006f), but it is difficult to know for certain the reason(s) for the Neanderthals' disappearance. The restricted logical abilities of the Neanderthals, which seem to have had more powerful intuitive capacity (Gooch, 2006), might have been one cause of their demise, as the advanced human consciousness possible among more modern humans could have provided superiority in adaptability and domination of the environment. Since a portion of the thinking skills of human beings is inherited strictly through women's mitochondria but no intelligence is sourced strictly from males (Rose, 2005), and since all non-African modern humans can be traced genetically via the mitochondria that is inherited only through women to an "Eve" in Africa existing about 70,000 years ago, it seems likely that such a thinking capacity that originated first in women is what differentiates modern humans from Neanderthals and other more archaic human species. Because an advanced analytical ability has continued to exist in the mitochondria of all men and women today, there isn't a large intelligence difference in intelligence between the human sexes, however.

The origin for intelligence in women 70,000 years ago might partially explain a natural human honoring of women of all ages that culminated in the eventual creation of matriarchal human societies about 35,000 years later when some older exiled women came to be regarded as superhuman, potentially by harnessing the powers associated with combining the advanced psychic abilities of Neanderthals with the logical thinking capacities of modern humans. As will be explained later, such a hypothesis clarifies the development of a culture of goddess worship 35,000 years ago that lasted until conquest by patriarchal barbarians about 6000 years ago, and it is consistent with the characteristics of today's human mind.

C. The Human Mind

At an elementary level, the human mind is powered by brain cells cooperating via electric and chemical impulses that operate like a group of bees, with decisions being made based on the number of individual neuronal impulses that are in the majority (Hooper, 2008). "Neurons, the specialized cells of the brain, can produce brief spikes of voltage in their outer membranes", and "these electrical pulses travel along specialized extensions called axons to cause the release of chemical signals elsewhere in the brain" (Eagleman, 2008). At synapses, whose construction may be partially dependent on normally folded prion proteins (*NewScientist*, 2009z), the axon of a neuron outputs signals to the nearby dendrite of another neuron, and these output signals are received by the dendrites of other neurons which use such received indicators to determine the frequency and magnitude of their own output firings that in turn provide signals to other neurons (Bernstein and Nash, 2008). The synaptic connections operate like electronic memristors that permit memory and learning through natural electromagnetic properties (Mullins, 2009c).

Fox (2007) provides an indication of how complicated the brain's neural signals are compared to computer signals, insofar as neurons have an average of about 10,000 connections with other neurons, can form new connections at any time, have variable output levels (not just 0,1), and utilize a firing code that has not been cracked yet. Each neuron acts like a complex computer that reacts differently to varied signals from connected neurons (Klein, 2003).

Neurons are genetically programmed to fire in response to electrical signals, which they receive from other neurons, each of which acts as a powerful minicomputer with a basic genetic code for processing over 60 different chemicals that permit the brain cells to learn to process proteins and fire like Pavlovian dogs (Klein, 2003). The probability of a signal being sent by any of the 100 billion brain neurons seems to be a function of a Bayesian learning procedure that uses past relationships between the signal impulses of other neurons to make predictions about the importance or relevance of current neuron signals (Huang, 2008). The neurons grow new connections to other brain cells when they are reinforced with dopamine, which is produced in the ventral tegmental area, and those rewarding pleasure chemicals may be spread among enough neurons in the rest of the brain to create a "high" or feeling of happiness (Klein, 2003).

Dopamine is apparently required in adequate amounts in the brain to prevent a perception of time moving too slowly and a feeling of boredom that can result in hyperactive behavior (Coghlan, 2009). This chemical is naturally produced at learning from new experiences and from other stimuli like anticipation of a reward that results in the brain's internal production of endorphins, which have effects which are very similar to both dopamine and opium (Klein, 2003). Given that people over 67 years old who tend to experience more sporadic electrical activity in the striatum reward area of the brain have been found to be less efficient at figuring out puzzles like investment games (*NewScientist*, 2010k), dopamine may be essential for rewarding analytical thought as well.

Electrical neural firings upon electrical signals by other neurons tend to create combined firings leading to patterns which are useful for further brain processes. Good decision-making is thereby chemically encouraged, whereas systems of neural firing that lead to erroneous information, false thinking, or counterproductive behavior gradually dissipate. This method of adaptive information processing seems to work on a system of self-organized criticality whereby neurons fire in a partially connected and partially random or unpredictable manner until a pattern is formed to create a stable locked state that is reflected in thoughts and actions (Robson, 2009). There even seems to be a productive brain process that sometimes increases random noise in neural firings which provide informational signals to other neurons in order to enable the amount of electrical signals that are communicated to other neurons to exceed the threshold needed for recognition of a faint stimulus to generate actions and thoughts (Spinney, 2008a).

The average number of neural firings that result from one neuron firing is just one, but the chaotic system often results in a cascade of firings that create instability until a meaningful pattern generates new stable thoughts and actions (Robson, 2009). Such neural firing avalanches, which work like earthquakes with smaller aftershocks, are thought to help the brain explore new ideas (Biever, 2009b). The amount of time between periods of stable neural firings is actually greater for more intelligent people because the extra milliseconds of insanity enable recruiting of additional neural firing possibilities for added information to deal with new situations (Robson, 2009).

This anarchic development of signals which tend to cascade with orderly efficiency functions according to a chaos theory that results in extreme speed, resilience, and memory capacity of processing (Graham-Rowe, 2008). Once a certain critical threshold level of electrical activity is reached in one region of the brain, brain cells tend to fire in perfect synchronicity across different areas of the brain as if they were quantum entangled, and this process enables thoughts that include a variety of different information inputs (*NewScientist,* 2010b). The system of electrical firings result in brainwave frequencies between 0.05Hz and 125Hz that are locked in phase during each period of stable synchronization, which tends to last for a slightly shorter time interval for more intelligent people (Robson, 2009).

However, very poorly synchronized brain waves or neural firing frequencies result in poor communication across the different parts of the brain, causing false categorization of perceptions and attribution of significance to random stimuli (Thomson, 2010a). Unstable synchronization of brainwave frequencies can even create "madness" like schizophrenia and epilepsy during excessive periods of instability in neural firings (Robson, 2009). Synchronization of the electrical neural activity at a specific frequency that optimally varies with the type of stimuli and thought processes does seem to enable a holistic awareness of the environment, and the synthesis and release of the neurotransmitter GABA internally in the brain seems to be crucial for such effective synchronizing of brain cell firings (Thomson, 2010a).

Due to both genes and the environment, some areas of the brain develop special functions, such as brain sectors that fire when particular stimuli (such as visual ones from the eyes) are received from the senses (Klein, 2003). The portion of the brain responsible for evaluating spatial locations actually has neurons organized into triangular patterns (NewScientist, 2010e), perhaps because such patterns are an efficient means of meaningfully categorizing the 3-dimensional space in which they exist, and which may reflect the 3-fermion nature of quarks.

In any event, waves of electromagnetic activity in different parts of the brain create discrete perceptions in time by taking snapshots of stimuli during the wave troughs (Fox, 2009b) that are apparently signaled to other areas of the brain during the peaks of electromagnetic activity. The neurons' genetic code apparently has them normally form connections in specified areas for efficient communication via the dendrites, but their actions are heavily influenced by learning (Klein, 2003). The system enables synchronized neural firings which cause "task forces" of groups of neurons to form that operate without interference from other neurons but that remain connected through a long-distance network that permits communication between any two neurons with no more than just 13 nodes of electrical signals (Robson, 2009). This mode of operation applies not only to the initial processing of stimuli, such as from the eyes, but also to the categorization of stimuli into symbols or words as well as to analytical processing of those signals.

For instance, information from the five traditional senses in humans and other organisms is sent via signals to higher thinking levels of the brain that in turn feedback information to the neurons nearest to the senses to enable more accurate comprehension of neural interconnections in a process that minimizes errors in judgment and neuron firings (Huang, 2008). The discrete time intervals over which stimuli induce snapshot frames of perception tend to vary with different tasks focused in different areas of the brain, and other areas of the brain process these signals to form a perception of continuity (Fox, 2009b). The electrical impulses or firings of the brain cells are apparently coordinated via "fancy editing tricks" of externally received sensory inputs to ensure that there is no discontinuity in perception even when one blinks one's eyes (Eagleman, 2008). For instance, the right inferior parietal lobe typically create new fixed images of the visual environment at the rate of 13 per second that create 13 hertz waves, while sounds are processed faster to create discrete perceptions at a higher frequency, but the brain is normally capable of adapting to variations in the frequency of perceptions and time lag changes in environmental responses (Fox, 2009b).

However, the brain can even be tricked into perceiving all sorts of events that don't happen, including out-of-body experiences (*NewScientist*, 2007zb). The ability to trick the brain with even routine visual illusions (Fox, 2009b) may result at least partially because of the subconscious guessing for the brief moments between stimuli perceptions. When several different possibilities exist for creating a complete picture out of incomplete information, different sets of neurons fire until one set dominates to form an impression (Griggs, 2010b). Conscious thoughts can be utilized to interpret the information, thereby leading to differing conscious perceptions that can be effectively selected by the neural firings associated with the thoughts.

Some electromagnetic waves that result from organized neural firings of individual brain cells reacting to signals from other neurons reflect the self-awareness of the neural firings of their own categorization of physical and emotional stimuli that are typically symbolized with words and language. This self-awareness permits conscious evaluation of all information, including from the emotional sectors of the brain that is also categorized into specific feelings, to draw deductions on the most rational decision(s) to be made. The electromagnetic waves that reflect the conscious conclusion(s) also incorporate an analytical forecast of changes in the categorized state of the electromagnetic waves that create one's own self-awareness. They are translated into action(s) by organized neural firings whose exact structure isn't consciously known even though information categorizing the consequences and successfulness of the action(s) is reflected in the electromagnetic waves of consciousness for further evaluation.

These electromagnetic waves of consciousness merely reflect the fields created by the organized neural firings that humans learn at an early age to enable logical decision making based on information incorporated into and categorized by those oscillating electromagnetic fields. The logical analysis of that electromagnetic information represents a person's consciousness, which is self-aware of its own categorization of the neural firings that cause conscious thoughts, deductions, and decisions. Within the environment of an infinite number of possibilities which may occur in the universe, and the virtually unlimited amount of information that can't possibly ever be known because of continued stochastic creation of new electromagnetic fields out of divisions of nothing, the rational processor called consciousness optimally evolved to utilize its own logical discretion to make choices using the sensory and emotional information that it receives about that environment.

Quantum physics may be directly involved in the process of an individual's consciousness making its choices. In particular, electrical signals in the form of calcium ions enter at brain synapses through channels that are less than a nanometer wide, which is the microscopic size at which quantum collapsing of probability states of waves of potential becomes noticeable, and so the collapsing of waves in those nano-sized channels via conscious thought in a process that spreads via entanglement is quite feasible (Radin, 2006). Quantum entanglement effects involve particles that have once been in contact with each other and cause a positive or negative correlation in the particle states long after they have physically separated, including in larger, directly observable matter (McTaggart, 2008). Conscious choice that collapses a single brain synapse might then cascade with enormous consequences.

In particular, the system of neural firings that create electromagnetic brain waves and that categorize the pattern of its own electromagnetic activity into symbols like language can collapse itself on a selected point of its own probability wave function by merely choosing a symbol which defines the chosen point. While information to the wave function of consciousness can affect the point chosen on its own wave function through that wave function's evaluation of the consequences of any choice, the wave function has the complete freedom to choose any point in the set of possible neural firing patterns. That feasible set is heavily motivated by chemical rewards to the individual neurons of the brain that construct any pattern of self-aware thought, but the system of such coherent firings partially determines those rewards, especially with respect to the neurons involved in the conscious decision-making which effectively does collapse its set of possible decisions on the point it chooses.

Any possible point on any wave function, including with respect to the feasible pattern of neural firings that cause conscious decision-making, will happen in some parallel universe. However, the probability wave function of any pattern of brain cell firings that create reflective self-awareness is unique insofar as it itself can select the point on that system of electromagnetic waves of neural firings that will happen in any moment. The electromagnetic field of consciousness thus has free will to make its own decisions.

For instance, an electromagnetic field of consciousness may perceive a superposition of 3 possible choices, and that field can choose to collapse the probability wave function on just one through its control of itself. The self-control of consciousness is exercised through its selection which then collapses the probability wave function of individual electromagnetic waves through light photons and other particles it sends through the neurons' microtubules that then affect and entangle others to collapse the aggregated coherent field of consciousness to the chosen decision. The nanosized openings in brain cells effectively enable the collapse of the superposition states of electromagnetic waves that pass through them that can then spread via entanglement to not only affect conscious thought but also the probability wave functions in the external environment. Interactions of electromagnetic fields involved in the chemical reactions are also part of the collapsing of the wave functions involved in making a decision or changing a train of thought. The process of will power can thereby cascade into macroscopic events that are observable. This mere scientific possibility, along with all the evidence on the potent impact of will power that is summarized in Chapter 16, indicates with 100% certainty that an individual's conscious thoughts can cause patterns in neural firings that enable choice of the person's thoughts. Free thought therefore exists through a mere conscious belief in the free will that creates it.

Each conscious person can choose to not exercise the individual's will power just by believing in determinism, and the person will experience an average of others' will power. If no one were to exercise the will power created by their system of neural activity which causes conscious thoughts, everything would collapse on the averaged wave of possibilities, which sum to zero and thus lead back to the pure nothingness from which they arose. Thus, reality indicates that at least one individual is exercising, or has exercised, will power. Anything without the ability or the will to consciously exercise freedom of choice has the points on their probability wave functions determined either by chance or by the conscious choices of people actually using their free will. This set of hypotheses is confirmed by the Chapter 16 evidence on the potency of a belief in positive thinking that is reflected in the individual's feelings which add to the power through the electromagnetic forces naturally emitted by emotions.

An individual's thoughts and thus will power is affected by the person's internal and external environment because their interactions with the electromagnetic waves of consciousness influence the probabilities of possible thoughts and actions according to the laws of quantum physics which merely summarizes the basic properties of the divided pieces of nothing that make up everything. With the self-awareness of consciousness having arisen from the incentives provided by nature to forecast, thoughts evolved to be especially motivated by predictions of the possible consequences of thoughts and actions that include events which provide positive or negative reinforcement for any deed. As a result, the probabilities of a decision or thought are affected by the perceived optimality for the individual although will power can collapse a thought on any point of its own electromagnetic field of possibilities. This self-discretion in collapsing its own probability wave function naturally evolved because the infinite amount of information about the present and future make optimality in decision-making too uncertain and varying to leave to a system of automated responses.

Human beings evolved to have one of the primary motivators for human actions to be the love of others in order to enable them to generate the collective power that allowed this species which is too weak to survive individually to continue to exist. Each person's consciousness therefore has large incentives to direct thoughts toward resonating sufficiently with the desires and needs of others in order to obtain the chemical brain reinforcement involved in love. Doing so greatly magnifies the potency of individual wllls through entanglement and other quantum forces. In contrast, thinking with a purely selfish intent can result in destructive interference in the electromagnetic waves involved in the exercise of will power. Incoherence in individuals' wills drains the finite will power of the individual whereas harmony between them magnifies it as benevolence naturally leads people to return the good wishes because of the innately cooperative nature of humans that has evolved.

Human spite for selfish exploitation is one important motivation for the natural tendency of people to reciprocate, as the fear of spiteful reactions inhibits purely selfish acts. Spite directed for purely selfish purposes, however, drains will power and therefore has very limited potency. In contrast, spiteful actions undertaken for the good of humanity as a whole draw the loving benevolence of others and harmonize their wills in the form of a sense of communal justice that enables each person's spite to feed on the others' desire to punish anyone for a perceived injustice. Humans' sense of justice reflects in an absolute form the innate drive of all humans to help make the wills of all people coherent in a process that maximizes their individual power and pleasure.

Loving intent for others, whether through benevolent deeds or via spiteful actions to cause others to redirect efforts toward the good of others, thus draws the energy of other people's will power. This supportive thinking and feelings by others reinforces each individual's own loving energy, as the electromagnetic forces resulting from the complex chemical processes of these coherent feelings resonate across people. The power of love may therefore be infinite.

However, a purely selfish love that isn't motivated by a higher purpose of serving humanity as a whole may be inconsistent with the wills of the others, including that of the beloved. The power of such love may therefore easily decohere into impotence. Even two people in love with each other have a mutual will that may be inconsistent with that of others and thus have their potency reduced. In addition, two lovers undirected by the higher purpose of serving humanity have difficulty in maintaining their coherence in wills because of the intrinsic emptiness of love motivated strictly by a selfish desire for ecstasy without utilizing that feeling to serve the purpose for which it evolved.

Selfish love not directed for the higher purpose of human beings can also result in a one-way surrender of a person's will power to another or others. In particular, the will of an unloving beloved can collapse the lover's own probability wave of will power on points the beloved chooses. Such a surrendering of an individual's will power to other people can have a short-term positive impact by reducing the disharmony caused by incoherence in wills while also having a relaxing effect through the saving of energy involved in trying to collapse others' wave functions on the individual's own selected points that may be futile if they aren't feasible. However, unless there is mutuality in the love that would allow mutual collapsing of the wave functions, the love becomes an unhealthy one-way surrender. In today's societies that encourage selfish exploitation of others as opposed to natural communal love for the good of all, people tend to only take advantage of people possessed by such myopic love or have contempt for them. The exploitation of such one-way surrender inevitably results in a person experiencing adverse feelings that lead to an energy-draining spite. Love consistent with the aggregate will of all humanity can remove the possibility of the fate of exploitation that is otherwise quite likely today.

The neural firings involved in self-aware human thought that are reflected in brainwaves, which themselves are merely probability states, are thereby highly motivated to be coherent with the probability states of the system of neural firings that make up other people's consciousnesses. Today, the prehistoric group consciousness among human beings that naturally resulted has been disrupted by contemporary societies born out of the selfish exploitation of people long deceased, but people continue to have strong incentives to return to that blissfully hypnotic state that is also reinforced by the chemical rewards in the brain that evolved to maintain that powerful ecstasy. Thus, despite the widespread existence of exploitation and selfishness today, it may not be too difficult for a single individual to collapse the superposition of all possible states of human beings into coherence with that of humanity as a whole. Since each individual will so directed can collapse the wave functions of others, the process could conceivably cascade into a return to universal group conscious rather quickly.

The details of the quantum process by which conscious focus not only causes a collapse of a probability waves on a point but also enables that focus to select the feasible point merits further explanation. Conscious concentration is translated subconsciously into the collapsing of a particular molecule's wave of potentials into a single state at the electromagnetic field existing at a brain synapse, which is merely a location between fields created by brain cells. Entanglement and further chemical reactions spread the effect into a pattern of neural firings that create thoughts which lead not only to actions but also affect others. The probability pattern of electromagnetic waves of potential states of consciousness thus represents a distribution of parallel virtual thoughts, which can be directed by a focused awareness of the brain observing itself through the conscious, and which can thereby self-selectively collapse a selected probability state through concentration.

Each brain cell firing in the collective pattern of a conscious thought is itself merely a chemical process associated with the electromagnetic attraction of various divided pieces of nothing for their opposites. Each firing results in the emission of electromagnetic waves like photons that then create coherent brainwaves when their firings are harmonized with others through entanglement and other quantum forces that include chemical reinforcement which naturally occurs in the brain for firing patterns that result in evolutionarily useful actions and consequences. Such coherent patterns of neural firings are partially influenced by the signals sent by other cells outside the brain, including those that are passed through in an aggregated signal from the gut and pineal gland which provide initial processing of the mass of information they received into summarizing signals transmitted for conscious analysis as gut feelings and intuitive hunches. All this information includes not only the electromagnetic condition or state of the individual cells and their component molecules but also their encounters with external quantum forces such as "invisible" electromagnetic waves like photons emitted by others, as well as physical force fields detectible by the five senses. The pattern of neural firings involved in conscious thought can, however, also ignore many such indicators by conscious choice, although there seems to be irrevocable chemical and electromagnetic forces internal to the brain that evolved to motivate conscious choices that are coherent with the collective will of humanity.

Consciousness is just this reflective self-awareness created by systems of neural firings that allow these systems to influence their own future firing actions. These firings result in future thoughts by the prior firing system merely choosing a particular self-aware experience, symbolic concept, or decision upon which it wishes to reflect, and subconscious processes then cause the chemical and electromagnetic activity necessary to enact the pattern of neural firings to create the selected thought(s). Consciousness thereby wills its own self-aware future thoughts, feelings, and decisions. This will power affects the environment, but the freedom of choice remains restricted by that environment, and so the entire process of conscious thinking remains at least somewhat impacted interactively.

The evolved genes for the prefrontal cortex of the human brain enable this self-determining system of neural firings that remains affected by the environment, including by the internal biological situation in the brain. The human genome causes the neurons involved in consciousness to need chemical reinforcements and signals to continue to obey the commands of the entire coherent system of firings that lead to thoughts. At death, there is an insufficient supply of chemicals to keep the firing system of consciousness going, and so it terminates then, with damage to the system beginning within 30 minutes after the last delivery of the neural nutrients through circulating blood. In contrast, during sleep, the system of neural firings involved in consciousness is merely relaxed for hours at a time to enable processing associated with repairs and development of algorithms for subconscious action, but consciousness can be restarted thereafter because the neurons involved haven't been starved to death for chemical nutrients. While the system of neural firings associated with consciousness is enabled by genes, the actual neural firings involved are learned via Bayesian conditioning that begins shortly after the birth and continues for life.

This complex interactive system of consciousness doesn't begin at conception as there is not even a brain in existence at that time. Instead, the biological capacity for conscious self-selection only develops a few months after birth. Even after the brain has developed as a separate electromagnetic force field that has some independence as a coherent physical entity, it remains integrated with the environment via quantum interactions with other forces. Consciousness therefore only slowly gains more independence from the situation external to the brain, and there never is any complete freedom from environmental influences.

For all individuals, "every experience contributes to the person they will become", and learning begins in the womb just 22-24 weeks after conception (Phillips, 2009b). Memory capacity exists in human fetuses only 30 weeks old (*NewScientist*, 2009zzv), but activity in the cortex is restricted to stimuli based on physical touch until 2-3 months after birth, and the frontal lobes aren't activated until a baby is 6-12 months old (Phillips, 2009b). Consciousness would only seem to be possible after then, as a fetus or infant would just be able to react subconsciously until that point like a mere animal. The wave function of potential situations for a fetus or infant is therefore either determined by the collapsing on a feasible point by actually conscious human beings or remains an averaged probability wave function of possible superposition states. Once there is a consciousness in an infant that can make self-choices for its own state, the set of actually potential states that may be selected remain restricted by the environment, which includes both the genetic characteristics of the person, the prior experiences, and the existing situation, which all affect the feasibility of any individual's actions and thoughts.

The human brain circuits become more integrated" at around the age of 18 months, "a sense of other people having their own minds" begins at age 3 to 4, and "children start to apply logic and trust and to understand their own thought processes" by the age of 6 (Phillips, 2009b). These facts seem consistent with some form of group consciousness developing by 1.5 years, some individual consciousness by 4, and a fully aware, functionally individualized consciousness by 6.

Although some new brain cells grow even in adults, most are created in the womb or early in life, and it is the chemical connections between the cells and their interconnected electrical activity that create thoughts (LeDoux, 2002). Brain electrical activity, which is key to understanding the brain, tends to occur for the first two years of a human's life at the low frequencies of delta waves (0.5-4.0 cycles per second or Hz), to exist at 4-8Hz (theta waves) in years 2-6, to typically be at 8-12Hz (alpha waves) in subsequent years, and move at the age of 12 to the standard state of consciousness that can be in the form of beta waves (12-35Hz) used in normal human adult activity and gamma wave activity (>35Hz) during peak brain performance (Lipton, 2008). While different brain cells can fire at varied frequencies simultaneously, it is the synchronized electrical activity of neurons at a dominating coherent frequency that enables the consciousness which molds awareness of the environment and thoughts (Thomson, 2010a). A capacity for consciousness exists for fetuses after 40 weeks in the womb that enables an integrated self-reflective awareness of information from the senses and predicting the future (*NewScientist*, 2010aaz).

There are of course variations in the ranges of typical electromagnetic frequencies for the different age groups, but they do provide some indication of the status and level of thinking that is normally occurring at any moment of a child's life. For example, since theta waves are associated with deep sleep, and since alpha waves tend to exist during light dreaming (McTaggart, 2008), the minds of children seem to be comparable to an altered dream-like state. The low theta waves that are typical in very young children are also experienced by adults in a deeply hypnotic state that causes them to be very open to suggestion (Lipton, 2006), possibly indicating conscious questioning of commands and stimuli isn't possible at such low frequencies of neural firings. Instead, the neural firings associated with carrying out the suggestions are reinforced without any of the reflective self-awareness of consciousness that occurs at higher, less relaxing frequencies.

The mutual hypnosis that characterizes people deeply in love and the group hypnosis that characterized much of prehistoric human nature may be associated with such low frequencies for electromagnetic brain cell activity. Subsequent rapid frequencies are enabled to carry out the tasks the group hypnosis suggests has to be performed without any interfering conscious doubts, which would disrupt the efficient, reflexive subconscious processing necessary to accomplish the task.

Lucid dreaming, which involves being consciously aware of a dream results in brainwave frequencies that at times lie within the area of consciousness at 40Hz, are somewhere between sleep-like and awake states (Hamelzou, 2010). The latter frequency mix may allow self-hypnosis by enabling conscious suggestions to one's own subconscious at low frequencies that won't normally be interrupted later by the rapid frequencies of conscious thought which would disrupt efficient subconscious processing to carry out one's own commands.

Such mixes of mental states indicates that the frequencies may vary from moment to moment and not be easily subject to exactly differentiated categorization. Hypnosis may be especially complex, since it seems to exist in different possible degrees of intensity and effectiveness that may involve various different brainwave frequencies. For instance, hypnotic suggestions made in the slow theta state would be less disrupted by the higher frequency firings of conscious thought, thereby allowing uninterrupted subconscious processing or training of the brain at the low frequencies. Such training at low frequencies still allows rapid neural firings as high as gamma speeds when rapid thoughts or actions are necessary to carry out the hypnotically induced commands. The individual may be consciously unaware of the disciplined training at the theta frequencies that caused or enabled the subconscious processing at slow or fast speeds, as long as conscious justification for the actions can be rationalized.

An individual's memory and analysis of thoughts and behavior caused by neurons interconnected to fire together at low theta frequencies may be difficult at least partially because such frequencies below the minimum necessary for consciousness indicate no higher frequency firings to enable a self-aware knowledge or memory of them. The more limited coherence of neuron activity across the brain at lower frequencies may also result in less connections to other neural firings that make it more difficult for the memory process to recall them because there are not conscious thoughts, memories, or stimuli associated with them that might spark their recall. In addition, lower brain wave frequencies may also mean less of the harmony between neural firings that may be necessary for the experience of the feelings, which would otherwise enhance memory by facilitating emotional connections to the systems of neurons firing that are developed or learned while in a theta state.

Most people remember nothing of the earliest times of their lives, and anyone who believes they remember something of those days may be merely imagining things based on an understanding and observation of what fetuses and babies experience. Such logically induced "memories" could also be valid because they might partially stem from some potential subconscious recall of early brain connections that the conscious determines are consistent with known facts.

Hypnosis can theoretically be used to assist in recall of forgotten experiences in low frequency states of brainwaves, but hypnotic suggestions often only motivate a conscious rationalization for beliefs induced by those suggestions that are accepted without question. As a result, instead of helping reconstruct actual past experiences from traces of neural firings associated with the events transpiring at the time, hypnotic trances can lead to "false memories", including relating to imagining all sorts of things such as being Martians or non-human, as well as being abducted by aliens. Out-of-body experiences, which can be induced via electrical stimulation of the brain's temporoparietal junction that processes sensual signals to indicate one's location (Ananthaswamy, 2009e), is only slightly different since those perceptions apparently result from electromagnetic activity in the brain that causes a mistaken categorization of basic signals from the environment because of inadequate information about them.

The human brain's higher thinking logic evolved at least partially to correct such editing imperfections involved in automated subconscious processing and categorization of perceptions. It can do so if it is motivated to change subconscious firing tendencies that are leading to undesirable effects. A recognition of past subconscious programming during childhood or while otherwise in a hypnotic state can enable conscious evaluation of feelings and urges in various circumstances, thereby permitting a reflective self-awareness of whether they should be continued, suppressed, modified, or redirected.

Only after reaching the age of 12 do people's brains seem to spend more time in a normal mental state of more rapid beta waves, possibly thereby enabling greater conscious focus as they become teenagers. However, at all ages, there are at least occasional deviations from the foregoing listed range in an awakened state. For instance, conscious recognition of words requires more rapid thought waves (Thomson, 2010a) than is normal for young children. All such states of reflective self-awareness at any age enable consciousness collapsing its own superposition of possibilities on points of its own choosing.

The frequencies of 13hz or higher that create perceptions of an individually differentiated self-awareness is above the normal frequency for very young children, but such individualized consciousness may start at the "terrible threes". At three years old, toddlers are developing a sense of individual consciousness that reflects itself in a refusal to exist at their normally obedient state of theta waves. Such uprisings against blind obedience may stem from the emotion of spite, whereby prior compliance to learned rules or behavior hasn't been rewarded and may have even been associated with a penalty. As a result, the child naturally revolts and even seeks revenge. The spite provokes higher frequencies related to emotions that are in conflict with the frequencies associated with blind obedience. Even higher frequencies are required to harmonize the conflicts into conscious thoughts in the process of logically determining the optimal action based on the stimuli categorized at that frequency into self-awareness. This hypothesis is consistent with the conjecture that spite and weighing the advantages and disadvantages of actions motivated by it compared to submission to the wills of others was an important factor in the development of individual consciousness in the prehistoric past.

The low electrical frequencies of the earlier ages are optimal for learning without conscious questioning, including from verbal teaching by others once language develops (Lipton, 2008). While the delta waves typical at the ages of 0-2 years may reflect the frequency of a system of brain cells learning to fire automatically in response to external stimuli, the theta waves observed in children 2-6 years of age apparently can enable integration of learned neural firings in more coherent patterns across the entire brain in a process that permits the use of symbols or language. Symbolic learning permit individualized thoughts and interpretations on how to creatively carry out suggestions.

Symbols also facilitate the conscious analytical judgment of the large amounts of information available. Inconsistencies in categorizations can result in some groups of learned neural firings being out-fired by others firing stronger and more rapidly. A re-categorization can then occur that can harmonize thoughts to make them depend more on the situational context. However, incoherent thoughts derived from the suggestion of a completely trusted individual can lead to human spite and an absolute rejection of prior learned behavior and thoughts.

The behavior of 3-year-old children may not be much dissimilar to that of adults who attempt to discontinue interconnected electromagnetic neural firings learned automatically in the theta wavelength because of a hypnotic state while in love or otherwise mesmerized. Conscious spite against anything associated in any way with an exploitative hypnotizing agent provides the strong motivation to engage in electromagnetic activity of the correct wavelength to out-fire and suppress the unwanted neural firing connections.

Since dopamine rewards are only provided in both children and adults for the neurons firing in the process of making decisions which lead to actions that are subsequently reinforced, theta-state suggestions about which an individual isn't consciously aware can also be overcome if the neural firings associated with it aren't reinforced, or if other firings that dominate them are reinforced more strongly. As a result, if theta programming into a person as an infant or under deep hypnosis results in inadequate satisfaction of human needs and desires, the individual will be strongly prompted to change or reprioritize the basic set of assumptions upon which the individual operates in order to increase levels of pleasure chemicals in the brain. These neural rewards, which might only sometimes be related to physical pleasures, can include those associated with feelings of self-fulfillment for changing goals or priorities to be more consistent with human nature and trying to satisfy the innately rewarded urge to serve humanity as a whole. Incentives to change prior hypnotic programming also exists in the form of consciously rewarding feelings of being human or "right" when externally induced programming was wrong or counterproductively manipulative.

Thus, while theta brainwaves facilitate group consciousness and cooperation, they can also lead to extreme frictions when there is a perception of manipulation for the benefit of some individual(s). Higher frequency alpha waves permit some integration of symbolic learning in conjunction with the neural firings associated with external stimuli in the present, but it apparently takes the beta frequencies for integration of neural firings across the entire brain in order to enable the consciousness that incorporates forecasts of the future.

The higher frequencies of beta electrical activity appear to permit more filtering of learning from stimuli and other knowledge using logic, thereby allowing individuals to exercise a self-aware judgment on the usefulness and validity of such information and suggestions. As previously mentioned, the beginnings of such filtering may begin as early as the "terrible threes" when children under six may start to experience conflicting choices that they have learned absolutely without question from others like their parents or guardians. Frustrated by such conflicts or by a lack of reinforcement in the past for obedience in some circumstances, they begin to resist automated learning externally imposed on them. They then slowly begin to recognize when certain learned behavior is appropriate and when other actions are more suitable. Such discriminatory learning requires the beginning of conscious filtering instructions so that they know when to follow the suggestions and when not to. Such resistance and filtering may be reflected in electromagnetic waves at frequencies higher than the typical theta waves experienced by kids under six.

While animals can survive via purely automated behavior that is specified for well-defined circumstances, human beings need to learn to decide on their own which behavior is appropriate and when in their complex interactive societies. Because it isn't possible to teach the optimal behavior in all situations, human beings evolved the capacity via conscious thought to learn to customize their reactions by a logical evaluation of information both from their own experiences and learning from others. This information includes subconscious inputs in the form of intuition and feelings that help weigh the more concrete thoughts in the working memory of the consciousness. Stories are very useful for the purpose of helping people generalize what behavior is appropriate and when, because it isn't possible enough for them to learn as much from simple trial and error, or merely watching others. To optimally adjust to the environment that includes complex human cultures which rapidly change it, humans have to learn self-discretion in application of learned behavior.

Most brains learn from mistakes in logic, deduction, and actions via neural firings that are concentrated in particular portions of the brain (*NewScientist*, 2008c), although there is some overlap between the functions of different sectors of the brain (Gynn and Wright, 2008). The specialization of brain cells may enhance multiple tasking while the overlap may assist in coordination of those activities. The electrical frequencies indicate the rate of brain neural firings which act as signals to other brain cells and which form into systems of electrical firings that prompt activities in the body through nerve connections to the brain. Conflicts between the signals given by the different sets of brain cells require resolution through some prioritization process that is determined by the thinking area of the brain in human beings. This area has access to information from the rest of the neurons and the entire body itself, thereby permitting it to make decisions with more complete information of the situation or self-awareness.

Continuous neural firings tend to dampen weaker signals, and the decision-making area of the human brain apparently has the capacity to generally out-fire any other area of the brain if it merely summons sufficient will power. By exercising this ability, the "thinking" system of neural firings generates greater coherence among the various neural firings to achieve the tasks it determines are top priority. Since the decision-making sector of the human brain receives signals from other areas of the brain, it takes more energetic conscious firings to overcome more urgent signals from elsewhere. In any event, the thinking system of neurons must decide whether to outfire external signals or let them take control of brain activity to help them maximize the chance of satisfying whatever need or desire they are signaling.

Conscious thoughts themselves therefore reflect a coherence in neural firings that is obtained by filtering out incoherent and random brain cell firings via neural firings at higher frequencies. The coherent pattern that can allow informational inputs on the state of the body and brain enables logical reasoning about that information, and this reflective process, which includes an awareness of both one's own self-awareness and logical thinking, represents the consciousness that is experienced by human beings.

The conscious thoughts formed by coherent neural firings are typically reflected in symbols like words, thus explaining the conscious talking to oneself that can even take the form of two or more characters representing different points of view. The monologue involved in consciously talking to oneself reflects the self-awareness which doesn't need stimuli external to the brain like actual spoken words because the words involved in consciously talking to oneself are internal to the brain. Such monologues may assist the brain in logically reasoning solutions to problems in the symbolic form of language, as well as enabling addressing subconscious desires and emotions by summarizing them in a categorized form for easier analysis.

The coherence of conscious thought can come at a cost because the process of judging what information to filter out before logically analyzing the remaining information slows down the more rapid information filtering and decision-making processes of the subconscious. The subconscious evaluates far more information and much faster because it makes deductions based on simple but fairly effective algorithms. Subconscious processing can take place at the same time as the faster electromagnetic firings associated with consciousness are occurring, but those firings related to conscious thoughts can disrupt the firings related to subconscious processing. The disruptions are sometimes even greater when the conscious thoughts are directed to a task also set for subconscious performance because more of the same neurons are then involved in the two processes involving different frequencies, thereby creating confusing noise in both.

Sleep may be the optimal time for subconscious processing since there is normally no conscious interruption or noise then. Because the lowest firing frequencies aren't fast enough for translation of simulated decisions into the analytical framework of language used by consciousness, the subconscious simulations of dreams are typically communicated in picture form, and the subconscious deductions that derive from sleep activity result in neural firings that are normally perceived by the conscious as intuitive gut feelings.

To avoid the problem of slow reactive speed and indecision involved in too much conscious thought, the conscious can decide to adopt a particular personality prototype, with different choices selected for different situations. The most efficient subconscious reactions to the environment are thereby allowed while still having some conscious control exercised through the conscious choice of the character type that remains subject to reevaluation upon the appearance of pre-specified negative stimuli.

The highest frequencies of brain firings, such as exist in gamma waves, may require a complete focus of the entire brain on a single task that enables subconscious filtering of all information deemed to be irrelevant to that task. Gamma waves represent electromagnetic activity outside their normal range during consciousness. Such electrical wavelengths are associated with highly focused thought (Motluck, 2009), and the frequency of discrete perceptions of the environment speeds up during periods of such rapid electromagnetic brain activity (Fox, 2009b). The conscious decision to surrender the filtering process to the subconscious enables the most rapid coordinated brain firings that occur when there is no conscious interference.

Conscious focus on a single thought can, however, also result in the slower alpha waves that promote creativity by enabling some less coherent firings that are reflected in the mind wandering as in day-dreaming and that can later be transformed into the more coherent conscious thought involved in beta waves. Maximum creativity has been found to occur with lots of slow alpha waves and few gamma waves, implying that having significant periods of less focused thought and more imaginative dreaming is better for creativity (Motluk, 2009). Even artificially induced increases in beta waves tend to inhibit activities such as simple, sustained brain processes that are more suitable with slower brain waves (*NewScientist,* 2009zzzt), as the latter may be difficult to coordinate with more rapid firings.

Because of the lower amount of conscious judging and filtering of information for minds in the alpha-wave and theta-wave states, memories of thoughts formed during those periods are usually very weak unless there is a rapid switch to beta wavelengths. Such changing brain states may be prompted by the more random and incoherent thoughts at the lower frequencies suddenly resonating coherently at higher frequencies in what then can be perceived to be an insight, which can be transferred to consciousness for more coherent evaluation. Without such coherency, emotions, which are necessary for effective decision-making (Gigenrenzer, 2007), may be weak or even nonexistent, as neural firings at lower frequencies are more automatic and less subject to conscious judgment that incorporates inputs from the emotions. On the other hand, the lack of emotional judgment at lower frequencies may assist in the process of allowing free thoughts without the inhibitions associated with experiences of emotional rewards and penalties.

A neuron in any place in the brain fires when signals from neurons to which it is connected indicate that its own activation minimizes its own error function with respect to providing appropriate signals to other neurons, with the entire system of electrical activity designed to minimize the predictive error in the complete brain system of all neurons (Huang, 2008). The total integrated framework is dependent upon dopamine and other reward chemicals that help motivate groups of connected firings that are useful in adapting to the environment. The coordinated neural firings are grounded in a Bayesian learning process that is similar to that used by the individual neurons in the system of firing because each brain cell is thereby motivated to take the most probable successful firing actions by combining learning from prior actions with innate firing connections that evolved in an organism's genes.

The neural firings in the brain result in electromagnetic waves because of the photon and other emissions caused by the chemical processes involved in brain cell firings. The electrical impulses of a neuron firing just reflect the light waves involved in the heat of the chemical reactions in the neurons that involve joining the electromagnetic fields of different atoms or molecular groups of atoms. Each such brain cell effectively acts like single-celled organisms seeking gratification in the form of reactions with chemicals like dopamine, and this process motivates the Bayesian learning as the cells adjust innate firing dispositions based on new information on the probability of dopamine rewards that come with new experiences. The coordination of electromagnetic chemical firings actually may also be facilitated by quantum entanglement of fields that remain affected by prior contact, as well as by chemical and electromagnetic signals.

Conscious human thoughts tend to reflect neural firings in the most advanced portion of the human brain called the cerebrum, with the extent of the neural connections between the two halves varying among different individuals (Klein, 2003). The left side of the brain tends to be geared toward linear logical thinking that categorizes information into concepts, while the right side both captures more realistic details and is essential for creative and multi-dimensional analysis (Gynn and Wright, 2008). Creative thinking "outside the box" generally reflects more right-sided brain activity, where the box has been formed by the left-side's linear classification system.

The extensive information processing in the rights side as well as its summary analysis of it in the form of intuition and emotions aren't normally well understood by the logical left side where limited capacity to analyze more than a few pieces of information or symbols at a time motivates it to simplify reality with assumptions that often deviate greatly from reality. The separation of the brain into two interconnected halves, however, may facilitate integrated thinking on two different levels that provide two different perspectives simultaneously. By thereby examining an issue or problem on both a conscious and subconscious plane or firing frequency, many more angles maybe be deduced that enable an efficient multidimensional analysis.

Since creative thoughts are associated with more activity on the right side of the brain (Motluk, 2009), and since feelings also tend to be concentrated in the right side of the cerebrum (Bernstein and Nash, 2008), emotions may be important to generating creative insights into phenomena. Although the left side generally dominates the right in most people's brains, and tends to become even more dominating as a person ages in contemporary societies, the right side actually appears to be more efficient in directing a person's present activities and is generally more functional when operating alone (Gynn and Wright, 2008). The right side seems to be more reactive to present circumstances whereas the left side enables reflection based on logical analysis of both the present environment and future forecasted ones.

Integration of the two sides of the brain tends to occur more frequently in women and can enhance speech, mathematical skills, hand dexterity, and imagination (Klein, 2003), as well as many other tasks. In contrast, people suffering from schizophrenia appear to have problems coordinating neurons across the two sides of the brain (Gynn and Wright, 2008). That difficulty seems to inhibit their ability to apply logical analysis to all the existing information, especially with respect to the emotions, to draw reasonable conclusions about the present and future. While brain "hemispheric differences are not well understood" (Flanagan, 2009d), it is possible to utilize existing research findings to facilitate a better comprehension of both sides and their integration.

Happiness is experienced more in the left side as a result of a combination of chemicals and electrical signals that are related to the extent of dopamine in various portions of the brain (Klein, 2003). The most pleasant experiences, such as spiritual highs, have heavy right-side brain activity (Gynn and Wright, 2008), but happiness seems to be dependent on the judgment of the left-side of the brain that considers input from the right-side with respect to emotions, pleasures, and memories. Those inputs evolved to motivate survival and reproduction of the human species.

Dopamine along with other internally produced chemicals like opioids create feelings of pleasure that lead to the natural human drive for activities that lead to stimuli which cause their production (Klein, 2003). Satisfaction of basic human needs is associated with the chemical reinforcement, but so are efforts toward meeting such urges. Evolution caused successful behavior to be rewarded with neural excitations or firings in pleasure locations of the brain that include the nucleus accumbens and ventral tegmental areas, which are more sensitive in some humans than others (Nettle, 2007). Activation of the nucleus accumbens apparently occurs even prior to initiation of risk-taking behavior in anticipation of a potential reward (Kuhnen and Knutsen, 2005). Self-forecasting is thereby reinforced via anticipation of future benefits upon undertaking various actions.

The same chemical system in the brain appears to reward satisfaction of all human desires (Klein, 2003). The same reinforcing pleasure has even been found to be associated with erotic as well as monetary rewards (*NewScientist*, 2008x), probably because money acts as a secondary or indirect reinforcer that can lead to, and is therefore identified with, more basic desires like sex (Bernstein and Nash, 2008). Humans are thus motivated to take actions that lead step-by-step to final rewards. Interestingly enough, even near misses on monetary success such as in gambling activates the reward system within the brain (*NewScientist*, 2009m), possibly because humans evolved to self-reinforce behavior that is perceived to represent good efforts at moving toward successful attainment of the final goals of satisfying human needs and desires (Klein, 2003). Many of humans' goals were secondary or indirect to the most basic needs and wishes even in human prehistory and hence were good efforts at a final goal, such as constructing a tool for digging up nourishing roots.

On the other hand, negative feelings are associated with internal and external stimuli that decrease the chances of human survival and reproduction. Negative emotions such as fear are caused by neural firings in the amygdala part of the brain that apparently are affected by a gene which can lead to greater anxiety, neuroticism, and depression in some people (Nettle, 2008). The amygdala tends to fire when there are possibilities of danger, especially those associated with unknown risks or things not recognized or perceived consciously (Griggs, 2010b). Such emotions tend to prepare an individual to fight or flee, thereby leading to stress that is unhealthy if it continues long-term.

Feelings of depression are partly caused by the internal release of cortisol that readies a person to deal with a threat (Bernstein and Nash, 2008). However, that chemical inhibits risk-taking and cognitive abilities (*NewScientist*, 2008za) as well as depresses the immune system (Callaway, 2010b). Depressed states of mind may have evolved to help humans in trouble save energy and hide from predators (Fisher, 2004), but they can also help an individual to change paths or goals, as would seem to be especially advantageous in the case of the failure of specific strategies or the futility of trying to achieve the impossible (Marshall, 2009). Depression also apparently can serve to display sad emotions that solicit help from other humans who are empathetic (Nettle, 2007), as may once have been useful in the more egalitarian, cooperative societies of human prehistory (Murphy, 2008). In addition, since the brain tends to remember events only if they were associated with an emotional feeling (Klein, 2003), activation of the amygdala such as during periods of sadness or other negative feelings also enhances memory (Marshall, 2008a). Remembering what made one sad reduces the negative feeling (Hamzelou,

2010a), and part of the offsetting benefit may be the satisfying pleasure related to the learning gained from the experience.

On the other hand, long-term depression is associated with inhibited neural growth and development (*NewScientist,* 2010aas) and even results in a dying off of brain connections during extended periods of such feelings (Klein, 2003). Sadness provides the incentive to discover new information or re-categorize existing knowledge to enable more optimal behavior to avoid distressed situations, as a failure to do so causes a continuation of the negative sentiment. Eliminating some old neural firing connections can be useful in the process of rethinking one's behavior, but if they aren't replaced with new ones, the depression continues. There can be a tendency for the brain to not only eliminate many existing neural connections but also not expend the energy necessary to develop new brain processes when attempts to rethink everything are repeatedly unsuccessful. Depression tends to result in a person giving up totally on any activities that can cause adverse situations (*NewScientist*, 2010aar) because the individual's existing and new thought processes aren't leading to actions that are reinforced by positive stimuli.

Negative emotions inhibit learning if they persist, as they create stress, which is associated with the release of noradrenaline, adrenaline, and cortisol (Klein, 2003). That stress may be at least partially related to the energy involved in trying to remove an adverse situation, and giving up on doing so may be one means the brain has evolved to help temporarily relieve such stress. However, if unfavorable circumstances persist, the brain remains stressed to take actions to avoid the negative situation even though it has "deduced" that inactivity is the best response. By remaining sad, there may be a hope of help from others that often enough may have been forthcoming in prehistoric human cultures more attuned to mutual assistance. Often enough today, however, giving up merely leads to a continuation of the depressing situation.

The tendency of an organism to give up can be reduced when the brain cells react with one particular internal neurotransmitter called serotonin (*Newscientist*, 2010aar). However, sufficient neural reactions with this naturally produced brain chemical that moderates stress don't take place in depressed people whose learning therefore continues to be inhibited (Klein, 2003). Brain serotonin, which promotes greater cooperation with others even when it hasn't previously improved an undesirable situation (Clark et al., 2008), may have evolved because it normally leads to benefits resulting from organisms acting as a group. On the other hand, a human being can consciously deduce today that any activity, cooperative or otherwise, is futile, especially in contemporary cultures where exploitation is encouraged instead of the mutual assistance that would have normally relieved stressful depression in prehistoric humane societies. In this contemporary environment, any actions benefitting others can be logically deduced to be counterproductive because of the likelihood of negative feelings relating to the frequent exploitation and social contempt that results, and artificially induced serotonin may lead to only a delusional impact that can have serious side effects.

The stress associated with depression itself can also have very negative side effects. Stress concentrates the body on fight or flight situations (Bernstein and Nash, 2008) that distract the brain from firing neural connections related to knowledge acquisition. Since connections between neurons that are not used frequently weaken or are turned off altogether to save on brain energy, and since stress hormones like cortisol concentrate the body and brain on preparing for emergencies to the neglect of other tasks like learning, extended stress that is associated with long-term bouts of negative emotions like depression reduces brain connections, knowledge, creativity, and productivity (Klein, 2003). The fact that people learn better when knowledge is input at a regular pace than when a large amount of information is crammed in a short period of time (LeDoux, 2002) may at least partially reflect the stress involved in cramming that inhibits learning.

There is another chemical, acetylocholin, which could contribute to negative feelings like depression. Acetylocholin is produced internally in the brain and inhibits experiencing the pleasure chemical dopamine (Klein, 2003). The genetic processes involved in the production of these chemicals that cause adverse feelings may have evolved to inhibit rewards relating to the anticipation of forlorn hopes. They can also serve to provide feelings of satiation related to fulfillment of needs such as basic food requirements that reduce incentives for hedonistic excesses like overeating. It could also lead more generally to unhappy experiences of emptiness and uselessness when a person has prioritized the goal of overindulging personal desires in lieu of serving the interests of humanity as a whole.

Acetylocholin, as well as depression itself, may have important adaptive purposes by motivating changes in behavior to discontinue the behavior that leads to the adverse feelings they create. However, such unhappy emotions can have negative effects if their duration is excessive, as can occur if there is a failure or inability to modify behavior or thinking sufficiently to enable replacing the negative feelings with positive emotions. For instance, it can contribute to a long-term state of depression that is counterproductive to the individual and humanity as a whole.

Anti-depressants like Prozac typically promote the presence of serotonin in the brain (Bernstein and Nash, 2008) that reduces stress and therefore enhances learning (Klein, 2003). Melatonin is produced in the pineal gland from serotonin, and this chemical increases production of endorphins and oxytocin (Gynn and Wright, 2008), Serotonin can therefore promote personal happiness through pleasure and social contentment through bonding (Klein, 2003) as well as learning (Gynn and Wright, 2008). Since melatonin can be induced with sleep, the benefits of the artificial promotion of serotonin by anti-depressants may be enabled with such natural behavior. More sleep along with setting a goal of merely trying to serve humanity as a whole could be a potential cure for depression, since it would end the frustration associated with the inability to meet any particular goal while also promoting learning and other pleasures.

The amygdala, which is the primary center for negative emotions, varies in size for humans with different levels of neuroticism or tendency for depression, but it can always be willfully overruled by the hippocampus and right dorsolateral prefrontal cortex (Nettle, 2007). The hippocampus is involved in recording memories and categorizing them according to time and place (Esch and Stefano, 2005a), including biographical ones (Phillips, 2009a). The right dorsolateral prefrontal cortex controls and processes emotional information sent from the amygdala and "is involved in the control of impulses, judgment, and decision-making" (Gosline, 2009). Klein (2003) has indicated that the brain's conscious thinking portion in the left side of the cerebrum can overrule the emotional right-side of the brain and thereby inhibit negative emotions, and that this domination by the left-side of the brain can be learned even by the 1/3 of all humans who tend to be more dominated by "right-side" thinking and hence normally unhappier.

Of course, the thinking area of the brain covers the sectors beyond the right dorsolateral prefrontal cortex. For instance, the anterior cingulate and orbito frontal cortices are "involved in more thoughtful and reflective responses" (Young, 2008a). Those two brain sectors assist the dorsolateral prefrontal cortex in inhibiting some behavior, such as attempts to seek immediate gratification at the potential cost of reducing the chance of greater long-term benefits (Nettle, 2007). The right frontal lobe regions that overlap with working memory appear to be important for humans in making decisions with respect to self-restraint to achieve higher goals (Phillips, 2008b). Intelligence may have evolved partially in order to permit evaluation of such tradeoffs, as might have been motivated by the need for decisions relating to the long-term relationships originally needed for human survival, especially those involved in mating and raising offspring with a long-term dependency.

Like humans, apes also apparently weigh in a consistent fashion the benefits from different actions such as eating or looking at ape pornography (Phillips, 2008a), but that evaluation may take place more automatically in those primates. In particular, ape "analysis" may consist solely of the subconscious evaluating the relative immediate benefits that have in the past been normally obtained by established responses to stimuli. Primates as well as rats have decision-making capacities similar to humans, but they appear to be based on more subconscious processes (Douglas, 2007) that wouldn't consciously evaluate future implications. Chimpanzees display even stronger self-restraint than humans with respect to the patience needed to achieve a higher reward (Phillips, 2008b), but this restraint may also be strictly automatic.

Subconscious decision-making processes are based on simple rules that are highly efficient (Gierenzer, 2007), but those heuristics can be manipulated by others. For instance, those who recognize and adapt to the patterns caused by subconscious decision-making processes can engage in optimal behavior by sacrificing near-term rewards for the sake of greater benefits resulting from exploiting such patterns, as is especially possible via logical thought and forecasting. Human beings have developed special abilities to consciously weigh both current and long-term rewards and penalties from taking particular actions and expected future ones.

Humans can even use such analysis to reprogram simple subconscious evaluation of alternative courses of actions, thus enabling behavior based on more complex analysis. It seems quite likely that the conscious self-restraint involves the massive firing of the brain cells in the right frontal lobes that swamp the more limited activation of subconscious brain cells indicating one should seek immediate gratification. A conscious determination can be made to reprogram the subconscious to have the cell connections whose activation motivate immediate gratification be always silenced under consciously specified conditions via the firing of subconscious brain cell connections that urge postponement of such gratification until the urge for the undesired short-sighted behavior ceases. Such programming requires reinforcing the connected firings associated with delaying gratification with positive feeling like pride and happiness at having decided to do so, even if the anticipated better rewards can't possibly be realized until far into the future. In any event, self-restraint consciously programmed into the subconscious is more efficient and more effective than a conscious reevaluation and suppression of each individual desire as it arises that requires extra will power and energy, besides distracting from focus on other matters. Consistent suppression of a short-term desire for gratification will result in the neural connections related to them to learn to fire less frequently and eventually die out if there are enough rewards in terms of actual or anticipated future benefits.

The process of evolution has given each person's consciousness a great deal of individualistic latitude in placing relative values on short-term vs. long-term rewards, as well as with respect to the satisfaction of different human needs and desires. However, a very effective reward system has evolved to motivate each consciousness to always act with an intent to serve humanity as a whole. This reward is consciously experienced in the form of feelings of self-fulfillment. It is so overwhelmingly strong that the satisfaction of almost any human need or desire can be postponed if actions for immediate gratification are perceived to be inconsistent with the core purpose of consciousness to think logically for the benefit of humanity as a whole. The incentive is so powerful that people are willing to die for causes that they at least subconsciously believe benefit humanity.

Subject to that overriding incentive for the direction of conscious thought, consciousness has great liberty in determining the relative value of satisfying different human needs and desires, as well as their relative value at different points in time. Consciousness, which rewards itself according to those values it set, also has complete freedom of choice with respect to deciding how best to achieve any goals. Individualistic service for a collective purpose is thus innate to human nature.

Conscious complex analysis, especially when combined with the empathy that facilitates forecasting others, permits manipulation of beings living strictly via simple subconscious decision-making rules. The conscious can even manipulate its own subconscious. For instance, negative feelings are generally stronger motivators than positive emotions, leading to aversions generally being more powerful in governing behavior than rewards, but it is possible for both positive and negative feelings to exist at the same time in humans and even for the conscious to decide to have the positive emotions dominate (Klein, 2003). The conscious thus appears to be able to program its own incentive system for long-term benefits, subject to the overriding evolved purpose of consciousness to serve humanity as a whole.

However, once set consciously or subconsciously, the effective application of subconscious heuristics is actually inhibited by conscious thought (Gierenzer, 2007). Conclusions can be reached subconsciously that defy existing cognitive knowledge and logic but yet be useful for decision-making purposes since many events happen that appear to be anomalous only because science hasn't developed far enough to enable their understanding (McClenon, 2002).

Humans adjust their conclusions based on new information from experiences and learning that incorporates an evaluation of the reliability and relevance of the information (Bernstein and Nash, 2008). Such learning occurs both subconsciously and consciously. That process involves Bayesian learning, which apparently derives from natural electromagnetic chemical reactions that enable even inanimate objects to act if they are "learning" (Barras, 2010). A complex type of statistical analysis called Bayesian econometrics (Judge, Hill, Griffiths, Luetkepohl, and Lee, 1982) has been developed that essentially evaluates the probability of both new and past information having been generated by a set of believed assumptions about the existing environment, and this methodology seems to be similar to the natural learning process of neurons. Brain cells appear to use new information on receipt or denial of a dopamine reward to reevaluate the probability of receiving one in the future upon firing in conjunction with a given set of other neurons.

Human decision-making in any situation is based on the same Bayesian process utilized by the individual neurons involved in creating conscious thoughts to determine when they fire. People often use heuristics to make assumptions and deductions that are based on the limited information that can be recalled from memories, which can come from both personal experiences or learning from others, and this human learning procedure permits deductions in the face of the typically incomplete information, as only some information is perceived to be worth the effort of obtaining (Bernstein and Nash, 2008). These educated guesses seem to be based on the Bayesian logic which allows learning to both make decisions and change them for the same situation as a result of new experiences or information. This adaptive process results in responses to stimuli that are more likely to lead to rewards in the form of good feelings like happiness. Happiness itself derives from a categorization of all the neurons being reinforced internally in the brain with pleasure chemicals.

The neural firings that lead to conscious thoughts and decisions, which themselves are caused by different organized sets of neural firings, are heavily formulated by conscious and subconscious evaluations of the reliability of the information obtained, and such judgments are utilized in making the assumptions employed to model the situation for optimal reactions to it. Emotions seem to be vital in that process, possibly because they permit more efficient subconscious evaluation and categorization of the available information, as well as analysis of the potential cost and benefit of obtaining more information that is often extraordinarily uncertain.

Bayesian learning seems to develop naturally out of the electromagnetic and chemical properties of nature, as even inanimate objects like drops of oil displaying behavior that is similar to that of live organisms learning to adapt to the environment (Barras, 2010). This process may drive a learning of emotions as well. In particular, the neurons firing to cause particular feelings are rewarded with dopamine when they lead to successful actions and are discouraged when they don't, thereby allowing any particular genetic tendency for any emotion to change.

Conscious thought is also based on Bayesian learning techniques. For instance, second guessing impulsive acts based on subconscious intuition that derive from emotions stems from drawing lessons from prior experiences that resets the probability of the feelings being appropriate in such circumstances. In addition, the fact that a person's decisions or acts can affect the individual's satisfaction with future decisions (*NewScientist*, 2010w) implies that Bayesian learning affects even internal judgments as a whole about the future success of the conscious in making decisions. In particular, a prior decision that leads to failure typically reduces self-confidence in even activities unrelated to a particular situation. Moreover, in the process of exercising the capacity for complex decision-making that allows complicated generalizations of experiences and knowledge for modeling the environment, Bayesian analysis provides a logical means for both consciously and subconsciously evaluating the reliability of information and assumptions utilized in categorizing any situation.

The subconscious and conscious processing involved in solving an individual problem is extensive, and enormous confusion in thoughts results from trying to address too many issues at once. To avoid conflicting and confusing neural firings related to different tasks, the brain typically learns to focus all of its capacities on a particular activity. The emotions and other subconscious signals can then be directed to decision-making because subconscious Bayesian probability analysis that modifies existing assumptions about the environment is concentrated on only information relevant to the task at hand.

Early human forms likely engaged in their first learning by trial and error. Behavior rewarded with more dopamine reinforcement for the neurons involved in engaging the behavior was repeated. For instance, going to a specific tree was rewarded with the pleasure of eating the fruit, and so the neurons involved in the system of firings that caused the individual would be motivated with a reward to do so again until there was no more fruit. On the other hand, penalized behavior such as walking near lions was avoided via similar brain processes. This conditioning consists of not only pleasant feelings caused by chemicals created naturally in the brain like dopamine and endorphins, but also dynophrin, which is produced internally for the brain as a penalty to motivate aversion to stimuli that can damage an individual's body, such as extreme pain, hunger, pain, and other bad experiences (Klein, 2003).

Even complex modern human reasoning is based on certain assumptions and deductions from them that follow similar "if-then" arguments. This process creates expectations about possible rewards or penalties in the future that depend on both the realism of the assumptions as well as the validity of the logical conclusions following from them (Bernstein and Nash, 2008).

Humans continue to have a strong tendency to have their attention directed to situations or events that violate their expectations, thereby leading to changes in beliefs or assumptions (*NewScientist*, 2009d), which can enable learning new neural firing connections to improve decision making. People are thereby driven to exploration in order to learn to fire their neurons for better reactions to stimuli. The mere association of stimuli with subsequent events or rewards and penalties causes people to seek meaning in all phenomena (Klein, 2003). This quest is taken to the extreme in the search for the meaning of life, the universe, and everything.

Besides satisfaction of basic individual human needs and desires that include the pursuit of happiness, learning itself is also a task rewarded with pleasure chemicals in the human brain (*NewScientist*, 2006e). The human conscious is therefore motivated to yearn to know for knowledge's sake alone. Knowledge can enhance an individual's ability to satisfy human desires and attain other goals, and hence the evolution of the learning pleasure chemical associated with it seems natural. The chemical pleasure process associated with learning may have actually developed out of an anticipated future satisfaction of desires that is forecast with a higher probability because of the improved knowledge. Learning causes one million new neural connections to be formed every second at the same time that other neural connections are weakened when they no longer seem to provide useful information (Huang, 2008).

D. The Human Combination of Empathy and Spite

Humans have evolved to experience pleasure at helping and cooperating with others (Murphy, 2008). There are actually chemicals in the brain that cause humans to experience pleasure just from such actions (Begley, 2007b). Human nature typically results in helping others to be rewarded reciprocally, even if the assistance isn't equal due to inequality in talents. Some of the pleasure in helping may relate to unknown but expected future rewards forecast subconsciously that reflect learning from past experiences as well as innate tendencies. Cooperation does tend to naturally promote reciprocal treatment by others that is mutually reinforcing, as once was necessary to allow humanity to survive a world with stronger and faster predators.

Empathy, or the natural tendency of humans to experience what they perceive others to feel (Bernstein and Nash, 2008), may be an important aspect of this human characteristic of cooperation, insofar as helping others permits greater happiness of others that is then naturally experienced empathetically. The tendency of children to experience empathy with animals and even inanimate objects (Bernstein and Nash, 2008) facilitates a reverence for everything, including oneself, that many Native Americans feel (Peat, 2005).

This tendency can be accompanied by nirvana feelings of oneness with the universe (Strassman, 2001) that apparently are focused in the brain's parietal lobe (McClenon, 2002). This brain sector, the right side of which is also responsible for mathematical abilities (Thomson, 2010b), is itself involved in integrating all mental information (Rose, 2005). Given the extreme ecstasy that people experience when they feel a trusting unity with all else (Strassman, 2001), it appears that striving to act coherently with everything is strongly motivated in humans. Doing so can certainly facilitate collapsing the probability wave functions of all on points of maximum benefit to a person. In fact, the blissful feeling may stem from a resonance with all else that can create both strong electromagnetic potency and pleasure in anticipation of accessing that power. Since people's empathy, which is processed in the temporal parietal junction, can be increased with conscious deliberate efforts to do so, such as via compassionate meditation (Richter, 2009), increased coherence with others' probability wave functions can be freely willed.

People seem to naturally purge empathetic feelings with respect to humans or others who don't reciprocate cooperative efforts or attempts to help. Uncooperative or exploitative actions therefore reduce empathy. Spite is often felt instead. Purely selfish individuals may even be perceived to be evil. Such generalizations of evil may be quite logical given the brain's natural process of categorizing information for purposes of making better decisions with respect to both cooperation and other activities. For instance, if a person's own needs and desires are not being fulfilled, empathy may be overpowered by spite, thereby resulting in a natural categorization of others as evil, if they aren't helping the individual in need despite their own needs and desires being fulfilled.

Thus, there seems to be ingrained into humans a desire for fairness and equality in treatment, as my own children taught me well when they were very young. The unique human feeling of injustice causes negative reinforcement to the conscious when a person perceives manipulation, exploitation, or otherwise unfair treatment. Spite results in such cases which create a somewhat offsetting reward that is often felt in the form of teaching the exploiter a lesson. Spiteful acts thus represent a form of providing learning to others that may even be experienced positively by the spiteful person through empathy. In the case of a failure of uncooperative people to learn their lesson or any perception that the others can't learn to be humanly cooperative, the offending people may be categorized as evil. The uncooperative people may therefore be perceived as inhumane, thereby making those "monsters" deserving of exploitation by humans without any inhibiting feelings of human empathy whatsoever.

Categorizing specific people as evil tends tend to be maximized in more capitalistic societies, where people are indoctrinated into believing they are intrinsically selfish and are trained to suppress empathy. Friction, exploitation of others, and violence are thereby encouraged, and so many people are naturally labeled as evil. Whereas spite might tend to cause mere benign attempts to teach others to cooperate in humane cultures, perceptions of evil people can lead to very inhumane acts.

Spite and other reactions to injustices can be inhibited in any environment through various artificial means. For instance, beliefs in religions such as Christianity that "promise" a system of rewards after death for "good" behavior during life, as well as infinite punishment for "bad" actions, can motivate a repression of spiteful feelings. The "good" behavior largely consists of obedient servitude that the ruling elite is able to exploit. The fear of punishment, including not only a horrible but fictitious one in an afterlife but also real punitive actions such as result from penal systems, seems to be rather effective in keeping people from acting spitefully against the contemporary societies that cruelly exploit them.

In addition, since people often don't compare themselves with people above them in terms of some performance criteria when they don't feel like they can actually achieve the higher levels attained by others (Bernstein and Nash, 2008), propaganda and ideologies that make an individual feel inferior, such as by birth, genetics, inherited skills, or possession of wealth, seems to inhibit spite by creating feelings of unworthiness. Natural feelings of injustice between equals can thereby be offset if people don't recognize they truly are equal.

E. The Special Nature of Human Consciousness, Happiness, and Love

Consciousness largely consists of a working memory that is used for decision-making and forecasting (Bernstein and Nash, 2008). The conscious processor receives information inputs from the senses, emotions, and memories and it applies basic logic to evaluate that information and make decisions. The information received by the conscious causes changing perceptions of the self that helps it make better decisions.

Conscious self-awareness enhances decision-making since it improves forecasting what an individual will do in the future. Self-forecasting is especially important for humans whose actions increasingly affect the environment in which they must make decisions. Because the environment includes other self-aware individuals, conscious self-understanding also enables better forecasting the behavior of others through empathy.

Since the process of forecasting optimally requires determining the motives of the entity being forecast, it seems natural that humans would seek to figure out what motivates themselves. Humans' yearning to know the purpose of themselves and their lives may therefore follow directly from the self-forecasting process. In particular, as human brains evolved to reward neurons for coordinated firings that predict the future more successfully, they were driven to try to forecast their own actions as well as that of others. A desire to know people's reasons for actions would have naturally followed from the drive to predict humans, and so there grew the incentive for humans to deduce a purpose for themselves.

Such a drive for knowledge could have developed naturally from forecasting in a mutually self-reinforcing process. In particular, predictions enhance decision-making with respect to actions to satisfy needs and desires that rewards brain cells with pleasure chemicals like dopamine. Forecasting itself enables imagining future anticipated satisfaction of needs and desires that are felt in the present to help in evaluating the relative optimality of any similar situation. In particular, the act of forecasting results in pre-rewarding neurons with dopamine that would be reinforced if the predicted situation happens.

Both the Bayesian learning process and evolution motivate the neurons involved in predicting to engage in forecasts that were more likely to result in successful satisfaction of needs and desires by withholding dopamine in the case of failure. Imagining situations that have been judged to be unlikely to occur is discouraged because frustration of previously anticipated pleasures cause negative feelings of unhappiness and dissatisfaction with oneself. Wildly unrealistic fantasies are more often predicted in children because they haven't yet learned what is actually infeasible.

Forecasting, imagining, decision-making, motivation, and actual stimuli appear to be so integrated into the brain's electrical circuits that planning an action and perceiving having already engaged in it differ with respect to neural activity only in terms of the relative magnitude of the firings of a particular network of brain cells. For instance, it has been found that electrical stimulation of particular areas of the parietal cortex make people want to initiate movement of specific parts of the body, while stronger electrical stimulation of that same brain sector create the feeling of having already moved those body parts (*NewScientist,* 2009zu).

Because understanding everyone's reasons for actions greatly assists in forecasting future behavior that affects the typical environment of people living in groups, humans have evolved a yearning to understand the motives of others as well as their selves. For optimality in predicting complex human beings, understanding human motivation requires knowledge deeper than mere recognition of simple stimulus-response incentives.

Empathy is very useful for this purpose, although a failure of empathetic feelings to result in correct prediction of some person, organism, or thing can create a natural incentive to turn off all such feelings. Uncooperative or exploitative people and animals can then be perceived as monsters that might be considered like objects or lions. Such categorization of things based on their behavior may have been the forerunner of systems of ethics that regard everything as either good when acting like oneself or evil when not.

Such categorizations not only helped in predicting but also in establishing moral guidelines for human beings to follow. Since people in prehistoric times were characterized by a group consciousness and motivated to help the human group to survive, people naturally learned to believe anything contributing to helping other humans was good and therefore to be given reciprocal aid. Anything harming humans was evil and therefore to be avoided or destroyed. Everything else was an object to be exploited. Such morals are innate to human nature, which evolved to perpetuate humanity.

For maximum depth in understanding themsleves and others, humans naturally developed a desire to understand their own purpose. Thus, the forecasting that led to consciousness also created a desire for a deeper understanding of the meaning of life and consciousness itself. In particular, the complexity of all the various stimulus-response incentives and the art of forecasting motivate knowing consciously the reason to determine the overall purpose for which consciousness and its forecasting powers evolved. Knowledge of this purpose can assist the conscious in utilizing its capacities for rational logic in decision-making that leads to thoughts and behavior which optimize the various dopamine and chemical rewards provided to the brain's cells that are experienced in a conscious feeling of satisfaction and self-fulfillment.

Knowing the purpose of existence therefore not only assists in forecasting, but it also can aid human beings in making decisions on their own behavior. In particular, in the quest for positive feelings, the conscious seeks to make decisions that lead to the maximum rewards, and codes of norms are naturally developed internally for such decision-making. These guidelines or rules for behavior learned from experience can be taught to children in the form of morals that define actions as good, evil, or neutral.

As previously explained, the innate human morals of cooperation that evolved create a natural tendency to define people exhibiting cooperative empathy as good whereas exploitative people are labeled as evil. The former are regarded as human beings while the latter are treated like animals. Those who don't exhibit natural human empathy are often regarded as subhuman objects to be exploited.

People are also widely categorized by labels like "good" and "evil" based on whether they obey a particular society's morals. While such cultural norms have been influenced by the innate morals of humanity, exploitative elites typically try to create norms for behavior that maintain their power, and so they distort ethical systems with religious fictions that include cruel codes of conduct that they actually label "morals".

Many divide consciousness into two parts: a conscience that is determined by morality and a rational consciousness that assists with decision-making (Kharkhordin, 1999). The Freudian superego, which represents rules and morals imposed on the individual by society and culture, and the ego, which corresponds to a rational consciousness designed to achieve the goals of an individual's unconscious pleasure-seeking id (Bernstein and Nash, 2008), seem to represent a similar separation of consciousness. While each human conscious is really an integrated whole that is created by the coherent set of neural firings in any moment, there can be information from different parts of the brain that create conflicting implications for behavior. For instance, artificial cultural norms brainwashed into a person may signal one action while innate human nature is pushing for another deed that is inconsistent with society's "morals", and consciousness must logically choose between the alternatives.

Conscious thought is affected by past experiences, learning, and the opinions of others, but each person can consciously set both goals and morals as well as the strategies and tactics to achieve the goals within the ethical guidelines that the individual establishes. Even codes of conduct hypnotically indoctrinated into the subconscious at an early age can be eliminated when they are recognized to stem from exploitative religions and lead to inhumane behavior like abstinence.

Consciousness evolved to not only engage in complex processing of information to consciously make decisions but to also set rules for the subconscious to follow that include establishing priorities for decision-making, action, and information filtering. While the conscious is free to select any goals it chooses, neuron firings signaling inadequate satisfaction of basic human needs and desires like nourishment strongly motivate the conscious to concentrate on meeting such human necessities, especially since it is recognized that setting the wrong priorities can result in death.

Normal conscious thought leads to neural firings throughout the entire brain that synchronize both the frequency and phase (*NewScientist,* 2009zg). This fact is consistent with the conscious view of the self having to be consistent with information about reality (Robinson and Ryff, 1999) that comes from the emotions, senses, and subconscious. Unless human needs, desires, and emotions set the values for the goal-setting process, the application of conscious logic with an alternative purpose, such as serve an elite in an inhumane hierarchy or act only out of hedonism, causes an unhappy useless sense of meaninglessness.

Like the subconscious and the rest of the brain, the conscious is designed to learn from the past, including from mistakes, via a system that rewards good decision-making. The fact that memories are consciously experienced from an external point of view (Ananathaswamy, 2009e) is consistent with conscious memory having evolved to permit logical analysis for decision-making purposes that is detached from the emotions involved in self-awareness. This detachment from subconscious feelings at the moment enables a more objective analysis of a situation that itself may have spurred an emotional response which was rewarded.

Human consciousness itself seems to be dependent on the functioning of the thalamocortical system of the brain, insofar as the thalamus acts as a filter of inputs from the rest of the brain to the cortex thinking sector, and all the interconnected neural firings and other activities within the cortex cause conscious thoughts (Nichols and Chemel, 2006). This system has the conscious, which seems to be concentrated in the prefrontal cortex of the brain, make its decisions based on inputs from other parts of the brain, i.e., the subconscious. The brain's frontal lobes, which are involved in organizing and categorizing information as well as complex reasoning (Owen and Highfield, 2010), can control information flow into it through the thalamus filter and thereby allow the conscious to also determine that it needs additional or alternative sets of information. The thalamus thus coordinates the coherence of neural firings in the subconscious with those of the conscious to provide information deemed relevant.

The self-aware processing system of consciousness experiences happiness at making "correct" decisions while unhappiness or regret is felt otherwise. Unlike the other brain processes, however, the conscious is the sole judge of what is "correct" or best in terms of undertaking actions to satisfy human needs and desires, as well as self-realization goals it also sets for itself. As ancient philosophers like Aristotle long ago hypothesized, happiness comes from one's actions, with happiness for one's activities being dependent on the person doing the best that is perceived to be doable under the circumstances (Klein, 2003). Happiness doesn't appear to be necessarily dependent on the success or outcome of the actions but merely on an ex-post conscious evaluation that one did the best one could in the situation encountered with the available information.

Since personal satisfaction lasting for weeks can come from just writing about a "positive experience" in one's life (Jones, 2010), it appears that the reinforcing benefits from making a correct decision can be rather long-lasting, despite intervening failures, if one just concentrates on past satisfying experiences where one exercised good judgment. Although the conscious appears to have a great deal of latitude in applying rewards and penalties for its decision-making success with happiness and unhappiness chemicals, respectively, input signals from the rest of the brain and elsewhere can influence and create boundaries for that evaluation process. It is therefore easier for the conscious to reward its decision-making with happiness when its decisions can be logically concluded to be responsible for situations that are physically and emotionally pleasant. On the other hand, it is more difficult, albeit not necessarily impossible, for it to perceive decision-making to be "correct" when standard "if=then" logic implies that it is to blame for adverse situations.

The greater confidence after making a conscious choice that exists in more introspective people who tend to have more "grey matter" in their right anterior prefrontal cortex (*NewScientist,* 2010aaf) may reflect the greater insights into the purpose of both making and judging decisions that such individuals could have developed through their introspection. As has been explained, there seems to be strong motivation for the intent of actions to be consistent with serving humanity as a whole, but the conscious has freedom of choice in determining behavior and evaluating it.

The internal "judge" called the conscience in religious and philosophical studies (Kharkhordin, 1999) is somewhat similar to the actual human conscious. However, the conscious judges more than just morals and intent insofar as it also evaluates success in the development, implementation, and execution of plans and actions. In addition, the conscious judges its own intent and even the set of morals it chooses to follow.

Even though designed to maximize satisfaction of human needs and desires, this reward/punishment system for optimal decision making seems to be more powerful and motivating than even the chemical incentives in the brain that motivate satisfaction of basic personal human needs and wants like nourishment and sex. In particular, because satisfaction of needs and wants can occur in many different periods in the future as well as in the present, and because the future itself is uncertain with forecasts being mere probabilities that the conscious itself must estimate, the situation is too complex for a simple reinforcement system. Instead, a scheme of weighting the various possible outcomes and their risks had to be established for evaluating conscious forecasting and decision-making. Consciousness, which evolved to assist in decision-making to meet future needs by forecasting potential changes in the environment, needed an incentive system for it to do so, and feelings of happiness or unhappiness are the evolved result.

Conscious analysis is strongly motivated by that system to serve more than an individual's immediate needs and desires. In particular, since consciousness incorporates a self-awareness of stimuli and emotions, its evaluation of decisions is affected by more traditional motivating systems like subconscious emotions and feelings of want or satiety. In addition, because human beings had also evolved chemical brain rewards to motivate cooperative behavior for the common benefit, consciousness had to address the task of best meeting those social or higher goals.

The complexity of trying to weigh the relative importance of relationships, including with both present and future members of the clan across all time periods, was made even more complicated because the relative unpredictability of the other human beings themselves increased the amount of uncertainty in prediction. The problem was resolved with the evolution of an ecstatic feeling of unity through the love of all human beings, including past, present, and future ones, who were also acting to benefit humanity as a whole long-term. While the conscious was given wide latitude in evaluating its own performance in this task, it is also discouraged by feelings of emptiness and lack of self-fulfillment at behavior that deviates from its purpose.

These logical deductions on the nature of happiness as a motivation system for the conscious forecasting and decision-making are based on a substantial amount of empirical evidence. For instance, happiness indicated by natural smiles is controlled by the logical thinking portion of the brain (Bernstein and Nash, 2008), and so happiness is clearly connected to conscious processing. The fact that young children typically indicate via crying far greater emotional responses from frustrated behavior than from failures to fulfill important needs and desires provides an early indication of the importance of being satisfied with achievement of tasks in being happy. Only later in life can people learn that just trying to accomplish goals can lead to happiness even without success.

Since happiness is experienced on the same side of the brain (the left) as the rational portion (Klein, 2003), a hypothesis that happiness serves as a motivator for optimal conscious analysis that includes both forecasting and decision-making based on those forecasts seems like a logical explanation for this feeling that everyone seeks but has so far escaped the grasp of most. This hypothesis can also help illuminate some otherwise counterintuitive findings. For instance, having too many choices reduces happiness due to buyer's remorse (Loftus, 2009), even though having some choice in decision-making is very important in creating happiness (Klein, 2003). It appears the conscious is more likely to be less than fully satisfied with a decision if there are more choices, not only because of greater expenditure of effort needed to decide but also because the probability of making the optimal selection is reduced by more choices. In particular, while the conscious isn't satisfied with itself unless it is employed to make decisions, it tends to be more self-satisfied with itself if there is a greater chance of making the optimal choice.

The fact that hedonism doesn't bring happiness (Parvin, 2008) also strongly supports the hypothesis that the conscious is motivated more by higher goals than mere satisfaction of needs and desires. Given that many humans are willing to sacrifice their very lives for others or for an idea to help humanity, the overriding importance of higher goals seems rather clear. Even though the conscious has wide latitude in setting goals without being penalized with feelings of emptiness from mistakes as long as the overall purpose is humane, the suicides of the rich, famous, successful, and hedonistic are good examples of the meaninglessness of purely personal goals.

Robert Stewart (2009), a Professor of Psychology at Oakland University, has stated that "Self-acceptance … is the best single predictor of satisfaction with life". Since the human conscious has a large degree of control over its own reward and punishment system, self-satisfaction can be consciously willed through self-acceptance under any circumstances, but such self-acceptance requires a feeling of self-fulfillment that can only arise from following the purpose of trying to serve humanity as a whole. Focusing on doing one's best in serving humanity represents the naturally evolved way to avoid the frustration associated with making the wrong choice(s).

Since all conscious decisions are influenced by inputs from the emotions, the satisfaction of innate physical needs, and other neural connections to the subconscious, the conscious setting of its own criteria for happiness is always affected by the subconscious. Conscious setting of goals must therefore be consistent with evolved human needs that include not only physiological requirements like food but also a feeling of self-fulfillment. A feeling of self-fulfillment seems to be dependent on attempting to serve humanity naturally as a human being, a trait that evolved as part of the need of human beings to cooperate in order to survive.

By setting a goal for oneself of trying to serve humanity naturally, one automatically generates sufficient dopamine to avoid behavior that one believes is inconsistent with that objective. When the goal is merely the attempt to serve humanity in a fashion that one thinks is best in any moment, failure in actual positive service of that objective doesn't create any of the frustration that can motivate returning to bad habits. In addition, establishing a goal that merely relies on intent and not on outcome enables rewarding humor from laughing at and learning from the failure of any actions, before, during, and after the fact. Actions to try to serve humanity as a whole therefore motivate uninhibited activity that enables the most efficient processing by the subconscious without interference by the conscious, which itself is thereby set for optimal operation by focus on the overriding objective of serving humanity at all times. Such focus also helps the person resonate with the human nature of all other people for expanded power, as will be explained in more detail later.

While the logical left-side of the brain that tends to establish purely "utilitarian" systems of ethics (Flanagan, 2009d) might "rationally" attempt to internally establish a goal of mere maximization of personal happiness or pleasure without regard to the circumstances, neural connections to the right-side of the brain and to the subconscious inhibit such happiness hedonism. Even without such connections, self-satisfaction itself seems to require a feeling of self-fulfillment and higher purpose that go beyond mere hedonistic feelings of happiness. In addition, since the human brain requires ever more dopamine to maintain a feeling of satiety or happiness (Klein, 2003), it doesn't seem possible to have the conscious merely will continuous happiness without meaning, just as narcotics can't provide long-term pleasure as ever greater amounts of them are needed to maintain the same "high". At some point, the dosage or the abundance of dopamine becomes toxic. Setting a goal of serving humanity seems to avoid that pathology by providing varying amounts of dopamine in an evolved tendency to optimally both reward and motivate further efforts at serving humanity as a whole better.

While feelings of happiness enhance creativity and problem-solving abilities, as well as promote a proactive attitude, positive moods can also cause a person to make more careless decisions, as well as make an individual both less logically persuasive and more gullible (Jones, 2010). It appears that too much dopamine reinforcement for decision-making results in less conscious effort being exerted, resulting in less optimal decision-making that may make one subject to manipulation, as can lead to especially adverse results in today's societies that glorify exploitation of thy neighbor. On the other hand, an inadequate level of happiness results in the reinforcement of rational decision-making insufficient to motivate the energy needed for creative thinking and will power, thereby leading to depression, passivity, and less rational thought for the opposite reason. In either case, inputs to conscious self-awareness from the emotions and the state of human needs are intrinsically designed to help motivate a more optimal amount of rational decision-making and creative thinking.

The human brain has evolved to optimally balance the feelings of happiness depending on the situation. The finding that humans tend to have a fairly stable level of happiness throughout their lives regardless of extremely beneficial or adverse events

(Dreifus, 2008) represents important evidence supporting this conclusion. The primary chemical determinant of happiness, dopamine, is generally released in the brain to maintain a normal mood, but an abundance of this dopamine tends to be provided as rewards for some actions, such as those that satisfy human needs of nutrition, while shortages of chemical reinforcers exist when there are unsatisfied desires that result in aversive feelings (Chemali, Chahine, and Naasad, 2008). Negative experiences include a feeling of emptiness from failing to serve humanity's purpose.

Happiness can come from merely thinking about how to change an adverse situation, as well as from a general positive attitude that can result from both feelings of gratitude and questioning negative feelings (Jones, 2010). While happiness can thus derive from mere conscious thoughts, they have to be directed for the purpose for which consciousness exists for long-term happiness.

The level of happiness is a function of one's own perception of the optimality of current decision-making as well as one's own evaluation of one's situation that reflects on the optimality of one's past decision-making. This system, which rewards decisions perceived to be correct with happiness and punishes decisions perceived to be incorrect with reduced happiness, is important for optimal decision-making.

As previously mentioned, long periods of inadequate happiness are reflected in depression that creates an incentive for figuring out how to improve decisions and change behavior. The conscious is also motivated to avoid the mistake of allowing the experience of too much happiness as it can result in a hangover effect. It can subsequently even lead to depression not only because it needs to grow to continue to provide satisfaction but also because it can actually cause failure during the period of gloating and subsequent unhappiness. In particular, selfish gloating distracts a person from the goal of serving humanity as a whole because it concentrates attention on oneself and not the overriding goal of a human being. Such imbalances that can result in carelessness and even passive depression may be avoided by mere focus on the innate human purpose of human beings that could innately prevent extremes.

It has been found that the brain functions best when it operates at an average 75% level of happiness (Jones, 2010). Oriental philosophies such as Buddhism that emphasize balance (Capra, 1999) seem to be partially coherent with this view of happiness. In contrast, the chronic depression existing throughout the U.S. (Weiss, 2005) implies that the Christian hierarchical belief system, which is so prevalent in that country, is not conducive to such balance. A person's internal belief system does appear to be crucial in affecting feelings of happiness (Bernstein and Nash, 2008), but a mere belief in balance without a higher purpose creates a feeling of meaningless emptiness that is itself warped and imbalanced.

In spite of these facts, it is interesting that mere existence of "belief … may stimulate naturally occurring health processes" in humans that are especially useful in reducing stress (Esch and Stefano, 2005a). Perhaps the mere conscious decision to believe in something with a perceived reward from that belief represents a decision and forecast for which the conscious provides its own rewarding happiness and satisfaction. However, beliefs inconsistent with reality eventually result in chemical incentives to change that are stressful. While stress can help organisms overcome unfavorable environments by enabling appropriate actions (Bernstein and Nash, 2008), can promote an organism's resiliency (Mattson and Calabrese, 2008), and can even enable genetic mutations that enhance adaptability in organisms such as bacteria (Calleman, 2009), the effects of stress can also be excessive enough to be destructive and counterproductive to an organism's survival (Bernstein and Nash, 2008).

Belief in something that will lead to a better situation may be a natural human tendency that evolved to reduce stress. However, a belief in a fictitious deity is guaranteed by reality to lead to frustration in achieving a desired improved environment. It can only lead to the hell of emptiness that has to occur no later than the moment of death. In contrast, a belief in humanity not only maximizes the chances of a better future environment while alive but also the eternal heaven of having served the true purpose of each human being.

Although the conscious normally attempts to balance feelings of happiness to inhibit excesses, it is also possible to develop seeming imbalances in this system. For instance, many believe that personal happiness is the purpose of life, and many different procedures have been determined to help achieve happiness, including by helping others and by concentrating strictly on the present moment (Klein, 2003). However, focusing solely on the present cannot alone bring a long-term feeling of self-fulfillment, as the human conscious has evolved to live for the future. In addition, helping others without a higher purpose or reason for doing so may only result in exploitation in contemporary societies either by the very people who are helped or those who exploit them.

Each person's conscious appears to have developed to be motivated to try to advance humanity into the future but allowing complete individual free choice with respect to the means of doing so. While advancing humanity can sometimes require self-sacrifice for the benefit of others or the future, selflessness can certainly be taken to a very unbalanced level. For instance, some religious or philosophical hermits claim to have found nirvana in deprivation and only minimal satisfaction of personal needs (Baring and Cashford, 1993). While such activities can lead to inspiring insights, it is questionable whether they actually found anything more than a hallucinating delusion that eventually leads to emptiness if they didn't come to the conclusion that helping humanity is their higher purpose in life.

The fact that there was no word in Germany for happiness until less than 1000 years ago (Klein, 2003) may be related to inhumane religious views. In particular, a conscious prohibition on happiness may have been related to a severe patriarchal culture, such as the Christian one which encourages passive suffering on earth, perhaps to avoid threats of a worse hell in an afterlife. People adhering to such cruel beliefs may only hope in complete futility for a better existence after death, as do some extreme religious cults in the USA today that discourage smiling as an indication of a sinful act.

Somewhat similarly, severely depressed individuals seem to give up on meeting most wants and needs, probably because of a learned helplessness caused by past frustrations of attempts to satisfy their desires (Bernstein and Nash, 2008). The only incentive to continue life for many depressed people is the hope for better in an afterlife. Depressed feelings of helplessness in the face of seemingly uncontrollable events in an overwhelming environment can be eliminated via feelings of spite or rage (Klein, 2003), but patriarchal religions such as Christianity typically try to brainwash and/or people, especially women, into accepting positions of powerlessness without resistance (Walker, 1987).

Imbalanced mental systems can be subconsciously created, as well as consciously. They can even be rewarded when there is greater satisfaction of human needs and desires with such imbalanced systems in place. For example, people may be severely disappointed or frustrated when they employ more balanced systems, whereas they may receive more attention, help, hand-outs, or even an expectation of heaven in a forecasted after-life when an imbalanced depression dominates their minds.

The current society's rewarding of extremely selfish behavior, such as psychopaths who tend to rise to the top of organizations (Nettle, 2007), is another example of imbalanced mental systems being promoted by inhumane societies. It is important to recognize that although many people try to use success in attaining larger amounts of wealth to offset "social shortcomings", having goals relating to money and associated status are negatively correlated with happiness (Buchanan, 2009b). The psychopaths could be driven by their unhappiness to seek ever more short-term rewards for themselves in a possibly futile attempt to be both happy long-term and purely selfish.

While the conscious itself acts as its own judge and jury of its own performance, considering only information it wishes to consider, there seems to be a subconscious process that leads to an inadequate amount of dopamine in some sectors of the brain unless there is the goal to help humanity as a whole. Selecting that purpose as supreme not only yields self-satisfaction but also creates a rewarding feeling of self-fulfillment regardless of the perceived short-term or long-term effects. Mere submission to that purpose also maximizes the chances of winning the blissful admiration of future human beings, who are also innately driven to that goal.

Thus, while the conscious has been given the authority by evolved human nature to judge its own individual actions and to distribute to itself a level of rewards of happiness that is a function of its own verdicts, human nature evolved to penalize it with empty unhappiness eventually if its decisions on actions or verdicts are inconsistent with the evolved goal of humanity. Service to humanity has thus been programmed into human beings as the only means to eternal bliss and happiness. However, if consciousness selfishly distributes too many rewards to itself without adjusting the level based on success and failure, it automatically creates a feeling of worthlessness, and so even recognizing and following the purpose of humanity results in varying levels of happiness.

Information concerning human needs and feelings are also well-known to affect conscious thought (Klein, 2003), and so it is a documented scientific fact that the conscious can't totally control a conscious feeling of happiness. As previously explained, the evidence on the unhappiness and emptiness associated with pure hedonism indicates that the feeling of happiness does involve much more than satiety with respect to basic human needs and desires for preservation and reproduction of the self. The needs and happiness of humanity as a whole must also be considered for true happiness, as is indicated by even patriarchal religious studies showing the importance of compassion in inducing feelings of happiness and self-contentment. The conscious is motivated to consider not only negative feelings such as pain, hunger, and negative animalistic emotions that are concentrated on the right-side of the brain but also a perception that one is serving a higher purpose which is consistent with the evolved human nature to perpetuate humanity. This happiness system is adaptable to many different environments, as is a great deal of flexibility in the system for balancing individual and group needs with those of humanity as a whole. While consciously making improvements in the happiness system represents one of the functions of consciousness, it is impossible to modify the overall framework without destroying the human being and all the related power in the human body and mind.

The concept of happiness, which "is surely the most coveted of emotions" and has been defined by some philosophers (such as John Locke) as the "Utmost pleasure" experienced by humans, derives from a Greek word for "good spirit" and can be translated in several European languages such as French and German as "good luck" (Chemali, Chahine, and Naasan, 2008). Feelings of happiness result in an activation of the brain's anterior cingulate cortex, which is also associated with a concentration on emotions and social interactions (Bartels and Zeki, 2000). This fact may provide a clue as to the innate balance that the happiness framework imposes on humans. In particular, while a person can consciously think oneself into a happy state, including through meditation, etc., that particular positive feeling naturally causes people to think about their social relationships and related emotions such as spite and love.

The left side of the brain that is less emotional than the right side tends to be more activated in happy states, including during meditation (Esch and Stefano, 2005a). Any meditation-induced relaxation of the anxieties associated with amygdala activation (Bond, 2011) enable greater focus on social relationships without inhibiting, counterproductive phobias. Such focus may motivate the individual to try to act for the common good of a group of humans, as Aristotle himself hypothesized people optimally should for maximum happiness (Kharkhordin, 1999).

Although the conscious establishes its own goals and behavior, humans have evolved to experience the highest pleasure from mutual love for other humans. This hypothesis is well supported by evidence indicating the importance of social relationships and helping others in achieving happiness, which itself creates a desire to give to others (Klein, 2003) that can create an eternal self-perpetuating cycle. Human love generally is also mutually reinforcing through empathy that helps ensure maximum satisfaction of all personal needs and desires. In particular, via empathy as well as chemicals in the brain that reward cooperative behavior, humans have evolved to seek to maximize the happiness of not only the self but also of others. On the other hand, human spite and fear thereof inhibit any individual from selfishly exploiting that tendency of humans.

The human balance of empathy and spite that helped motivate the balance of selflessness and selfishness results in humans naturally seeking to maximize the satisfaction of the needs and wants of humanity in the aggregate. Chinese philosophers tend to emphasize the balancing of some other set of characteristics like yin, or "female" intuitive wisdom and receptivity, and yang, or "male" strength and aggression (Capra, 1999), but such merging of opposites would seem to result as easily as using the left and right hands if an optimal balance between selflessness and selfishness is in place. Such a balance motivated by empathy and spite may follow very naturally in a harmonious society. As previously explained, empathy and spite are merely two sides of the same altruistic coin that promoted the continued existence of humanity. Even overtly selfish behavior that isn't unfair to others promotes humanity by enabling oneself and one's offspring to help humanity further.

Close (2009) has compared collisions of yin and yang to the self-destructing of matter and antimatter upon contact that produces gamma rays, which in turn induce the creation of new matter and antimatter out of a vacuum. Such self-destruction may be analogous to the male and female destroying their separate selfishness and even their separate selves, which must change to harmonize their partnership through their mutual surrender to each other. Their egos must merge for the good of the spark-filled union that can produce offspring of a combined or different nature. Since the physical separateness of the mates remains, however, a male and a female in romantic love may act physically more like a particle and an antiparticle attracted to each other that are then balanced by the attraction to a third fermion, which is analogous to the child that is formed from their virtual union, to form a quark of three tightly bonded fermions moving as a unified whole despite the pulls in opposite directions through time. Here the magical creation of the third fermion needed to form a bonded quark out of nothing may have originated in forecasted conceptual form upon the pair's initial mutual attraction, since love, sex, and offspring are three sides of the same procreative coin.

However, such balanced love and feelings are very difficult to achieve in today's sick societies whose foundations are ideologies of pure selfishness. Since indoctrination systems cause people's innate tendencies for selfless service to others to be contemptuously manipulated for exploitation by others without any intention of reciprocity. Examples of such exploitative ideologies can be found in the lies of today's patriarchal religions, which solicit behavior which is labeled selfless but is clearly motivated by each individual selfishly seeking personal rewards and punishments promised after death. Religions also employ both emotional and habitual rituals to "keep the flock in check" (Douglas, 2009e). The true purpose behind the religious propaganda systems that solicit "selfless" behavior on earth is to keep people docilely working for the benefit of the rulers.

People can be made to think they are purely selfish and to act without concern for others if their basic current needs aren't being met. Since humans naturally think about the future, the same selfishness holds when a human being has fears with respect to meeting future human needs, as are maximized in non-caring capitalist societies. However, most people's lack of knowledge of what really does happen at the end of physical life, concerns about existence after death create fear that is manipulated to the maximum by cruel religious leaders to obey their dictates or suffer a horrible afterlife.

When an individual's own current or future needs appear unsatisfied at the same time that the needs of others in the group are, human nature tends toward pure selfishness. In today's cultures, the necessary physical environment with adequate nourishment available to all now and in the future doesn't typically exist to avoid selfishly motivated actions, nor does a healthy spiritual environment exist that requires trust, communality of purpose, and equality with other human beings. Although humans naturally seek to serve an entire group of humans and to work together for the good of humanity as a whole, they are also innately hard-wired to do so only if everyone in the group is doing likewise. The selfishness and spite induced by the inhumane contemporary cultures are actually harnessed by the "superior" patriarchal rulers to have the people act exploitatively against each other, thereby dividing them and weakening any resistance to or spite against the exploitation by the elite.

In addition, while a society showing little or no concern for individual human needs naturally induces spite, vindictive actions against contemporary cultures are inhibited by fear of the hierarchal rulers as well as their fabricated afterlife. Deceitful propaganda about this system being "naturally" the "best" one also contributes to turning off people's natural spite. Human action against the exploiting rulers is further inhibited because the deprival of human desires for nutrition, health, and sex in the cruel patriarchal societies causes people to focus on attaining satisfaction of such basic desires, whose satisfaction is made partially dependent on serving the elite and the societies they administer.

Concentration on personal desires is magnified because of the propaganda of selfishness that succeeds in fragmenting so many humans into artificially selfish individuals, with whom natural human cooperation is counterproductive. The propaganda that motivates selfishness in many who exploit those who don't believe in selfishness is crucial here. Such exploitation provides "proof" of the validity of the propaganda that wins ever more adherents to the doctrine, thereby supplying further motivation to focus on personal needs to minimize both undesirable exploitation and deprival of one's own personal physical and emotional needs. Social pressure further widens the belief in this inhumane philosophy.

Fortunately for those living in perverse societies consisting largely of selfish robots incapable of selfless love, even self-love can evoke happiness, especially if it motivates seeking and striving for the love of other humans. A self-loving person has the advantage of being less subject to manipulative exploitation by others via the false complements and related promotions of self-image that are typified in feigned love. Feigned love or respect is often employed for exploitative purposes today and typically results eventually in spite and long-term unhappiness. A self-love that brings self-respect and long-term happiness therefore can't be dependent on others but must instead derive from a feeling of self-fulfillment from serving humanity as a whole as best as one can.

A complex combination of actions, which are both selfless and selfish, and which are often seemingly "irrational", is motivated by the fundamental human desire to maximize the love between oneself and others that is both timeless and priceless and that can create and maintain maximum human happiness. The respect and trust required for any love, including self-love and mutual romantic love between two partners, may only come with a common adoption of the overruling purpose of serving humanity as a whole. Serving the latter purpose ensures at least the love of future human beings.

Although mutual human love has evolved to represent the highest human pleasure, and although such love provides infinitely more love or "utility" than the satisfaction of any personal desire or need, human happiness probably can't be reduced to an equation that maximizes the expected value of current and future love, or to any other utility function for that matter. In particular, human happiness is dependent on far more than a simple rational attempt to maximize the collective utility of all or of some group. It requires the evolved common purpose of all humans to serve humanity as a whole to experience the feeling of self-fulfillment and true happiness.

There are many routes to contributing to the good of humanity and thus to happiness. There is also never any permanent equilibrium feeling of happiness, as an ever-changing environment would never have allowed humans in a state of constant satisfaction or bliss to survive. As a result, the conscious appears to have evolved to be motivated to always seek new ways to achieve happiness. This hypothesis seems to be strongly supported by the fact that happiness is dependent on a person engaging perpetually in new activities (Klein, 2003) to avoid boredom. However, novelties can only bring positive feelings of self-fulfillment long-term if they are directed toward the higher purpose of serving humanity in ever more creative ways. Exploring the new without purpose must eventually lead to feelings of empty meaninglessness for it all.

Many people set an idealized static form of romantic love between sexual partners as the supreme happiness that they seek but that is very difficult to attain (Morrell, 2005). While romantic love certainly brings feelings of bliss, it is like an internal narcotic that must be appropriately managed. To maintain the ecstasy of mutual romantic love requires that both partners maintain it through continuous effort, sacrifice, and trust, which themselves can be exhilarating and sexually exciting when they originate from selfless reverence for the partner and are appreciated by that partner. A mutual belief in a partner being capable of anything, regardless of past failures and mistakes, can also help, especially since it spurs mutually loving partners into the power of believing in themselves. The empathetic love toward serving their mates contributes to the mutual bliss, but such ecstasy becomes pointless unless the higher purpose of human beings to serve humanity as a whole is followed.

Effort and sacrifice represent unhappy chores for individuals brainwashed into believing they should act strictly for their own selfish benefit. They are therefore often undertaken strictly as a bargaining chip in the standard trading involved in most marriages and sexual friendships. In that sense, most sexual relationships operate more like capitalistic businesses, and it isn't at all surprising that most end up being dull or outright disasters. Marriage typically only creates a legal obligation to trade sex for money and artificial "shows" of love that inhibit true mutual reverence and trust, which motivate continued efforts to improve for the benefit of the loving partner and all humanity.

Within the U.S. environment, the feeling of romantic love rarely lasts more than a year even in persisting relationships (*Wall Street Journal*, 2005), and sexual activity tends to decline with the term of the relationship regardless of age (Rathaus, Nevid, and Fichner-Rathaus, 2008). The root problem may be a transformation of romantic love into a cold business relationship by the selfishness of one or both partners. Many people in the U.S. marry because they think it will make them happier, but the high divorce rate there (Rathaus, Nevin, and Rathaus, 2008) indicates that it often does not. Reich (1936) found even more persuasive evidence that marriage doesn't bring happiness in other times and countries.

Enlightened spiritualists realize that love and heaven are just a state of mind (Morrell, 2005). One's own neural firings and electromagnetic chemical processes can indeed create a desired feeling through their own communal selection of coherent firings that are reflected in the appropriate thoughts to do so. The thoughts that evolved to bring happiness are those that reflect a desire to serve all of humanity.

Self-deception about oneself, especially with respect to one's future situation about which there is less conflicting evidence from stimuli because of the very uncertainty of the future, is a typical human trait which is positively associated with anticipated future happiness (Robinson and Ryff, 1999). This tendency reflects a natural human optimism which can be a self-fulfilling prophecy because of the power of positive thinking that can collapse wave functions of possibility on points that are desired. The mind thus naturally engineers its own happiness to be integrated with its will power but only if it is serving its evolved purpose that creates internal and external harmony.

A happier mood, and even greater optimism, can even be promoted in a person's mind by simply having that individual move the right side of the body, such as with a smile that raises only the right side of the mouth, because the experience triggers more activity in the brain's left side which specializes in more positive emotions (Lawton, 2007). However, the smile generally must be accompanied by an internal feeling, as faked smiles that are exhibited with different muscle expressions don't necessarily reflect or affect an emotional state (Chemali, Chamine, and Naasan, 2008). Although the entire perceived universe is merely a construct of one's own mind (Strassman, 2001), and although the human brain is prone to generate experiences that don't reflect all external factors (McClenon, 2002), that system is affected by the reality that is internal and external to the individual.

Mere concentration on some trivial thing or aspect of nature can create happiness (Klein, 2003) and is actually one means of attaining pleasant meditative states (Capra, 1999). Recognition of the fact that everything is connected through quantum forces that may reflect deeper forces could be useful for such concentration to make it more enjoyable and enlightening. However exclusively enjoying the present experience can lead to an unhappy feeling of emptiness unless one recognizes and acts upon a higher purpose that engages the conscious in its evolved tasks of forecasting and making decisions based on the predicted future.

The conscious evolved to be able to make decisions, including in setting goals, in a process that includes forecasting the future to better plan for it. A failure to so employ the conscious supplies it with an unpleasant feeling of uselessness even if concentration on experiencing the moment can be temporarily relaxing by giving consciousness a pausing rest from tiresome activities that can enable more subconscious processing on other matters. However, mere concentration on the present long-term would make one's conscious worthless, as the subconscious operates most efficiently in the present. Extended focus on the present would eventually transform oneself into a reactive vegetable leading to an unrewarding feeling of emptiness.

Meditation, which is often initiated by concentration on the present, has a similar relaxing effect that is temporary, as does hypnosis. However, hypnosis, including not only meditative self-hypnosis but also the hypnotic suggestions to toddlers and others in the infantile state of learning without conscious analysis and processing, also allows the conscious to later exercise its powers to forecast and make decisions subject to the purpose(s) set by the hypnotist. This process can be rewarding long-term just as meditation may also be if it enables new insightful learning for purposes of serving humanity. On the other hand, since a hypnotist's purpose may or may not be consistent with the innate goal of serving humanity as a whole, serving the wishes of the hypnotist, especially one motivated by selfishness (including a parent), can lead to extreme unhappiness and feelings of worthless if the individual figures out that the hypnotist has provided misleading direction. Spite can result if the individual feels exploited.

An obsession or continuous single goal system that results from hypnosis or other suggestions to worship and serve any deity, feeling, or idea is inconsistent with serving humanity. Pursuit of such a myopically motivated goal contrary to the innate purpose of human beings, like worship of an external deity, enslaves oneself to it, and humans haven't evolved to be slaves to anything but their evolved higher purpose. Goals like going to a fictitious heavenly afterlife, believing in a particular external being, experiencing romantic love, or satisfying the sex drive, can lead to a one-way worship of a being presumed correctly or incorrectly to have the power to enable attainment of the goal. That being can be fictitious deity, in which one has been programmed to believe, as well as in a person with "magical" powers, and the worship can even be directed at a source of the desired goals, as in the body or genitals of another. Frustration of a person's attempts to achieve an objective is typical and can even magnify the worship if externally programmed beliefs or ignorance inhibit a natural spiteful reaction.

Frustration is especially likely for people not directed by the innate human drive to serve humanity as a whole since other human beings at least subconsciously perceive such people as subhuman if they act according to a goal other than their naturally evolved purpose. The contempt of others can therefore inhibit the satisfaction of a misguided intent. In fact, since humans are motivated to take advantage of all non-humans or subhumans, exploitation can also result from the contempt, which then typically leads eventually to spiteful reactions on the part of the exploited.

The one-way worship of another being actually makes mutual human love impossible, insofar as human love is only possible among equals who are devoted to something other than a love of human nature and service to all humanity. While seeking or striving for a specific subsidiary goal can be temporarily enjoyable both as a challenge and as a pleasant process, a perceived obligation to continuously seek or strive for it leads to slavery to it that leads to a lack of respect from others that can destroy one's own self-respect.

Even if a particular achievable goal like romantic love or satisfaction of the sex drive is attained, it can lead to an unsatisfactory feeling of emptiness that eventually causes unhappiness unless it is recognized to be a component of the innate higher purpose of a human being. Strictly striving for a particular goal other trying to serve humanity as best as one can leads to the same feeling of emptiness, at least partially because there isn't the mutual feeling of security associated with serving humans' evolved purpose.

Human minds evolved to be capable of both setting goals and changing them as the situation and newly learned knowledge merit it, and they are driven by human nature toward doing so in service to humanity as a whole. As Strassman (2001) has indicated, a deeper understanding of reality can permit one more easily to let go of particular goals, and an understanding of each human being's innate purpose enables optimal timing of goal switches.

Excessive focus on a particular goal, even on the maximization of happiness, is largely inconsistent with human nature and human happiness. The human conscious evolved to dynamically set priorities among multiple goals that include survival and procreation as well as "higher" purposes. Continuous focus on just one matter does result in the creation of dopamine in the brain that leads to feelings of happiness and optimism (Klein, 2003) that may be related to the power of concentration, self-hypnosis, and positive thinking (McClenon, 2002), but humans had to evolve to be able to deal with many different matters that include complex social issues involved in the creative group cooperation necessary for humanity to have survived through the ages. The human brain seems to become quickly bored with one particular goal and to be inspired to greater productivity with changing activities and goals (Klein, 2003). However, merely varying actions and goals can't bring long-term happiness any more than narcotics can.

Humans seem to have evolved to have just one overriding purpose of serving the human group or humanity as a whole. Focus on that meaning to life provides them with feelings of happiness and self-fulfillment, as well as direction that facilitates decision-making. Human beings developed a capacity and even an urge to solve more than one subsidiary goal in pursuit of their higher purpose.

Regardless, for purposes of attaining some goal, it is optimal to set intermediary goals that can maximize the chance of reaching the final destination while providing positive reinforcement on the way (Klein, 2003). The establishment of multiple steps toward a higher goal can also permit deviation from the final goal to other objectives along the way.

Humans' pursuit of varied goals may have a feedback effect. To the extent that striving for some set of goals makes an individual happy, the person will be motivated to continue to pursue those goals. However, when a set of goals no longer provides happiness, humans have an incentive to change paths, or at least adjust the order of priority for objectives. Modifying goals can also be motivated by frustration or boredom and may be especially appropriate for objectives that seem unlikely to be achieved or that appear to demand more effort than they are perceived to be worth. Goals which are more consistent with human nature, such as ones that promote the survival of humanity and that enhance human enlightenment into serving humanity in the best way they individually can, are the ones that will most likely make humans happiest. The specific subsidiary goals of life, the universe, and everything vary not only between individuals but also over time and across environments, as is consistent with the differential abilities and situations of each person. Such variation contributes to the wonderfully diverse structure of the universe.

F. Conscious Focus and Intelligence

It is an important task assigned to the conscious to determine the priorities between the various activities and goals that may be selected. The ability to filter in only information relevant to a particular problem upon which the brain is consciously focused is actually a key determinant of intelligence (Motluk, 2007). Human consciousness is geared for maximum efficiency to focus on one thing at a time (Bernstein and Nash, 2008). In contrast, an overwhelming mixture of unfiltered sensory information can lead to disfunctional mental processes like autism and schizophrenia (Trivedi, 2008).

An information overload can be caused by insufficient conscious focus and filtering as well as by inadequate categorization of informational inputs. For instance, schizophrenia can result from an inability to distinguish between imagination and reality due to inadequate barriers between the subconscious and conscious that may lead to excessive signals from the subconscious caused by strong psychic powers (Gooch, 2007). Since schizophrenia is also associated with less "grey matter" in the brain cortex (*NewScientist*, 2010l), it is also possible that this brain disease results from an inability of the thinking area of the brains to properly categorize and filter such signals. Similarly, autism, which has a higher incidence in children exposed to higher levels of testosterone in the womb (*NewScientist*, 2009c), may actually be caused by excessive receptivity to the incomprehensible noise of non-integrated sensory perceptions that can inhibit optimal brain functioning (Szalavitz, 2008b).

Autism, which has also been found to be related to excessive amounts of glutamate in the brain that ramps up neural activity (Coghlan, 2010b), and which may be related to a breakdown in neural connections during brain development (*NewScientist*, 2009zo), results in an individual being so overwhelmed by information that the individual tunes out most everything except an overspecialized focus on some particular matter that isn't relevant to the person's goals to satisfy basic needs and interact socially (Szalavitz, 2008b). While some autistic people can utilize their extraordinary focus for the accomplishment of incredible mental tasks, such as learning a foreign language in a week and memorizing 20,000 numbers in a row (Tammet, 2009), autism often results in poor social adaptation, at least partly because of inadequate amounts of the social bonding agent oxytocin in the brain (Szalavitz, 2008a). This mental state can also result in "savant" or genius abilities with respect to particular tasks (Biever, 2009a) may result from an excessive focus on a particular task in order to block out the excessive information or noise that they are unable to integrate or categorize.

However, while autistic people are indeed able to discern well-ordered patterns better than others, their ability to deal with the uncertainty associated with some randomness in signals makes them much less able to react to most environments (*NewScientist*, 2010). Their excessive focus may have a negative effect because it filters out information and analysis that might often be useful, especially in complex situations that are characterized by uncertainty, as with social relationships. Thus, while anyone can potentially accomplish extraordinary mental feats by simply consciously drawing strong enough associations between words or numbers that are remembered subconsciously via a categorization process by which all people learn languages (Tammet, 2009), the concentration necessary for such tasks may make it difficult to be able to deal with other situations.

The process for accomplishing incredible tasks typically does require an organization of information into associations that can be categorized by the brain. For instance, one useful way to increase memory of a string of numbers is to identify something with each and then make up a story about the objects identified with the numbers for easier recall (Robson, 2010b) The extraordinary psychic feats of the schizophrenic (Gooch, 2007) may also be accomplished by anyone with sufficient focus and self-confidence to categorize the huge amount of information that the human brain is capable of absorbing in a meaningful manner. The abilities of some autistic and schizophrenic people provide an indication of the extraordinary capacities of the human mind if it is properly focused.

A conscious focus of the entire mind on a particular task consciously deemed to be important can enable the subconscious to rapidly consider as much information as possible and sort out what can be creatively applied for conscious evaluation. A concentration of attention is reflected initially in the activation of a particular area of the brain that apparently determines the subject of focus, and this concentration is followed by the firing of special brain connections related to the focus of attention that swamp and eventually weaken any continued firing of other brain connections relating to prior subjects of focus (*NewScientist*, 2008zzze). However, the brain is capable of switching focus rapidly, as can be useful when a particular problem has been resolved, when a more pressing concern arises, or when more information or analysis is available such as after subconscious processing during sleep.

Since focus enables the brain to concentrate on solving a particular problem, it enhances applied intelligence for cognitively processing informational inputs in accomplishment of mental tasks. Focus is very important for filtering into consciousness only sensory inputs, emotions, memories, and other information that are relevant to analyzing a problem. Intelligence is also related to "the capacity of short-term memory" (Eagleman, 2008), as that capacity determines the amount of information that can be evaluated in conscious thought. However, focus may enable the organization of information into symbols that can help overcome the capacity constraint of conscious working memory.

Memory itself is divided into short-term and long-term capacities within different portions of the human brain. Gynn and Wright (2008) have indicated that the emotional right side of the brain is involved in storing long-term memories, while the logical left side is more involved in storing short-term memories which typically are maintained in verbal or language form. It would appear that the conscious tends to categorize events via words, and it recalls them using such categories.

Short-term memories are typically sent to the hippocampus via either conscious thought or emotions arising out of the amygdala (Bernstein and Nash, 2008). The hippocampus, which fires at various frequencies in different parts of that sector of the brain in response to both present stimuli and past experiences (Thomson, 2010a), is generally involved in tying together information from various areas of the brain, such as from "vision, memory, and imagination", to form "cognitive maps" (Berdik, 2009). The hippocampus is located in the center of the medial temporal lobe of the brain next to the perihinal cortex section of the parahippocampus that provides feelings of familiarity for remembered phenomena (Phillips, 2009a). The human hippocampus has "specialized neurons dedicated to sense of direction" just as that area of the brain does in other animals, but it tends to function less effectively for this purpose in human beings because it has evolved to be usable for additional purposes that allow humans to make better sense of their environment using experience and reasoning (Berdik, 2009).

Conscious thought has a working memory that retains information for less than 20 seconds (Bernstein and Nash, 2008). An emotional association or connection is therefore important in enabling conscious recall of longer-term memories, as well as for more intuitive subconscious use of experiential learning that is based on feelings. The mere feeling of familiarity can be sufficient for recall of some memories. The fact that memories are separated from feelings of familiarity in people with déjà vu, often due to brain diseases or damages that distort the normal connections within the temporal lobe (Phillips, 2009a), provides an indication of the importance of emotions in memories and logical analysis of them.

Information in short-term memory is discarded in a matter of hours or days if it is not applied or doesn't appear to be needed or connected to anything relevant, but it is otherwise integrated into subconscious long-term memory for possible later conscious use if appropriately categorized for easy recall via categorization or connecting cue (Bernstein and Nash, 2008). Memories are contained in systems of "synapses, which are the tiny connections between brain cells" (Eagleman, 2008) that are formed using the chemical acetylcholine (Bernstein and Nash, 2008). Although the proteins in the brain that form the connecting synapses are constantly being replaced, long-term memories are preserved in a process called methylation, which creates methyl caps on cells' DNA to control the expression of new proteins by the cells (*NewScientist,* 2008zzzb). Higher levels in the brain of one particular protein, alpha CaMKII, appear to block or eliminate the effects of painful memories or connections (*NewScientist*, 2008zzx), but the neural connections involved may thereby be only inhibited or weakened as opposed to totally erased once they have been created with a firing code embedded into the cellular DNA.

Long-term human memory may have a virtually unlimited capacity and might never be totally lost even though it can be weakened (Bernstein and Nash, 2008). For instance, there is evidence of people having extraordinary memory of past events when hypnotized that they wouldn't consciously be able to remember in a normal state of mind (Gooch, 2004). The latter fact implies that the brain categorizes memories in forms that enable easier conscious recall of a chosen few, which it deems more important, but the synaptic connections formed by all memories may remain in at least some weakened state forever.

Re-introducing external stimuli that were originally associated with an experience, sensory input, or learning when that information was initially placed consciously or subconsciously into long-term memory can be very useful to facilitate recall (Bernstein and Nash, 2008). While memory "mostly encodes the relationships between things" that can be called up with "no external stimulus" (Eagleman, 2008), the more associations with a memory, the more easily and likely it can be recalled.

Memories are not precise but are instead interpretations of fragments of recalled information (Bernstein and Nash, 2008). Since the brain expends a great deal of energy while daydreaming and analytical learning from experiences (Fox, 2008), it is reasonable to hypothesize that the brain routinely utilizes these processes to create more and stronger neural connections for information evaluated to be more important. The strengthening of those connections is enhanced by more frequent recall, including when idly imagining and dreaming. However, the brain normally only keeps a processed form of memories that both facilitates recall and enables remembering without time-consuming re-evaluation. Storage of memories in processed form also minimizes brain capacity use. However, some very weak brain connections may exist for all the details of one's entire life experiences that themselves can be strengthened subconsciously via dreams.

As reported by Phillips (2006), recall is not only inexact, but the brain also sometimes fabricates memories that are falsely deduced in the face of incomplete and fragmented remembered facts, which can even result in completely false memories. In particular, memories are kept in connected neural firing systems that reflect individual aspects of remembered phenomena, and each connection provides only a fragment of sensory or other information relating to a memory. These pieces of information pieces are optimally categorized by people for easier and faster recall. The conscious can assist in the task of organizing memories and other information, and such categorization is also important in trying to logically evaluate which memories are likely to be real and which are distorted by the weak signals and random electrical noise in the brain that can lead to false memories.

The older connections are generally very weak, unless they are often re-fired through frequent conscious remembering those thoughts, and so they may be difficult to distinguish from the random firings that occur in the brain. False memories may develop from an inability to distinguish between such informational static and weak neural connections. Out-of-body experiences, such as in cases of near death, represent examples of what happens when the brain is starved for oxygen (Fox, 2006) and may represent a form of false memory that could partially result from some neurons in a firing system not having the elemental chemicals needed to fire.

When old neural connections fire in the process of remembering something, some other neurons that were related to those connections may also be re-activated. However, the conscious can inhibit further remembering by conscious neural connections firing to swamp unwanted ones and thereby indicate changes in the train of thought or focus.

The subconscious can be creatively programmed for the suppression of painful or unnecessary memories from consciousness and sometimes does that even automatically (Phillips, 2006). Such programmed memory suppression, as with other mental behavior, may be a trait learned by the subconscious itself, but it can also be set consciously. Repression of memories from conscious recall, however, doesn't necessarily inhibit them from affecting subconscious processing. Some subconscious processes lead to actions and intuition that are realized consciously to be suboptimal, and dreams may assist in recognizing such behavior to enable recognizing and changing the unwanted tendencies

It is possible to reverse memory repression, so that suppressed memories can indeed be remembered. Doing so can facilitate changes in unwanted subconscious tendencies. In particular, while learning from experiences or otherwise is often transferred to the subconscious for automated and more efficient tasking (Douglas, 2007), the resulting subconscious algorithms can sometimes be counterproductive and hence merit recall of the forgotten events that prompted the undesired activity in order to be able to consciously rewire the associated neural firings. Accomplishment of this feat is sometimes difficult if a prior subconscious or conscious order to forget a repressed memory has caused the brain to have the neural connections associated with the memory swamped by other neural firings whenever the interconnected neurons of the memory fire sufficiently to make their way past the brain's thalamus filter to consciously recognize them. Nevertheless, the filter may be opened with sufficient conscious concentration on possible associations or via hypnosis when necessary to consciously recall past events.

Information is normally transferred out of conscious memory into longer-term memories to free up the conscious for analysis of other knowledge. Conscious thought would be impossible without such transfers which prevent mass confusion (Marshall, 2008a). Once information about an event leaves consciousness, it can be subsequently recalled by the brain's process of categorization that has the same network of brain neurons firing that were initially excited upon experiencing the event (Marshall, 2008a). Since the conscious can hold only a few separate concepts in working memory at any one time, the categorization of thoughts, ideas, and information seems to be essential for optimal brain functionality and thinking (Bernstein and Nash, 2008).

While intelligence is a function of both the "ability to quickly resolve cognitive conflict and the ability to store stronger associations between facts" (Eagleman, 2008), allowing some connectivity between unrelated concepts and categories of ideas enhances creativity that can be important in many tasks (Nettle, 2007). Some seemingly unrelated brain connections that some people have, such as greater connectivity between the portions of the brain responsible for the senses, may even enhance memory and numerical skills (Robson, 2009b). Such connections help them categorize information and data better for easier memory and recall. However, excessive connections that aren't organized for meaningful interpretation or recall can result in the extreme neural noise that inhibits social functioning and other activities involving large amounts of information from many sources, as they do in autistic people.

Within this context, it is interesting that those with higher intelligence quotients (IQ) are more creative when metabolism is higher in the anterior cingulate, but others are actually less creative when there is less activity in that area of the brain (Geddes, 2009d). Since the anterior cingulate is responsible for filtering out irrelevant information flow from conscious thought (Gruzelier, 2006), it seems likely that more intelligent people tend to already have open minds which allow for enormous thinking outside the box, and they need to filter out more information in order to develop usefully creative conclusions from it. However, other people require reduced restraining of neural firings to be able to categorize excessive information when trying to generate meaningful creative ideas in order to allow sufficient information into consciousness for creative thought. Those with lower intelligence may have a smaller conscious memory capacity to integrate a large number of factors into meaningful ideas, and they may tend to filter out far more information automatically under normal circumstances as a result. Since further filtering activity just reduces the information flow that is needed for more creative thinking for them, such individuals need to open up the "box" of their conscious more to enable greater creativity, as opposed to increase the filtering out of information.

"Operating at the edge of chaos" may be necessary for efficient brain functioning in a coherent "self-organized criticality" that is somewhere between randomness and order and that can be strongly affected by small changes in inputs (*NewScientist,* 2009zl). For instance, under stress, people often "revert to irrational thought patterns" and "child-like reasoning" (Brooks, 2009b) which reflect the inability of the existing neural connections of the conscious brain to deal with a problem. Such thought processes enable new connections, or the reconnecting of old ones that are re-fired, to creatively deal with the stressful situation.

An optimal balance between order and chaos in the brain seems to be enhanced by focusing one's attention on one particular matter at a time. A belief in the brain's subconscious and conscious abilities to optimally address any particular issue can be useful in directing concentration of all the brain's resources on successful analysis and decision-making by inhibiting diversion into making contingent plans in case of failure.

Research indicating that "beliefs about the future self can create their own reality" (Robinson and Ryff, 1999) is consistent with a hypothesis that a conscious forecast by an individual result in neural connections in the subconscious portion of the brain that cause the individual taking the actions necessary to maximize the probability of validating the conscious forecast or will. Therein may lay the power of belief, which functions like positive thinking and will power, as will be explained in more detail in Chapter 16.

Placebos, which are chemically impotent pills that have a positive effect in treating diseases in 1/3 of all cases (Lipton, 2008), represent an example of a means of inducing such powers as they induce a belief that the useless pills will be effective. For instance, placebos given to deaden pain have been found to affect neural activity in the left dorsal horn which is located in the portion of the spinal cord in the neck (*NewScientist*, 2009zzzw). Placebos have been shown to significantly affect neural electrical activity in scientific brain scans (Lipton, 2008), and the different biological responses of people to the same shock to their bodies (*NewScientist*, 2010h) may conceivably be controlled by electromagnetic waves stemming from belief and will power. The power of placebos were long ago recognized to be related to hypnosis (Radin, 2006) although hypnosis has been found in one study to have an even stronger effect (Jones, 2009b), perhaps because it induces a stronger, unquestioning belief.

However, beliefs in lies can lead to incoherencies in many cases that make success less likely and thereby eventually create doubts about future success. The power of belief is thereby reduced in the future as increasing evidence of the invalidity of the "faith" is discovered. Beliefs based on known facts are far more powerful and useful in promoting success and thus maximizing belief. Nevertheless, a willingness to change beliefs upon discovery of information that definitely contradicts old beliefs is necessary for maximum long-term power of belief. Making plans for operational task failures prior to engaging in focused attention can help maintain effective connections if they are specified to be implemented only upon the occurrence of specific situations that might merit a change in assumptions or beliefs.

Chapter 12
The Evolution of Belief

Humans naturally make simplifying assumptions about the world in order to categorize information into organized concepts about the complex and uncertain environment to facilitate predictions of the future. Putting any situation into a category with others makes it possible to deduce forecasts based on the past probability of such generalized environments to lead to a future result. This process, which represents a simplification of the universe, makes it possible for the limited human working memory to analyze an extensive amount of information with only a few variables. Since the primary goal of the organization of stimuli into groups is to predict how each situation confronted in the present will develop in the future, the brain naturally attempts to ascertain the particular aspects of any environment that cause future results. Assumptions about cause-and-effect are therefore made that can be based on a mix of assumptions learned from others and Bayesian probability analysis based on the individual's own past experiences.

Learning from another person can take the form of accepting as fact whatever the other individual says, as with infants or hypnotized individuals, but it can also incorporate some other information, such as proof of an assumed cause for a particular effect. More information and analysis is often required when a suggesting person is less trusted, and a Bayesian evaluation of the trustworthiness of the suggestions based on past history may be an important component of trusting process. Bayesian probability analysis based on a person's own experiences is inevitable in any case, as even suggestions made under hypnotic trances can cause the hypnotist and the suggestions to be disbelieved if they conflict too much with the reality that has been or is subsequently observed or deduced.

Assuming something without proof is similar to the concept of belief that is a core component of religion. Without complete insights on nature of the universe, humans have adapted to make decisions using assumptions that they learn from others until they have more information. The assumptions lead to deductions on the probabilities of different environments unfolding in the future, and the failure of assumptions to predict well can result in questioning the assumptions as well as the people from whom the individual learned them. On the other hand, assumptions that are difficult to disprove, such as a beliefs about existence in an afterlife, may not be subject to such analysis. Religions that require an unquestioning faith in a fictitious deity, which is assumed to offer better treatment in an afterlife in return for such beliefs, minimize the chances of trying to analyze whether the religion's assumptions are valid.

The staying power of belief in the absurd assumptions of religions like Christianity seems to largely derive from the fear of repercussions after death for any normal Bayesian analysis by the brain to reevaluate the probability of the religious assumptions being valid as new information is received. With at least conscious doubts therefore inhibited, many religious believers maintain their faith for life, and some become emotionally fanatical about temptations to question them. However, the subconscious innately operating according to Bayesian logic continues to draw its own conclusions as new information is learned that can surface into conscious doubts or reinterpretations of religious doctrine. A belief that some religious assumptions either are mere generalizations with exceptions or are merely symbolic in nature often results.

Any belief in a set of assumptions can enable the brain to make sense of the enormous amount of information that does exist through the process of filtering out irrelevant inputs and categorizing the rest. Believed assumptions can thereby promote focus and efficiency in both subconscious and conscious brain functioning, even when they are invalid, although more valid assumptions of course lead to both more effectiveness and less efficiency-reducing doubts resulting from failure of the assumptions to hold.

Religions and ideological belief systems relieve individuals of the daunting task of determining a set of assumptions or beliefs on their own (Lovelock, 2009). Most religions today assume the existence of a superior being or beings and mandate worship of the deity or deities. However, not all belief systems assume the existence of superior beings, as assumptions that there are deities began only in the last 35,000 years when matriarchies began to impose on humanity their beliefs in the superiority of females and goddesses to be worshipped without doubts. This first such religion was followed about 30,000 years later when males overthrew the pompous matriarchs and established their own cruel hierarchy based on a belief in the superiority of males and worship of male deities.

While a reverence for others is innate to human nature and likely began with romantic love, a one-sided worship of a superior being is not humanly natural. Since there was a belief in equality among humans for millions of years, religions based on worship of "superior" beings conflict with basic human nature and are therefore inhumane. Unquestioning belief in a "superior being" is also a far less adaptive process than establishing beliefs based on the straightforward Bayesian logic that is innate to human brains.

A. Development of Superstition, Magic, Science, and Belief Systems

The simple carrot-and-stick approach to learning that is intrinsic to human brain analysis, which evolved from stimulus-response logic, is useful in developing focus, consciousness, and intelligence. However, such rationality no doubt resulted in many initial superstitions over time. For instance, chewing on a twig while in transit to a particular tree might become a ritual if initially, and later periodically, reinforced by eating bananas there. The power of positive thinking also probably reinforced many behaviors that served no purpose other than to increase humans' confidence, which enhances humans' courage, initiative, and success (Howells, 1986). By enabling efficient subconscious processing without interfering conscious doubts, belief even in false assumptions can have powerful effects.

In addition, some of the seemingly irrelevant superstitious beliefs like omens, as well as tarot cards, etc., might have prompted various thoughts to enter into consciousness that would not otherwise have been considered. Many of these ideas may have been creatively useful. Even random "signs" could have actually assisted in decision-making if the appearance of a particular omen prompted a relevant analysis from a new perspective, including rational as well as intuitive interpretations.

Moreover, as previously mentioned, some superstitions such as providing music to crops may have been rewarded for specific scientific reasons that are only now being understood. There is also some evidence that humans can even consciously affect otherwise random event generators (Nelson, 1998), possibly because of photons and electromagnetic waves, which all people emit. Such emissions affect outcomes in what is often referred to as the power of positive thinking, as will be explained in more detail in Chapter 16, and the resulting chances of the future unfolding as per the superstitions lead to reinforcing faith in them.

Some seemingly superstitious beliefs and rituals can have more direct foundations in reality. For instance, since people's hearts emit an electromagnetic field that extends out 5-8 feet (Braden, 2007), it is possible that some people can see and interpret that field, which they might even perceive as some type of aura. Auras, which reflect electromagnetic fields (Childress, 1999), are perceived as having varying colors by some people, and those colors might indicate something about the emotions of people since they would reflect, at the very least, the frequency of the heart beating that creates an electromagnetic field. Auras have indeed been detected with special instruments that were widely used in the USSR for medical diagnostic purposes, and humans themselves can indeed effectively exercise psychic powers using such forces (McTaggart, 2008).

Ancient people appear to have utilized psychic forces more often (McClenon, 2002), as they were considered intrinsic components of their spiritual belief systems (Radin, 2006). The neural impulses in the brain have evolved over enormous periods of time like hundreds of millions of years to have special properties within the environment of this universe, and the purpose and usefulness of many of the interconnections are awaiting discovery.

In any event, as previously mentioned, the foundation of superstition or religion is a set of assumptions that categorize events and things in the world. Some of these categorizations may simplify and enhance people's understanding of events, such as an assumption or belief that the universe is structured as if it were created or controlled by a particular deity. In particular, while it is possible to ascribe some events with unknown causes to random chance, as many Native Americans do (Peat, 2005), many religions such as Christianity attribute unexplainable phenomena to the will or acts of a supernatural power or deity (McNamara, 2006).

The beliefs and morals that one attributes to a deity are processed in the same area of the brain that is activated upon thinking of one's own religious beliefs (*NewScientist*, 2009zzzzj). A belief in a supernatural being is thereby integrated into one's own set of assumptions about reality by essentially believing in a deity thinks as oneself does. Since the portion of the brain that is activated when people try to imagine how a deity feels is the same part that experiences empathy (Geddes, 2009b), it appears that "believers" are effectively ascribing human characteristics to the power or deity that is assumed to cause otherwise unexplainable events. In particular, the same brain patterns are "triggered" when "people communicate with each other" and when humans offer "prayers" to a deity (*NewScientist*, 2009zm). These findings imply that religions merely adapt natural human brain processes that evolved over millions of years to deal with culturally induced beliefs today, as opposed to any innate human religious belief in an external deity.

The area of the brain involved in deciphering metaphors is activated when people are thinking about their deity (Geddes, 2009b). The deity thus acts as a symbol for some higher power that may once been humanity as a whole. Actual prayers to a deity activate the same part of the brain that is used "to process desire and consider how another individual—in this case God—might react", whereas the same neural activity doesn't occur with respect to entities like Santa Claus that are well-recognized to be fictitious (*NewScientist,* 2009zm). Deities thus act not only as symbolic representations of reality but as entities considered to have empathized human qualities despite their powerful nature.

The overall evidence is consistent with a hypothesis that human beings once believed in a higher power that was human but had extraordinary power that might even was used for the benefit of humanity in some way. The essentially innate human belief in humanity as a whole was redirected first by matriarchies into a faith in females and goddesses that was later transformed about 5000 years ago into an assumption of the superiority of males and gods. The extra reverence for such powerfully inhumane entities was then exploited by such selfish entities, thereby leading to religious worship and unquestioning servitude at hypnotic suggestion.

Religious or other assumptions can provide a simpler and reasonably effective framework for understanding, categorizing, and predicting the world than blaming random chance. For instance, an assumption that the universe behaves as if there were a deity can lead to the same conclusions as an actual belief in a deity. A fanatical believer in a fictitious deity may therefore evaluate and predict events in a manner similar to an atheist with just a belief in the assumed analogy. The assumption that the universe operates as if a deity with human qualities is in charge is the same in each case. Belief in a deity with human traits may not be that different from the prehistoric belief in humanity as a whole. Given the virtually omnipotent power of humanity that will be explained in subsequent chapters, faith in an omnipotent deity whose actions are motivated by the tendencies that evolved in human beings may reflect reality better than a belief there is no higher power. Religions defining the deity with more of the characteristics of human nature therefore have greater predictive accuracy.

In any event, an individual assigning unexplainable events to a single cause facilitates the process of categorizing phenomena. Categorization enhances memory and recall (Bernstein and Nash, 2008), which are important in predicting the future (Marshall, 2008a) and can be useful in developing intelligence (Eagleman, 2008).

Everyone makes assumptions about things and therefore believes in something (Urban, 2007). A system of beliefs represents one of the most important components of religion (McNamara, 2006) that harnesses that human tendency. Children begin to form theories about the world at an early stage in life with very limited empirical evidence or testing, but acceptance of "counterfactual" beliefs actually increases with age (Dow, 2006). This seemingly incredible phenomenon could result from the inhumane religious requirements for unquestioning faith in order to achieve a fabricated heavenly result in a make-believe afterlife, about which people are uninformed. This ignorance subjects them to fear-induced manipulation which causes them to reinforce absurd beliefs through conscious repetition that more deeply ingrains their faith in the lies.

Another reason for belief in unverified phenomena that often results in people making assumptions that are "wrong" is the desire of people not to appear "stupid" in the process of going through the trouble of testing out hypotheses (Hammond, 2009). The rejection of a hypothesis that a person tests is possibly the most embarrassing aspect of experimentation. In particular, both thinking of a failed idea and then actually believing it enough to try it out makes a person unhappy with the individual's own conscious decision-making. Many people therefore lose some confidence in themselves as a result of such experimental failures. They may also suffer negative social consequences, as well as lose focus by feeling a need to expend extra brain energy trying to explain the failure.

Most people therefore often prefer to live in ignorance rather than go through a learning process that can require stressful conscious analysis, which diverts brain energy from more pressing tasks of life. In addition, such stressful doubts create unhappy feelings of insecurity that reduce the self-confidence associated with the real power of positive thinking. Without an alternative for categorizing information, it also becomes difficult to evaluate and react to all the stimuli in the environment, and indecision or poor decisions often result. A failure to have the same faith in what others believe can reduce social acceptability, thereby providing a further incentive not to doubt culturally induced religious expectations.

While one can adopt a social style that permits experimental stupidities, as well as self-program oneself to enjoy learning from both successes and failures in experimentation to offset any negative aspects, the aforementioned costs of that trial-and-error approach motivate most to believe in existing religions or explanations of reality. A mere belief in something, even if it is truly false, such as a belief in the power of a "magician" or shaman that is induced by a sleight-of-hand trick, has been shown to have positive psychological and physiological effects, including for curing diseases via hypnotic suggestion (McClenon, 2002). Some of the magic employed by shamans may actually have some scientific basis in quantum physics, but much of it may have been merely superstition (Radin, 2006). Even if the assumptions of a religion are unrelated to reality, faith in them can benefit human beings because of the power of positive thinking.

Both the human capacity to be tricked and the power of belief are not that difficult to understand given that "normal" human perceptions themselves are a distorted categorization of the true underlying quantum reality of the universe. Human perceptions are useful for adapting to the existing macro environment, but they don't actually reflect the underlying quantum reality of things. Some knowledge of reality may not be extremely useful for most people in many situations. For instance, the fact that all solid objects are really characterized by electromagnetic fields which cause spatial areas empty of anything but forces of attraction and repulsion to be perceived as impenetrable doesn't normally have any practical implication. However, such information, which can seem irrelevant for immediate employment, may hold some possibility for application at some point if it is properly organized for timely recall when needed, such as in providing useful perspective for creative thought. At the same time, the generalized effects that typically occur on the macrocosmic level can still be recognized via the normal method of categorizing human perceptions.

In any event, the fact that belief in anything by itself has positive health benefits for humans with respect to relieving unhealthy levels of stress (Esch and Stefano, 2005a) alone makes it possible to better comprehend why humans have a tendency to believe in religions in the face of counterfactual information. Beliefs, whether in a deity, deities, or assumptions, reduce uncertainty and fear of the unknown and the related stress. They can also aid in the process of categorizing phenomena for purpose of forecasting the future and less mental exertion.

Religions serve to satisfy the widespread human need to see the purpose and worth of everything when things go right and feel reassured in adverse environments

(Lovelock, 2009). They thereby meet a human requirement for happiness to regard others benevolently while also comforting people with a prediction of in a better future when there are unsatisfied personal needs and desires. While a belief in humanity and the power of future human beings is more realistically useful for attaining those benefits, most people today are brainwashed into ignorant faith in cruel deities invented by elites long ago to manipulate them into slavery, such as via a belief in a religious fairy tale,.

Some of the earliest sets of beliefs were probably naturalistic ones, where people empathized with all other organisms and even Nature itself. Such belief systems, which were likely similar to those of the Nature-revering Native Americans (Peat, 2005), seem consistent with the fact that people, especially children, naturally ascribe human feelings to others, including to inanimate objects (McNamara, 2006). Such empathetic beliefs are consistent with the fact that all live and inanimate objects are made of the same sort of electromagnetic fields. This fact of life could possibly be perceived by the senses before the brain categorized raw stimuli into macro objects for practical survival and procreation purposes.

Ancient people may have had at least some subconscious insights into the meaning of life, the universe, and everything about which most are unaware today, but that ancient knowledge itself was clearly incomplete. Otherwise, there never would have been the matriarchies and patriarchies which are so inconsistent with human nature that they have even resulted in the suppression of known insights via religion and ideologies based in lies.

The first religious systems probably included a belief that the Earth was a sort of Mother that provided nurturing and regeneration of life (Sjoo and Mor, 1987). Continuing with that analogy, a cultural belief that the sky impregnated the earth with rain and sunshine might have also existed. A similar absurd assumption that females self-generate offspring might have even developed in the religions invented by matriarchies which were based on simplistic naturalism. Such absurd categorizations of reality might have been one of the cornerstones of religions based on worshipping femininity that began 35,000 years ago and continued for about 30,000 years until the conquest by patriarchal barbarians. Whether ancient humans had either been unable to detect the facts of life from prior patterns, or whether they were brainwashed into ignorance through means such as hypnosis, an "elite" of inhumanely selfish people was able to use people's beliefs in lies to motivate slavery to them.

Nevertheless, even some earlier human beliefs can have some usefulness. For instance, in today's modern world, regarding the Earth as if it were a living organism provides a useful analogy for purposes of understanding the need to stop pollution and global warming. In particular, if it is assumed this planet is a sort of Mother Earth which nurtures humanity and all other life dependent on it (Underwood, 1965), it can be concluded that treating it contemptuously like an exploitable object and trash bin is likely to result in Mother Earth wrecking revenge by discontinuing the nurturing of humanity. A catastrophe of immense proportions could indeed occur if pollution-induced global warming causes the ice caps to completely melt, as is very possible, that would cause sea levels rise by 70 meters (Baxter, 2008).

However, it's also possible to take such belief systems too far instead of as a mere analogy applicable to limited circumstances. For instance, one Native American tribe actually refused to till the ground to grow crops because their believers felt it would represent painful incisions into the revered Mother Earth (Underhill, 1965). Such behavior may reflect an excessive amount of reverence for something that may actually constitute a belief in a superior entity or even worship that very unnaturally places one's own needs below that of something else, thereby creating an unhealthy imbalance. Some extreme environmentalists today apparently have a similar problem, as may have many in matriarchal societies for tens of thousands of years. In contrast, a scientific understanding that all things and individuals are mere processes that change, as opposed to being permanent objects or organisms, can make any worship of anything or of any individual utterly absurd, since everything is merely a part of an aggregate whole (Capra, 1999).

In any event, ancient human experts, who were especially good at knowing what led to rewarded behavior and what categorizations of assumptions or beliefs best characterized the world in the context of existing knowledge, were probably followed and possibly honored by others seeing their success (Peat, 2005). Unverifiable and even seemingly "counterfactual" beliefs of such leaders would have been accepted if they led to successful behavior (Dow, 2006, 2007). The result was the emergence of the first recognized magicians, who represented the earliest scientists to the extent that there was in fact either a true or apparent cause and effect behind their theories (Peat, 2005). This social process apparently helped lead to the development of many technologies, including a sophisticated medical system that has been documented in the case of ancient Egypt to have led to many medical treatments which have been proven to be effective today (Pain, 2007). For instance, acupuncture has been found to be successful in treating some injuries by promoting a stress response that inhibits inflammation and thereby improves sensory and motor functions (*NewScientist,* 2010s). Even the ancient use of leeches on wounds has been found to produce positive medical benefits (*NewScientist,* 2009zzzzt).

B. Hypnosis and Belief

One early form of medical practice involved hypnosis (McClenon, 2002), which apparently has long been utilized by shamans to help heal people with diseases (McNamara, 2006). Hypnosis, which continues to be widely used in aboriginal human cultures (Howells, 1986), and which has been documented to be relatively effective for medical applications in more modern societies with the most advanced technology (Bernstein and Nash, 2008), seems to be integrated with the power of belief.

In particular, hypnosis causes people to carry out various tasks simply by suggesting them (Hewitt, 1996). Suggestions can promote belief without questioning or proof if the trust of a hypnotized subject can be obtained, as is crucial for the successful use of hypnosis to maximize the power of suggestions (Gruzelier, 2006).

The power of hypnotic suggestion is at least partially related to the placebo effect, in which the mere belief in a curing agent increases the chances of healing even when the hypnotic agent itself has no effect (McClenon, 2002). The placebo effect is named after the useless pill that is given to some patients in medical tests of developmental drugs to see if a new drug actually has an effect over and above just the power of suggesting a treatment is supposed to cure them (Brooks, 2008e).

Belief in something, even in a placebo, causes a reduction in stress hormones like cortisol, which negatively affects physical and psychological health, and it thereby relieves the directly negative impact on health of stress hormones as well as contributes indirectly to health benefits by reducing the unhealthy impact that stress has on lifestyle

(Esch and Stefano, 2005a). Psychological variables are well known to affect humans' immune systems, and so the therapeutic power of hypnosis in directing the body's internal resources to relieve symptoms and cure diseases is not incredibly magical, given the ability of a person to control blood flow through the appropriate brain neural firings (McClenon, 2002). Belief in a placebo or anything else has been documented to result in a release of nitrous oxide in the brain that reduces stress by creating relaxed mental states (Esch and Stefano, 2005a).

It has been shown that individuals can adjust their behavior via changes in brain neural connections just by thinking about doing so, as in an athlete rehearsing future sport actions in that individual's own mind (Bernstein and Nash, 2008). Such self-hypnosis may also be related to the power of positive thinking. In particular, any hypnosis likely induces brain reactions which cause biological and physiological changes in the human body that are at least partially related to hypnotized individuals believing that a suggested action or event will come true (McClenon, 2002).

The effects of hypnosis and positive thinking can be enhanced by specifying the exact contingencies for the brain to act on a suggestion. Such signals prompt the subconscious release of hormones such as adrenaline for seemingly "superhuman" actions and strength if necessary. The ability of people to swim in icy cold waters that would normally kill a person and that is achieved by self-mind training which apparently results in the production of adrenaline or epinephrine (Graham-Rowe, 2009) is just one example of the application of the extraordinary power of hypnosis.

Focus tends to be essential for hypnosis (DeLaurence, 1910). Since women tend to concentrate more on the moment than men who have minds that wander more between matters (Deida, 2005), females may be able to better access the powers of hypnosis.

Focus on successful completion of a particular action may even lead to psychic waves that influence others. Given the travel in time by antiparticles that are emitted by the human body all the time and that can also be spurred into creation by emitted light photons, concentrated focus might also possibly impact the past, as will be explained in Chapter 16. The quantum fact that a future event can collapse the wave function of particles in the past could be a related phenomenon.

Emitted antiparticles can not only collapse past wave functions of other particles but also affect the wave functions of the particles and, through annihilating contact, even impact the very existence of some wave functions. In both cases, however, affecting past wave functions or their collapsing on particular points doesn't preclude the existence of those functions of potential in other forms in other unconnected universes that exist parallel in time. In other words, it isn't possible in the present to affect unconnected past universes, but it is feasible to connect past potential universes to the present by collapsing the past wave functions of particles in the past on selected points or otherwise affect those functions in a way to make present and future events happen or be more likely to happen. Frequent efforts to affect the past can result in a powerfully reassuring belief that one is helped in the present by such psychic waves that will travel in time from one's future focus. A belief in something may free the subconscious to carry out the neural firings without the interfering firings associated with conscious doubts to enable such results.

A recognition that the mere collapsing of a present wave function on a particular point can disconnect some past possibilities even without time travel by antiparticles or light photons can also help maximize the power of such positive thinking through belief in scientific facts. For instance, it has been 100% verified that collapsing a particle wave on a particular point in space through focused observation excludes not only other contemporaneous occupation of other possible points in which it would normally exist simultaneously but also precludes past travel on some paths that are impossible according to quantum mechanics. Those past paths that would have been connected to the present reality are thereby disconnected into separated parallel universes from having ever occurred in the reality where the focused observation took place. Such observation also affects future possibilities.

Hypnosis is also based on focus that enables impacting people's future behavior. In particular, the focused attention of a person is essential in hypnotizing the individual (DeLaurence, 1910). However, the belief and willingness of a hypnotized subject in the ability to carry out the suggestion appear to be crucial to the success of hypnosis. There may have to be a coherence in the belief and other probability wave functions involved in carrying out hypnotic commands to enable sufficient collapsing of those waves on the intended points.

While experts remain unsure how hypnosis actually works, it has been well established to be effective when a person believes and expects a suggestion to come true, as can occur more easily when the barrier between the conscious and subconscious is weaker (McClenon, 2002). The weakened impediments to a free flow through of the hypnotic suggestions without the normal conscious filter permits efficient subconscious activity to carry out commands with no conscious interference by the hypnotized subject. In particular, hypnosis allows the creation of subconscious neural brain firings and connections with little or no conscious firing by the thalamus to distract or stop them. Such suppression of the thalamus filter is apparently made possible by a person trusting the hypnotist or hypnotizing agent (Gruzelier, 2006). Since stronger will power in a hypnotist maximize the effects of hypnosis (DeLauren, 1910), it appears that sufficient belief of an individual in an outcome can overcome the conscious filter of a weaker individual's mind. Because the effects of positive thinking are grounded in quantum reality, strength in will power has to be a function of the coherence of an individual's beliefs with innate human nature, which itself evolved according to the laws of quantum physics.

The brains of people who are more hypnotizable have been found to be characterized by thinness in the culture-dependent gating barriers between their conscious and subconscious that normally restrict the spontaneous flow of images and information within the brain (McClenon, 2002). The power of the subconscious to accomplish seemingly amazing tasks when there is no conscious interference (Gigenrenzer, 2007) may help explain these characteristics of hypnosis.

In particular, hypnosis enables the efficiency of unrestricted subconscious action. People who are more easily able to open up their cognitive boundaries can more easily eliminate distracting doubts and fears as well as prevent conscious interference in subconscious activities. Amazing feats are thereby possible for hypnotized individuals (Gynn and Wright, 2008).

For maximum effectiveness of hypnosis, whether in curing a disease or enabling walking on hot coals without burn pain or blistering through subconscious control of the body's immune system and blood flow, conscious focus on an external force or deity to help one achieve a particular goal can be useful, but it seems most important to have all other brain functions, especially conscious ones, turned off (McClenon, 2002). This focus apparently permits the subconscious to use all the brain's resources to achieve a desired goal with confidence that excludes interference or noise from unrelated sensual perceptions and from conscious doubts.

It has been found that tasks involving complex physical coordination are optimally achieved totally subconsciously by just focusing on the task, as in keeping one's eye on the ball when catching a baseball, and conscious attempts to interfere or help achieve such goals are counterproductive (Gigenrenzer, 2007). A focus or belief on an external force apparently helps keep the conscious from intervention that might disrupt the subconscious's efficient procedures for handling many tasks. A belief in an external force may eliminate fears of failure that can result from relying on one's own conscious direction. Fear and doubt create neural noise which distracts from concentrated subconscious activity and even diverts the subconscious into actions to prepare the body and brain for a possible failure. As a result, a mere statement to oneself that failure isn't an option may have an effect even more powerful than faith in a fictitious deity that has to be at least subconsciously deduced to be subject to doubt.

Within this context, it's interesting to recognize that the ancient Druids believe deities are mere intermediaries between humans and Nature (Paxson, 2005). Belief in such fictitious deities enables the subconscious to harness electromagnetic forces to accomplish consciously directed goals. However, the actual deities here seem to be the conscious thoughts or beliefs that make the subconscious carry out a task without conscious doubt or interference, as opposed to actual real entities. Because of the fictitious nature of any deities, the wills of those believing in actual quantum forces are far more powerful, regardless of whether an intent is willed through self-suggestion or via the suggestions of others. A belief in anything apparently enables the conscious to surrender its command function to the subconscious for uninhibited functioning to carry out the perceived will of the entity of the object of the faith, and believing in real quantum electromagnetic waves focuses directly on the actual forces that are involved in any macrocosmic phenomena without any false deity intermediary that has no actual potency.

Since suggestions can have a hypnotic effect on the brains of all humans (McClenon, 2002), medical applications of hypnosis have been found to be especially effective today in treating numerous diseases that may be related to the psychological well-being of an individual (Bernstein and Nash, 2008). Hypnosis has also been scientifically confirmed to cure many illnesses relating to factors that can have psychological causes and remedies like ulcers (McClenon, 2002). Many diseases that have bacterial, virus, or other causes may be avoided through hypnosis or the power of belief, at least partly because it reduces unhealthy stress (Lipton, 2008) that arises from worries or doubts which would divert the subconscious from the actions necessary to maintain health into preparations for a disease and thereby cause neural firings to permit an illness in an effective self-fulfilling prophecy of sickness.

The standard procedure for inducing hypnotic effects seems to function by initially disassociating an individual from everything but a particular item upon which the hypnotist suggests should be focused (Bernstein and Nash, 2008). For instance, hypnosis in pain management appears to work by the disconnecting of sensory perceptions from consciousness (Gruzelizer, 2006) that seems to come with hypnotic focus on something else. However, hypnosis has been found to have therapeutic value for many illnesses by focusing greater concentration on a task, problem, or disease (DeLaurence, 1910). For instance, the focus involved in hypnosis can even enhance the mobilization of natural killer cells that are effective in fighting diseases (Gruzelier, 2006). Hypnosis can also be used effectively in conjunction with other medical applications like herbal remedies (McClenon, 2002).

It has been alleged that hypnosis applied with a strong focused will is able to cure any sickness (DeLaurence, 1902). Unfortunately, difficulties in testing the effectiveness of hypnosis, as well as inadequate incentives within capitalism for funding such research that can't be owned and profitably exploited using conventional patents, has made hypnosis a more fringe medical treatment despite some evidence of its efficacy, and stage hypnotic theatrics may have contributed to an unprofessional view of this powerful procedure (*NewScientist,* 2008zzr).

While suggestions can be effective even when made to fully conscious people, the effects are most powerful if suggestions are made while a person has been put into a hypnotic trance (McClenon, 2002). The trance-like form of hypnosis represents a physiological state resembling sleep (DeLauren, 1910). This form of hypnosis that is characterized by sleep-like paralysis and dreamlike perceptions appears to result from "the firing of cholinergic neurons" in the brain (McClenon, 2002). Those neural firings seem to result in a reduction in the conscious logic that exists in awake alert states. They enable imagining and believing many suggestions that are passed without the normal conscious skepticism to the subconscious, which therefore unconsciously sets simple but effective rules to make the suggestions come true.

Hypnosis actually has many characteristics that are similar to sleep (Hewitt, 1996), when the pineal gland, which is located in the middle of the brain and appears to be involved in healing and dreaming, is especially active (Strassman, 2001). The pineal gland controls other hormone-releasing glands (Stray, 2005), and it produces health-enhancing pinoline and melatonin that is normally maximized during sleep but that is inhibited by night lighting and by television viewing (Gynn and Wright, 2008). Since all the functions and capacities of the mysterious pineal gland are unknown (Strassman, 2001), it seems possible that hypnosis involves that particular gland. The allegations that hypnosis works on iron magnetism (DeLaurence, 1902), that psychic powers can be enhanced by hypnosis (McTaggart, 2008), that those with greater iron content in their bodies have greater psychic powers (DeLaurence, 1913), and that greater consumption of iron in food enhances the psychic powers of the pineal gland (Stray, 2005) are consistent with the hypothesis of a connection between the pineal gland and hypnosis, as well as with psychic powers in general.

However, despite similarities between sleep and hypnosis, with the possible involvement of the pineal gland in the magic of both, hypnosis is substantially different from sleep. In particular, there is extreme focus in hypnotic sleep that is the exact opposite of normal sleep with its wildly wandering dreaming, and a hypnotic concentration is also on an external influence, except in cases of focus on one's own conscious thoughts and words in self-hypnosis (McClenon, 2002). Whereas normal sleep typically causes the thalamus to filter out ordinary stimuli from the environment that would otherwise cause conscious perceptions of the external situation during sleep and lead to reactions to it (*NewScientist*, 2010au), hypnosis apparently allows information from a hypnotist's voice to be processed and acted upon.

Nevertheless, since hypnosis is typically thought of as a person consciously programming the subconscious of another, or of oneself in the case of self-hypnosis, sleep may actually represent a form of the subconscious self-programming itself. As part of the rewiring and repairing of connections that definitely do occur during sleep (Erb and Erb, 2003), sleep could actually be a form of self-hypnosis by the subconscious. In particular, the subconscious seems to be programming itself during dreaming to prepare for future simulated possibilities. It certainly can train the neural connections to fire upon prompting by various types of situations that are similar to past ones but symbolically general enough to incorporate many different future environments that might be encountered. Conscious interference in this process would be counterproductive to both the training and any actual carrying out of subconscious reactions, and so the conscious is largely shut down during sleep, as rational though might detract from both efficiency and creativity if it reflected on everything.

It has been found that one can suggest a topic to dream about and the more hypnotizable people are more likely to actually dream as suggested (McClenon, 2002). Since the same effect can be achieved when one goes to sleep thinking about some issue or problem without an external suggestion, sleep does indeed have aspects of self-hypnosis and often enough results in the resolution of a problem.

There are, however, different degrees of sleep that include being partially awake and partially asleep (Ananthaswamy, 2009e). Since deeper external hypnosis in the form of trance-like sleep has a more powerful effect (McClenon, 2002), it might be concluded that deep sleep is the most potent form of self-hypnosis. The dreaming that tends to occur during deep sleep may very well be designed to prepare the subconscious to act upon various simulated situational signals to carry out the goals that were previously set, or allowed to be set, by the conscious prior to falling asleep.

Cognitive functions are still possible under hypnosis but require extra effort (Gosline, 2004). The brain is run mostly via subconscious activity during deeper states of hypnosis (McClenon, 2002), and so hypnotic effects are maximized then. While under the influence of hypnosis, an individual has reduced capacity in planning and associating appropriate actions in the face of external cues, and it tends to focus on the current situation (Gruzelier, 2006). Without conscious logic designed to forecast and logically analyze the probability of future scenarios, the subconscious just reacts to the perceived environment, including ones simulated during sleep. Even though the subconscious merely reacts to contemporary stimuli, hypnotic suggestions and dreamed events can apparently enable it to solve and train neural firings in preparation for the present stimuli of hypnotic commands and dreams that can provide future signals for further action and processing when they are triggered by the indicated stimuli.

Deep forms of hypnosis probably originated in animals hundreds of millions of years ago as an evolved psychological state of mind that results in an organism playing dead in the presence of a predator to reduce the chance of detection (McClenon, 2002). This type of hypnosis seems to depend on an extreme focus of the organism's controlling organ on a particular task, such as escaping detection in the case of playing dead. There may be even a subconscious recognition, perhaps aided by invisible electromagnetic waves received, that failure to move might make it difficult for a predator to notice the endangered being.

Motionless behavior in the face of a predator might be inhibited by distracting conscious doubts about the ability to do so or by making contingency plans in case of failure that might require at least movement relating to heavy breathing. Perhaps as a result, in organisms that can consciously think, belief in a suggestion without the typical conscious questioning evolved to be fundamental to the success of hypnosis in having it carried out. Greater belief caused by deeper forms of hypnosis that result in less conscious doubts or distractions do indeed increase the chances of success. Suggestions are therefore most powerful when made during a hypnotic trance because the conscious is inhibited from questioning a suggestion.

In the case of a hypnotist providing instructions while the subject is in a hypnotic trance not to remember a suggestion, the hypnotized individual normally doesn't even realize the suggestion has been accepted by the subconscious to be followed without conscious thought or doubt. Commanding no memory of the trance can inhibit conscious recall because all associations with the related neural firings are immediately disconnected from reflective self-awareness through the command that becomes an automated subconscious process itself. Suggestions are then carried out automatically without being questioned and without realization of the true reason for carrying it out. Nevertheless, when informed not to remember a suggestion, the conscious may later try to rationalize reasons for carrying it out, or at least for trying to follow the suggestion.

Recognition of hypnotic suggestions having been made can prompt conscious doubts about them that may also result in conscious overriding of them. The brain does so by consciously directing the firing of systems of neurons that weaken and eventually silence the subconsciously programmed system of connected neurons firing. As a result, suggestions that a person is told not to remember tend to be more likely to be followed and generally will be as long as the individual never realizes hypnotic suggestions have been made, as long as the individual doesn't consciously have alternative neural firings inconsistent with them silence them. By itself, analytical questioning all subconscious tendencies and actions that seem artificial or make little sense with regard to innate human nature can eliminate them, and even connections of neural firings resulting from suggestions made during childhood can be thereby broken.

Deep hypnotic trances in modern times are generally induced by telling a person to focus on something until the subject becomes tired, such as via eye strain, and then instructing that person to fall asleep (DeLaurence, 1910). Focus on something can induce such excessive activity in the visual neurons that the entire brain becomes exhausted and willingly accepts a suggestion to sleep. The hypnotist then takes control of the subject's subconscious activities during the trance through a command to focus only on the hypnotist's voice prior to inducing the sleep state.

Normal self-induced sleep shuts down conscious thought to enable the subconscious to run on its own to try to evaluate and solve problems that the individual has encountered or may encounter in the future. However, when the sleep is induced externally at the instruction of a hypnotist, the subject's subconscious is set to follow the instructions of the hypnotist without conscious interference. In particular, the hypnotist, as opposed to the subject, has turned the subject's conscious filter off and has therefore taken command of its processes. The subconscious can, however, revert back to working according to some innate, habitual, or self-commanded sleep process in some circumstances, such as due to lack of direction from the hypnotist, or conflicting suggestions that eliminate trust and cut off the connection between the hypnotist and the subconscious.

Hypnosis begins with heavy firing of the right side of the brain (McClenon, 2002), which is involved in negative emotions such as fear that normally inhibit behavior (Klein, 2003). Since the earliest form of hypnosis was the freezing of all movement in a sleep-like trance at the frightful sight of a dangerous predator that many animals exhibit (McClenon, 2002), the initial heavy firing of the area of the brain involved in fear when entering into a hypnotic trance may be related. In particular, while hypnotized people don't normally experience fear, the extreme focus on an outside stimulus like a hypnotist's voice may distract them from feeling the activity in the right-side of the brain because the focus is so strong, just as extreme life-and-death focus on a predator does for many animals. In particular, the original "playing dead" hypnotic trance, which can save prey from being observed by predators, is initially induced by extreme focus on the predator that results in ignoring heavy right-side brain activity, which normally would result in following a potentially suicidal instinct of fearful flight. With no chance of escape by any known means, at least one such prey was frozen into indecision on taking any action, and the resulting life-saving inactivity was passed on to subsequent generations.

The playing-dead form of hypnosis appears to have developed further in primates into a similar process of focus followed by heavy right-side brain activity that isn't consciously experienced but actually triggers a hypnotic trance. For instance, the hypnotic effects of apes grooming each other (Mcleonon, 2002) probably originated when the sexual urges of non-alpha male apes left open no feasible opportunity for satisfaction of their irresistible desires via the normal forced form of ape intercourse. The male was subconsciously driven to just pet a desired partner to get closer to the wanted sex object. The resulting grooming created such a pleasant feeling in the partner that the mate relaxed into a submissive hypnotic trance to allow more of the same. The focus on each other then enabled harmonious fulfillment of their mutual sexual urges.

As explained in much more detail later in this chapter, humans developed a deeper form of mutual hypnosis via eye contact that eventually became the well-known love at first sight. With mates being consciously clueless as to how to have their irresistible lust satiated, they froze in a gaze that allowed them to evaluate the electromagnetic "psychic" signals passed between them to harmonize their actions for the mutually desired sexual intercourse. This process required each partner to do what the other wanted, thereby collapsing their probability wave functions on a coherent point to eventually allow the irresistible procreative act. The blind, unquestioning surrender later developed into blind, unquestioning romantic love in anticipatory rewarding of the sexual bliss as human learned to forecast the future. This ecstatic feeling results from each mate's internal rewarding of the gaze between the eyes with pleasure chemicals in the brain that was self-reinforcing and became so overwhelming that no thought of the consequences of failure could be considered.

The hypnotic focus on a human partner's body and genitals developed in a similar fashion during the sexual act itself as each mate strove to please the other in a mutual submission that harmonized their particular sexual preferences. They merely interpret signals from the body movements and sounds as well as from "psychic" evaluation of the electromagnetic emissions from the partners. A sort of mutual body hypnosis resulted that required a complete turning off of all conscious doubts or inhibitions to attain the necessary resonance for mutual fulfillment of their irresistible sexual desires in orgasmic bliss that effectively represent the attention of their divided sex cells for their opposite halves.

Likely much later in prehistory, the sight of a predator from which flight was suicidal and in the face of which motionless inaction unthinkable because of attachment to nearby loved ones that may have included both mates and offspring resulted in a freezing of humans in fighting groups. Being trapped in a cave might have further promoted battling instead of trying to flee. At any rate, when attacked by the predators, the humans responded instinctively by each defending against the attackers, as would have been futile alone but was effective in a hypnotically unified group. Spurred by internally produced adrenaline, they were able together to accomplish incredible physical feats that might have also included creative defensive acts such as throwing objects, tools, or weapons against the predators. With such communal reactions reinforced with survival, the trait was passed down to subsequent generations, and an innate group consciousness was born. This form of mutual group hypnosis may have been prompted by the group noises involved in chanting.

Thus, humans evolved to do more than "play dead" while in a hypnotic trance. The group hypnosis via chanting that is widespread among aboriginal peoples (Howells, 1986) represents only one of the original forms of hypnotic trance that is special to humans. However, human ancestors freezing together as a group in the face of predators that allowed them to overcome the natural instinct to flee probably was likely the origin of the more general form of group hypnosis that included more than just sexual mates. Warning signals would have easily been learned in the form of chants that enabled them to prepare for battle as a unified whole upon the sight of a predator in their vicinity, and the resulting group hypnosis enabled greater cooperation between the defenders. This tendency, which probably did originate from their loyalty to other humans with whom they lived that included young offspring who may not have been old enough to "play dead" at the sight of predators, gave humans the focused determination to protect each other in fearless acts of bravery that were successful only because the humans stood together in a large enough group that hypnotically stood and fought together. Grunts or chants like "fight, fight, fight" might very well have developed out of such inclinations, enabling humans to overcome much more powerful animals which normally would have eaten them. The hypnotic capacity was then selected genetically through sexual selection of those who were characterized by putting the group above the individual as well as by the enhanced survival of groups that did so.

Hypnotic trances may thus have evolved in a special form in humans as a way for them to band together bravely as a group instead of dying individually. As humans become more intelligent, they may have learned to harness the powerful effects of hypnosis by inducing extreme focus even in other situations. Any sort of rhythmic beat, music, or dance can hypnotize people into an altered state (Gynn and Wright, 2008), and a group consciousness based on empathy and chemical reinforcement for cooperation might have even made at least a mild form of group hypnosis a normal situation.

Today, individual hypnosis typically requires that a person trust a hypnotist enough to "let go" and give up control, and, by inhibiting thoughts of emotional consequences, it enables both embarrassing actions at the command of stage hypnotists and clinical recall of suppressed memories (Gruzelier, 2006). Since frequent back-and-forth eye movements can result in a change in focus away from painful memories (Bernstein and Nash, 2008), and since watching a hypnotist's swinging pocket watch is one of the stereotyped methods of hypnotizing people (Hewitt, 1996), concentrated focus on the visual sense and the voice of the hypnotist that is often utilized to induce deep hypnosis appears to disconnect consciousness from other sensory perceptions and normal brain functions (Gruzelier, 2006). The fact that the attention necessary for learning from others is often optimally obtained for both children and dogs with eye contact and visual gestures, as well as vocalization (Douglas, 2008), provides a further indication that eye focus can be useful in inducing hypnosis. Any learning from others results from a form of hypnosis in terms of following the suggestions or instructions of others, but it is enabled more easily with deeper hypnotic trances that eliminate most if not all conscious filtering of instructions.

The focus involved in just relaxing meditation on simple sensory perceptions (Bernstein and Nash, 2008) usually follows professional hypnotists' instructions to close the eyes once attention has been attained, and a command to imagine a pleasant passive environment results in a suspension of "reality testing and critical evaluation" at the same time that it causes the hypnotized person to surrender executive functions to the hypnotist (Gruzelier, 2006). The result seems to be that the normal subconscious processing involved in sleep and dreaming isn't set to run according to the individual's own conscious needs to prepare for the future and solve problems but instead is directed toward the commands of the hypnotist. The concentration on one particular phenomenon that requires imagining may help in opening up the normal thalamus barrier of the conscious to the subconscious, thereby enabling unconstrained subconscious focus on the hypnotist's commands and resulting in the training of connections of neural firings that cause obedience to them unless they are later counterbalanced by conscious thoughts that exclude them.

Relaxing meditation tends to result in slowing brain activity into alpha waves that characterize light sleep (McTaggart, 2008). It makes sense that such an altered state, which is of the same frequency as that normally exhibited by impressionable children between 6 and 12 years of age, would be conducive to hypnotic suggestion with little resistance. Self-induced meditation, which is often induced by relaxing focus on some particular item, can thereby lead to self-hypnosis that is just hypnosis under one's own direction.

The power of positive thinking appears to also be somewhat similar by involving focus and avoiding all thought of failure. Positive thinking can be as powerful as hypnosis when an individual is so motivated that a failure to believe in success is unthinkable. Such states of total concentration are often naturally induced when the outcome of a failure to focus is perceived to be too adverse to have any thoughts or fears of anything but how to succeed, as may likely have happened in prehistoric times when human groups had to face down predators.

As previously mentioned, the sleep-like trance involved in today's hypnotic practices is typically prompted by the focused attention of the hypnotized subject on the hypnotist that is eventually accompanied by a suggestion to relax (Hewitt, 1996). A suggestion to imagine a pleasant situation provides reinforcement for continuing to listen to the voice that prompted that feeling, thereby enhancing the trust which motivates following the dictates of the hypnotist once sleep is induced. The closing of the eyes then shuts down visual perceptions and permits complete concentration on the hypnotist's voice. The deep forms of hypnosis that result in a trance in modern humans typically involve surrendering control of one's will to a hypnotist, as is typically accomplished via relaxing suggestions by a hypnotist who believes in his or her abilities to hypnotize through strong focus of his or her own attention on that task (DeLauren, 1910). Such externally induced trances may not be that much different from the group hypnosis that once occurred upon a warning chant indicating the need to prepare for battle against sighted predators. Without any other thought than a complete and unquestioning trust for each other, they actually relaxed them into fearlessness because of the confidence created by the power of cooperative mutuality in intent and action.

Pleasure chemicals like dopamine, endorphins, and related nitrous oxide that can be created by an imagined pleasant event upon which the hypnotist instructs focus to be directed, probably cause a very relaxingly soothing feeling that lead to a sleep-like state. Brain chemicals like DMT, which inhibits depression when present in the right hemisphere of the brain (Gynn and Wright, 2008), and melatonin, which affects neural connections while a person is asleep, and which stimulates the endorphins that creates pleasure and the oxytocin that promotes social bonding in the brain (Gynn and Wright, 2008), might then be released once a sleep-like trance is induced by the hypnotist. DMT, which can cause psychedelic experiences (Strassman, 2001) that are associated with sleep-related imagination (Gynn and Wright, 2008), may prompt creative thoughts after the combined suggestion to relax, imagine, and sleep. These various chemicals combined might provide impetus for relaxation of the barrier between the conscious and subconscious due to the brain's tendency to associate melatonin and DMT with such a disconnect that often occurs during sleep and that facilitates subconscious learning without interference.

DMT itself may assist in learning by disabling neural connections that are otherwise consciously inhibited because they haven't seemed to contribute to survival and reproduction in "normal" environments but that reflect a deeper reality which is symbolized in the form of remembered dreams and seemingly irrelevant experiences. The disconnect between "normality" and the deeper reality might enable the subconscious to figure out how to accomplish tasks that seem impossible from the perspective of the perceived "normal" reality. The incredible feats of hypnotized individuals can thereby be explained, as subjects subconsciously perceive how to carry out the suggestions of the hypnotist.

On the other hand, in the normal sleep that is internally induced and therefore causes the subconscious to attempt to figure out how to carry out the objectives of that person's own consciousness, the person's subconscious tries to utilize DMT effects and dreaming as a means to that internal end. That subconscious processing can be constrained, however, by the conscious doubts which existed prior to sleep that were induced by the person experiences in the "normal" reality. In contrast, such interfering doubts may not exist with a self-confident hypnotist, thereby enabling a person induced into a trance by that hypnotist to carry out the conscious will of the entity that induced the sleep.

An effective hypnotist believes so strongly that the commands can be carried out that it isn't just a hoped-for prediction but a known event that will happen. The same belief is involved in the success of "positive thinking" that can enable the subconscious to figure out ways to accomplish tasks that seem impossible and that is reflected in the similar process of self-hypnosis, including during normal sleep. The hypnotic process is indeed enhanced by the hypnotist having extreme self-confidence in the ability to hypnotize and a strong will (DeLauren, 1902).

At least an initial indication of desiring to help the subject that is implied alone by the initial suggestions to relax facilitates transference of the person's trust from the individual's own conscious will to that of the hypnotist, whose own conscious doubts would adversely impact the power of the suggestions then made. The unshakeable faith necessary for the most effective hypnosis is helped if the hypnotist is more hypnotizable and more able to experience anomalous psychic events and fantasies that are associated with thinner cognitive boundaries (McClenon, 2002). Such traits, which enable a greater blending of the conscious and subconscious that sometimes results in schizophrenia and an inability to distinguish fact from imagination (Gooch, 2007), seem to enable confidence in the accomplishment of tasks that might seem incredible to those believing only in a "normality" that is characterized by a distinction between reality and imagination. A more unified nature of the subconscious and conscious also allows less conscious interference in subconscious communication between people that can be useful in creating resonance between their brainwaves to merge their wills into one.

The fact that people who are better able to hypnotize are also more hypnotizable can create a weakness in them and in psychic abilities in general, however. In particular, hypnotists or psychics may be more susceptible to being manipulatively hypnotized by others with hypnotic powers. They are thus exposed to "invisible" revenge by others who utilize the services of other strong-willed hypnotists to inflict harm on them. This possibility inhibits them from selfish exploitation due to human spite if they use their powers for "evil" purposes, with spiteful actions taken including not only "invisible" psychic ones but actual physical violence. The murderous battling against "witchcraft" and shamans perceived to be "evil" that has persisted for so long in the last few thousand years (Howells, 1986) seems reflective of the counterproductive nature and deadly weakness of selfish use of hypnosis and psychic powers in the course of human history.

In addition, because of anticipated regret or fear at being exploited, it is also possible for people to resist hypnotic suggestions of any agents, whether the hypnotist be an individual or a group of people. This capacity no doubt has evolved more as there has been more manipulative use of hypnosis over the last 30,000 years. Any hypnotic suggestions perceived to have been exploitative can result in spite that not only motivates reversal of the suggestion but also vindictive acts that can be dangerous to the hypnotizing agent.

As previously mentioned, reducing the conscious questioning and inhibitions to following others' suggestions can be achieved with various levels of trust. However, complete commanding control is unnecessary to have some hypnotic influence. Hypnosis in its various forms appears to occur in groups of social animals like primates as a means of relieving stress that likely preceded its use in humans (McClenon, 2002), but its use for selfish manipulative purposes that can be counterproductive to the goals and wishes of the hypnotized subject is probably a fairly recent development.

C. The Hypnotic Characteristics of Human Nature and Manipulation

The unquestioning surrender to an agent in hypnotic trances can lead to one-way worship when it isn't mutual. While psychic feelings indicating non-mutuality are easily detected in eye contact and thereby inhibit the trust necessary for the selfless surrender of deep hypnosis, the hypnotic characteristics of submission to a mate's sexual desires through focus on the genitals don't necessarily provide for mutuality because the concentration is then on a particular body part as opposed to a human whole. The complete obedience involved in this hypnotically blissful process can be manipulated by another who isn't already unified via hypnotic love. One-way worship can therefore follow and even be perverted into total slavery to the manipulator. Because women' more complex sexual needs require greater hypnotic focus on their bodies by the males for sexual ecstasy, the first able to harness the inhumane power of one-way hypnosis and worship were likely selfish females.

The first matriarchies arose from this tendency about 35,000 years ago, as explained in Part I. Such exploitation eventually led to violent spite when males became so sexually, emotionally, and spiritually frustrated by matriarchal cruelty that they overcame the natural bliss of selfless hypnotic submission. Patriarchal rule enforced by violence therefore eventually replaced the matriarchies 30,000 years later. With both women and men still susceptible to hypnosis due to the group consciousness that arose from the hypnotic chanting that enabled them to collectively fight off predators without question, the patriarchal elite were also able to utilize hypnosis to maintain their own power. However, the unequal hypnotic tendencies with respect to sexual desires also continue, and females have been able to exploit this inclination to re-assume greater control within the framework of patriarchal rule.

Only a conscious decision to return to mutual group hypnosis can prevent further pathologically vicious battles between the sexes. Progress toward that end can begin via the naturally ecstatic group hypnosis between mates in love. All that is required is a mutual trust which is consciously allowed after all doubts have been removed, and these relationships can easily expand to a more general group consciousness eventually. Human nature drives all human beings toward this end at leas subconsciously.

As will be explained in more detail in Chapter 13, human consciousness originally had communal characteristics, and humans cooperated closely in groups in order to survive an otherwise overwhelming environment. In this group consciousness that characterized early human culture (de Quincey, 2005), people felt as if they were part of a group whole, as in different limbs of the same organism, instead of as separate independent individuals. The brain's perception that a tool in one's hand is an extension of the hand (*NewScientist*, 2009zzo) may provide an indication of the feeling of group consciousness that once existed.

As previously indicated, the attention of groups of humans could be focused on a communal activity in prehistoric times with a form of group hypnosis that was achieved via chants, as well as by drumming and music later (Howells, 1986). Hypnosis was long used by Native Americans to enhance the fighting skills of their warriors (Underhill, 1965), and drums can slow brain waves down to trance-like states even today

(McTaggart, 2008). Since groups of soldiers have been found to fight better and more effectively when there is music or drumming, it seems likely that the capacity of humans to be hypnotized so easily evolved originally in order to focus group attention on communal fighting by disassociating distractions like fear and thoughts of fleeing. In early human prehistory when only sharpened rocks and sticks would have existed as potential weapons, group hypnotic focus would have been vital in fearlessly cooperative battles against stronger predators like lions against whom only groups of humans totally trusting in each other's fighting abilities had a chance.

A belief in the omnipotent power of such harmonious human groups might even have arisen from the first such successful encounter that encouraged further application of that omnipotence and might even have led to a religion of believing humanity as a whole was a deity, as it may indeed be. This belief might have even enabled humans to overcome their innate fear of fire, thereby leading to further human powers, including the creation of campfires that permitted them to spend the night in the open far from trees with the assurance that the fires would provide a good deterrent against predators. Wild animals naturally fear fire (Kenneally, 2008b), and its initial harnessing by humans may have been primarily motivated by its usefulness in defending against predators, with the power of humans being magnified by using hot flames as a weapon in a group of humans hypnotized into cooperation. Human hypnotic capacities themselves might have actually developed further around the campfires (McClenon, 2002), which also enabled human ancestors to adapt and migrate to colder climates over 1.5 million years ago (Adovasio, 2002).

Therapeutic use of hypnosis might have actually developed later in human culture. Ancient magicians, shamans, and religious leaders frequently used hypnosis for healing and other purposes and were thereby recognized leaders (McNamara, 2006). Shamans have long utilized hypnosis to facilitate childbirth and ease labor pain (McClenon, 2002), and it is likely that, in the prehistoric matriarchal societies, one of the first uses of the technique on an individual collectively hypnotized by each other, as opposed to a group of individuals, was for that purpose. Knowledge of the use and power of hypnosis would therefore likely have begun with women, whose magical aura would have thereby been further enhanced, resulting in even greater reverence and potential for control.

The pineal gland, which was used more in prehistoric times and appears to enable psychic powers (Stray, 2005), likely played a significant role in the group consciousness and communication, especially before the existence of well-developed languages. The relaxing of the barriers between the conscious and unconscious that occurs in hypnosis and that is associated with anomalous or psychic happenings (McClenon, 2002) may enable the pineal gland to operate without the interfering conscious skepticism that makes it ineffective (Stray, 2005). Thus, the pineal gland may have played a significant part in group hypnotic power. In particular since chanting leads to effects that include blissful surrender to an even larger benevolent whole that connects everything, it seems plausible to assume that hypnosis, at least in its deeper forms, activates DMT and the pineal gland. The connection between psychic abilities and hypnotic powers (McClenon, 2002) provides evidence in favor of this hypothesis.

In modern individualistic societies, human psychic powers and the pineal gland tend to calcify by adulthood (Stray, 2005). The finding that psychic capacity and hypnotizability are related phenomena that peak at the age range of 10-12 years (McClenon, 2002) is consistent with this inclination. Societal skepticism, discouragement, and ostracization of psychic powers probably contribute to their lack of use and deterioration later in life. Another cause of the decline in a more psychic group consciousness and communication may be related to a perception that human language provides more precise communication. However, some people, especially aborigines, continue to able to communicate better in silence via meditation that permits a more basic understanding of each other's thoughts without the necessity of translating them into words that represent mere symbolic categories for the true detailed experiences of reality that are distorted by languages (Gynn and Wright, 2008).

Greater intellectualization and language learning after the age of 5 apparently do contribute to the suppression of psychic abilities, which seem to actually enable moving objects with energetic imagination in some people (Gooch, 2007). Nevertheless, humans continue to be susceptible to the power of hypnosis, including self-hypnosis, and those able to harness this focused power can perform "magic", such as harnessing seemingly superhuman strength, and this control extends beyond any medical effects (DeLaurence, 1902).

As previously indicated, the fact that the pineal gland produces the mind-relaxing chemical melatonin might also help explain the relaxed nature of hypnotic trances. In addition, unusual sensory perceptions and a relaxed feeling of oneness with the universe that can be induced by DMT may also emanate from the pineal gland (Strassman, 2001). The feeling of "blissful surrender" experienced under the influence of DMT may be partially related to a natural release of pleasure chemicals in the brain such as the nitrous oxide that might accompany the DMT-style "hallucinogenic" effects in the natural bonding process of humor at the irrational submission to each other.

In any event, group chanting not only facilitated better communication among members of the group but also enabled the group consciousness to be better focused and more efficiently unified for important tasks. In particular, group chanting works to focus attention on one particular activity, such as fending off predators as a group with no thought or conscious evaluation of other courses of action like trying to flee. It essentially accepts the group chant as a suggestion to be followed without rational analysis of alternative courses of action and without any fear or preparation for possible consequences, thereby enabling complete focus of the brain on the particular action for which the group is chanting. As with any hypnotic suggestion, it is more effective when there is less skepticism as to whether it is the right course of action or even absurd and suicidal. Trusting in the entity making a suggestion is essential for it to have maximum power, and since the group is the entity making the suggestion in group chanting, the group consciousness that existed in prehistoric times was very useful in causing the chanting to prompt hypnosis due to the mutual trust that always existed. Group consciousness may even have developed as a result of, or concurrently with, group chanting, as may have been initially prompted by the emotional grunts that were communicated empathetically between humans in their first group defense against stronger predators.

In group chanting, everyone hypnotizes each other, and no one controls another person. Although someone could certainly lead a group chant, others would follow only voluntarily. However, everyone following a leading signal, such as someone indicating the approach of a predator, might be spontaneous in a society characterized by group consciousness where everyone acted for the good of the group, including in unison when faced with immediate danger. The mystical feeling of possession by a spiritual force or by Nature itself that so many people occasionally experience even today (McClenon, 2002) may reflect a natural human tendency to want to return to this benevolent communal consciousness. Psychic abilities would have been well developed in such an environment (North, 1997), at least partially because of the absence of well-developed languages for alternative communication.

The fact that people in socialist countries tend to experience more anomalous psychic events (McClenon, 2002) could possibly be related to their greater communal experiencing of the world that incorporates a larger degree of group consciousness, trust, and confidence, which are less inhibited by the "rational" skepticism imposed by individual consciousness, selfishness, and manipulation. Evidence of group consciousness being reawakened does exist. For instance, in the socialist USSR, members of a collective were found to voluntarily forego rewards in order to keep an individual member from suffering for a mistake (Kharkhordan, 1999). Such tendencies could certainly contribute to the psychological well being of individuals in the group as well as enhance psychic communication.

Group support systems reduce individual stress and enhance health (Bernstein and Nash, 2008), and prehistoric group consciousness, care, and focus would have therefore provided excellent therapy for sick and healthy individuals alike in such an environment. The evolution of humans to be susceptible to hypnosis for healing purposes (McNamara, 2006) would have further contributed to group mental and physical health in this communal environment. The capacity to be hypnotized by a group of individuals would have therefore been selected by evolution.

Given the extreme power that can be harnessed via hypnosis (DeLaurence, 1902), and given that humans have long employed hypnosis, it is probable that it played an important part in the survival of early humans as well as in their later domination of other organisms. However, group hypnosis, as opposed to hypnosis of others and the group by any individual(s), was likely the prevalent form of prehistoric application of hypnotic "magic". In particular, evolution would have selected for the human groups that hypnotize themselves communally as the more united form of a group consciousness. Such groups are far more powerful than groups of manipulated zombies hypnotized by any individual(s). In fact, except in extraordinary circumstances such as in the face of predators, mere suggestive power is often better than the complete commanding control of a hypnotic trance induced by a single hypnotist since it enables diverse opinions that include better analysis of what is optimal given the needs, desires, and abilities of each individual in the group.

While individuals could have early on harnessed the power of hypnosis for themselves, a true human being, who naturally seeks to serve the entire group due to empathy and fear of spite, may not want to exercise it for any purely selfish purpose. Since hypnosis requires a resonance between brains, it may additionally be difficult to utilize it to the detriment of others. However, deceptive manipulation of trusting people like naïve people, who voluntarily relinquish control of their normal conscious interference out of curiosity, out of a desperate desire to be helped by a hypnotist, or out of a subconscious wish to return to the ecstasy of group hypnosis, may not be extremely hard, just as it isn't difficult to manipulate children who haven't developed a conscious defense against exploitation by others.

The evolution of individual consciousness that came with individual self-forecasting, as will be explained later would have contributed to a decline in the use of group hypnosis. In particular, forecasting for the group was more effective using each person's prediction of the individual self that had access to more direct information about that self. A more separate individual self-awareness was thereby spurred that led to individual consciousness as a useful component of the group consciousness. However, at some point, such individual consciousness prompted some to perceive the goal of the self to be separate from that of the group, and such a selfish individual, which may have been some sort of a mutant cross between an ape and a human or simply an old hag brainwashed by Neanderthals, decided to try to exploit the group trust for purely personal benefits to the detriment of others and the group as a whole. Such a pervert might have normally been eventually subjected to human spite by the group and exiled, but the monster could have eventually been worshipped as a superior being instead if it learned some special technology or "magic" such as individualistic, one-way hypnosis for exploitative purposes. A more individualistic use of hypnosis by special priestesses or shamans places the hypnotist in control, thereby creating a hierarchy and setting the groundwork for one-way worship that may have begun when someone consciously learned the hypnotic process which previously may have been only subconsciously generated spontaneously.

People under hypnosis display strong activity in the temporal lobes (McClenon, 2002), which are involved in hearing sounds or speech (Bernstein and Nash, 2008). The fact that electromagnetic stimulation of the temporal lobes can lead to spiritual and hallucinogenic experiences, which may be strongly felt through activation of the amygdala (McTaggart, 2008), implies that there may be a connection between hypnosis and religion that has the spoken word accepted as a sacred, factual representation of reality. McClenon (2002) has indicated that increasing language abilities over time actually enhanced hypnotic powers because they raised the level of symbolism that could be applied, with symbols being useful in eliciting emotions that are associated with them and that can affect physiological processes in the human body like the immune system.

The left side of the brain, which is involved in language comprehension (Bernstein and Nash, 2008), is also activated in people under hypnosis after activation of the emotional right side, and there is indeed communication between the conscious and subconscious in hypnotized people (McClenon, 2002). The shutting down of the conscious filter to the subconscious apparently keeps active any conscious thoughts necessary to understand the words of a hypnotist who has commanded sole focus on those words. With human hypnosis having originally developed through group chanting that included words, such a finding isn't surprising.

Part of the continuing power of hypnosis may also lie in the fact that language was originally developed to provide information, just as the chanting was designed to prepare people for common actions to deal with a particular situation like the approach of a predator. In particular, a communal system with group consciousness probably perpetuated the use of language for strictly honest, beneficial, and non-deceptive purposes, and so humans naturally developed a belief in spoken words that were designed to serve the human group as a whole. As a result, statements about the future were likely received as facts and hence caused the human brain to do what was necessary to make those forecasts consistent with the "prophecies" that were stated to happen in the future.

In a similar fashion, things told to children when they are very young have a deep, long-lasting effect that is ingrained into their subconscious because everything said then was accepted as true due to their "naiveté" resulting from lack of knowledge and experience. It is therefore difficult, albeit not impossible, to change beliefs learned as children no matter how absurd (Baring and Cashford, 1993) because the suggestions then are accepted without any conscious doubts or associations that can be recalled for the analysis needed for skepticism. Gruzelier (2006) has actually observed that the heavy activation of the emotional right-sided brain area under hypnosis, despite some activation of the language portion of the left side of the brain that allows understanding the voice of the hypnotist, is similar to the relationship between a parent and a child whose more limited development of the left side of the brain results in the bonding "resonance" being concentrated on the emotional right side. Group chanting may have also harnessed such emotional resonance that is still reflected today in the pleasant feeling of acting for the common good of all that made human beings so powerful.

While naiveté can therefore represent a weakness, for both children and adults with respect to manipulation, it can also represent a strength if a person is intelligent enough to use it via positive thinking and self-hypnosis. Naiveté also reflects an openness to new ideas that haven't been polluted by false assumptions and beliefs. It can also reflect a willingness to learn that hasn't been as inhibited by the manipulation that breeds skepticism, negative feelings like pre-emptive spite, and unhappiness.

After at least a few individuals started consciously using deception to manipulate others at some point in time, human consciousness began to develop some defenses like skepticism against lies and manipulation. However, hypnosis goes directly to the subconscious after the words are heard, processed for understanding, and believed without conscious filtering of the statements for validity. It thereby avoids the skepticism defenses, while it focuses the subconscious to efficiently carry out whatever has been said under hypnosis. Malevolent uses of hypnosis were therefore very likely in human prehistory (McClenon, 2002), probably beginning 30,000 years ago with the first manipulative religions involving worship of "superior" beings.

Hypnotic control can be used for any purposes, including for the benefit not only of the subject, or the subject's group, but also for the benefit of the hypnotist (McClenon, 2002). It can be utilized to deceive or manipulate people, especially with respect to making fools of hypnotized people for the entertainment of others that can be perceived by the hypnotized to have the possibility of bringing desired attention and even empathetic affection (McNamara, 2006). Because hypnosis enables a person to return to the blissful feeling of group consciousness, it is easily used to motivate an individual to provide entertainment and benefits to others because the letting go of conscious doubts gives the conscious a rest. Nevertheless, the evolution of the human capacity to be hypnotized that some other primates also have, albeit to a far lesser degree, implies that it has provided more benefits than disadvantages (McClenon, 2002), at least for much of distant past.

While there are many different views on the origins of human religious beliefs (McNamara, 2006), the power of hypnotic suggestion probably contributed significantly to the development of human belief systems (McClenon, 2002). In particular, hypnosis can shut down doubts to even the most incredible fairy tales. Under these circumstances, evolution might select for religions whose benefits from belief more than offset any reduced predictive power resulting from blind faith in religions based on falsehoods. Skeptical atheists wouldn't benefit from any such effects, and since non-believers might also decrease the power and benefits of hypnotic beliefs for others by promoting their skepticism, they might therefore have been perceived to contribute negatively to group welfare. Societies may therefore have sought to silence and possibly eliminate those with contrary opinions, especially in desperate situations such as facing down a predator with unquestioning group courage.

However, the realism of skepticism was apparently evolutionarily useful enough to enable skeptics to persist extensively today. The lack of blind faith enables an atheist to recognize simplifying assumptions may not always hold true and to therefore be prepared for exceptions. The ability of atheists to better explain and predict exceptions to some assumed laws of nature might have provided sufficient advantages to balance the placebo, psychological, and physiological effects of faith.

Advances in scientific knowledge, which was not available to early humans, only increase the utility of skepticism against absurd religious lies. Atheists of prior eras without the knowledge of the power of belief in something may therefore not have had a relative advantage until current times. However, scientific knowledge has now reached a point when one can confidently believe realistically in the evolved purpose of human beings to serve humanity as a whole. Despite their current minority and continued status of being largely silenced and ostracized, non-believers in today's "mainstream" religious lies can potentially exert enormous influence if they realistically recognize and believe in the power of human beings of the future.

An egalitarian belief in humanity as a whole seems to be an innate trait that easily results from the evolved characteristics of human nature and therefore likely was the norm for millions of years at least subconsciously. However, this natural human religion began to be replaced by hierarchical religions based on blind faith in an external deity and total suppression of skepticism about 35,000 years ago. Mutual reverence among humans as a whole was thereby replaced by worship of a fictitious deity that served as a proxy for a ruling elite of selfishly cruel people. These sorts of artificial belief systems developed across the globe, often harnessing all sorts of revered entities such as Nature, animals, and celestial bodies to serve as the deities which were utilized to manipulate the masses into slavery to the elite. Some religions like Judaism and Christianity even fabricated a patriarchal deity with human characteristics that served as a sort of omnipotent and omnipresent king to enforce obedience. Such bizarre fairy tales, many of which may have been spread via various forms of hypnosis, are from natural to human beings, but they do successfully manipulate human traits to turn people into robotic slaves to an elite.

One existing aboriginal group which lives entirely in the present has no religion and finds the concept of blind faith in something ridiculous (Everett, 2008).The fact that a group focused on living in the present has no blind faith in fairy tales supplies circumstantial evidence that religion or belief systems may be related to some people being able to consciously forecast the future. Those who could utilize hypnosis to influence future human behavior and those who could predict patterns in celestial objects and other events in nature were apparently able to convince others of their superiority at some point in time, and faith in the lie of an external deity may very well have grown out of such manipulative exploiters of humanity.

The feeling of nirvana that exists even in a European able to live completely in the present without any conscious thought to the future (Blech, 2008) may actually reflect, at least in part, the elimination of the effects of religion and worship that enables a mere acceptance and wonder at what is. On the other hand, one of the primary reasons for the development of many religions may have been the lack of knowledge about existence after death, and the human drive to prepare for the future including after the death of the self is indeed exploited by many religions through empty promises of eternal punishment after then for disobedience while alive.

In any event, women's abilities to predict their "magical" reproductive cycles by the moon (Shlain, 2003) could have been used to enhance their mystique among uninformed men without that "mysterious" forecasting power, thereby prompting males to also believe in the ability to predict other phenomena, including with respect to what happens after death. Just as sleight-of-hand magic tricks and deception enhance the power of hypnotists to have their suggestions believed and followed (McClenon, 2002), the initial menstrual cycle predictions by women that probably developed into relatively superior forecasting abilities for them, especially during their "mysterious" bleeding periods, would have enhanced females' ability to use hypnotic suggestions. The fact that women were widely used exclusively as oracles or predictors of the future even under some patriarchal religions with male gods (Barings and Cashford, 1993) is consistent with a continued belief in women's special forecasting powers even after imposition of patriarchies.

The first magicians, scientists, and religious leaders in most cases were therefore likely to have been women, whose special biological creative and nurturing powers would have made them even more easily revered (Sjoo and Mor, 1987). Powers that can be exercised for the good of the group would have been especially revered, and it would have even been possible for selfish people with exclusive knowledge or powers to extort an entire group of people into one-way worship and servitude under the threat of withholding those powers. Demonstrations of inexplicable magic might have contributed to one-way worship of such people since they increased the uncertainty as to the extent and source of the powers.

Magicians seem to exert substantial control over observers' minds that scientists are having difficulty understanding, and although many magic tricks are known to depend on attracting the attention of observers in order to distract from sleights of hand, the effectiveness of some magic has not yet been fully explained by science, such as why an observer selects a playing card simply because it has been seen more frequently (Powell, 2008). Perhaps, a natural tendency to forecast based on past patterns leads to the prediction of that card that the subconscious takes as a command, thereby enabling it to select the card through well-registered observation of it as it is being mixed. It's plausible to hypothesize that the earliest leaders integrated magic and hypnosis for maximum effect and influence, and they may have thereby also learned the art of deception and manipulation necessary to attain one-way worship of themselves.

It has been shown that religious prayer is more effective in turning off the prefrontal and anterior cingulate cortex areas of the brain involved in judging the truth and importance of what people say (Coghlan, 2010a). Through this natural mental process, the mysterious power of magicians might have easily led to greater trust in them, as could have been magnified by hypnosis, which might even have promoted the tendency. Such trust would have facilitated the magicians' ability to hypnotize people. Since the power of hypnosis that can be applied to groups of humans is a natural characteristic of humanity that goes back millions of years (McClenon, 2002) to a time when there was group consciousness, it is probable that the earliest magician leaders learned to harness the power of hypnosis that is maximized by repeated suggestions. Repetition, including in the earlier group hypnosis involved in chanting, apparently reinforces the brain connections necessary to carry out the intent of the suggestions or chanting.

D. Reverence and the First Deities

After the death of magician leaders, reverence for them, especially those whose magical science and rituals continued to be successfully practiced, would have likely persisted, leading to the first deities (Howells, 1986). These deities were likely females at least partially because the first leaders were probably women, as previously indicated. Females are often the leaders of other herding animals like elephants, reindeer, and sheep anyway for various reasons (*NewScientist*, 2009o), even when magic is not involved. In any event, it did not take a great leap of faith to go from revering non-living ancestors to deification of them over time for their important characteristics such as magic, science, and general creative power (Davidson, 1990).

Initially, reverence for ancestors, their memory, and their deeds may have just represented a form of nostalgia. Nostalgic reverence serves multiple purposes, including partially offsetting a feeling of loss, or depression, with a pleasant memory. It is also possible to use the rewarding feeling of nostalgia for learning purposes, such as to encourage one to try to bring back desirable characteristics and practices that led to a particular past situation that was pleasant.

Given humans' capacity for reciprocity and keeping track of social "debts" (Miller, 2001), reverence for ancestors may have also represented an attempt to reward ancestors for their actions while they were alive. It may have also provided an educational indication to living humans of a good role model to follow that would enable anticipation of rewarding reverence not only in life but after death.

Ancestor reverence is enhanced by out-of-body experiences, the perception of ghosts, and other hallucinations that many humans experience and that create a belief in an afterlife and a soul independent of the living body (McClenon, 2002). While the form of apparitions of dead people varies across cultures from apparitions to totems, analysis of electrical emissions in the portion of the brain involved in sensory perceptions have verified hallucinations to be consciously experienced and therefore perceived as real (Gooch, 2007). Since all such experiences can be induced with administration of a chemical that is also created by the human body (Strassman, 2001), they appear to be mere creations of the human mind. In fact, they are likely mere images created by the subconscious that can be enhanced by psychic energies that freely cross wide spaces (Gooch, 2007).

However, such hallucinations may have symbolic value that create evolutionary advantages with respect to encouraging humans to seek to act optimally for the benefit of others so that they are optimally revered in survivors' memories. The mere thought of being positively remembered creates rewarding feelings in human minds. In addition, the thoughts of deceased loved ones still being around can both create a sense of security and reduce sorrow.

If people believed that ancestors could influence the present, a positive thinking or placebo effect might have resulted in the feeling of reverence for ancestors persisting for some time. Moreover, as is explained later in Chapters 16 and 19, ancestor reverence might have actually enabled stronger psychic powers being carried from the past by the ancestors. Nevertheless, it must be emphasized that reverence can continue only if there is a perceived mutual reverence, and so the reverence might have been relative to the perceived efforts made by the ancestors for future humans.

As long as people recognized the humaneness of the ancestors, reverence for them would never deteriorate into the pathological condition that comprises one-way worship. Ancestor reverence can grow into worship only if a revered entity is perceived to be superior. The one-way relationship involved in worshipping another is normally prevented by human spite. Spite is typically motivated by inequality among humans that is perceived in an innate human sense of egalitarian justice that evolved under the communal societies of prehistoric times. However, normal human spite in the face of human inequality can be discouraged by a belief that another is superior. Reverence for an ancestor would therefore likely have grown into worship only in cultures where there was a belief in an ancestor's superiority, perhaps because of a perception of "superhuman" powers as implied from special capacities, such as great creativity, harnessing new technology, special powers of persuasion such as via individualized hypnosis, and physical strength, etc. Such powers might have been perceived as "magic" especially if the "superhuman" kept the origin of such power secret from others.

It has been documented that many scientific advances, such as steam engines, advanced geared devices, robots, music powered by wind, vending machines, and machines for making complex astrological predictions, were known for thousands of years before they were used for more than demonstration purposes (Marchant, 2008). Such fairly advanced forms of technology that might very well have existed in prehistoric times might have enabled people demonstrating that "magic" to be superior beings or deities. The social, sexual, and emotional benefits derived from exclusive control of them might have motivated their nature being kept secret.

Some of these technologies, such as music which can be naturally relaxing and pleasant as well as have more focused effects on the left-side of the brain (Esch and Stefano, 2005a), may have hypnotic effects by themselves that could have been exploited. Special music bringing back vivid memories of nostalgic or reverent feelings about ancestors who were associated with the pleasant sounds could have alone enhanced the power of the seemingly "superior" beings. Accurate astrological predictions that might have appeared to be even more miraculous, impressive, and incomprehensible than the ability to forecast women's menstrual cycle by the moon phase would have created an even greater aura of mystique, thereby further promoting the astrological sciences and forecasting in general. Those able to benefit society with their inventions and knowledge would have naturally been rewarded with at least special reverence, as well as more sexual opportunities.

Animism, or the reverence for animals and nature (Howells, 1986), may have also developed from humans' tendency for reciprocity. For instance, the tendency of Native

Americans to thank their hunted prey prior to eating it (Underhill, 1965) makes perfect sense in terms of attempting to compensate the dead animal with at least gratitude for the eating pleasure and nourishment it provides. Such reverence in no way indicates worship, however, as there is no belief therein that an animal is a superior being. It represents a mere reverence for perceived equals.

There may have been a particular power or magic associated with some animals that led to a special reverence for them. For instance, images of wild boars at Neanderthal graves 100,000 years ago that were perpetuated by later peoples who appeared to see that animal, which is a scavenger of dead animals, as having some supernatural relationship with or control over death (Gimbutas, 1989). Such animals with big stomachs could have also symbolized the creative powers of pregnant women. However, wild boars could have also served as a source of food and thus been revered for that purpose alone in animist religions.

Totemism, or the identification of a group or clan of people with a particular animal (Howells, 1965), may have originated because of reverence for a particular animal out of reciprocity or admiration. Given the desire of Native Americans not to eat the animal symbol of their clan and given the predator totems of some tribes (Underhill, 1965), it seems that an animal symbol for a group of humans derived more out of admiration. A totem might have possibly developed out of a clan exhibiting a reputation for having characteristics or acting like the symbolic animal, such as tough as a bear. The latter hypothesis is consistent with a Native American tradition of having people named after symbols of their reputation or accomplishments (Sandoz, 1992).

Since the totem symbolizes a clan, totemism may have represented one of the earliest forms of religion insofar as it represents a reverence for a group's own collectivity, and totems could also have been used as a deity to define the specialized virtues of the group (McClenon, 2002). Reverence for the group is a natural instinct ingrained in humanity over millions of years, and animal symbols may have been utilized to help individual members act virtuously using the animal as a role model in different situations. Some human groups may have become known for special capabilities in some particular virtues and thus became labeled after the animal symbol for those virtues. Such reverence wouldn't have corrupted into worship of a superior group because of inter-mating between clans.

Some totems might have originated from a particular magician or leader of a group of people having some characteristics like an animal. Given that women have long utilized masks for appearance purposes (Gimbutas, 1989), it seems quite likely that female leaders wore animal masks to cause others to draw a greater connection between them and revered animals, resulting in more mysterious admiration or reverence.

As is consistent with ancient stories of old women turning into attractive females at night (Baring and Cashford, 1993), masks could have been especially useful for older females, who may have wanted to disguise their age or unattractive looks in order to continue to be able to attract those of the opposite sex. Masks might have also served to disguise the humanity of the wearers of the masks so that others might perceive the masked leaders as something mysterious. Such an effect would have been magnified if other artificial effects and magic were employed. A perception of superiority and a mystical nature, perhaps representing a mix of the traits of a human being and a revered animal, might have been corrupted into a deification of the female leaders even before their death.

E. Romantic Love as the Predecessor of Religious Reverence

It is likely that the predecessor of today's deities actually occurred prior to ancestor reverence, totemism, animism, and the first widely recognized magicians and scientists. In particular, given the large similarities between romantic love and religious worship, the capacity for religious feeling may have originated with the adoration of mates, as explained in Part I. That capacity, however, was based on a mutual reverence for each other as equal but different deities.

According to ancient Taoist philosophy, pure romantic love is devoid of judgment or thinking and is pure feeling that is represented as two hearts united with nothing between them (Chia and Wei, 1994). The surrender to a sexual partner who is physically and spiritually beautiful (Deida, 2005) may enable both mates to experience divinity (Chia and Wei, 1994). Even scientists recognize that love can lead to a different level of consciousness or altered state that feels so good as to be "out of this world" (Esch and Stefano, 2005a).

The magical pleasure caused by romantic love, as well as the ecstasy of mutual surrender during sex (Deida, 2005), would have easily led to reverence of mates as magicians. Security and support, which create trust, belief, and even "meaning and spirituality" that motivate positive behavior (Esch and Stefano, 2005a), is endemic to the environment of mutual romantic love and could therefore have easily grown into a religion.

The ecstatic state of heightened concentration that is the basis of all mystical experiences and that is most strongly felt during sexual arousal and orgasm (Lai, 2001) would have, along with offspring resulting from sex, contributed to partners perceiving each other as procreative deities. The surrender to a partner in romantic love therefore likely originated in an atmosphere of mutual awe for each other. With mates selecting each other based on relative desirability, the mere entry into mutual surrender would have created a feeling of proud happiness with oneself that is reflected in a feeling of self-fulfillment. Such surrender, which creates a heavenly feeling, has absolutely nothing to do with the one-way, servile, self-depreciating surrender to somebody or something, like a deity, as a submission of a person to another without mutuality is an unnatural self-depreciation of the worshipper that only solicits contemptuous exploitation from the worshipped and creates hell on earth.

Concentration or focus, which tends to be extreme in romantic love (Esch and Stefano, 2005a), is the means to hypnosis, which leads to submission and control by another (DeLaurence, 1910). When the submission and control are mutual as in romantic love, a sort of unified consciousness arises. That unified or group consciousness likely began with romantic mates and then expanded to larger groups of humans, including offspring, relatives, friends, and other friendly humans in general. Only much later, with the sexual and emotional frustration of human beings about 35,000 years ago, did any fetishes arise that served as a substitute for the sexual and emotional union with other human beings. Such one-way worship or exploitative hypnotizing leads to inhumane hierarchies and counterproductive cruelties that couldn't have characterized humanity over the millions of years when only the individualistic creativity of mutual cooperation in an egalitarian reverence and sense of purpose could have allowed the species to survive an otherwise overwhelming environment.

The effects of deep hypnosis are, in general, typified by a loss of control, brain fluidity, a loss of a sense of time and space, a change in meaning and experience, and rejuvenation (McClenon, 2002). These feelings are so similar to those of romantic love that it seems almost certain that romantic love represents a form of hypnosis. The fact that the brain's anterior cingulate, which is involved in evaluating emotional outcomes and responding to errors, is activated in both people under both hypnosis (Gosline, 2004) and in love (Bartels and Zeki, 2000) provides some additional support for this hypothesis.

Hypnosis results in an emotional reduction in cognitive brain barriers or inhibitions in order to surrender to suggestions (McClenon, 2002), just as romantic love tends to result in blindly ignoring faults in a loved partner (Etcoff, 1999). The heavy emotional activation of the anterior cingulate in both states may represent a silencing of the inhibitory function of the error response task of that brain area. Hypnosis has been found to interfere with the detection of deeper conceptual mistakes at the same time that it doesn't affect abilities in correcting mistakes in current activities being performed (Gruzelier, 2006). This characteristic of the cingulate, which is involved in learning, may result in the uninhibited carrying out of actions suggested by the object of focus, which is the loved one in hypnotic love. The activity in the anterior cingulate does change significantly when a person goes into a hypnotic state that exhibits much different brain oscillations in electroencephalogram measurements of spectral patterns (Fingelkurts, Fingelkurts, Kallio, and Revonsuo, 2007).

It has been found that activation of the anterior cingulate is involved in creating altered states of consciousness such as in a hypnotic trance by blurring the distinction between imagination and reality (Nash and Benham, 2005). Belief can induce seemingly impossible hypnotic feats people can't otherwise perform, such as walking on hot coals without pain or blistering, and such a capacity to do things humans have previously learned from the experiences of normal daily life and the cultural environment (McClenon, 2002) seem to result from cutting off conscious doubts about imagined outcomes. The inability to discern what normally can't be done appears to derive from the activity in the anterior cingulate. A belief in imagined phenomena or suggestions might be enabled by the anterior cingulate turning off conscious doubts because of the belief in the hypnotizing agent, be it a hypnotist, a loved mate, the self, or a human group.

There is a disassociation between the thinking portion of the brain, i.e., the prefrontal cortex, and the anterior cingulate, i.e., the thalamus filter to consciousness, under hypnosis (Gruzelier, 2006). The dorsal anterior cingulate, which is very active in cases of paranoia and obsessive compulsive disorder, is involved in error detection and conflict monitoring (Geddes, 2011), and so the disconnect between this area of the brain and reflective thought during hypnosis may make it possible for suggestions to flow through to the subconscious without any interfering doubts and thus be implanted without question. Some feelings are easily mixed up in the brain even in a normal state of mind, such as physical pain being felt upon social exclusion due to the same dorsal anterior cingulate portion of the brain being activated during experiences of both physical and emotional pain (Giles, 2009), and so distorted views of reality may certainly be induced during hypnosis. A similar mix-up or disconnect between imagination and "normal" reality may be an important cause of the mysterious phenomena of love that is so magically powerful.

As in the original form of self-hypnosis that evolved to cause some organisms to freeze or play dead in the face of predators or other dangers, hypnotic love often results in a trance-like state that focuses only on dealing optimally with the situation that represents a matter of life and non-life for the potential offspring of that love. In that focused hypnotized state, an individual may be motivated and able to perform all sorts of irrational actions besides playing dead in the face of danger, depending on the individual's subconscious evaluation of the situation.

The most powerful hypnotic focus may actually be the mutual romantic love which enslaves two people to their mutual attraction that normally creates a desire for sexual procreation of offspring. Such mutually hypnotic submission reinforces itself and may thus magnify the effect.

Hypnotic surrender between mates is not a rational trade for satisfying each others' sexual, emotional, or other needs. Instead, mutual romantic love is a hypnotic obedience to each others' suggestions that surrender the two to cooperate to make a unified whole, which is greater than the two individuals separately. This process serves the primary purpose for which it evolved, i.e., to procreate and raise diverse adaptive offspring. It is a bonding designed by evolution to create offspring, who are both the product and the object of the parents' selfless love. It is characterized by a unity of purpose that maximizes each other's pleasure through the mutual feeling that persists only as long as there is unity of purpose. The feeling motivates the partners to treat each other well to help each in creating and raising offspring, and the sex drive component of it can be thought of as the pull of the future children for their creation. The commonly loved offspring naturally feels reciprocal love, at least until after the children reach puberty and are then driven by human nature to selflessly give of themselves in romantic love in the eternally recurring process of continuing the existence of humanity.

Romantic love loses power, feeling, and pleasure when it isn't shared or mutual. Love thus inhibited can even result in spite if the submissive obedience associated with it is exploited. Without the mutual feeling and unity of purpose, relationships lead to irresolvable inconsistencies and conflicts that are resolved coldly based on the relative bargaining position and will power of the partners. Such relationships are more like business trades where a constant conscious evaluation of fairness is required based on an exchange of benefits instead of love. To the extent that there are any hypnotic effects in such relationship, they may be more akin to one-way hypnosis. One-way hypnosis to another is slavery that is unnatural for humans, as humans evolved to feel spite if they feel exploited.

The capacity of human beings for one-way hypnotic trances may be related to the prior group hypnosis that enabled human beings to obey and surrender to a group of humans. Abuse of hypnotic trust is widespread today by individuals who choose not to submit to the group, and these selfish individuals abuse others' trust in humanity for their own exploitative purposes. Whereas group hypnosis maximizes coherence in relationships, one-way hypnosis effectively collapses the wave function of another on a point desired strictly by the hypnotizing agent without regard to the wishes of the hypnotized victim. Spite and a refusal to submit further generally results if the victim figures out the exploitative game.

Since there may be attempts to totally inhibit the fulfillment of purely selfish wishes of the hypnotizing agent and act spitefully at any time, there are natural incentives for humans not to exploit one another in a humane society that frowned on exploitation. Such risks would enhance the trust required for hypnosis because of the known repercussions resulting from abusing such trust. However, the risks of vindictive actions are greatly reduced in contemporary cultures that glorify exploitation and actually make it difficult to undertake any revenge. The trust necessary for mutually hypnotic love is therefore difficult to find today, but the ecstasy of hypnotic surrender to others is so strong that many submit to one-way worship.

One-way hypnosis in the unnatural hierarchies of contemporary societies generally involves servitude to another person or to an externally imposed religious system and deity for the purpose of exploitation of the masses by an elite. Spiteful feelings are prevented only as long as the hypnotic effect or brainwashing continues. The ignorance of the exploited in believing in a mutuality can only occur if there is a belief in the superiority of the exploiter, as can be facilitated with a belief in fictitious deities that are fabricated to motivate servitude to an elite.

Despite its abuse by selfish exploiting agents today, hypnotic love is an extremely powerful motivator and can still lead to incredibly productive deeds through the power of hypnosis. For instance, the conscious belief involved in hypnotic focusing the subconscious and other parts of the brain on attaining a task enables incredible accomplishments by human mating partners. The power of hypnotic love may like other forms of hypnosis be related to the power of positive thinking in a sort of self-love that is reinforced and magnified by the reverence of another.

Hypnotic love might clearly have been selected by evolution not only for the survival advantages relating to miraculous feats and powerfully harmonious behavior but possibly also with respect to subconscious choosing the mix of genes involved for ever smarter and more adaptable offspring which culminated in contemporary humans. Since it has been found that females of at least several organisms unconsciously help select the sperm that fertilizes their eggs by chemically inducing sperm with more diverse characteristics to move faster both inside and outside their bodies (*NewScientist,* 2008zx), the emotional quality of a hypnotic human love relationship may indeed influence the genetic characteristics of human offspring.

It has been discovered that human mates do tend to choose those with more diverse capacity for immunity to disease, making for offspring that are more resilient in the face of medical threats (*NewScientist*, 2008zzj). This beneficial selection process, which is promoted by scents given off by people that are not consciously detected (Potts, 2008), appears to be related to subconscious forces or feelings that influence the existence of romantic love. Due to the long prehistory of speechless cohabitation among humans, subconscious detection of non-verbal signals from other humans remains more effective than long periods of verbal communication (Buchanan, 2008e). Subconscious signaling might be expected to be most powerful with respect to the deepest, most basic, and most important human relationship of romantic love.

Romantic love may therefore have been both the earliest deep form of hypnosis among humans and one of the primary causes of humanity's evolutionary success. This theory is supported by the mysterious magic of love at first sight. In particular, because eye focus frequently induces hypnosis, eye-to-eye contact can actually lead to the hypnosis of one or both parties that can bond into a deep unifying love (DeLaurence, 1902). Thus, eye contact between partners can apparently lead to a love at first sight that may actually be a form of mutual hypnosis. With the mutual hypnosis involved in lovers' eye contact, there would have been little fear of exploitation to inhibit the hypnotic effects of that contact. The permanent relationships that often develop from that mysterious initial moment of eye contact between partners and that can be important in the bonded raising of human offspring may be explainable in this context.

The ecstatic high of mutual love can indeed be initiated at first eye contact through the hypnotic effect. Love at first sight could have easily evolved in the social environments before human speech developed and thereby led to the bliss of romantic love then. If there was mutual hypnosis through maintenance of the mutual gazing, the magical characteristics of mutual romantic love would result.

Many people experience the ecstasy of love at first sight several times a day, but almost invariably one of the partners of the opposite sex discounts the feeling and finds an excuse to be distracted from the gaze (Braden, 2007). The effect therefore dies when the hypnotized recognize no mutuality.

Although the mutual hypnosis in eye contact between mates is a type of group hypnosis with only two people in the group, it differs from shamantic hypnosis, which is one-way control that can lead to abuse. In distant prehistoric times, a spiteful loss of interest would, sooner or later, result if hypnotic love was not mutual. Similarly, a perception that efforts and sacrifices to serve a group of two or more that are not appreciated and reciprocated can end the effects of any group hypnosis.

Since animals generally don't do any gazing into the eyes of their sexual partners, it is likely that the extremely hypnotic nature of both love at first sight and romantic love exists only among humans. Hypnotic love provides far more motivation for greater things than just the bonding brain chemical oxytocin that is also involved. Oxytocin alone seems to be sufficient to bind prairie vole pairs for life (Fisher, 2004) after their first sexual encounter (Klein, 2003), but the complex hypnotic nature of romantic love seems to do far more than that.

The potentially transitory nature of romantic love often enough only lasts less than a year (*Wall Street Journal*, 2005), thus just long enough to provide incentives for a mate's care of a mother and offspring through delivery. The effects themselves enable humans to fall in love multiple times in their lives. Human love therefore creates the opportunity for both more adaptability and more diversity than exists in prairie voles.

Most of the other small minority of species that tend toward monogamous relationships, as opposed to sexual free-for-alls, are also characterized by serial monogamies that permit mating with others in a fashion similar to that for human beings (Blum, 2009c). Such facts are consistent with the needed care of offspring having evolved through systems other than purely dual parenting. Some form of group childcare, as opposed to strictly parental raising of children, could therefore have existed in prehistoric human cultures.

Nevertheless, pledges of mutual love until death that are motivated by the high of romantic love can bond human pairs for life, even without any legal or moral obligations (Morrell, 2005). Mutual hypnosis is an important component of this tendency that is spurred by chemicals like oxytocin. The blissfully surrendering aspect of love is typically felt most ecstatically if it is expressed to be forever, and so pledges of eternal love also motivate human pairs to stay together as long as it is mutual because of the natural human inclination to be honest unless information arises that creates a perception of exploitation.

Within this context, though, a mutual agreement for breaking up is also quite possible biologically for human beings at some point, as was evolutionarily useful upon the death of a partner. Break-ups have also thereby been facilitated if a relationship ever collapsed from being a hypnotic state into a business relationship trading benefits as they are so typically in the exploitative cultures of today. New states of hypnotic love are then possible for the human beings that enable new romantic loves which can provide more motivation for greater acts for the benefit of others as well as more genetic diversity in any resulting further offspring.

Romantic love and sexual desire are often but not always tied together in humans (Aron, Fisher, Mashek, Strong, Li, and Brown, 2005). The characteristics of human nature that permit sexual desire separately from romantic love even for those in a lifelong relationship between mating pairs does enable greater genetic diversity. As a result, some raising of non-related children in human groups likely developed. While a system of purely egalitarian group childcare could conceivably have resulted, such a system might have made the continuation of a hypnotically bonding romantic love superfluous. Since there hasn't been such a transformation of human nature that could have made human cultures similar to the matriarchal ones of the totally uninhibited free group sex of bonobos living strictly for the present moment, it seems likely that parents always maintained a special devotion to this offspring product of their mating love, at least until puberty.

The natural attachment between parents and their biological children indicates a special relationship has always existed between them. Some preferential treatment for a person's own offspring may have easily existed with the heterosexual matings of older people with younger ones of other human groups at puberty. These relationships could have persisted for long periods of time depending on the loving commitment of the partners, but they might have often been terminated early by the death of the older people, and so some group childcare seems likely.

As is discussed later in Chapter 17, longer-term romantic pairings of the old with the young seems to be most consistent with human traits. It is therefore the most likely probable culture in which those characteristics developed in prehistory. Romantic love bonding between older men and younger women would be useful for the long-term care of offspring in such societies. In particular, under the influence of romantic love, the older partners would have been more than willing to provide care not only for the younger mates but also for their offspring even if they didn't know who the biological father was and even if the connection between sex and procreation hadn't yet been realized. Even though tight sexual bonding wouldn't be selected by evolution between younger men and older female caretakers assigned while they were in menopause, the latter pairs could still enjoy sexual relationships as well as strong emotional bonds at least as powerful as between a mother and her biological son.

Such old-young sexual relationships may have long been the norm among human ancestors who may have also experienced romantic love in such a naturally caring, mentoring environment. Care of unrelated dependent offspring might even have been spurred by the indication it provided of beneficial mentorship relationships that might have facilitated attracting younger mates in a self-reinforcing process.

The development of romantic love could actually be related to the evolution of the ability of humans to forecast. The mere desire or intent and pledge of lifetime attachment to a partner that is motivated by romantic love indicate thoughts for the future that only humans may have. Given that humans typically feel more pleasure in anticipation of a possibly positive stimulus than when actually experiencing it, the extreme pleasure of romantic love, which is caused by similar brain chemicals (Klein, 2003), could very well have developed from the prediction of the possible pleasure of sex and devotion to the offspring in the future. The sense of self-fulfillment love seems to provide to humans itself may also be related to a subconscious prediction of future benefits arising from past and present actions by the partners that entitles them to the bliss of love. Other organisms don't have such feelings because they don't consciously forecast themselves and therefore have no need to develop a framework for determining their own self-satisfaction as a means of evaluating and planning for their future selves.

Romantic love and love at first sight may also run deeper than mere eye contact on a physical level as well. For instance, the pineal gland, which was once the third eye of organisms but is now buried in the middle of humans' brains (Strassman, 2001), has a direct connection to the genitals (Chia and Wei, 1994) and appears to be involved in psychic senses (Stray, 2005). It might therefore be a factor in the love-at-first sight phenomenon that could essentially surrender the partners' conscious command function to the innate urge for union of their sex cells.

The pineal gland, which is near an emotional center of the human brain and may react to various electromagnetic waves that permeate the surrounding brain tissue, especially ones with a resonance of frequency (Strassman, 2001), could potentially pick up the brainwaves of energy from others that are invisible via normal human visual senses. Everyone and everything emit electromagnetic fields (Radin, 2006), as well as radioactive waves (Segre, 2002), and the pineal gland could be involved in picking up coherent signals in such waves emanating from others in a psychic sense that may be imparted in first impressions during love at first sight. The nearness of the pineal gland to the emotional center of the brain may explain why psychic senses are "felt" through intuition as well as partially clarify how first impressions can be so powerfully experienced. Since it is also possible for humans to perceive meaningful patterns in entangled quantum particles that are undetectable by the normal five senses (Radin, 2006), the pineal gland, whose functions have long been mysterious, could have a special capacity for both initiating and interpreting such correlations in microscopic particles that could motivate attraction.

Given that people make a "snap decision about qualities such as trustworthiness, likeability, and competence" of another within 1/10 of one second of just seeing them (de Lange, 2009), it seems likely that there is some subconscious information processing involved in first impressions and love at first sight. Regardless of the human body parts that are responsible for subconsciously sending and receiving signals that are consciously undetected, people do in fact emit light waves that reflect the state of every physiological system in their bodies, and those states can be intentionally sent to others with love to create a pleasurable synchronization of electromagnetic brain waves that leads to a joint consciousness (McTaggart, 2008). Love at first sight may reflect such phenomena and thereby incorporate far more than normal visual looks. Much of such information could flow visually between the eyes

Other information may also be exchanged through other senses in the process of love at first sight. For instance, unconscious scents known to affect both human and animal attraction (Williams, 2008b) could also affect love at first "sight" via undetectible smells. Since future procreation of offspring might later occur as a result of love at first sight, there can indeed be an infinite eternity in that mutual "look" involving all six senses that is only partially motivated by the normal sexual desires and drives that all animals have.

Whether enabled at first sight or later in a relationship, the magic of mutual romantic love and hypnosis is still possible for all but skeptics whose doubts restrain the fluidity of the brain to utilize all information to attain the blissful enlightenment. Hypnosis enables a shutting down of the distorted perceptions of reality that have been learned by each individual in order to adapt to the macro environment that normally make different electromagnetic force fields impenetrable. A lack of skepticism about alternative realities is required for this empowering process to function, as is a trust in a hypnotizing agent that removes the fear of exploitation by the agent. Whether induced by romantic love or otherwise, hypnosis enables not only greater power to affect the environment but can also enhance an understanding of the true underlying reality of all things that human beings are all connected and that humans' normal perceptions don't reflect the true nature of things.

Any type of focus permits observation of one particular state instead of the averaging of states normally perceived. It thereby permits observing much more incredible possibilities. The mutual focus of romantic mates in love not only collapses their probability wave functions, it also does so on points desired by both partners. It thereby creates the opportunity to experience what each desires because of the submission of each partner to the other's wishes. This process is analogous to the collapsing of the common wave functions of the two into a single desired feasible point on those two interacting waves. Through the hypnotic focus on each other that is involved in mutual romantic love, a sort of group consciousness arises that results in each partner blissfully and powerfully serving the other.

Romantic love often results in a feeling of being born again (Rathaus, Nevid, and Fichner-Rathaus, 2008). This feeling may actually reflect the incredible new potential that is made possible by the common collapsing of the loving partners' wave functions that opens up a whole new set of possibilities. Rebirths in romantic love represent a form of surrender of each partner to the other that motivates the lovers to recreate themselves in images perceived wanted by their mate. In mutual love, the blind belief in the wonder of the other partner (Etcoff, 1999) causes a loving mate to attempt to be worthy of that faith through change if the loved image isn't yet reality. The mutual attempt of the loving partners to justify each mate's belief in the other only magnifies the love, faith, positive thinking, and striving to fulfill the dreamed-for images. The extreme focus of each partner for the other can result in a form of hypnosis that leads to seemingly superhuman abilities for the benefit of the loved mate.

Thus, the partners create each other and, in that sense, are true deities, who are rewarded with happiness and the highest high that represents living in heaven. This bliss requires a surrender of a person's self to the other, thereby enabling a form of unity which may have been the precursor to a more general group consciousness. Such surrender is based on empathetic mutuality, where an individual's own happiness is determined by empathetically feeling others' happiness and therefore doing whatever it takes to make them happy.

The unification of two members of the opposite sex in the group consciousness of romantic love was, in the distant past, expanded to unifying all human beings in an effective consciousness of all past, present and future human beings, as it incorporated both reverence for ancestors and care for offspring. This consciousness of humanity is fragmented by the hierarchal societies of today but human nature is driving humanity back toward the harmonic unity of human prehistory.

Marxism, whose core belief is the unification of opposites (McClenon, 2002), is a theory of economic and social relations that prophesizes such a return to human communality in the future. Marxism recognizes the original human cultures were communist ones which only later changed into societies based on slavery, feudalism, and capitalism that are now in the process of a struggle to return to a communist unity that is geared toward a group or common satisfaction of all human needs and desires (Lotta, 1994). As such, communist economics helps harmonize human beings and their wills back toward a group consciousness that enables true human love. Belief in such an ideal often results in an ecstatic feeling of being born again into the new potential that appears to reflect the collapsing of the wave functions of groups of people on a commonly desired point for each. Such a possibility would be impossible if people continued to act selfishly alone, as only an averaging of the possible points on other wave functions of potential is feasible without the group unity of purpose.

Being born again is also felt in religious rebirths under Christianity and other hierarchical religions when people are "possessed" by a belief in a unity with a "superior" being. This unification of opposites cruelly enslaves the believers into a slavery of one-way servitude as there is no mutual submission on the part of the hierarchical deity. While hypnotic states that result from Christian "rebirths" can open up new possibilities, it does so by the collapse of the hypnotized victim's wave function on a single point set by the dictates of the religion. It thus represents one-way hypnosis that enables exploitation by others, including by an elite which uses the religions to maintain their hierarchical power. The fact that systems based on exploitation have been shown to be economically inefficient and lead to far less happiness than communist ones (Murphy, 2000) provides one indication in an aggregated sense of the relative impotency of hierarchical one-way hypnosis compared to egalitarian group hypnosis.

The hierarchal societies based on selfish exploitation are, according to communist theory, destined to return to naturally human communal societies (Lotta, 1994). Even within such inhumane societies, emotional, spiritual, and physical unity continues to be achievable, such as via the eye-to-eye hypnosis between members of the opposite sex. Such basic relationships may be a primary force that indeed drives humanity away from pure selfishness and back to a group consciousness and communism.

The blissfully unique eye contact between human mates can also occur during sexual intercourse. The ability of humans to have face-to-face contact during sex enables the partners to look into each others' eyes to maintain mutual trust as well as the mutual hypnosis that facilitates the partners serving each other sexually without words beings spoken. With orangutans, which may be humans' most direct ancestors, being one of the few other species that engages in face-to-face sex (Schwartz, 2005), this characteristic of human love may go back over ten million years.

To spur female sexual excitement, manual and oral stimulation of the female genitals, which are the most sexually excitable parts of the female body (Rathaus, Nevid, and Fichner-Rathaus, 2008), could also occurred naturally long before language developed through the mutually hypnotic eye contact that may be unique to humanity. Since women's most sexually arousable parts are near the opening of the vagina (Rathaus, Nevid, and Fichner-Rathaus, 2008), initiation of sexual intercourse between two human partners was likely motivated reflexively to revolve less deeply inside the female, with only later or occasional deep penetration, requiring substantial self-control by the male for impregnation. Some Eastern Taoist sexual traditions maintain this tradition (Chia and Wei, 2003), and hypnotic eye contact during slow, repetitive, and long sexual activity would have both facilitated speechless communication for this purpose and enhanced bonding in this context for males able to maintain sufficient self-control. While touch is also useful for bonding (Esch and Stefano, 2005a), the hypnotic and blissful effects of love would seem to be more powerful.

Evolution through sexual selection would also have led to males experiencing greater pleasure by maintaining their sexual excitement longer in order to please their mates who required more stimulation and passion, thereby reducing their motivation to seek other sexual partners. The increased pleasure by prolonging the time to orgasm would have led to an evolved capacity for longer erections in males and multiple orgasms in females. Spiritual Taoists advocate males never having an orgasm before their female partners (Chia and Wei, 2003). The desire to surrender between partners in sex probably originated in the chemical pleasure experienced by each mate at surrendering to each others' innate desire for a mutual orgasmic experience, and such mutual submission typically carried over into a lasting romantic love.

In particular, the surrender of mates to each other during natural human sexual intercourse that creates the maximum pleasure (Deida, 2005) would have easily existed in a state of mutual hypnosis during such sexual activity that would have culminated in the procreative act. The ecstasy of this high, along with the hypnotic trance involved, would have helped motivate the longer-term bonding that the bliss of romantic love promotes. The surrendering aspect of romantic love may actually represent in part the anticipatory pleasure of this mutual servitude to each mate's sexual desires.

If offspring result from romantic love, a brand new mixture of the lovers is created. The creation of life represents an act of deities, which isn't surprising given the power that the lovers impart to each other with their mutual reverence as deities. The created life eventually develops independent free will and an ability to also become a deity in a potentially endless process of procreation. The natural mutual love between parents and offspring, while significantly different from romantic love, results in a further development toward larger group consciousness. Such sexual and emotional pleasures may drive humans back to the spiritual human love, whose own blissful hypnotic power creates further motivation for a return to that state with all.

The powerful placebo effect that is based on the power of positive thinking (Brooks, 2008e) may have actually been originally related to the mutual belief in other humans that characterizes mutual romantic love. As a result, accessing the full power of positive thinking may require a return to a mutual belief in and reverence for other human beings. The power of positive thinking does tend to only function effectively if there is a belief in something other than oneself, possibly because a person's wave function of potential is collapsed by such a belief.

If there is belief in someone or something else that isn't mutual, only the believer's wave function is collapsed by the wave function of the believed-in entity in a fashion similar to one-way hypnosis. While extraordinary acts are possible via one-way hypnosis, those acts are chosen by the believed-in entity. When the believed-in entity is a fictitious deity, the acts are selected by the deity's human fabricators and promoters. In any event, the lack of mutual selection in the collapsing of the wave functions typically results in doubts about the believed-in entity that reduce the power of positive thinking. Spite can even result if the trusting faith is perceived to be exploited by others, thereby leading to a collapsing of the wave functions of the hypnotized on a point least desired by the hypnotic agent. One-way belief is therefore impotent in comparison to mutual belief, and evolution should drive humanity as a whole to move back toward the more powerful form of belief.

The high from the combination of a spiritual unity induced by hypnotic mutual surrender and the physical pleasure facilitated by sexual excitement also provides strong motivation for individuals to select mutual belief in each other over one-way faith in the superiority of a manipulating elite or their deities. Only the repression of love and sex by contemporary cultures, along with a separation of them into two components that don't require mutuality, inhibits such incentives from being overwhelming. The matriarchies that preceded today's patriarchal societies likely utilized such repression and divisions of innate human desires. For instance, with many males having evolved to perform cunnilingus during sex, a natural mutual reverence among lovers for each others' bodies appears to have been manipulated under matriarchies into vulva worship. Such one-way worship was likely only possible in humans because of repression of male sexual urges for exploitative non-sexual purposes that divorced mutual love from sexuality. Hypnotic female body movements and parts, as opposed to the mutual hypnosis involved in eye contact, might have been an essential ingredient of that separation in the inhumane matriarchies.

In particular, since sexual excitement is dependent on the shutting down of the inhibitory portion of the brain (Portner, 2008), a complete surrender to a partner is required for maximum excitement. Sexual repression by a culture suppresses the natural mutual surrender that exists without the inhumane cultural influence, and it thereby transforms the mutual reverence into a one-way worship and obedience that creates contempt in a partner and motivates exploitation as opposed to love. Sex and love are then divided instead of drawn together. Also contributing to the separation of love and sex could have been the matriarchs' use of masks that may have inhibited the eye contact involved in mutually hypnotic romantic love. One-way worship, whether of the vagina, the female body, or women in general, returns hypnosis to its purely animal form based on purely primitive subconscious processes that would seem to only lead back to more primitive life forms. While the patriarchal conquerors replaced the matriarchal repression, the patriarchs' own spiteful repression of women's sexual pleasure motivated women to renew their frustration of male sexual desires in a process that seems to be returning society back toward matriarchal imbalances, as will be explained later in Chapter 17. Only mutual hypnotic love can create harmonious balance in human sexual relationships.

As human groups grew larger, dancing could also have later evolved in groups of humans, who might have been hypnotized by the repetitively moving bodies. Dancing in the dark, perhaps in caves or at night near campfires, might have enhanced the focus for maximum hypnotic effect by reducing some of the distractions of a more lighted environment that would have more visual stimuli.

If a mate provided other magic, such as child bearing and nourishment as only women are able to do, the reverence would have been even more strongly promoted. The greater sex drive of men (Rathaus, Nevid, and Fichner-Rathaus, 2008) might have also led to them to believe themselves to be more "dependent" on the opposite sex. The greater desire by men for younger women because they are more fertile (Etcoff, 1999) might have led to greater competition for them and enhanced this effect, besides motivating the male human ancestors to seek longer-term relationships than exist in other animals.

Greater reverence for women's bodies helps explain why the triple vulva symbol "6" may have come into fashion in the matriarchies. However, prior to the worship of a triple goddess in the matriarchies where "666" denoted the female genitals of the maiden, mother, and crone that are unified in that triple goddess symbol (Walker, 1988), it's quite possible that the "6" also once represented both female and male genitals, and that a third "6" symbolized the resulting offspring. The drawing of "6" to indicate the genitals (Acharya, 1999) isn't apparent from compilings of prehistoric artwork, but such art did include symbols for both the male and female genitals, thus indicating some deep psychological reverence for them (Gimbutas, 1989).

In any event, prehistoric reverence for the opposite sex and their genitals might have even involved a form of mutual body reverence that may have even developed into a mutual hypnosis via moving or dancing bodies of the opposite sex. Religions such as Taoism, which persisted through matriarchal and patriarchal times (Jiayin, 1995), and which may have preceded both systems, stress the harmonious unity between the opposite sexes that naturally flow into spiritual and physical ecstasy (Chia and Wei, 1994) and are therefore consistent with this hypothesis.

Until the establishment of the first matriarchies, romantic love was never worship of a superior being or entity, but of a different being who provides enormous pleasure that may be enhanced by the greater passionate desire which can come with body reverence. Spite would cause it to be discontinued if it weren't appreciated and rewarded, or if it were exploited in any way other than for mutual sexual pleasure. While overall mutual reverence between partners rewarded each other with the deep human pleasures of love and sex that could have easily lasted a lifetime, the normal state of communal bliss was disrupted about 35,000 yeaas ago because of inhumane selfishness.

F. Worship and Manipulation by Others

The complex and unequal sex drives and procreative/nurturing abilities of humans of the opposite sex might have eventually helped lead to the hierarchal relationships that are involved in most religions. The first such religion was therefore likely a goddess worship. The natural analogy of the Earth being a nurturing mother for humans no doubt contributed to the first deities being construed to be female ones. Generalizing women's reproductive cycles to the cyclical characteristics of the universe (McClenon, 2002) like the moon cycle would have provided confirming evidence of a female nature for celestial deities. The fact that women are stronger emotionally and sexually, as well as physically with respect to endurance (Chia and Wei, 1994), might have further promoted female worship. Thus, it seems very probable that the first religions to develop were matriarchal in nature, and that goddesses came before gods.

Goddess worship exists today even in the patriarchal culture artificially imposed on humanity. Some males even advocate worshipping the female as a means to greater happiness, sexual pleasure, and surrendering love (Deida, 2005). However, the cultural phenomenon of worship could never have been innate or evolved into human genes. For instance, most northern Native Americans (i.e., those "Indians" living in what is now the USA) never worshipped anything and never made sacrifices to deities perceived to be superior to themselves (Underhill, 1965). The fact that Native Americans attempts to explain the world seem to be more consistent with reality than Western religions (Peat, 2005) may not be surprising in this context. For millions of years, human mates revered each other mutually, and a perception of higher beings that resulted in deification and worship is an artificially new development in human history.

Submission to another or to an imagined concept is an essential component of all religions (Lai, 2001). Religions can therefore also include any belief in natural human love as a feeling or concept. Such belief is hierarchical by nature because an emotion or concept can't love back. It also separates the feeling from the person as a whole, making the rest of the individual hierarchically inferior. It also diverts love away from real people and thus may make mutual human love impossible.

All such one-sided hierarchal worship of a superior deity, as opposed to mutual reverence between egalitarian mates or human deities, appears to have begun only about 35,000 years ago. Since human happiness is very negatively correlated with obedience and powerlessness (Klein, 2003), the obedience to superior deities, who are assumed in most hierarchical religions to have supreme power over each individual, is obviously a somewhat new phenomenon for which humans didn't evolve. Just as the obedience of ants to their queen can even be used by other species to enslave them into meeting their own needs (*NewScientist,* 2009q), unquestioning worship leads to blind obedience that can be manipulated by others besides the worshipped deity, as robotic slaves to a master are easy to predict and control.

A relationship of hierarchical worship and obedience is actually somewhat similar to that created by one-way hypnosis that permits exploitative manipulation by others besides the hypnotist. Religious worship and one-way hypnosis reduce human beings back into being manipulated externally instead of mutually controlling their environment. Hierarchies of worship and servitude thus reduce the adaptability of humanity in comparison to group hypnosis that permits maximum individualized control of the environment through the harmonization of wills.

The original motive to solicit worship may have actually developed out of a desire by some selfish, possibly mutant or mentally deranged individuals to manipulate others for sexual purposes. It could have developed concurrently with the fetish of leader worship and worship of deities in general. Those individuals manipulating others into worship would have easily developed contempt for the worshipping people and their ignorance. That contempt would have prompted the worshipped to believe ever more contemptuously and arrogantly in their own superiority, motivating demand for ever more subservient sacrifices for their "inferior" subjects, including for sexual exploitation.

The natural mutual reverence that humans normally feel for each other was thereby transformed into an unnatural one-way worship by merely persuading, tricking, or hypnotizing other people into believing in their own inferiority relative to some superior being(s) who deserved unconditional reverence.

To the extent that some people in a group were sexually frustrated, worship could have easily been extorted as a condition necessary for sex. The natural human bliss of mutually surrendering reverence was thereby transformed into a perverse servitude to the deity. To the extent that sex was tied to the one-way exploitation of the deity or its representatives, sex became may have become associated with humiliating slavery, resulting in masochism. The servitude to a superior necessary for satisfying any need or desire by itself creates feelings of unworthiness. Allowing sex only for slavery may also have been the forerunner of paying for sexual release that has continued today. To the extent that sex was dictated by the religion to be evil, there would be guilt with personal gratification that would lead to associating painful feelings like humiliation with sexual desires.

To the extent that sexual or other favors might have initially been dependent only on pledges of worshipping subservience, feigned statements of worship or love might have begun to be made dishonestly to get some of those favors. The inhumane religious leaders would have eventually begun to figure out the understandable ruse and required tangible "signs" of subservient worship to "prove" their "inferiors" weren't just saying they believed in the elite to counter-manipulate them. An assumption that sacrifice was necessary to prove belief (*NewScientist,* 2009zw) could have easily developed thereby, as the religious leaders required evidence of sacrifice to prove their servants' subservience and worship.

Thus was born the concept of religious sacrifices to a deity. Today, in societies like the USA where women have regained much power over their men through sexual extortion, men are also often forced to perform a form of sacrifice by their female sexual partners to "prove" their love. Such required sacrifices essentially allow female exploitation of males for economic and other benefits that are similar to those of a matriarchy, as will be explained in Chapter 17.

In this exploitative brainwashing system, the act of worshipping and serving, as well as mere existence, may have been learned to be perceived as an ecstatic right that may have become habitual and addictive via subconscious brain connections. In addition, prohibitions against youthful sexual intercourse inhibit natural human connections to sexual excitement from initially being formed at puberty, and perversions grounded in suppression of the sex drive become the habitual norm. Without a conscious recognition of these facts, it is possible for all sorts of bizarre fetishes to develop and continue. In particular, the frustration of the desire for natural sexual and emotional unions between human beings can even lead to a one-way worship of the opposite sex's genitals and other body parts, as well as other objects that are identified with them such as underwear. However, since surrendering to such objects can't provide emotional mutuality, fetishes provide nothing but temporary sexual relief.

The members of the opposite sex that the frustrated slaves worshipped would have been female during the tens of thousands of years of matriarchy. Then, tit-for-tat, the conquest by patriarchs resulted in thousands of years of an even more hierarchical and cruel system which enslaved all the vanquished of both sexes that are supposed to believe their only purpose is to serve a male elite. To help motivate servitude to hierarchical rulers, the male elites originally imposed upon the masses one-way worship of them through a symbol of a long-deceased barbarian conqueror, who was to be loved above all else in lieu of the mutual reverence between sexual partners. A combination of worship and sexual repression led to the perversions that continue today like a disease that is passed on from one generation to the next.

In particular, many of the various patriarchal religions promote worship of a fictitious deity as a means of becoming worthy of entering a heaven after death. Sex on earth may be frowned upon by these belief systems in many cases, if not outright proclaimed to be evil, at least partially in order to motivate a complete lifetime of surrendering service to the deity in return for sexual relief after death in heaven. Some of the patriarchal religions, such as Islam, have even fabricated a story of the reward of virgin women for Moslem men dying in holy wars for their particular patriarchal deity.

In the meantime, the creation of the fairy tale that the invented deity loves its inferior worshipping slaves can provide people with a brainwashed feeling of being loved, thereby sometimes creating a religious "high" and making it difficult for the people to feel spite at having to wait for an afterlife to "earn" relief from their suffering toil. Many people become addicted to using the deity as an emotional fetish at the same time that they are addicted to some sexual fetish while they await their reward in a make-believe afterlife. With an emotional surrender to a fairytale deity, sexual relationships become divorced from the natural emotional surrender of human beings, and members of the opposite sex themselves become mere fetishes or sex objects and thus less than human.

In contrast to Western religions, ancient Taoist philosophy asserts that all people are equal gods and that sex between willing partners is not only extremely healthy both spiritually and physically but even holy for all (Chia and Wei, 1994). This belief is likely to have been the norm for millions of years until humanity was enslaved with matriarchy 35,000 years ago and then with patriarchy another 30,000 years later for most people.

Within this context, it is interesting that most Native Americans' belief systems include a reverence for sex that motivates them to grant children permission to receive sex education in the most natural manner by watching adults whenever they wish (Underhill, 1965). The mystique and taboo typical in other cultures related to sex is thereby removed, making frustration and exploitation of sexual drives more difficult. Since there wasn't a requirement that Native American marriages or matings be permanent, and many were very short-term in nature (Sugden, 1997), a system of natural human sexual freedom existed, thereby helping to explain the absence of any worship among Native Americans. A philosophy of natural human sexuality is in stark contrast to the vilification of sex in the patriarchal religions today that promote only perversions through bizarre taboos (Eisler, 1995).

The original matriarchal religions that initiated sexual repression may therefore have been grounded in selfish perversions. In particular, older women might have been frustrated by the reduced desire and love exhibited for them as they aged, and some of these females might have been motivated to trick or coerce men into submitting to their sexual and emotional desires. Such females, who might merit a label "old hags" because of their manipulatively unattractive natures that itself may indicate the origin of this derogatory term, and who might even have had perversely mutated genes that caused their inhumane selfishness. They could have used the "magic" of any knowledge that they learned and that caused people to think they were "superior" beings to be obediently worshipped and served. The fact that until the last few decades people often experienced nightmares with a terrifying Old Hag (Gynn and Wright, 2008) is consistent with such a negative view of the matriarchal goddess.

The self-transcendence which enable hypnosis as well as psychic abilities (McClenon, 2002) can be enhanced somewhat by inheritance of specific genes (McNamara, 2006), as is consistent with the utility of this human capacity having evolved over time. At least one old hag blessed with this gene by humanity could have used that gift for the selfish purpose of tricking or extorting human beings into granting them the special reverence and sexual attention they received when they were younger. Some humans may have even developed special extra sensory perceptions that enhanced their forecasting powers (McClenon, 2002), possibly via the pineal gland (Stray, 2005), which remains largely incomprehensible to modern science (Strassman, 2001). Such powers might have been perceived to emanate from or be caused by a superior being. In any event, paranormal mental powers, which are concentrated in the right-side of the brain that is inhibited by the logical left side (Gynn and Wright, 2008), would have contributed to greater reverence for such individuals (McClenon, 2002) that might have culminated in the hierarchal worship of femininity that characterized the matriarchies.

Psychic abilities, which can be activated by shutting out conscious inhibitions against thinking without logical interference, can be initiated by performing seemingly illogical rituals (McClenon, 2002), which characterize religious worship. Psychic powers can also be exercised by acting playfully (Gynn and Wright, 2008). Either process may help return a person to a childhood state, thereby helping to eliminate the consciously learned skepticism that inhibits those powers, which are partially affected by heredity, just as are capacities for hypnotic powers (McClenon, 2002). As a result of a genetic component of psychic powers, a potential worship of paranormal "magicians" in some societies might have easily taken on a hereditary nature that might have resulted in holy rulers somewhat similar to kingdoms in patriarchal times.

There has long been a widespread belief in women's predictive abilities capacities such as the "Fates" that were thought to result in inevitable future events and that continued even into patriarchal times (Baring and Cashford, 1993). One of the primary extrasensory powers of women that could have led to the creation of matriarchies may therefore have been related to an ability to forecast the future. Those powers might have been over and beyond simpler logically deduced ones involved in detecting patterns in past associations such as between lunar and menstrual cycles. They may have even involved more than women's special intuition that reflects their tendency to allow more psychic information to go through the conscious filter and affect their thoughts. In particular, since the words for forecasting in some languages, such as "pre-dictating" in English and "sagen voraus" in German, imply the mere statement of a prediction makes it a future factual occurrence, hypnosis may have been one means by which the female "prophets" made their statements about the future come true, as the hypnotic commands affected future human behavior.

Some of women's forecasting accuracy might have actually been derived from others' naïve belief in the "soothsayers'" abilities. In particular, the mere statement of a "prophecy" could have resulted in the believers' subconscious acting to make it come true through the power of suggestion without inducing a hypnotic trance. Such powers that even appear incredible today might have caused a belief in a superior being that led to the worship of such in matriarchies and later in patriarchies.

Religious worship of old hags may have been facilitated during the matriarchies as older women used masks to hide their unattractive looks and appear as non-human. That appearance, combined with psychic "magic" and hypnosis, could have been used to enhance their mystique as supernatural powers at the same time it might attract males, who could be controlled to at least provide oral sexual stimulation to the unattractive hags without the manipulated males even experiencing sexual desire. The earliest representations of an apparent deity had a female human face or mask attached to either a rounded female body that represented an earth goddess or to an Old Hag figurine whose aged phallic appearance represented a Death Goddess (Gimbutas, 1989). It therefore seems very likely that the original worship was related to old females who purported to symbolize or represent Nature and the regeneration of life, and males may have been tricked into sacrificing themselves into reverent oral servitude to the old hags to cause the regeneration of life.

Since even children today can be fooled with a mask into thinking a cat is a dog (Bernstein and Nash, 2008), it is certainly conceivable that masks could have been used by some women to make males believe that they were somehow superior beings to be worshipped and sexually or otherwise served, especially when used in conjunction with the employment of magic and hypnosis. The evidence that ugly Old Hags symbolizing death and appearing like stereotyped witches continued to be feared into the Middle Ages for being perceived to have extraordinary powers (Gimbutas 1989) is consistent with the hypothesis that it was indeed older women who initiated the matriarchal religion. Such "crones" were once considered to be the wisest of all women (Walker, 1988), perhaps because of their special insights that may have derived from meditation and narcotics, as well as because of their use of hypnosis and psychic "supernatural" powers to cause people to believe they had superior knowledge. It has been established that goddesses were originally revered women who were reputed to have performed all sorts of magical feats (Jiayin, 1995), and such "magic" could have been combined with masks to hide their faces in order to create an image that solicited worship and obedience.

Given many of the figurines, found from prehistoric times are of fat, plump, or allegedly pregnant women (Gimbutas, 1989), and given the worship of pigs in the matriarchal religions (Baring and Cashford, 1993), it seems likely that many overweight women were leaders of the religion of goddess worship. The reverence for wild pigs or boars and their assumed supernatural powers that predates matriarchal religions (Gimbutas, 1989) could have resulted in people drawing a connection between fat women and the revered boars, which were assumed to have some supernatural power over death because of their scavenger nature that resulted in them often being observed near to dead animals. That connection might have actually encouraged some women to take on additional weight to motivate or enhance reverence for themselves.

As women developed their forecasting powers further, people may have become increasingly concerned with what happened after death, and a big belly that was identified with the wild boar scavengers of corpses may have helped make their prophecies more credible in the superstitious environment. It wouldn't have taken a great leap of faith from there for people to believe the fat women had supernatural powers over life and death, thereby motivating further worshipping servitude to them because of the increasing concern among human beings after death as their forecasting capacities motivated them to consider such events ever more.

Since there was also a prehistoric worship of both pigs and pregnant women with big bellies in Asia (Jiayin, 1995), as well as in Europe (Gimbutas, 1989), it is also likely that old women's desired obesity in order to be revered as pregnant women were. The hags' fat bellies could have easily been increased in size because of the extra food that may have been made available to the revered creatures who also didn't engage in any physical exertion to work off the extra food because they were meant to be served just as pregnant women might have been.

Even today, the innate tendency to treat pregnant women with reverence persists, typically resulting in special treatment of them. The fact that depression in pregnant women harms their future offspring (Wilson, 2009) is consistent with a hypothesis that the females of human ancestors were always treated with reverence by other humans when they were of child-bearing age, as otherwise they would have evolved an ability to tolerate the mistreatment, lack of care, and abandonment that can lead to depression. While many pregnant women today imbibe fetus-harming drugs like alcohol, narcotics, and cigarettes, they may do so only because they don't receive the ancient cultural reverence in which they evolved. In any event, the innate human reverence for pregnant women might have been manipulated by fat old hags into a worship of their fat bellies.

The reverence for pregnancy and a round Earth Mother were indeed among the earliest representations of the goddess (Gimbutas, 1989). That reverence could therefore easily have contributed to fat females often dominating the matriarchies. Older women not pregnant or with offspring might have had been motivated to take on extra weight to try and gain such reverence.

When older women were young, they had been the desire of all men, and when they were pregnant, they had been revered and served by all. Once older and infertile, they may have continued to be revered for their experience and wisdom. However, with men having evolved to be attracted to women of child-bearing ages (Miller, 2001), the special sexual desire of men would have been replaced by a special lack of lust, and older women may have desired more than just reverence for their knowledge.

Given that women, including those with characteristics most attractive to males, feel more jealousy toward taller and more powerful rivals (*NewScientist,* 2008p), it seems likely that some of the older fat women employed force in some manner, at least with respect to daintier female rivals. Since women also tend to be more jealous of prettier females (Rathaus, Nevid, and Fichner-Rathaus, 2008), such coercion might likely have been applied on more attractive younger women in order to require them to help fat old hags to seduce men into religious servitude to women in general. Some extra weight might have been useful in any such application of force or threat thereof. Thus, older women might have yet another incentive to take on more weight, insofar as it might help them control the younger and more slender women. As human cognitive abilities developed, some older women may no doubt have figured out the innate tendency of men to surrender to young women, and they could have been motivated to utilize that knowledge to manipulate male sex drives to selfishly satisfy their own sexual desires.

Prior to 35,000 years ago, the pre-matrarichy culture of sending youth at puberty to unrelated custodians of the opposite sex in other clans, as explained in Chapter 17, would have provided older women with the opportunity for some sexual pleasures, especially if they stayed in shape, offered the younger males in their custody the knowledge and kindness they would naturally want to provide, and used their experienced knowledge of sex to seduce the youthful males. However, some selfish older females may not have been satisfied with going from being the most desired and served to becoming the least. Feelings of jealousy may have arisen in these older females, especially as they saw younger men, including those in their custody, lust for and mate with younger women, including even their own daughters. Many of these older females may have also longed for the attention of some older men who had surrendered to them completely to their beauty when they were younger. The development of make-up or masks would have helped somewhat, but some may have gone beyond that into a purely selfish manipulation involving tricking males into believing in their superiority.

To the extent that older selfish women were entrusted with custody of male teenagers, they may have utilized their control of the stronger boys to force the younger women to submit to the old hags' wishes, even offered to let those boys have the younger women in return for frustrating the older men into wanting the older hags again. In addition, the older women could have brainwashed their daughters as young children, and also later via hypnosis, to withhold sex from men, thereby frustrating males into more sexual interest in the mothers. The old hags may have even suspended sending away their daughter at puberty for this purpose, thereby stopping the inflow of new young women they couldn't control and keep from sexually freeing their male slaves.

Thus, both unnatural manipulation and violence may have been employed by the old hags to impose their sick hierarchy on humanity. Regardless of the process by which one-way worship of women entered into human culture, it transformed heaven on earth for all into pure hell for all but the mutant old hags. The selfishness of the older women may also have given rise to unhappy feelings of emptiness, guilt, and fear of spite in the matriarchs themselves, thereby resulting in their own inhumanely designed heaven for themselves being a hell for them as well.

In any event, selfish, older women might have had the incentive to engage in the unnatural and cruel manipulation of male sex drives in some way, and their greater experience with males would also likely have enabled them to be the first to figure out how to do it. The greater enlightenment or intelligence of some older women might have been an important ingredient in the development of the mystic of superiority, deification, and worship more generally, as their counsel was sought in many matters, including those of sexual relationships.

One possible reason for greater enlightenment of some older women that may have provoked feelings of superiority might have been related to a withdrawal of at least some of them to the fringes of society after menopause. Such removal could have occurred due to a exiling from society because of some inhumane behavior like violence against a younger female out of jealousy. Some older women may have left without societal pressure if they were no longer attractive or revered for their prior services and no longer served the community with child-rearing and education. In that situation, such females, like the philosophers and hermits who are able to generate special insights when away from everyone else (Baring and Cashford, 1993), may very well have figured out some psychic magic that likely enabled them to act like the "witches" of medieval times were reputed to do, as explained in Part I. The continuing sexual drives of older women

(Rathaus, Nevid, and Fichner-Rathaus, 2008), the extensive prehistoric existence of figurines portrayed by Gimbutas (1989) that look like dildos, the fact that spiritual and other insights can be inspired by orgasms (Deida, 2005), and the medieval witch practice of using dildos for sexual stimulation (Stray, 2005) all combine to imply that masturbation could have been one "inspiring" practice of such hermits.

The infamous "Tree of Knowledge", which was part of a matriarchal goddess religion before it was used in the Bible to symbolize the forbidden fruit of individual consciousness (Baring and Cashford, 1993), may have involved a herbal hallucinogen that promoted greater awareness (Stray, 2005). Drugs may thus have been involved in the enlightenment of the matriarchs. That "tree" may also have been a phallic symbol that created life-inspiring enlightenment for those women who inserted it into their earthy female bodies during masturbation, just as inserting the tree-like penis into the vagina creates new life.

According to Chia and Wei (1994), ancient Taoist philosophy asserts a Tree of Life is the male just before ejaculation into a woman in a process that is actually a form of meditation which isn't thinking but commuting with the divine through surrender to one's partner. The Tree of Life, which is referred to as the World Tree by the Vikings, is symbolized by an ancient cross sign that actually reflects the basic structure of the centrioles that direct living cells and that is considered to represent the source of existence in microcosmic fractal form (Calleman, 2009). Since the centriole is actually contributed by the male sperm to direct the female egg cell that creates life, the phallic symbol does appear to represent a microcosmic form of the male genitals that may have predated the "6" and that is an essential ingredient in life.

The Tree of Life was probably once related to the Tree of Knowledge, whose fruit enabled tasting enlightenment for both sexes through the spiritually meditative aspects of sexual intercourse. Both Trees are vilified by patriarchal religions like Christianity, which refers to a forbidden Tree of Knowledge of Good and Evil. This tree that is so disparaged by Christianity probably reflects that religion's repression of both sex and knowledge that are connected in the Tree. Because the subjects of the patriarchs' conquest may have questioned whether it was humanly "right" for them to be subjected to slavery in an inhumane hierarchy, the cruel patriarchs imposed a system of good and evil on them that included a morality of sexual repression and ignorance to motivate slavery to a deity. The true fruit of the Tree of Knowledge that was forbidden by Christianity was actually the enlightening pleasures of acting according to one's human nature, such as in sexual intercourse. The patriarchies were thereby able to impose their own ethics internally on each human being to transform the natural tendency to serve humanity as a whole into an exploitative slavery to the elite that was labeled as "good" while all else was categorized as "evil", including human nature itself.

Since sex can certainly enable alterations in the mind that can generate inspirations and insightful knowledge (Deida, 2005), it may very well have been a key force in the development of special inspiration that advanced humanity but that also eventually resulted in the imposition of matriarchies when old hags didn't share their insights with everyone else. However, other mind-altering activities may also have developed, including imbibing drugs. The fact that today many Buddhist monks, who normally rely on meditation or self-hypnosis through focus on repetitive acts like breathing, initially were inspired to the religion through the use of psychedelic drugs

(Strassman, 2001) implies that such drugs could have preceded individual conscious application of individualized hypnosis to reach the magical enlightenment of altered states of minds for many. The very drug-induced special state might have helped spur individual consciousness, since it deviates from the frame of mind of others not so drugged.

Alcohol, which has been drunk by humans for at least 9000 years, may also have played a significant part in the development of much of human culture and religious beliefs (McGovern, 2009). Given the long-term importance of tobacco in spirituality within Native American cultures (Underhill, 1965), it seems that different cultures may have also long used different narcotics for spiritual inspiration. The connection between creative intelligence, painful suffering, and lung problems in the main Nordic god in ancient times (Paxson, 2005) indicates that both smoking something and suffering were used to generate creativity. Since smoking cannabis stimulates the pinoline and relaxing melatonin which is normally produced by the pineal gland during sleep (Gynn and Wright, 2008) when creative dreaming associated with DMT occurs, narcotics and suffering may both spur creativity through chemical processes that result and prompt new ways of thinking.

Under the matriarchies, the widespread existence in prehistoric times of female bird deities (Baring and Cashford, 1993) that were considered in ancient Egypt to represent the "seed-soul consciousness" (Stray, 2005) may have symbolized the "flights" of medicine women or priestesses during hallucinogenic trips. While narcotics may have been widely used to attain special "powers", self-hypnosis can also be employed in such "trips" that have long been widespread among shamans of both sexes and that can also be prompted by the sensory deprivation that is natural to the deep parts of caves where many prehistoric paintings, perhaps inspired by such "trips", have been found (McClenon, 2002).

It's also possible that some of the knowledge of the original matriarchs stems from their living with Neanderthals, who were still around about 35,000 years ago. There is strong evidence that there were some sexual relationships between Neanderthals and modern humans in the distant past that left only a very tiny imprint in today's human genome (Gibbons, 2010). It's possible that many of the old hags and other dissidents from the prehistoric human cultures of modern humans left to live with the Neanderthals and drew inspiration from them. Given that the Neanderthals treated their people well, were very matriarchal, and also may have had strong psychic abilities (Gooch, 2006), it seems very likely that the Neanderthals would not only have welcomed the old hags but also honored them. They may very well have provided the older revered women with unique insights into developing psychic skills and other "magic", besides supplying the sex the hags really wanted. Since the hags may have been too old to bear children, they would have left few if any offspring, and any they did produce may have stayed with the Neanderthals until they finally died out about 35,000 years ago.

When the Neanderthals disappeared from existence, some of the old hags may have returned to human culture and so impressed modern humans with their "magic", which they had learned from the Neanderthals, that they adopted the Neanderthals' perverse matriarchal religion. Such a very tentative hypothesis is based on the circumstantial evidence of religion among modern humans first appearing when the Neanderthals died out. It's also conceivable that the pathological matriarchal religion caused the final demise of the Neanderthals, just as patriarchal religions may lead to the eventual demise of modern humans unless they are changed.

Some of the old hags may have actually defected back to modern humans prior to an extermination of the Neanderthals, bringing with them knowledge of psychic powers that helped modern humans defeat the Neanderthals. Such aid might have been offered only in return for worship. The worship with which the old hags may have been spoiled while living among the now extinct Neanderthal breed of homo sapiens might have motivated them to make such an exploitative trade. Such help certainly would have induced greater reverence for these witches among contemporary Cro-Magnons in any event.

It's conceivable that Neanderthal support for a continuous inflow of old hags could have started to become a burden for them, possibly even contributing to their disappearance. Such a situation might have reduced the Neanderthals' reverence for the hags, some of whom might have therefore been motivated to return to modern human societies, which they helped with the "magic" they learned while away. That "magic" was then used to push on modern humans the matriarchal religion, which they might also have learned from the Neanderthals, who had been driven into extinction by that pathological belief system.

In any event, all hermits and refugees might have voluntarily returned to society if they attained sufficient enlightenment to develop magic, including new primitive technologies as well as hypnosis. Such tools could have enabled them to dazzle others into renewed reverence. The voluntary departures may have been only temporary to think in an environment without being as impacted by "normal" societal views that require diversionary compromises and efforts involved in social interactions. These hermits might very well have utilized alcohol, narcotics, hypnosis, and special knowledge learned from Neanderthals to supplement the mystique of their intrinsic knowledge in order to dupe others into believing they were superior beings with superior powers.

The evidence that music and special sound effects created in ancient buildings and caves induce strong emotions and an "otherworldly" feeling (Brooks, 2008f) implies that prehistoric humans may have eventually developed more elaborate manipulative methods like tonal "magic" to induce particular emotions. The naturally greater psychic abilities of females on average (Gooch, 2007) would have contributed to women being the most effective creators and users of such "otherworldly" capacities to solicit awe and worship.

Since clothing covering the genitals creates a mysterious aura about the opposite sex (Reich, 1936), it is likely that hiding the female "private" parts was an important component of the female "magic" of matriarchy and vulva worship. Because clothing may not have been developed until about 75,000 years ago (Holmes, 2008a), matriarchies likely only followed some time after that when it was discovered that covering the genitals could be used to create a mystique about them. Such hiding of the genitals, along with unnatural sexual repression in general, is a large contributor to human sexual perversions that create de-motivating feelings of unworthiness, disincentives to cooperation, violence, and energy distracted from productive non-sexual tasks (Reich, 1936). It therefore seems probable that the perverse matriarchal hierarchies began well after the development of clothing and other advances in human technology that was utilized by the perverted matriarchs.

The older matriarchs might have found especially good use for disguising their faces and bodies with clothing to make it difficult for men to recognize whose clitorises they might be tricked into sexually stimulating. The development of the triple goddesses of maiden, mother, and krone as a unified whole under matriarchies (Walker, 1988) might have even been related to a more overt form of such manipulation. For instance, old hags could have used their daughters and granddaughters as bait to attract males, who naturally felt a desire for complete surrender to the maiden but may have had to provide sexual stimulation to the mother and krone in order to prove they would love the maiden when she too became a mother and krone.

By wearing a new mask, luminaries or prophets could have even been unrecognized. Unrecognized people, as well as those who moved from other groups or cultures, might have been especially revered if they had special talents like insights and magic. In particular, the unfamiliarity with immigrants' actual human birth and childhood might have helped spur a belief that such outsiders were superhumans to be worshipped. Such a belief might have broken the power of empathic group consciousness by causing the "superhumans" to feel superior and contemptuous of others and their feelings.

The ability to deceive others with masks and other magic might thus have contributed to the breaking with group feelings and equality, as it would have led to a perception of their inferiority. Older women feeling such contempt as well as spite at less reverent treatment than in the past when they were younger might have promoted the exploitation of this deception that would be undesirable with true empathetic equality. However, even that perversion of human nature might still create some empathetic happiness in the "inferior" people, who partially experienced the resulting bliss of the masked leaders, who were perceived to be entitled to greater ecstasy because of their "superior" nature.

The use of masks to promote special worship might have later developed into fetish worship of images, such as artwork depicting a human wearing a mask. Images of women with the heads of other organisms have indeed been found to have been widespread in prehistoric cultures (Baring and Cashford, 1993). Later, such fetishes could have fragmented into worship of only portions of the female deities such as just the body, breasts, or vulva. Such focus on particular parts might have been useful for old women in distracting attention from their unattractive faces or non-human masks to attract males, who might have been further fooled into sexual excitement. The younger female assistants, who may have willingly cooperated due to the excessive attention they receive from males, acted as preliminary masked bait that was then replaced by old hags indistinguishable from them because they were wearing the same masks in a secretive dark environment of goddess temples or caves.

The extensive female figurines and vulva symbols in prehistoric art (Gimbutas, 1989), as well as the recognized secretive nature of the prehistoric goddess religions (Sjoo and Mor, 1987) that seems counterproductive to humanity as a whole, do provide support for this general hypothesis on the origin of image fetishes. The prohibition on the worship of any images that existed in some patriarchal religions such as Judaism after the overthrow of the matriarchy (Baring and Cashford, 1993) may also possibly be explained by this theory, as spiteful males fed up with female manipulation of images for hypnotic control purposes wanted to end the worship of artificial symbols or parts.

The first evidence of worship of images relating to a Goddess religion were vulva cave drawings of vulvas that are 30,000 years old (Gimbutas, 1989). Because human clothing had existed by that time for nearly 50,000 years, it is indeed likely that the hiding of the sex organs behind clothing was instrumental in the road toward greater reverence for women from males, whose greater sexual drives possibly resulted in them being more frustrated and mystified. The then incomprehensible magic of female pregnancy and birth-giving would have only magnified the effect.

Given the relationship between the women's reproductive cycles and the moon (Shlain, 2003), vulva worship would have likely led to lunar fetishes that were connected to females. The other-worldly nature of alternative mental states induced by the hypnotic effects of religious worship, animal lust, and narcotics (McClenon, 2002), as well as by human love, could have also induced a belief in otherworldly entities like celestial bodies. Given a reverence for all of Nature (Peat, 2005) that may have characterized humanity through millions of years, the widespread religions focused on the moon, sun, and stars, as well as on animals and humans masked as animals that developed (Jackson, 1992), would have followed logically from an initial fetishism based on female genitals whose emissions were observed to be associated "magically" with lunar cycles. The connection of the old female shamans to animal masks and powers as well could have easily led to an association between celestial bodies and the shaman's animalistic spiritual images.

Prehistoric art datings seem to indicate a worship of a Goddess as part animal, such as represented via a human body and an animal mask, before there were any astrological connections (Gimbutas, 1989). This finding seems consistent with a hypothesis that clothing and masks were initially utilized by old women to mystify males.

Goddesses were sometimes represented as snakes because of the regenerative nature of those reptiles shedding skin (Walker 1988). Since ancient snake coil artwork seems similar to that of labyrinths (Gimbutas, 1989), it is possible that the ancient sacred labyrinth even originated from the snake coil symbol and developed further as new insights into the spiral nature of the cosmos unfolded. While the labyrinth and its insightful symbolism could have preceded the Goddess religion, there is no evidence of that chronology of events.

In any event, the worship of a superior being, human, image, object, or body part makes a person less than the object of worship by definition and therefore a mere tool of the worshipped entities. The regard of women in the goddess religions that men were mere exploitable object for women (Sjoo and Mor, 1987) is consistent with this hypothesis. Even if a worshipped entity is recognized as merely human, the worship of that entity makes an ideal subhuman.

Regardless, the hierarchy of worship motivates slavery that may be natural for insects like bees but is inimical to human nature because of spite. The fact that one of the key symbols used in goddess worship was the bee (Gimbutas, 1989) may be indicative of the type of society the matriarchy was striving to create and to which goddess worship was attempting to lead humanity. The fact that bees actually try to copulate with the flowers they pollinate because of the pheromone chemicals released by them (Rathaus, Nevid, and Fichner-Rathaus, 2008) can provide an example of the manipulatively frustrating direction in which humanity as a whole would have gone if the matriarchs and their frustration of male sexual desires to create disgusting fetishes had continued.

Fortunately, human nature includes characteristics such as spite and revolt against inequality, injustice, and exploitation that prevented humans from de-evolving back into more primitive life forms.

Although Eisler (1995) has asserted that the matriarchies were egalitarian, the very nature of belief in the superiority of women made them hierarchical and allowed the exploitation of men by women. Besides the cruelty to men that appeared to exist in matriarchy and that reflected their lower class status religiously and politically (Gimbutas, 1989), there is also some evidence of unequal economic treatment (Baring and Cashford, 1993).

It may be useful here to speculate on the means by which the smaller sex was able to maintain 30,000 years of keeping the stronger sex subdued. Besides actual use of hypnosis and technology kept secret from males, religious indoctrination was probably crucial. The matriarchal doctrine of females being "superior" would likely have been ingrained into the brains of both sexes starting as infants when everything is accepted without question. A two-class society with the physically smaller sex ruling the stronger would have been seen as natural due to the long-term hypnotic effects of learning at that age. The propaganda would have continued throughout life, and the entire cultural system was likely designed to reinforce such learning. Differential treatment of male and female children was probably the norm, and there would likely have been rituals at puberty that may have not only honored the females with extraordinary reverence for their attainment of their special powers of procreation and sexual attraction but also clearly indicated males' roll as one of complete one-way submission to women that was clearly tied to sexual relief. Such rites would help explain the spiteful treatment of menstruation as vile after the spiteful patriarchs' overthrow of the matriarchies a few thousand years ago, along with their vindictive casting women into the new role as being evil slaves to men.

In any event, the worship or regarding of another or others as superior leads to unnatural hierarchies that are inhumane. Even among women in the matriarchies of prehistory, there appears to have been a religious hierarchy that put a High Priestess at the top (Gimbutas, 1989). A belief in the superiority of magicians, shamans, women in general, or symbols thereof, was the unnatural characteristic of matriarchy that led to the counterproductive hell involved in matriarchal worship of special individuals, images, and parts.

Some might suggest that overpopulation could have caused sexual repression that in turn led to the creation of matriarchies as the greater male sexual urge was frustrated into servitude of females in a competition for satisfaction of that drive. This hypothesis, which is examined in more detail in Appendix D, doesn't explain the worship of old hags and fat women, nor does it explain why the goddess included a pregnant mother and a crone as well as the more desirable maiden. In contrast, the theory presented in this section on the origin of the matriarchies and worship in general seems to explain all the known facts about the past and current human circumstances quite well.

G. The General Characteristics of Religion

Regardless of speculative hypotheses on the origin of worship, the original religions appear to have been matriarchal, as explained in Part I. The "magic" employed by the leaders of the new religions that were typically called priestesses did include rituals and fetishes with respect to an animal or deity image that may have included application of some technological tricks unknown to others (Sjoo and Mor, 1987). Since some of the ancient rituals such as those with music or chanting have hypnotic effects (Howells, 1986), and since hypnosis has long been powerfully used (DeLaurence, 1902, 1910), it is very likely that hypnosis was one of the technological tools employed by these religious leaders. The spiral shapes found in the eyes of some ancient goddesses (Gimbutas, 1989) could reflect some of the matriarchal hypnosis employed that may have relied on eye contact. Spiral shapes that resemble labyrinths are still sometimes used to focus people's eyes in modern applications of hypnosis, as many of these techniques were later adopted by the patriarchal conquerors and their religious elites in those succeeding cultures.

In any event, shamans of both sexes have long utilized hypnotic powers to attain high status, and hypnotic religious rituals are motivated by the very hypnotizability of humans that not only enhances their abilities but also helps avoid the psychological problems such as depression that is often associated with increased hypnotizability (McClenon, 2002). The association today between hypnotizability and depression may relate to the greater innate desire for the hypnotic ecstasy of group consciousness that exists in such individuals but that is frustrated by contemporary individualistic societies. Deceitful religions are designed to fulfill that need in a manipulatively artificial manner.

Religious experiences themselves today seem to be related to a chemical DMT that is released naturally into the brain and that can result in a blissful experience relating to an irresistible desire to unite with other beings (Strassman, 2001). It can also result in a "bad trip" when an individual resists the desire to surrender (Gynn and Wright, 2008). The feeling of interconnectedness that is involved in a hypnotic subject submitting to a hypnotist (Urban, 2007) apparently occurs in the middle of the brain at the pineal gland, which also is involved in the deepest form of sleeping and dreaming (Strassman, 2001). This feeling is affected by the release of DMT and other hallucinating chemicals by the pineal gland (Stray, 2005), and the power of hypnotism may be partially explained by this connection.

Within this context, it is interesting that people who have been put under the influence of the natural hallucinogen DMT at some point have indicated they actually gain many benefits when they let themselves go, including not only enhanced enlightenment and creativity but also greater control (Strassman, 2001). Such phenomenon may be understandable if it is recognized that the decision of whether or when to let go or surrender to others and the environment always remains completely up to the individual. Allowing oneself the freedom to let go at times frees oneself from the very restrictive bondage of robotic self-control that may be both constraining and counterproductive to what a person wants to do.

Hallucinogenic brain chemicals such as DMT can be induced with reverence for a mate, others, or the entire universe as well as via a "modern religious experience". The pleasure of complete mutual surrender experienced during natural human love and sex (Deida, 2005) might also be related, as may a group consciousness that once was empathetically invoked to maximum happiness via public chants, music, and rituals. These phenomena can still have an effect on many individuals and are both quite natural and adaptive for humans. For instance, the fact that babies born in the normal fashion experience a large dose of DMT at birth may be especially useful for humans, as people who have been delivered using Caesarian sections encounter psychological and spiritual problems later in life, possibly because of a lack of that drug in Caesarian births (Strassman, 2001). Hallucinogenic experiences can, via chemicals causing the conscious to ignore normal external stimuli, lead to a better understanding of one's own nature, including one's emotions, memories, and personality, and it may also lead to special insights into reality, via special stimulation of the novelty-detection portion of the brain that is called the locus coruleus (Nichols and Chemel, 2006). Such insights, which are normally inhibited because they are unessential for normal daily tasks, could also utilize psychic phenomena, cosmic rays, and electromagnetic waves, especially to the extent that the pineal gland is involved (Stray, 2005).

Narcotic drugs can also be utilized to induce special insights (Narby, 1999). However, the terrible harm to human fetuses caused by pregnant women imbibing natural or artificial drugs, alcohol, and chemical-laden cigarettes in the very early stages of pregnancy (Bernstein and Nash, 2008) indicates that humanity as a whole hasn't evolved to fully adapt to such external influences. Pregnant women have even evolved to have a subconscious aversion to some imbibed materials that would harm the fetus (McClenon, 2002), and so it seems likely that any use of narcotics by women would have been restricted to only those who weren't females of child-bearing age. In particular, with women being unaware of when they are pregnant, narcotics were probably only utilized by females after menopause. While the failure of the fetus to evolve a defense against narcotics is consistent with a hypothesis that narcotics haven't been employed by humans for an extensive period of history, it's also possible that older women may have long used herbal hallucinogens after menopause to enhance their special knowledge, which they utilized along with a secret understanding of herbal medicines to eventually promote worship of themselves.

In any event, natural production by the pineal gland of some pleasurable, hallucinating chemicals is enhanced by fasting (Stray, 2005). Many Native Americans and shamans seek out visions via some self-sacrifice that includes hunger (Underhill, 1965). Hallucinations, which can promote insights and a vision of reality not normally perceived (Narby, 1999), could have been induced with the sense of hunger to promote creativity in dealing with particularly adverse environments that caused human suffering which could potentially be relieved with a novel idea. In particular, by helping the brain develop new neural connections between seemingly unrelated ideas, hallucinations may enhance the "normal" senses which evolved to deal only with typical situations, so that people could adapt to "abnormal" environments. The hallucinogen DMT does seem to be produced by the body in extremely stressful situations (Strassman, 2001), thereby explaining the success of fasting and suffering in promoting spiritually inspirational experiences that can promote creativity.

Systems that promote occasional fasting seem to provide some benefits, as the food thereby saved is available for starving humans. Periodic bouts of fasting are also healthy to the person not eating, especially if that person is overweight, besides perhaps providing some creative inspiration. Sacrifices out of mutual reverence or loyalty to a group or idea can also enhance group cohesion and survivability (McNamara, 2006). In addition, it appears that the brain has evolved to associate some pleasant experiences and useful insights with the sacrifices that are often involved in mutual reverence for a mate, loved one, or human group. For instance, the mutual love between mates that results in an idealization of the partners by each other may not only motivate longer-lasting relationships among mates (Rathaus, Nevid, and Fichner-Rathaus, 2008) but also likely results in partners striving ever harder to be the person revered by the other no matter how seemingly impossible.

In the very distant past, personal suffering likely resulted in empathetic sympathy and help (Nettle, 2007), and so extremely stressful personal conditions would have generally been experienced only then. During periods of group privations, internally generated DMT might have had vital use. Some individuals believing that they weren't contributing sufficiently to humanity to justify their own sustenance, or who thought they could contribute more to the group of humans than they were if they could direct themselves better, might also have left the group to exist in a state of want to help induce enlightenment toward this end. However, a belief that pointless personal suffering is necessary in order to attain spirituality and a pleasant religious experience is an "ego-based" concept that has no validity (Chia and Wei, 1994), and it is a counterproductive, evolutionary dead end.

One-way sacrifices due to worship of beings perceived to be superior are especially pathological. Such suffering serves no adaptive purpose except to enable manipulation by the "superior" deities or their priest and priestess handlers. Humans exist to satisfy their needs and desires, not to suffer needlessly (Chia and Wei, 1994) or for the exploitative benefit of others. Suffering in servitude to superior beings is encouraged by most of today's religions because they promise eventual rewards, such as in an afterlife, for such exploitative slavery. People sometimes inflict harm upon themselves, such as fanatical adherents of perverse patriarchal religions like Christianity that glorify pain, suffering, and self-sacrifice for a selfish, sadistic, fictitious deity and that vilify pleasure (Eisler, 1995).

The personal gain from suffering in the absurd christian religion appears to be to not only gain some spiritual inspiration but to also appease a fabricated god through self-sacrifice. It essentially represents a pathetic attempt to atone for sins against artificial morals allegedly established as universal truths by a fictitious god to permit entry into a make-believe Heaven. It probably also derives from some absurd logic of the deranged Christians that they can selfishly draw sympathy and attention from their imagined deity. This aspect of Christianity reflects the religion's absurdity as well as the infantile mental state of those afflicted with that pathological disease.

The pathology of deliberately imposed suffering caused by religious manipulation of ignorance may have originated in the unhealthy system of worshipping female deities, femininity, and women. That religion seems to have resulted over 10,000 years ago in a belief that springtime required a one-sided male sacrifice or surrender to symbolize the planting of seed in the earth goddess and that autumn harvesting had to be accompanied by a gruesome shredding of male bodies (Gimbutas, 1989). Such sacrifices apparently developed into a widespread tradition of an annual sacrifice of a son-lover of a queen (Baring and Cashford, 1993). That ritual might have involved an old high priestess taking sexual pleasures with a young man for a year, perhaps in return for granting him some privileges and power before having him gruesomely disposed of and replaced. Such traditions were apparently literally carried out (Sjoo and Mor, 1987).

It is possible that the victims had "volunteered" for such fates, but the mere fact that they did so indicates the cruelty of the society in which they lived. For instance, an extremely adverse social or cultural setting, such as one characterized by the physical or emotional torture that is so prevalent today, can certainly motivate some to commit suicide. In a matriarchy, an enforced sexual abstinence for at least some males could have served as a primary tool of controlling more powerful males, and the resulting frustration of male sexual drives might have represented frustrating torture for them. They might have been further motivated to extreme self-abuse and self-sacrifice through women's use of hypnosis and other forms of brainwashing.

Possibly as a result of such atrocities, some call all systems of worship a "darkened consciousness of the soul" (Baring and Cashford, 1993). One-sided worshipping relationships create a system of inequality and injustice that artificially separates the worshipping from the worshipped promotes contempt, frustration, and spite. Such relationships are in conflict with humanity's evolved need for mutuality and integration with all others and the universe in order to survive an overwhelming environment. In fact, the inequality and injustice of a worshipping relationship results in unnatural hierarchies that must promote ignorance in order to maintain the injustices against spiteful resistance. Because the natural drive for just equality in human beings requires each to act to try to offset feelings of inferiority with a perception of superiority to others, the spite induced by slavery to deities and exploitation by others deemed superior can be manipulated via brainwashing into violence and war against other brainwashed people.

Humans could never have survived such conflicts created by worship when their technological level left them endangered by stronger and faster predators. However, once humans had advanced to a technological stage where they dominated all other life forms, the need for human balance in order to survive a dangerous environment would have seemingly disappeared. Then, humans would have been able to expend time and efforts on counterproductive activities such as worship and related behavior like slavery and war. A rough guess that religious worship began about 30,000 years ago (Dow, 2008) seems reasonable within this context. While the actual manipulation of other organisms for human exploitation began with conscious baiting of hunted prey or planting of seeds and culminated in shepherding and agriculture, actual widespread manipulation of others for selfish purposes didn't begin until about 35,000 years ago.

Prior to the establishment of the first religions, a balanced culture must have existed, or humans would have already evolved into honey bees serving a minority of procreators, or been dominated by alpha males like their chimp relatives. In either event, they would have perished when they had to come down from the trees due to weather changes that confronted them with overwhelming predators.

At least one group of people in China, the Mosuo, have continued to maintain a form of a fairly balanced society today, albeit a matrilineal one (Jiayin, 1995), and so there is evidence of such cultures existing despite being moderately impacted by some residual influences of the matriarchal hierarchies that once dominated the globe. In addition, many, such as northern Native Americans who demonstrate a continuation of rather balanced views by honoring grown crops like corn and thanking their hunted prey for allowing them to eat them, still see a long-term need for balance in Nature that would well serve their offspring many generations into the future (Peat, 2005). Within the context of human group consciousness, such balance and thought to future humans is quite natural. Within those cultures, attempts to exploit others and solicit one-way worship would have been unthinkable, especially since doing so would have probably resulted in spiteful ostracization and even exile.

As previously mentioned, most Native Americans who live in what is now the USA don't worship anything although many revere their ancestors, nature (Underhill, 1965), and the universe (Peat, 2005). While some categorize as a religion any belief system that holds something sacred (Dow, 2007), a more narrow definition of religion, which requires worship and sacrifice to a perceived superior being, would not apply to the beliefs and reverence of most northern Native Americans. Such "non-religious" reverence for other human beings and the universe likely characterized human culture for millions of years until the matriarchal religions were established among modern human beings.

The purely selfish individuals, who were able to manipulate others into religious worship about 35,000 years ago due to insights they possibly drew from Neanderthals or drugs, may have been prompted to do so because of mutated genes that created their pathological mental illness of selfishness. In any event, once people had been manipulated or tricked into religious worship, they could easily be exploited sexually and otherwise by selfish leaders who would be motivated to impose a hierarchy, as previously mentioned. Since a belief in an all-powerful and all-seeing deity can discourage "sins" against the imposed morals of that deity and also motivate cooperation among its adherents (McNamara, 2006), widespread belief in a religion would have enhanced obedience to the leaders by inhibiting attempts to stop the inhumane exploitation. Religious leaders of a hierarchy with assumed superior beings could utilize the submission and violent tendencies promoted by worshipping those superior beings to force groups of "inferior" humans to conquer other peoples, especially if the "enemy" were propagandized as even more inferior and subhuman. The cultural phenomenon of worship and sacrifice to superior beings, including via fighting in unnatural wars between humans, was thereby widely spread to become dominant in the unnatural world order of today.

Some have asserted that belief in imagined deities provides a social bonding system that enhances group cohesion (Brooks, 2009a). However, humans evolved to bond naturally through chemical processes in the brain that motivate group cohesion without any artificial worship of fictitious beings. Belief in a set of established lies actually seems to create violent frictions through disagreements on what exactly should be imagined. The requirement in most religions to believe without question or proof in a set of assumptions, and to continue to believe even in the face of contradictory facts, appears to be a self-defensive mechanism for perpetuation of a faith in a deity (McNamara, 2006) that is based on lies. While such faith may inhibit vindictive human spite against the manipulating elite that would arise if the deceitful manipulation and lies were uncovered, the psychological brainwashing of human beings into ignorant slavery merely channels the hate into mutually destructive wars without end.

Some have suggested or implied that religious worship can provide actual human benefits, such as satisfying some hypothesized desire for a perfect parent-figure (Dow, 2008). However, there is absolutely no evidence whatsoever that humans would prefer such dominating relationships to the mutual reverence of egalitarian relationship between unrelated sexual partners of opposite ages that exist in a humane society. Even with respect to the relationship between parents and their offspring, non-worshipping Native

Americans act more like servants to their children than dominating dictators (Underhill, 1965).

Believing in an all-knowing and all-seeing deity motivates obedience with limited actual monitoring or enforcement (McNamara, 2006). Manipulative religious leaders therefore modeled their invented deities as parent-like super-beings in order to cater to infantile desires for everlasting nurturing from "benevolent" and "knowledgeable" guardians that seemed to replace the mentorship system involved in the natural young-old relationships between unrelated sexual partners that existed before 35,000 year ago (as explained in Chapter 17).

In addition, the "powerful emotional relationship" supplied by worship (Dow, 2007) is much more productively and less spitefully provided by mutual reverence. Even the fantasy of mutual love had to evolve to be more humanly satisfying than worshipping a fictitious being or anything with imagined superiority. Fantasies of any kind seem to open up the barrier between the conscious and subconscious and therefore make a person more susceptible to hypnotic and psychic control (McClenon, 2002). However, imagined events that serve a group of equal humans also facilitate group hypnosis that is controlled by the group for the benefit of the group as a whole. When fantasies are so directed for the good of all, they harmonize the subconscious group psychic power for the benefit of humanity in a process that is far more powerful and ecstatically rewarding than one-way worship designed for the selfish exploitation by a cruel mutant elite, as explained later in Chapter 16.

While it has been suggested that one of the reasons religious worship developed is because it could possibly enhance sexual attractiveness (Dow, 2008), it is unclear how such a tendency might evolve. In particular, worship of some deity takes reverence away from a partner, and religions like the Christian one that demands the love of a fictitious deity to be the top priority can make it difficult for a true believer to fall completely in love with a real mate or to even love anyone at all. Christianity, which vilifies all humanity as sinning slaves to a god that alone is supposed to be worshipped creates self-contempt which restricts event the self-love that can attract a loving partner.

It has been hypothesized that belief in a deity somehow makes an individual seem more trustworthy to members of the opposite sex and society in general (McNamara, 2006). However, given the evidence that atheists are at least as honest and ethical as religious believers (Shermer, 2004), a dogmatic religious leap of faith is needed to think that belief in a set of manipulating religious lies enhances trust.

Religions can provide set answers to complex questions such as the meaning of life and thus allow people to avoid the effort involved in investigating such questions on their own (Urban, 2007). However, contemporary religious answers seem to be motivated by attempts to manipulate and exploit people via deception and are therefore less likely to have any degree of validity or usefulness than a random fictitious cartoon. For instance, even fantasy worlds like Pokemon that were created more for entertainment than for manipulation permit people to dream up their own fantasy worlds, thereby spurring creative imagination. Such naïve fantasy worlds have to be far more useful for the individual human being, as well as for humanity as a whole, than slavery to religious lies that were promulgated to enable ignorant and unquestioning servitude to "superiors".

H. The Inevitable Death of Worship or Humanity: Which Will It Be?

In today's technological world, worship seems to be pathological, not only because of the scientific evidence in conflict with religions but also because of the destructive nature of the modern weapons. The development of technology enables the creation of weapons to destroy all of humanity, including some that may be used without detection of the source, such as those that affect the weather and the environment (Smith, 2006). They therefore don't have an effective retaliatory deterrent, and rivalry between selfish elites in different countries, combined with the religions promoting wars for their separate deity fictions, encourages their employment. Violent tendencies against other humans are promoted by beliefs in "superior" beings and other lies that are spread through the use of invisible weapons like hypnosis and brainwashing which also don't have an easily detectible source. While reverence for others and everything else is a natural human trait that enhances humanity's ability to adapt and survive the environment, taking it to an extreme of one-way worship that implies inferiority and hierarchy is an evolutionary dead-end, and human beings must realize this fact before it is too late.

Sharks, cockroaches, and bacteria that have survived seemingly overwhelming creatures like dinosaurs and horrible environmental disasters like asteroid impacts may eventually inherit the earthly paradise if humans continue with their ultimately suicidal tendencies. The fact that enormous human cooperation will be required both for dealing with the catastrophic effects of global warming that may alone kill billions of people in a few decades (Vince, 2009) and for trying to minimize those effects (Brahic, 2009) is just one other example of the need to discontinue the individualistic tendencies that religions fostering worship of superior beings promote. Besides the weaponry and environmental destruction that is known to be applied today, there appear to be many other dangers from the millions of "control freaks" that aren't even publicized (Smith, 2006), and only a return to the prehistoric belief in human commonality can save humanity.

Although "modern" religions can create a "high" at times in some people, that effect is somewhat similar to that of a narcotic to which people selfishly addict themselves. Humanity appears to have evolved so that pure pleasure-seeking activities are eventually self-destructive even for the individual, as with a drug addiction for which the brain chemicals and pleasures are often quite similar if not identical (Chemali, Chahine, and Naasad, 2008). There are natural processes in the brain that turn off the yearning for ever more pleasures brought by actions like eating, sexual intercourse, and even love, they don't work for artificially imbibed drugs (Esch and Stefano, 2005a). Similarly, there aren't any innate antibodies against the new addictive disease called religious worship that has spread so widely like a plague.

Both external narcotics and artificial religious worship have only recently been introduced to human bodies, and so no natural defenses against an addiction to them have yet evolved. However, those addictions are only possible for human beings not enjoying the ecstasy of serving humanity as a whole that includes oneself. Such diseases can therefore easily be eliminated with knowledge of this cure. The natural internal administration of opiates to the brain that has characterized humans for millions of years has in fact been crucial to motivating actions essential to the survival of humanity. Artificial stimulation of these pleasures through narcotics or worship is addictive only if there are unfilled needs, such as the human need for feelings of self-fulfillment that only comes with devotion of one's life to humanity as a whole.

Human nature has evolved over millions of years to motivate actions in service of not only the individual but all human beings, and it does so in the aggregated sense through the drive for a happy feeling of self-fulfillment that can only come from serving a higher purpose in which the individual is a part. The pleasures of love, sex, and happiness are caused by internal biological chemicals, and those pleasure chemicals would have led to the extinction of humanity if attained artificially and pursued endlessly (Chemali, Chahine, and Naasad, 2008). The human body's natural production of optimal amounts of such chemicals at the right times in conjunction with others, such as cortisol, norepinephrine, and vasopressin for humans in the initial stages of human love, results in productively aggressive yearning that is absent with artificial drugs which promote only unproductive behavior such as doing nothing but taking more of the narcotic (Esch and Stefano, 2005a).

While the drive for more to offset the lack of self-fulfillment from taking narcotics eventually results in individual destruction through an overdose, the drive for the ultimate heaven in religions like christianity is directed toward the elimination of the human race. The pleasures created in the brain by narcotics do motivate taking toxic overdoses (Chemali, Chahine, and Naasad, 2008), at least partially because they only lead to the same feelings of worthlessness as religious worship, and so they can't provide any happy long-term feeling of self-satisfaction. Any case where reverence is transformed into a worship of a "superior" seems to have similar effects and can even lead to the bizarre trait of suicide wishes for those unfortunate enough to believe in the lies of a better heavenly environment in an afterlife. In lieu of a conscious recognition of their purpose in life to serve humanity as a whole in the way each individual chooses, human biological processes haven't evolved to adapt to the poison of the "worship" drug any more than they have for the external imbibing of standard chemical narcotics. The increasing tolerance for artificial narcotics that promotes ever-increasing use (Bernstein and Nash, 2008) should provide humanity with an indication of the suicidal nature of both hedonism and worship, both of which require ever more to be satisfied.

The first worshipped entities that originated 35,000 years ago were therefore potentially motivated partially by a temporary, artificial "spiritual high", perhaps with the help of both narcotics and hypnosis that the old hag deities could have learned from the Neanderthals. Since opiates can easily lead to behavior and beliefs that are irrational (Esch and Stefano, 2005a), the introduction of the narcotic of deity worship could certainly have spread rapidly like an epidemic of alcoholism. Although a belief that all humans can actually be thought of as sacred deities (Sjoo and Mor, 1987) could have served as an antidote, the old hag pushers of the religion that made themselves superior deities apparently enjoyed the temporary hierarchical "highs" of being worshipped too much to think of or promote that cure for the pathological disease they spread.

Prior to 35,000 years ago when the matriarchies established an earth goddess above humanity and placed female priestesses at the top of a hierarchy, the earliest magicians surely recognized that they were not all powerful deities. Their humanity probably encouraged them to promote the reverence for the unknown in the form of Nature and the entire Universe as the source of their special powers, thereby explaining the development of reverence for Nature and the Cosmos in many ancient cultures that preceded the pathological religious worship of particular entities. Such a non-worshipping reverence probably existed through most of human prehistory. Magic and superstition were components of the belief system and remain popular today among humans, who apparently derive benefits from it, not only in terms of self-confidence and positive thinking but also in terms of a feeling of happiness, contentment, and security (Howells, 1986). All humans are like deities insofar as they are able to create and jointly procreate, thereby enabling their genes live into eternity through their offspring. People can be rationally adored as such, as they indeed are when there exists the state of group consciousness that occurs even today for mates in mutual romantic love with each other. The adoration only becomes unhealthy to humanity if it grows into a worship fetish that makes the object of adoration more than human.

One-way worship of a leader, deity, or anything else thus doesn't seem to be at all intrinsic to human nature, and, being inconsistent with evolved human nature, it only leads to disharmony and unhappiness among human beings. For instance, about half of all residents in the USA suffer from depression each year (Weiss, 2005), and the primary cause of that symptom of mental illness is likely to be the disease of Christianity that afflicts so many of those unfortunate people.

The original mutual reverence that well served humans living in isolated pairs had to exist or evolve with an overriding tendency to serve the perceived best long-term interests of a larger group. That trait was needed in order to allow humanity to survive in the hostile world of stronger and faster predators. The bliss that people experience when they feel one with other humans and the universe (Strassman, 2001) may represent an evolved reward for such a group consciousness to which humanity must return or face extinction.

Chapter 13
Group Consciousness

De Quincey (2005) has indicated that humans originally developed a group consciousness without much independent individualized thought. Empathy, or experiencing the feelings of others through the brain's simulation of what it believes others feel (Bernstein and Nash, 2008), would have made group consciousness quite natural. Human empathy is so strong that recognition of another person's emotions naturally results in the premotor cortex area of the brain responsible for moving facial muscles to mirror that person's feelings (Douglas and Robson, 2010).

However, group consciousness goes beyond mere empathy. Even in "the competitive paradigm" of contemporary society, it has been found that "bonding is... what makes us happiest" (Bekoff, 2009). Those who increase their level of altruistic behavior actually indicate a higher level of self-satisfaction (Hamzelou, 2010e), implying that human beings evolved to experience pleasant feelings upon helping each other.

The bliss that people feel when they take the drug "ecstasy" and thereby feel like they are friends with everyone else in the world (Klein, 2003) may be indicative of what group consciousness originally was like. Drugs like ecstasy not only create a feeling of euphoria but also induce sociability, talkativeness, attentiveness, and relaxation (Fleming, 2010). The chemicals that ecstasy induces in the brain are internally produced naturally, including serotonin (Ainsworth, 2002) and oxytocin (Gynn and Wright, 2007), and so it seems quite feasible for consciousness to have once been characterized by such an ecstatic feeling of cooperative group unity. Since ecstasy has the side effect of inhibiting verbal learning but actually enhancing some forms of geometric and locational analysis (Ainsworth, 2002), this mental state may very well go far back in human prehistory before there was any language communication.

Such ecstasy would have been selected by evolution since it promoted the necessary cooperation for early hominids to survive an otherwise dominating environment. The human brain, which naturally creates similar sensations when perceiving physical and psychological warmth in the insular cortex area of the brain, and which creates feelings upon social exclusion similar to that of physical pain (Giles, 2009), seems to have evolved to make social bonding as important as a warm and safe physical environment. In the face of stronger and faster predators, humans would have to evolve such innate tendencies or be eaten individually after they were forced out of the safety of trees by a weather-induced decline in jungle vegetation. In treeless environments, humans had to develop characteristics that motivated them to serve the group of humans, as those that didn't were exterminated by the stronger and faster predators.

Group consciousness could have initially been spurred or at least enhanced by group hypnosis as in group chanting. The mere state of group consciousness is an ecstatic high in and of itself that promotes a unity of purpose to maintain this blissful feeling. Group consciousness is characterized by hypnotic effects that are made more powerful through the empathy of feeling what others feel. People are therefore motivated to do what they perceive others to want them to do at least partially to experience their happiness. Greater psychic abilities in earlier humans (Gooch, 2007) might have also developed from and contributed to enhanced group thinking, especially in times when language was primitive or nonexistent.

In addition, because mating across human clans had to develop to avoid the weaknesses associated with procreation by incest, serving the human group meant serving future ones in other clans who might one day be united by sexual relationships. As result, the original human group consciousness naturally had to include all of humanity.

Human relationships and group consciousness in the prehistoric past were originally so strong that no meaningful individual consciousness existed because everyone's own opinion of themselves was formed by others, (de Quincey, 2005). Such a situation is consistent with a harmoniously integrated society that once was vital to humanity's survival, and may still be. Even in the very individualist society of the U.S. today, people in a group setting such as in a crowd or demonstration often lose their "sense of individuality" and are "submerged in the group" in a process called "de-individuation" (Bernstein and Nash, 2008). In prehistoric human societies, there likely was no differentiation between the good of the group and the good of the individual, as every human in the group was needed, and everyone needed the group.

It has been well demonstrated that any activities done in unison with others promotes loyalty to the group (Robson, 2009a). The "psychological processes that can transform hundreds or thousand of individuals into a unit" may be motivated by almost any identification with a group of people having a shared belief or common purpose (Bond, 2009a).

The innate "shared intentionality" that exists in humans because of the evolutionary advantages it provided to social groups of humans (Wade, 2009) is at least partly spurred by chemicals produced by the brain internally. Natural chemicals in the human brain do reinforce behavior perceived by the individual as good for the group (Begley, 2007b). People's most cherished moments in contemporary times remain characterized by an emotional elevation that can be prompted by a feeling of being part of a group with an important mission and that leads people to be more trusting and open to others (Griggs, 2010a). Even simple activities such as exercise generate more pleasure from neural endorphins when people do them in a group than when they are alone (*NewScientist*, 2009zzzn).

A natural inclination of people to help each other starts in infancy (Wade, 2009). Buddhists assert that compassion represents the only path to happiness and that acting with the intent of helping others facilitates all activities because it enables their energy to flow through to oneself (Bogle, 2010). If everyone was experiencing what others felt, people would naturally act as one to maximum everyone else's happiness, so that they could all co-experience happiness.

Humans' recognition of and attraction to observed "kindness" to others begin even when they are less than a year old, and babies early on also exhibit an aversion to the "bad guy" who harms others (Bloom, 2010). Empathy seems to be required for this early learning of "good" and "evil" that otherwise might only occur with respect to the direct treatment of an individual by others. Although the human bonding experience is at least partially caused by increased amounts of the chemical oxytocin in the brain (Griggs, 2010a), empathy may be primary driving force in group consciousness and mutual assistance.

Empathy derives from the "mirror" portion of the brain that also exists in other primates to promote learning by copying, but humans evolved to be able to use that portion of their brain to do much more, including to predict the intentions of others (Slack, 2007). Feeling what others are experiencing facilitates predicting their actions and was therefore naturally chosen by evolution as human beings began to utilize their unique process of forecasting. Such experiences also promote acting for the benefit of others in order to avoid empathetically feeling their negative experiences, and so empathy was also crucial in promoting the human tendencies for long-term altruistic cooperation.

Empathy results in the premotor cortex mirroring the emotions of others, although the actual facial expressions exhibiting the emotions typically aren't actually exhibited today (Douglas and Robson, 2010). It appears that individuals can consciously inhibit both the actual physical mimicking and deep feeling of others' emotions, and they are certainly encouraged to do so by contemporary cultures that encourage selfish individualism through both propaganda and capitalist incentives. However, the innate neural connections in the brain that cause empathy remain, and children not yet acculturated into the philosophy of acting without concern for others exhibit the altruistic effects of this continuing human trait more openly.

For instance, children between 4 and 8 years old today have a tendency to refuse treats if they are receiving more than others (Whitfield, 2009). This innate altruism may be motivated by the negative empathetic feelings associated with injustice to others that seem to outweigh tangible benefits to the individual. The chemical reinforcements in the brain that result from altruistic actions may have evolved from that incentive created by empathy.

It has been documented that people's measures of self-worth, especially early in life, are heavily influenced by the impressions of others, and individual identities typically don't completely form until after the teenage years (Bernstein and Nash, 2008). Some research indicates that human "brains are geared to mimic our peers" (Robson, 2009a). Even personal styles, including for obesity as well as for clothes, are passed through many social connections (Bond, 2009a), and it's not really possible to "fully understand people without understanding the links between them" (Bond, 2009d).

People's moods are so greatly influenced by the opinions of others that some scientists have concluded that individual human beings have no significant amount of independence (Bond, 2009a). For instance, one person smiling is so contagious that it has the effect of significantly increasing the happiness of others (Chemali, Chahine, and Naasad, 2008). Laughter, which helps bond humans to each other, is especially contagious and can even lead to mass hysteria when people are put in the same stressful situation (Douglas and Robson, 2010). The latter fact may reflect an especially strong tendency for group consciousness in the face of a stressful environment that unites people together to provide them with some security from the situation through the very bonding which reduces the stress of the situation.

The contagion effect that spreads beyond those in proximity with a person goes beyond happy feelings or stressful situations, however. For example, it has been discovered that the emotional state of people unknown to a person strongly affects that person's own feelings through its impact on mutual acquaintances that passes through (Bond, 2009a)

All human beings even have an innate desire for oneness with humanity (Calleman, 2009), The Mayan greeting of "I am another yourself" (Stray, 2005) may represent a good example of the ancient group feeling of oneness with the human community. The prehistoric "shared consciousness" may also have resembled modern hippie culture (Miller, 2001) that promoted peaceful egalitarian cooperation through harmonizes their brains on the same wavelength. Chemicals likely evolved naturally to make such feelings of unity pleasant, thereby promoting the cooperation that was necessary for the relatively weak and slow human species to survive.

The fact that the same chemical reward centers of the brain, the nucleus accumbens and caudate nucleus, are activated upon giving to charity as when seeing pictures of loved ones (Loftus, 2009) implies loving and helping others is innate to human nature. Humans actually have a tendency to want to assist others and act for the common good more when they are happy (Klein, 2003). Thus, the natural tendency of humans to experience reinforcing pleasant chemicals in the brain upon helping others or acting for the common good (Begley, 2007b) has a magnifying effect that grows the tendency ever stronger. "Being concerned with another person's welfare at a cost to yourself" actually "can show measurable health benefits" (Richter, 2009).

There are indeed a whole list of human psychological reasons why it is natural to conform to the norms of the group, including the natural inclinations of people to want to be liked, to reciprocate, to do what is perceived as "right", to enhance one's feeling of self-worth that is affected by the opinions of others, to copy others who are doing likewise, and to avoid social punishments (Bernstein and Nash, 2008). It is not a question of acting for the good of the group or for the good of oneself: good for oneself is good for the group and vice versus, including for one's own preservation, optimal functioning, and pleasure.

As Peat (2005) indicates, one does not give up freedom by joining a group but instead gains freedom through being part of the group which actually forms oneself. The benefits in capacity to accomplish more things as a member of a cooperating group represent only one advantage of being a member of a group, however. There are also large gains in feelings of self-worth, self-contentment, and happiness from being a member of a group. This phenomenon is understandable within the context of human empathy alone. Everything and all are one, and this knowledge can provide a wonderful feeling.

The "identity shift" that causes the people in a group to act in the best interests of all those with the common goal upon which they are focused results in cooperative behavior that is much more efficient than masses of people acting for individualistic purposes (Bond, 2009c). Within this context, it's also interesting that people have been found to receive a negative reinforcement in the form of an "error signal" in the brain upon expressing an opinion which is contrary to the view held by a majority of others that creates pressure to conform (Robson, 2009a). Given that the majority opinion tends to be the best one (Surowiecki, 2004), this human trait may have evolved over time to promote both optimal and harmonious group decision making.

Although many people today think that human beings are intrinsically selfish, such false beliefs stem from contemporary hierarchical societies like capitalism that promote them in order to divide and fragment the people into being "rationally" obedient slaves to an elite. In spite of cultural encouragement for "rationally" acting strictly in one's own interest, people continue to behave in ways that seem irrational from the perspective of personally selfish rationality but that are perfectly logical from the point of view of the group as a whole (Frith, 2008).

The "universal theme" in literary fiction of the triumph of people working for the collective good over selfish private interests is so widespread worldwide that it is thought to exist in humans' "evolutionary roots" (Shetty, 2009). The deeply innate nature of group consciousness seems to be so entrenched that capitalism and hierarchies haven't been able to disable it in people's imagination, emotions, and ideals. Positive social moods remain identified with words like "unifying," "liberating," "joining", and "tolerant", whereas negative moods are related to "fragmentation," "separation," "restricting," and 'bigoted" (Casti, 2010).

A. Cultures of Group Consciousness

Evolution would have played an important part in the communal nature of human beings, as it would select for group cohesion. Such cohesion would result in stronger societies than an alternative possibility of each individual exclusively adoring herself or himself, and perhaps one significant other, without any cohesive feelings held in common with the group. Only those who have an innate tendency to empathetically promote the interests of other people as well as oneself survived, as groups of such humans would thrive in comparison to individualistic ones where people have no compulsions about taking actions in their own best interests to the detriment of the group.

The synchronization of women's menstrual cycles (Shlain, 2003) may have both contributed to the communality of group consciousness and evolved from it. In particular, an original group consciousness may have led those females who had menstrual cycles outside the norm to be outcast socially as being alien to the group. In addition such "abnormal" females might have even been formally expelled from the group for the danger their bleeding might have brought to it by attracting predators at more frequent times that might even have been more difficult to predict. Once the menstrual synchronization existed in a group of humans, forecasting the cycle with the lunar phases would have been facilitated, thereby enabling, within the context of group consciousness, the all-important human development of self-forecasting that likely started for the group of humans as a whole and naturally grew into group consciousness.

In this situation, communal sharing of everything was not only natural but likely the only conceivable option in prehistoric times. Although there are reduced incentives for work in societies that encourage exploitation when there is any enforced material equality, such problems likely didn't exist in the substantial portion of human prehistory that was characterized by a group consciousness. Besides losing the innate chemical pleasures in the brain that resulted from cooperation, failing to work for the common good meant a loss of the blissful unity in purpose that existed. The affection of other humans through cooperation was also once extremely motivating because of social stigmas against selfishness that compare starkly to the inhumane incentives and encouragement for selfishness that exists in capitalism. An innate greater attraction of mates to more productive contributors to a group, as well as empathy, social pressure, and human spite against those not maximizing their efforts for the group as a whole, would have provided further incentives for activities beneficial to both the group and the individual than material ones.

Substantial evidence of the greater productivity of more modern communal societies with fewer disincentives for selfishness is provided by Murphy (2000). The fact that friendship in the anti-capitalist Soviet Union was normally an end in and of itself without any regard for relative benefits provided by the friends as long as they trusted each other (Kharkhordin, 1999) may be reflective of the innate nature of human relationships that can actually exist in more collectivist societies. In particular, true human friendships are naturally valued for themselves and not for any personal benefits that result, and such relationships would have characterized everyone in cultures of group consciousness.

Despite the innate group consciousness of people, humans evolved to have different opinions and to persist with them in the face of other views, with such diverse opinions in group-oriented societies being held for the good of the group (Peat, 2005). For instance, even in artificially hierarchal, selfish capitalist societies, innovation by cooperating individuals tends to be more successful (Surowiecki, 2004). A diversity of opinions has been found to contribute to more successful decision-making (Oskin, 2009).

The fact that individuals do seem to have a natural tendency to want to differentiate themselves in spite of encouragements to act for the benefit of the group (Kharkhordin, 1999) implies that human individuality is innate even within the context of a group consciousness. Human overconfidence contributes to an ability of individuals to push minority viewpoints (*NewScientist*, 2007i). In addition, since collectivist cultures normally tend to attribute successes and failures of others to external events or to the group as a whole as opposed to casting blame or labeling heroes (Bernstein and Nash, 2008), there are fewer risks associated with forming and expressing individual opinions in a culture of group consciousness.

The aggregation of diverse individual opinions, as opposed to a group-influenced consensus, does tend to result in better decisions (Surowiecki, 2004). The democratic decision-making by monkeys may be illustrative of human group unity of action and purpose. In particular, any individual monkey of any age, sex, or status can lead a group of monkeys, and the existence of different leaders with different actions developing at any moment results in everyone, including competing leaders, following the leader who attracts the most followers (*NewScientist*, 2008zzc). The group memory and collective decision making of bees, whereby individual bees indicate their own opinions to others but are ignored by the group if there are more contrary opinions in a manner similar to the way neurons in the brain actually function (Hooper, 2008), may also represent a primitive form of the original human group consciousness.

A conscious belief in the usefulness of such a democratic decision-making process can enable both the diversity and cohesion that humans likely needed to survive their hostile environment millions of years ago. In particular, evolution seems to have programmed humans to serve the group but not in a robotic way. People are happier if they have a democratic say in decision-making processes, and greater happiness promotes good health and productivity (Klein, 2003).

Since the pleasure reward in an individual's own brain is dependent only on the person's own perception of trying to help the group, creativity and diversity in helping the group is enabled at the same time. Satisfaction of personal needs is still motivated with brain chemicals that reward behavior sufficient to keep the individual going (Esch and Stefano, 2005a). Subject to satisfying minimums for self-preservation and perpetuation via offspring, the brain pleasure chemicals for service to the group can motivate less purely self-serving behavior while also maximizing individuality.

Orgasms have even been found to be more frequent in women who are in better touch with both their own emotions and those of others (60 Seconds, 2009), thereby providing maximum reinforcement for a group consciousness in those who are responsible for bearing and nurturing offspring. With group consciousness and empathy, sexual competition and frictions might have been minimized. Sexual intercourse enhances the health of individuals in a group (Farnham, 2003), and sexual acts in prehistory would have likely been regarded as beneficial to all by improving the group members' health, numbers, and diverse capacities. It would therefore have been natural for everyone to want everybody else to be mated, and a human tendency for such group conscious wishes would have naturally evolved. People's wishes and efforts might have been especially strong for those doing the most for the group, thereby reinforcing individuals engaging in activities beneficial to the group with the most extreme reward. A group desire for those having the "best" genes, as indicated by their contributions to the group, to bear more offspring, as well as a widespread wish among the opposite sex to mate with such people, would have naturally developed into a special interest in the love and sex life of such "celebrities". A desire to mate with such "celebrities" would have also easily evolved since it both rewarded behavior benefiting others and also increased the likelihood of offspring who would be able to do likewise.

Most humans remain concerned about the welfare of others in a grouping of people and even continue to show an interest in the mating of others as indicated by extensive match-making assistance and gossip, especially about celebrities (Miller, 2001). Such a fascination with others' sexual lives is consistent with a hypothesis that the natural human cultures of group consciousness cared for all people's needs, including for sex. Sex might actually have been the primary motivator for motivating maximum individual efforts to contribute to the group, as individuals acting like selfish apes or like leeching tics without concern for the group would have been faced with an unpleasant life without love and sex with others. Gossip and ostracism remain effective tools to motivate fairness in treating others (Whitfield, 2009), and so the tendency to be concerned with others clearly continues to exist despite the incessant propaganda and lies in modern societies about humans being strictly selfish beings.

The disharmony between the members of the opposite sex and generational gaps would likely have been eliminated in cultures of group consciousness through the encouragement of young-old matings. As explained in Chapter 17, partnerships between unrelated members of the opposite age and sex are innately the most harmonious and were very likely to have been the norm in the distant past. As explained in that chapter, the battle between the sexes that is the cause of so much of the misery and violence in the world today can thereby be replaced with harmonious ecstasy. While the fact that people in contemporary societies tend to be more affected by the opinions and mood swings of those of the same sex (Bond, 2009a), such unnatural identification with the same sex no doubt stems from the lack of understanding and empathy between the sexes due to the paucity of harmony in mating relationships today that cause emotional and sexual frustration which leads to intolerance for the feelings of the opposite sex as soon as people reach puberty.

Matings between unrelated young-old pairs are especially useful for teenagers after they reach puberty in order to avoid the artificial dictatorship required in today's "modern" family system based on artificial slavery to biological parents with whom there can be no harmonizing sex. As indicated by natural reactions against authority that begin in childhood, no human wants to be controlled (*NewScientist*, 2007k), but the "modern" family system appears designed to force learning to live in an artificially hierarchical dictatorship.

All societal cooperation had to be voluntary originally. As previously explained, voluntary cooperation is encouraged by positive chemical reactions in the brain that result from trying to help others (Shermer, 2004) even without any social or other rewards for helping the group. Without patriarchal pressure and propaganda on selfishness to create mental illness, group interconnection would have originally been seen as normal from children's earliest years. The relationship between a parent and child continues in contemporary society to be dependent on the emotional right side of the brain that relies on empathetic processes (Gruzelier, 2006). Thus, empathy, love, and group consciousness seem to remain buried in the subconscious area of humans' brains despite a conscious belief in selfishness into which people are brainwashed to believe.

The ancient practice of having all the elder people in a clan labeled as "fathers" and "mothers" (Jackson, 1972) seems to be consistent with a communal upbringing of children. The fact that oxytocin is produced in women when their breasts are stimulated by anyone and the fact that offspring become attached to any female breast-feeding them (Klein, 2003) would have helped all women and children to bond with each other within such a group care-taking system for the young. Group childcare continues to be observable today in other intelligent mammals such as whales (*NewScientist,* 2009zzf).

In the distant past, children were raised together by each group or clan as a whole as opposed to separated into individual family units (Sjoo and Mor, 1987). Segmentation into families fragments the general feeling of empathy that would exist with communal group upbringing of children (Else, 2006). Since empathy has been found to reduce violence, the large increase in violent behavior that continues under the unnatural family system (Gosline, 2008) may actually be traced to "family values".

The division of society into individual families has definitely contributed to the replacement of some of the group consciousness with individual consciousness. In spite of that fact, cooperative learning still seems to continue to work best for children today (Bernstein and Nash, 2008), and children continue to prefer to imitate those who follow social norms (Douglas, 2010).

Without fragmentation into families, group consciousness would have seemed more normal. In addition, communal raising of children in an environment of equality and sharing would have likely led to children's actions being judged rather than separate worth or judgment placed separately on each individual. Such a child-rearing policy is extremely beneficial in personality development, since praise of an individual leads to conceit (Reich, 1936) and since condemnation of a person leads to a lack of self-confidence that is even more counterproductive. Although negativity may be more clearly harmful for children (LeDoux, 2002), conceit actually inhibits future action or improvement in individuals.

The Native American custom of rarely punishing children (Underhill, 1965) would seem natural and healthy within this context of group child rearing and group consciousness. The fact that it has been empirically shown parents spending more time with their children don't even help their offspring's performance in school or work even within a capitalist society like Germany (*Der Spiegel*, 2008a) provides evidence of the natural efficiency of communal raising of children, probably because of the communal characteristics that are encoded in humans' genes.

In prehistoric times, education of children could have been largely the responsibility of older people. The reduced physical strength of the elderly would have been less important in those tasks, while their mental capacities might have been ideal. Even sick, older people with limited physical abilities could help the young with the knowledge they learned from a greater wealth of experiences.

B. The Role of Spite as an Enforcer for Group Consciousness

Even if some, such as the very sick and elderly, were perceived to contribute less knowledge and other benefits to a society than they took in nourishment and care, the risk of human spite would have inhibited the creation of human societies that discarded such people. In particular, as soon as an individual feared the onset of disease or old age that might result in his or her abandonment despite all the individual had earlier done for the group, human spite would have been the reaction that would have led to counterproductive conflicts and manipulation. Empathy, respect, and a knowledge that all would eventually be faced with such a situation would have further ensured care for the old and incapable. Even far back in human prehistory, care has been provided for old and disabled people even when those "weak" individuals may not have had much potential for contributing much to those societies (*NewScientist,* 2010aao), but normal human emotions would have also caused the weakest to think of ways for them to help the group as best as they can.

Even in the fragmented society of the U.S. that encourages individualism so much, behavior tends to be more driven by the expectations of others than by a person's own feelings and logical reasoning (Bernstein and Nash, 2008). A continuous expectation that everyone would contribute to the best of their abilities to the greater good probably existed in the prehistoric past, and this expectation would have led to a self-fulfilling prophecy. Since spite would have been the natural reaction to anyone trying to take advantage of the communal system, negative reinforcement also existed to promote the group unity of purpose.

While social systems providing rewards for cooperation among human beings have been found to be more effective than ones penalizing free-loading, the effect is greater when it is recognized that the decision on using rewards or penalties is made by a human being (*NewScientist,* 2009zzzk). Such tendencies make sense within the context of a natural subconscious tendency to fear the spite of other people.

Evolution thus seems to have selected human traits in a process of "social selection" that would enable them to "reach a mutually satisfactory solution that is better than if each merely tried to satisfy its own selfish interest" (Holmes, 2009b). The tendency of selfish individuals to try to take advantage of others' operating more for the good of the group (Holmes, 2009a) is inhibited not only by this rational logic but also by the evolved traits of human brains that include human spite. In particular, spite promotes a fear of future revenge from exploited individuals that may result in feelings of guilt in individuals for selfish deeds. The very discovery of selfishness may create the negative feeling of shame in a guilty individual who has violated the trust of others that is likely to cause a spiteful withdrawal of love at the very least.

The deception, free-loading, and exploitation which are endemic to today's societies built on artificial foundations of selfishness would naturally have been inhibited in ancient times by the characteristics of innate human behavior that include spite and the adverse consequences thereof. The more people were able to forecast future possible scenarios, the more they would learn to fear others' spite for "wrongs" done to them, at least subconsciously, and a feeling of guilt certainly could have developed out of this context. In particular, exploitative actions would lead to such fear, and self-blame would have caused the negative experience of guilt feelings to evolve to inhibit selfish actions that harmed others.

Although there can be a tendency for selfish individuals to take advantage of any system based on cooperation (Holmes, 2009a) and to use deception if cheaters are penalized, the human feeling of guilt at cheating others provides intrinsic inhibitions against doing so. That feeling may be innate to human nature, having developed out of fear of possible future revenge from spiteful victims of their cheating.

The negative experience of guilt not only inhibits the individual from cheating, but it can also lead to external signals that can unravel the deception. For instance, the feeling of shame that humans uniquely feel and communicate to others via blushing, such as when a person is self-conscious about violating some social morals like honesty, may be related to an evolved human tendency to allow others to more easily recognize attempts to deceptively manipulate (*NewScientist,* 2009l). In particular, shame and blushing tend to occur upon self-conscious recognition of failing to meet some group expectations like honest attempts to act for the good of the group. It may have originally served as a form of primitive communication before words were created to express embarrassment and may even represent a subconscious method of apologizing for not trying hard enough to serve the group as a whole.

The feeling of shame may have evolved to help detect pathological, selfish cheaters who might take advantage of the cooperation of others without being caught and without therefore incurring spiteful revenge. Human groups that have cheaters without this trait may have been subject to excessive friction and suspicion that would have resulted in groups of human beings with the trait being the only ones that thrived and survived. Human beings are much more attuned to discovering problems associated with cheating than due to random errors (*NewScientist*, 2010t), and it is likely that this trait also evolved for the very purpose of maintaining the trust and harmony associated with group consciousness.

Empathy may also contribute to discouraging exploitation as the adverse feelings of others being exploited are empathetically experienced. For example, listening to depressed voices makes others feel depressed (Hutson, 2008), and that experience would seem to provide a disincentive to causing such feelings that might rise from exploitation. In this context, sadness and even suicide seem to represent a form of spite inflicted on empathetic loving others. In the prehistoric past, anyone showing signs of depression would have naturally been helped to avoid negative empathetic feelings (Gigerenzer, 2007).

Thus, pleasure feelings at cooperation, empathy, spite, guilt, and shame all once served to promote the development of group consciousness, which co-evolved with the natural honesty of humans that continues today, as long as people perceive they aren't being manipulated or tricked. The honest communal benevolence of human beings requires no positive enhancement by artificial morals imposed on them as obligations, as empathy, spite, and other innate human traits alone are sufficient motivators, at least in a humane society.

Human spite depends at least partially on people's sense of justice that is influenced by the social mores and system of justice in place. As a result, revenge provides little danger to selfish exploiters in today's capitalist societies that actually encourage taking advantage of others as long as it is legally allowed and disguised. Capitalism and other economic systems promoting selfishness are in conflict with the cultures in which human nature evolved, and human spite would have been a bigger deterrent to selfishness in prehistoric times.

It has been established that pursuit of selfish goals does not make people happy (Klein, 2003). In contrast to the allegations of Richard Dawkins and others that altruism must be taught (Douglas, 2009c), altruism is actually innate to human beings, but selfishness is learned from today's inhumane cultures.

Deception also originates from contemporary cultures, not from an innate human tendency for it. Exploitation and selfishness breed deceit just as cooperation and empathy motivate honesty. Although an ethics of honesty is strongly promoted in today's sick societies, the purpose of such "preaching" is to exact obedience to the inhumane hierarchal systems, which naturally evoke spiteful repayment of exploitation by the hierarchical elite with deceitful attempts to cheat the elite whenever possible. There remains a widespread desire to return to the humanitarian societies in which humans evolved (Lovelock, 2009) and which promote honesty, but that natural inclination is frustrated by the powerfully brainwashing superstructure of the existing economic and political system, which inhibits people from even recognizing their innate nature, much less taking action to change return to such cultures.

C. Group Consciousness and Capitalism

Human brains evolved to get pleasure for doing what is good for others, as long as it isn't taken advantage of. This trait is partially reflected in humans' desire to be liked or admired (Nettle, 2007) that is now distorted by money, for which no one likes to be liked (Pizzigati, 2004). For millions of years, money did not distort relationships because it didn't exist in the communist societies in which humans evolved to live (Baring and Cashford, 1993). Humans during those millions of years likely chose their mates based partially on their contribution to society (Jiayin, 1995). However, today success is measured by the amount of money a person has to a large extent, albeit in a distorted way, with wealth being a more important criterion of success when there are greater economic inequalities and insecurity (Pizzigati, 2004).

Money has been shown to be so important to people today that they value the same item more when it is priced higher (*NewScientist*, 2008zm). Money has therefore become a sort of fetish that, by the very nature of people having differential amounts of it, creates inequality, hierarchy, and the possibility of exploitation, as well as the drive for such things (Lotta, 1994). Such phenomena are all inimical to human nature. The mere existence of a differential power causes people to "objectify others, viewing them as disposable pawns rather than emotional humans" (*NewScientist,* 2010), thereby making inhumane manipulation as well as exploitation possible.

Within the context of capitalist exploitation, individuals' continuing desire for connectivity is often manipulated for the sole benefit of others, often for money, via deceit and hypnotic persuasion. As a result, the mutual trust needed to be earned for a true friendship sought solely as an end in itself, and not as a means to personal monetary, sexual, or other gain, is difficult to attain in a capitalist society that reveres exploitation (Kharkhordin, 1999). Friendship in capitalism is transformed into a cold bargaining process that effectively values each friend as a shallow monetary amount for business and personal gain, thereby turning human relationships and human beings into commodities.

Nevertheless, the fact that computer games sell much better when they enable social connections and when they don't rip people off with high pricing (*NewScientist,* 2009zzzm) seems to indicate the innate force of human nature. In particular, the popularity of games in capitalism indicate that human beings still prefer cooperation as long as there is no exploitation involved. The fact that there remains widespread positive sentiment for and even idealization of self-sacrifice for the good of communal cooperation (Cartledge, 2006) is also very illustrative of innate human nature.

As previously mentioned, one's own perceptions and evaluation of oneself and the external world continue to be formed and influenced by the perceptions and evaluation of others even in the very individualistic societies like the USA today (Bernstein and Nash, 2008). However, the effect of the contemporary opinions of others is less with older people who have already been molded by other people's prior opinions, including those based on individualistic selfish philosophies. The enormous increase in the number and variety of connections with others for older people usually leads to an individual consciousness that can be less influenced by the contemporary opinions of others.

Many indoctrinated by the selfishness of today's patriarchal, capitalist society may have difficulty understanding the concept of group consciousness that is extremely pleasurable and spiritually engulfing. The ecstasy of spiritually uniting the ego with another person, with a group of other individuals, or with the entire cosmos can never be understood by those who believe purely in their own self-interest (Sjoo and Mor, 1987). However, the religious and out-of-body experiences that occur when DMT is released into the brain may still be triggered by a perceived conscious union with a benevolent external being or beings (Strassman, 2001).

Such ecstatic feelings are inhibited by selfishness and a conscious decision to maintain one's chains of separateness. Just for instance, if one is trained by capitalism to expect an external reward for "selfless" behavior or service to others, any pleasure created in the brain will be reduced by the fact that the "selfless" act was not truly selfless. The result is less and shorter-lasting chemical pleasure, if any at all, from it. In addition, such behavior undertaken to obtain a predicted tangible reward will lead to unhappiness through disappointment, frustration, and anger if the external reward doesn't materialize. Moreover, any external reinforcement capitalistically conditioned to be expected also requires all payments of such cosmetically "selfless" behavior to be at the cheapest possible price. As a result, even attainment of an external reward for a "selfless" act may therefore lead to unhappy feelings of regret for having paid too much for it, or for not having received enough benefits. Within the capitalist framework, helping others or the group represents an emotionless trade that has nothing to do with altruism.

Helping others for the purpose of receiving something in return at the same time, or in the future, is a form of capitalist reciprocity that doesn't create the same health benefits as true altruism from the heart (Richter, 2009) that comes from innate human nature. The reciprocity of true human altruism has nothing to do with trades of equal value. Altruism merely requires mutuality based on unified purpose of action as a group that also seeks to help the individuals in the group. In the past, such group consciousness probably included greater social honoring or reverence for those who contributed more to the group. However, in today's society that glorifies selfish exploitation of others, people often hold selfless individuals in contempt if they aren't seeking rewards. Natural human altruism is thereby motivated to be transformed into artificial shows of altruism that are a perversion of human nature.

Even in a capitalist society, people still want the group consciousness that comes from communally serving a group's goals, and they continue to seek the thrill of being loved by a group in the form of admiration or popularity. People generate pleasure from both loving others and being loved, and more pleasure is derived if the love is mutual, thereby still providing some pleasure for mutual love. However, in today's society that strongly promotes selfish individuality, much of their natural interconnection is no longer a pleasant end in itself but is instead used by people to help them attain their personal goals. The interconnectivity is therefore less rewarding as well as even negatively perceived, due to exploitation and resultant spite.

Despite the fragmentation of group consciousness, attempts to serve the common good are reinforced in a particularly strong way, even within the capitalist framework. In particular, a person's reputation for seeming to help others altruistically can make one more attractive to the opposite sex.[4] While the latter phenomenon might represent a sort of handicap effect (Zahavi and Zahavi, 1997), evolution has clearly chosen humans with greater chemical pleasures for serving the mate, offspring, and the group, and mate selection has and probably continues to provide important reinforcement for such altruistic behavior. In particular, groups of humans preferring to mate with individuals who seek to serve the group, as well as the mate and offspring, would be even stronger and more cohesive than other groups. The evolution of this mate selection criterion was therefore natural, especially among women who represent the primary controlling gate in the reproductive process.

Mate selection for altruism has special benefits given that mating only requires attracting one other individual intensely, thereby making it unnecessary for more than one person in a whole group of people to respect the individual's actions. This situation continues to permit and even encourage diversity and creativity.

Diversity in pursuit of communal goals is built into humans' most basic instincts. Across sex, there remains a tendency to be attracted by partners who are different due to the evolutionary advantages of diversity in the gene pool especially with respect to disease immunity (*NewScientist*, 2007b). While it is actually possible for females to create identical offspring without a male, as with a 2007 shark (*NewScientist*, 2007o) that apparently was able to fuse the polar body by-product of egg production with an egg cell (Brooks, 2010c), and while cloning humans is becoming ever more feasible, reducing diversity through the creation of clones or other means is unlikely to lead to a very adaptive species. It also isn't innately satisfying.

According to Laszlo (2006), individualism is good to the extent that it serves the whole, but excessive fragmentation, selfishness, and thought-wave incoherence are evil. Since such "evils", which have permeated humans only in the last few millennia, bring only unhappy emptiness, it is clear that they remain unnatural perversions of human nature, which continues to push people for group consciousness. Although an inner communist desire to benefit the group in each individual's own creative way can bring personal rewards, it remains its own reward by itself. The brain chemical pleasures are experienced when a person thinks of ways to help the group and tries to implement ideas are felt whether others agree with the methods or not. Thus, helping, cooperating, and loving, including in the romantic love form, seem to be some of the evolved purposes of human life that the exploitative, hierarchical societies haven't been able to eliminate from human nature.

An illustration of some continuing human tendency for combining individual freedom with a form of a group consciousness even within capitalist society can possibly be seen in trends of consumption styles. For instance, despite a multitude of consumer goods, most consumers choose consumption items that are very similar to the consumption items selected by others, resulting in the development of systematic fashion that are rather fickle and not wholly predictable (Webb, 2008). The latter fact may represent the robustness that individualistic cooperation provides to the human races, allowing them to change quickly and without easy manipulation.

Other evidence of continuing human tendencies for group-conscious behavior even in exploitative societies also exists. For instance, group pressure remains more powerful than rational arguments even in today's individualist society (*NewScientist,* 2007i). The fact that today's term "common sense" derives from the concept of knowledge held in common (Shlain, 1998) provides some indication of group knowledge. However, "common sense" has been perverted by today's hierarchical cultures into "group" knowledge of how to exploit others without being exploited oneself. Today's "common sense" is so tainted by counterproductive lies and brainwashing that it encourages selfishness as opposed to natural human collectivity.

The socialist Soviet Union appeared to have made strides back toward greater group consciousness, with experiments even indicating that members of a Soviet collective wouldn't strive for rewards for themselves if by doing so others in the collective were penalized, as striving for group goals was highly motivating (Kharkhordin, 1999). People in collectivist societies tend to worry more about reputation than individual self-worth and actually engage in less social loafing or free-loading than do those in more individualistic societies like the U.S. (Bernstein and Nash, 2008).

Capitalism and the pursuit of self-interest actually lead to a superficial conformity that suppresses divergent views (Huffman, 2004). In particular, cultures of selfishness motivate concentration on exploitation and preventing being exploited, and capitalist "common sense" pushes people into similar methods of exploitation and ways of avoiding it. Regimented service to one's self-interest is directed with a strong reward and penalty system toward slavery to an elite with a particular set of expectations on behavior. Deviations from the norms of slavery to an elite while exploiting others are discouraged with social contempt and other penalties, thus creating robotic commodities out of individual human beings and replacing individuals creatively cooperating with each other for the good of all with identical servile behavior.

Thus, capitalist societies inhibit productive innovation as well as true cooperation (Lotta, 1994). Such exploitative societies tend to result in more violent solutions to frustration and other problems (Bernstein and Nash, 2008), as opposed to creative cooperative problem-solving. Capitalist societies based on the ideology of selfishness are also naturally characterized by a hierarchy that involves obedience to the rules and the expectations of those at the top that is not conducive to innovation or productivity.

In spite of the perverse effects on human behavior motivated by capitalist cultures, humans still prefer cooperation over selfishness when such is possible without exploitation. For instance, psychological tests and games indicate that people today still tend to prefer to cooperate as long as others also cooperate, with such cooperative behavior still leading to much more productive and successful outcomes (Surowiecki, 2004). However, individuals also experience spite, and they act spitefully if they believe other humans or the group as a whole are treating them unjustly or unequally (Buchanan, 2008b). Societies like capitalism that encourage selfish behavior are natural promoters of the human feeling of spite, which is prompted by observation of behavior motivated by selfishness as opposed to natural human tendencies for altruism for the good of humanity.

Human group consciousness requires mutual human love, which consists of a natural balance of selflessness and selfishness that is motivated by empathy to maximize the collective utility of all those in a group with that purpose. Excessive selfishness or selflessness is characterized by repression of one or the other and therefore represents a perversion of human nature that inhibits or prevents real human love.

Excessive selfishness represses the natural human urge for selflessness. It thereby constrains an important aspect of humanity: the desire for a selfless unity of purpose with others. In the face of frustration of needs and desires like sex, the repression of this and other human desires can lead to violence against other humans as opposed to cooperation with other humans to attain a mutually desired end.

On the other hand, many people may be affected by such a strong desire to return to the bliss of communal love that they overemphasize only the selfless aspect of love. This selflessness may be a conscious or subconscious attempt to attain mutual love that would have naturally resulted in the communal societies in the prehistoric past but is taken advantage of in societies that brainwash most people into believing they are purely selfish beings. Some excessively selfless people develop a philosophy that love should be totally selfless and feel guilty at the pleasure that love provides. Taken to the extreme, selflessness can even result in so much pleasure that it results in a desire for more selflessness and hence more pleasure, as in an addictive drug. To attain more pleasure requires more selflessness and attempts to avoid any pleasure, thereby sometimes even leading to masochism. This vicious cycle can be stopped by a person deciding to adopt a healthy balance between selfishness and selfishness, with the balance determined by establishing a goal of contributing to the continuing existence of humanity as a whole that requires avoiding selfless actions if they are perceived to lead to exploitation without any benefit to oneself or the entire group of human beings.

Natural human love was never totally selfless but was instead composed of a unique combination of selflessness and selfishness, with the selflessness only partially motivated by the pleasure it created. As the ancient Taoist philosophy of a unification of opposites indicates, it is important to also please oneself (Chia and Wei, 1994).

Group consciousness results from mutual human love. It is not a trade of any sort and not even an offer to give love in order to receive love. It represents more of a unity of purpose in a humanly balanced combination of selflessness and selfishness that is irrational from a purely selfish point of view but evolved in humanity to unify human actions for the mutual welfare of all. It thereby enabled humans to thrive in an environment that would have overwhelmed them if they had acted from pure selfishness or pure selflessness.

C. The Individual Within Groups

With group consciousness, there is no differentiation between serving oneself, others, and the group. It is service in one's own way, not slavery to others' demands, that promotes the greatest creativity and the most powerful positive feelings, which include pride in both oneself and the human groups with which one identifies. This human characteristic results in the most productive service for all. It also has the advantage of being devoid of pressure and disappointment for the individual because there are no purely personal goals. Instead, there is only the content feeling of self-fulfillment and happiness at trying to contribute to humanity in one's own creative way.

Each individual's own personal needs and pleasures are satisfied to the maximum extent possible in a natural human society based on group consciousness. In particular, attainment of personal goals are recognized to be good for the group by enabling and motivating the individual to contribute to the group, thereby maximizing one component of group happiness, whose utility function natural human cultures once sought to optimize.

Although human cognitive abilities seem to have difficulty in evaluating relationships in groups of over 150 people, larger groupings of relationships can be facilitated with technology like facebook and the internet in general (Fisher, 2010). The typical group and group consciousness in the prehistoric past that probably was long restricted to less than 200 can be expanded to much larger groupings, as is indicated by nationalist and general humanistic feelings. This capacity for commonness of purpose across larger groups probably stems from prehistoric matings across clans that strongly motivated cooperation with all human beings.

Helping others and the group is not at all an obligation or slavery to the will of the group or others. It is instead merely a natural, internally rewarding process. Having been cared for as helpless children, and having lived for millions of years with the group taking care of the weak or downtrodden, humans have also developed an expectation for others and the group to care for their own unfulfilled needs. As a result, an innate sense of justice and expectation of mutual care has developed in humans.

The unique human trait of spite (Barras, 2007a) has also evolved in humanity that can result in fighting against others, the group, and the social system upon perception of injustices against oneself. Given the great desire of people to penalize free-loaders in a system or society (Frith, 2007), that feeling of spite appears to serve an important purpose in inhibiting both individual injustices and free-loading in a cooperative system that relies on trust (Surowiecki, 2004). Spite seems to be extremely well suited for purposes of discouraging manipulation and exploitation, which destroy the socially cohesive trust that makes groups of individual humans so powerful.

Feelings of injustice to oneself can also lead to symptoms of what is often called paranoia, insofar as people often believe they are being persecuted when they perceive unfairness in actions by others. In the selfish individualistic societies of today that encourage exploitation, a feeling of paranoia is often quite justified since injustice is the foundation of exploitation.

In any event, paranoia often leads to feelings of spite, whether justified or not. Spite may be an especially useful trait in today's society that encourages purely selfish behavior to the detriment of others and humanity that motivates the vindictive feeling. For instance, the cruel patriarchal system continues to create negative feelings of frustration and disappointment in people, especially by making satisfaction of basic personal needs and wants, including sex, difficult. Recognition of this fact helps permit people to replace feelings of sadness and low self-worth with justified spiteful hatred of the inhumane system. In past human societies, others and the group of humans as a whole probably assisted anyone who felt hurt or disappointed (Nettle, 2008), but sulking or showing signs of depression cannot be expected to result in truly sympathetic caring in today's cruel system.

Uncontrolled spite against an unjust system and powerful rulers, however, may be suicidal. People may therefore optimally forego immediate overt action against the unjust, placing instead revenge on the back burner until it appears to be less futile. In the inhumane societies of today, spite against individuals is optimally redirected into behavior intended to change the cultures that causes the injustices. Starting or joining dissident groups can be useful here, not only to increase the chance of successful and less dangerous actions against the inhumane cruel manipulators but also to provide a legal outlet for a person's spite. Such groups also can provide some of the psychological and other benefits of the communal systems that were so natural in human history.

Spite not only has an effect on individuals in a group but also on society itself. For example, the elimination of a communal Nature-worshipping matriarchy that occurred thousands of years ago (Stone, 1976) and that happened shortly after the occurrence of horrible natural disasters like worldwide flooding and earthquakes, as explained in Part I, could have been caused, at least partially, by spite. In particular, it would have been very human to feel spiteful against the matriarchal system that did not care for the people who had offered their service and worship of the very Nature that inflicted so much harm on them. Males in the matriarchal society might have been especially prone to feelings of spite if they felt persecuted under rule by female priestesses. The worldwide matriarchal civilization of prehistoric times was in fact eventually replaced a few thousand years ago with barbaric patriarchal cultures that imposed a hierarchal society to the benefit of the male conquerors (Stone, 1976). One root cause of the imposition of a cruel domination by an "elite" of ruling males may have been not only spite against matriarchal rule and its Nature deities but also spite against individual women for any past frustration of men's sexual drives under matriarchy that was generalized for application to all women.

Spite could also help explain the differential resistance to patriarchal rule in Europe and Asia. For instance, widespread resistance continued against patriarchal rule in Europe and the Middle East after conquest by barbarian invaders (Eisler, 1987). This situation contrasts with a form of patriarchy that was self-imposed in Asia without foreign conquest after local males were able to partially inhibit the effects of the worldwide natural disasters and that was therefore initially less resisted than the patriarchies imposed on Europe (Jiayin, 1995). While global flooding and catastrophes resulting from the ending of the most recent ice age may have represented a pre-condition for imposition of patriarchal rule in both areas of the world, as explained in Part I, the fact that the patriarchy was imposed by outsiders in Europe and the Middle East may have enhanced the resistance to it. In particular, the peoples of Europe and the Middle East were able to direct a portion of their spite and blame for the human suffering caused by the natural catastrophes against the foreign patriarchal invaders, whose tyranny increased the suffering of the people. In contrast, it was not possible to do so in Asia since the patriarchs had actually mitigated the negative effects of the global flooding.

Hence, Asians were more willing to accept replacement of the Nature-worshipping matriarchal system because of the matriarchs' relative inability to control the very Nature that the matriarchal religion was supposed to control through religious rites that involved the sacrifice of men. Because Asian women may have therefore submitted to male domination more readily, a system of patriarchal gods demanding obedience was never needed in eastern Asia. In particular, except for parts of western Asia like India, which was conquered by patriarchal Europeans, Asia remained free of conquest by Kurgan, Aryan, or other barbarian tribes. Having discredited the matriarchal goddesses by the ability of men to overcome the natural disasters the goddesses were unable to prevent, thereby uncovering the deceit and magic of the matriarchal priestesses, it was possible for people to readily accept a philosophy that had men be superior to both goddesses and the women who had pushed the sick matriarchal religion upon them. On the other hand, the Asian men, who were humble enough to recognize that they couldn't completely control nature, never felt the need to trick people into thinking of themselves into gods, as they were already revered enough for their human accomplishments and capacities, including greater physical strength than women, that they didn't need for people to worship fictitious deities to maintain power.

Interestingly enough, the patriarchal culture of men dominating women didn't even penetrate into the lives of most of the common people of Europe until the Inquisition (Walker, 2000). That christian slaughter murdered millions of women in order to break the resistance and make significant inroads into the sexual relationships of the masses whose spiteful resistance against absolute patriarchal rule continued into medieval times (Walker, 1988). In contrast, as the harshness of the patriarchy increased over time in Asia, as shown by harsher Chinese laws for female crimes like adultery and the cultural spreading of female feet bonding, women actually supported greater repression of themselves (Jiayin, 1995).

Despite the most recent few thousand years being characterized by patriarchal rule which emphasizes self-interest in a non-egalitarian hierarchal society (Stone, 1976) and to which humans with the "traditional moral values" of "authoritarianism, conservatism, and religiousness" are best adapted (McNamara, 2006), human nature remains more suited to a group-interest system that evolved over millions of years, as explained in Part I. Even today, selfless attempts in the individual's own way to serve others are their own pleasures themselves even when the help is not successful because of the pleasant brain chemicals that result. Humans tend to resort to counterproductive and pathological selfishness only when trust in the mutual cooperation of others in a group is destroyed (Surowiecki, 2004). Despite the relatively recent weakening of group consciousness that results from selfishness, the innate human tendency toward group cohesion may reverse that trend.

D. Group Consciousness Within the Context of the Larger Cosmos

The evolutionary phenomenon of encouraging cooperation through pleasure reactions in the brain is promoted by the overall nature of the universe. In particular, the characteristics of the universe's most basic essences, consisting of vibrating particle waves are consistent with communal bonding and human desires. Pieces of matter are merely vibrations which, like even huge celestial bodies themselves, harmoniously reflect patterns in that energy as in musical notes (Peat, 2005). The most basic quasi-particles discovered, such as phonons, excitons, polarons, plasmons, and dressed electrons that are merely excitations, which can act like particles in a collective system that is more powerful when their waves are coherent, cease to exist outside a cooperative system (Peat, 2005).

A natural quantum bonding exists among everything via entanglement (Aczel, 2002). Entanglement essentially causes particles once in contact with each other to affect each other instantaneously despite being so distant from each other that any communication would have to travel at over 10,000 times the speed of light to create the same effect (*NewScientist*, 2008zu). Complete entanglement can continue over extensive space when there is limited interference in the particles (*NewScientist,* 2008v). The human desire for harmony may reflect in macroscopic fractal form the innate characteristics of quantum entanglement that coherently correlates the quantum states of particles that were once in contact with one another.

DNA itself evolved from bonding and attraction among molecules, atoms, protons, and electrons through the attraction of opposite electrical charges that initiated life as bacteria, one form of which later turned into the non-independent cellular mitochondria of animals and plant chloroplasts (Hooper and Khamsi, 2006). At extremely low temperatures, matter becomes more unified and less fragmented (Szego, 2002), and even particles with like-charges such as electrons no longer repel each other at the lowest temperature. Even in an energetic environment, the spin of electrons determines whether they repel each other, and dissimilar spins of electrons can create a frictionless pair that characterizes a superconductivity of electricity at even some reasonably high temperatures in specific metallic environments (Zandonella, 2008). Such phenomena may reflect the natural attraction of positrons and electrons for each other, insofar as similar spins of nearby electrons may indicate a similar pull on the same positrons in the past or future that causes them to repel each other in pursuit of the basic fundamental unity toward different positrons.

It therefore seems natural in a cosmic sense to want to be liked, to cooperate, and to help others. The trait of desiring the opposite sex is an especially natural one, to which primitive life such as viruses and bacteria naturally adapted via development of higher forms of life eventually (West-Eberhard, 2003). As indicated by Peat (2005), it's not a matter of bland, conforming equality but bonding according to individual needs and requirements that is innately natural, and it is also not survival of the fittest but a balanced cooperation with the environment that is coherent with human nature. The most productive cooperation helps mold and actually causes some randomness, which is neither good nor bad but merely creative.

The particular universe in which humans live developed out of a vacuum via quantum fluctuations, as has been proven mathematically by Magueijo (2003), and empirical evidence indicates that all those particles or vibrational waves are interconnected (Aczil, 2002). Experiencing the unity of the diverse universe has widely been considered to be blissful to human beings (Capra, 1999). Thus, greater understanding of the entanglement of all can enhance the overall coherence of human beings with the waves of the universe. Since light travels without time, it may even be possible to go to another world, universe, and time when a person may partake in reality in a wider way (Peat, 2005). Even in the contemporary societies that unnaturally fragment people today into artificial separated machines designed for exploitation by an elite, human beings can always just ride the wildly fascinating waves of quantum physics and life whose foundations involve a coherent unity, including with actual human beings of the future and past.

North (1997) has suggested that humans once had more communal instincts that were more deeply connected with each other and with the universe, and such feelings would have helped enable use of human psychic powers. The much greater free time that people have in more primitive societies that are engaged in simple hunting and gathering activities than exists in modern society (McClenon, 2002) would have helped humans to utilize such psychic powers better in ancient times. However, the development of technology caused people to think in a more compartmentalized fashion that resulted in the filtering out of a massive amount of information that still can be tapped but is not frequently accessed because it is unessential in a technologically advanced society (North, 1997).

Nevertheless, it has been documented that humans still have some hidden senses left that are utilized. For instance, people react to some scents, such as of the opposite sex, that they cannot consciously smell (Etcoff, 1999). These consciously unsmelled odors may even lead to homosexuality when the gene responsible for that activity doesn't function (*NewScientist,* 2007zk). However, it should also be mentioned that homosexuality is encouraged by repression of heterosexual relations (Reich, 1936), and so it is possible that any genetic tendencies toward that type of sex might have been irrelevant when repression of heterosexual relationships weren't artificially inhibited.

The ability of DNA to emit and receive electromagnetic photon psychic waves (Narby, 1999), which appear, among other things, to regulate some intra-body connections via cilia receptors for electromagnetic vibrations (Mercola, 2008), is another example of consciously unperceived stimuli. The human capacity to powerfully act, know, and influence via hypnotic focus (DeLaurence, 1902) also reflects continued "supernatural" human powers. The home sports team advantage, which has been found to be approximately the same across all teams without regard to stadium size, potential noise, or other differences (*NewScientist,* 2008o), may be a reflection of such psychic energy in group form and indicates its power regardless of the size of the group or external disturbance. Psychic effects are even more powerful when normal sensory perceptions are shut down, such as when sleeping (DeLaurence, 1910). There appears to be a continued connection of human consciousnesses through the force of quantum entanglement (Braden, 2007), although that connection is only fragmental.

The excessive fragmentation of Western societies that are built on Christian selfishness is therefore incoherent with the true nature of the universe. Many psychiatrists and psychologists find those who believe in religions like the Native American ones to be living in a fantasy world, but it seems that the Western system is the unreal one (Peat, 2005). For instance, Western psychology has a tendency to errantly fragment the human mind into separate components, such as cognition, emotion, and motivation (LeDoux, 2002), and Western thought tends to ignore the integrative approach of Native Americans to all processes that is more consistent with reality (Peat, 2002).

Western psychiatrists and psychologists just reflect the values of today's psychopathic patriarchal system that is divorced from reality and is inconsistent with human nature and the true quantum environment. Evidence indicates Western psychiatric and psychological assistance provide little benefit to humans seeking to cope in the world and may even have a detrimental effect on people (Bernstein and Nash, 2008), thus indicating how unreal the Western system is. The Western system is so incredibly unnatural that some medical experts have even labeled human "Happiness as a pathological" condition (Chemali, Chahine, and Naasan, 2008).

The fact that four in ten patients going through psychotherapy are worse off after they began such therapy than before (Begley, 2007e) provides strong empirical support against the relevance of the views and superstitions of classical Western psychologists and psychiatrists in dealing with reality. Years of psychiatric treatment have been found to be no more effective than placebos or useless pills and are worse than simple advice for self-analysis, as can be effected by a person merely writing down the positives and negatives of the individual's own life (Klein, 2003). The evidence indicating that ancient shamanic practices of hypnosis are more effective than psychotherapy regardless of the amount of training of the psychologist (McClenon, 2002) also seems to imply the ineffectiveness of Western philosophies and religions that appear to be the ones truly divorced from reality.

It has been stated that uneducated lay people understand human psychology better than trained Western "psychologists" (Burling, 2004). Although there are some efforts to have groups of lay people or even entire communities help individuals to adapt to their situation in the world, it is widely recognized that over 30% of all Americans each year suffer from some mental disorder (Bernstein and Nash, 2008). Evidence reported by Weiss (2005) indicates the percentage in the USA is actually closer to 50% and is higher than any other country in the world, with extended bouts of depression being the leading mental disease in that prosperous country.

Since prescription drugs for depression like Prozac boost levels of serotonin, which also leads to herding behavior in individualistic locust insects (*NewScientist,* 2009n), and which inhibits spiteful behavior in people (Clark et al., 2008), it seems plausible that depression is widespread in today's individualistic societies because the rampant exploitation there promotes spite, which is, however, penalized by the exploitative cultures. People are thereby left feeling hopelessly frustrated and helpless. In particular, both sadness and spite are normal reactions that would provide momentary pleasure, but they are inhibited by a conscious realization that acting sad or spiteful will only serve to make oneself more unhappy in the future in contemporary cultures that penalize both. Feelings of hopeless depression naturally result today in this environment that attempts to make robotic slaves out of human beings.

Scientific evidence indicating that anti-depressants like prozac initially increase serotonin levels in the brain cause the thalamus and nucleus caudate to fire more in rhythm with each other as opposed to in conflict with or in reaction to each other's firing (Klein, 2003). Depression thus seems to be partially caused by the thalamus filter inhibiting some conscious thoughts, which may include feeling sorry for oneself as well as possibly spite, to deal with the unhappy state. The nucleus caudate pleasure center of the brain motivates such natural thoughts to adapt to an unhappy environment, but a conscious recognition that sadness or spite will both be penalized results in hopeless despair. Serotonin may actually cause people to rationalize away the unhappy situation created by the cruel societies by learning to make the most of the adversity that other human beings are also having to suffer. It thus serves to make people into obedient slaves actively serving the elite.

Serotonin levels decrease with use of anti-depressant drugs over extended periods (Geddes, 2010b), perhaps because the conscious eventually learns from the early effects of the drugs, which initially increase serotonin in the brain, that the feelings of depression that used to lead to inactivity are less likely to result in a change in unpleasant conditions than the cooperation that serotonin promotes. Once a person has consciously arrived at this conclusion, the individual no longer needs the serotonin, which can also have adverse effects by prompting herding behavior that can be exploited. The fact that psychotherapy without any drugs has similar effects to anti-depressants, including with respect to only initial increases in serotonin levels (Klein, 2003), provides supporting evidence that conscious thought alone can overcome sad feelings. Since serotonin promotes social bonding (Gynn and Wright, 2008), which also reduces stress (Bernstein and Nash, 2008), it is also feasible that the initial rise in serotonin levels increases the friendship and support circle of other people, and with this reduction in stress comes a reduced need for the brain to produce serotonin.

Within this context, serotonin produced in the brain, including when prompted by prozac and other anti-depressants, may just enable happy acceptance of the unpleasant situation without stress. Serotonin, which motivates more crying at sad situations (*NewScientist,* 2010a), does reduce stress (Klein, 2003), perhaps via a process of allowing uninhibited emotions and empathy that can enable the good feeling of being part of a group, even if only a group of slaves.

Artificially increased levels of serotonin, however, inhibit the ability of human beings to act naturally for the benefit of the group as a whole when it would involve directly hurting someone (*NewScientist*, 2010aaj). It may thus reduce spiteful actions against an exploiting elite, and it thereby enables the transformation of people into robotic insects in service of an elite. Serotonin is responsible for mindless herding behavior by insects (*NewScientist,* 2009n) that may benefit only a few. People may even consciously rationalize such an unjust situation due to some hope that their spite-less cooperation will be rewarded at some distant point in the future, such as after death.

Beliefs in religions such as Christianity may have a similar or contributing effect by indoctrinating people to passively accept hopeless, exploited, or otherwise depressing situations on earth in return for the anticipation of pleasure in an afterlife that can be especially enhanced by being member of a group of similar slaves. Even though they won't be rewarded no matter what they do, they do continue to exist and allow themselves to be exploited in the ridiculous hope for a reward that can never be fulfilled.

Serotonin may be involved in assimilating the Christians to their slavery, and a communal fellowship involved in a belief in a fabricated deity offering a fictitious reward after death, can provide a pleasant feeling especially if there is involvement or identification with a particular church group. Obedient service to an elite is thereby reinforced even with reduced levels of serotonin, and the existence of a communal support group of similar believers in the Christian fairy tale can be an important motivation for continued obedience, as can fears of losing friends upon straying from the "flock" of sheep.

In addition, in overwhelming Christian societies like the USA, the leaders promote exploitation through propaganda of selfishness for the masses, who are thereby programmed by the entire culture and media to be slaves to an elite. Legal penalties are also levied for most spiteful behavior. Whereas prehistoric human cultures limited exploitation through empathy and spite and thereby reduced long-term feelings of sadness or despair, depression is a natural result of the frustrating situation of being exploited without recourse to spite.

Since the masses of slaves in Christian society are trained that they should be happy with their position in life, depression is generally met only with contempt, especially in the capitalist form of Christian societies like the USA that encourage selfish exploitation between slaves. Even being part of some supporting subgroup of humanity like a church group may not be sufficient to compensate for the negative feeling associated with being exploited without being able to do anything about it. In addition, exploitation inhibits natural human empathy, which in prehistoric times led to helping those feeling exploited or despondent, and any limited help from Christian subgroups may be recognized to be insignificant, artificial, and insincere insofar as it is motivated not out of human nature but out of a desire for a reward in an afterlife. A meaningless, helpless frustration often results that continues the depression in today's societies whereas they were very temporary conditions in prehistoric times. Western tendencies to negatively stigmatize people with psychological disorders only make curing the problems more difficult (McClenon, 2002), as the empathetic care for the sad that existed in prehistoric societies has been replaced today with socially shunning of them, thereby further magnifying their problems. Because the contempt of others is naturally sensed, a feeling is promoted in themselves that their own inadequacies are responsible for their sad state. Capitalist cultures that emphasize individual responsibility further contribute to such perceptions that only make the despondent more pessimistic about their ability to solve their problems. The fact that about 50% of Americans suffer from depression (Weiss, 2005) shouldn't be surprising in this overall social framework that prompts ever more despair among the sad.

Since many aspects of Native American spiritual cosmology that are seemingly the most fantastic are actually consistent with the reality of the universe that runs according to the patterns defined by quantum physics (Peat, 2005), it might be useful to provide education in such beliefs as a component of community help to the mentally ill Americans. However, it might be more optimal to merely have people be more aware of human nature and to act as consistent as possible with that nature within the legal confines of a society that is recognized to be extraordinarily inhumane. Instead of trying to behave in the way that the cruel and exploitative society of today expects them to, people should seek legal ways, such as through the permitted political processes, to remove the system that causes so much harm to so many human beings. Such attempts for the benefit of humanity are not only rewarding in terms of feelings of self-fulfillment but can also lead to bonding relationships with other human beings striving for the same goals.

One of the most important facets of today's cultures that needs to be eliminated for greater human harmony is the condoning of dishonesty. Dishonesty fragments the human unity which was so efficient that it enabled a weak species to dominate the earth. Attempts to justify dishonesty as a means to a good end are based on the false assumption that the liar's end is superior to human nature itself. The concept of "white lies" that are permissible because they hurt no one is itself a pure lie. Lies destroy trust and merit a total disregard for anything a person says. While tact and diplomacy, including exaggerated comments and jokes, are certainly something innate to human nature and promote bonding, outright lies, even with respect to complements, are incoherent with human nature and destructive to humanity and the heaven of group consciousness.

It might prove especially useful to recognize that group consciousness and the related mutual empathy represent characteristics of more evolved forms of life, and that people acting strictly for their own selfish interests is a step backward toward single-celled organisms. Humans no doubt evolved a group consciousness partially for the very reason that it reduces stress and promotes physical health, as well as for the many other benefits it provides.

For the same reasons, apes themselves might very well evolve over the next few million years to replace their alpha male cultures with the more advanced form of life that involves empathy and group consciousness. Just for instance, the few apes who actually develop close friendships as opposed to engage in the power struggles that set everyone against each other in alpha male ape societies have already been found to experience less stress and be physically healthier (Klein, 2008).

Evolution seems to be driving group consciousness even if the last few thousand years represent a temporary step backwards. However, the individual consciousness that has developed more completely during that time may in the future permit an even stronger and more beneficial group consciousness of which people are individually more self-aware. Individual self-awareness long existed, but it originally served to ensure that each individual's needs and desires were appropriately addressed, as was natural since each human body had more direct information on that body's needs and wishes. However, satisfaction of those needs and desires about which individual consciousness made each aware was guided by an overriding group consciousness that made all the individual needs coherent with those of the group. In prehistoric times, group consciousness may not have even been consciously perceived to be different from each individual's consciousness, especially because of empathy. Today differences in mental states are perceived and even maximized, but basic human characteristics like empathy and chemical reinforcement strongly motivate people to consciously choose to integrate into a group consciousness by individual choice as opposed to mere instinct. Greater individuality in serving humanity as a whole may therefore be enabled in a process that could enable increasing humans' resonating power to control the environment for the common benefit of all.

Chapter 14
The Forecasting Origin of Human Consciousness and Communication

Consciousness represents an analytical process for decision-making based on an awareness of the self and other stimuli that are perceived by an individual organism. This self-awareness, which includes an awareness of the decision-making process itself, perceives other things besides the self from the point of view of that self-awareness. This consciousness may be of an individual or group of conscious individuals.

Consciousness appears to require a synchronized firing of brain cells at the same frequency to enable a coordination of sensory perceptions that can be integrated into thoughts (Thomson, 2010a). One's self is formed from synaptic interactions between the various systems of the brain (LeDoux, 2002) that create consciousness when the awareness of the self arising from neural processing of stimuli internal and external to the individual are combined with logical thoughts. This coherency in neural firings enable decision-making that is aware not only of the self but of the conscious decision making itself. Self-awareness itself represents being aware of being aware (Hamzelou, 2010b), and the perception of the self is just this conscious self-awareness in the present.

Self-awareness enables conscious forecasting people's own future actions, thereby facilitating decision-making in the present since humans' future actions can impact the environment they might experience. Because a consideration of one's current self is required to contemplate one's future self, as well as one's past self (Robinson and Ryff, 1999), self-awareness facilitates predicting one's own behavior. In particular, through an empathetic understanding of why one felt and acted in different situations in the past, an individual can better predict how one will react to future possible situations. Decision-making can then be made based on such forecasts of oneself in conjunction with all the other information about present and potential future environments that are obtained from all the body's senses.

Consciousness "arises from the coordinated activity of the entire brain" that is characterized by higher electrical voltage levels and greater frequency and phase synchronization of neural firings "in different parts of the brain" than otherwise (*NewScientist*, 2009zg). This fact implies that the self-awareness of consciousness is dependent upon coherent processing of all internal and external stimuli received by the neurons involve in firing to create conscious thought. As will be explained in Chapter 15, an individual's conscious can at least partially control the very brain processing of stimuli that determines what information is utilized in forming a perception of oneself. It does so by out-firing emotional neural firings with the activities of neurons in the thalamus that are trained by firings in the commanding decision-making sector of the brain. The conscious is only aware of perceptions based on neural firings that aren't so suppressed, and this "reality" is organized in generalized categories in order to facilitate predictions of one's own self in different future situations.

Research has shown that mere "beliefs about the future self can create their own reality" (Robinson and Ryff, 1999) by apparently training or preparing the brain's synchronized neural firings to enable the future thoughts and actions to produce the intended result. A person's will reflects such intent. An individual's will power, which is formed by the neural firings of conscious thought that create intentions, is, however, like a single wave in an ocean. One can will oneself, albeit only within the context and constraints of the environment that includes other wills. A person can also partially affect others through will power but only to the extent permitted by the quantum environment. Many of the effects of will power are not consciously perceived because the processes are carried out subconsciously.

Consciousness is the most perplexing characteristic of the human mind, but a reasonable understanding of it is indeed possible within the context of its evolution in the environment in which human ancestors lived. The human mind evolved from subconscious reactions and learning that is based, as with all animals, on simple if-then logic and probabilities estimated from experience. Human consciousness developed from the application of that logic to forecasting oneself, as part of a group or individually, as a means of improving decision-making. Self-forecasting may have began with women's menstrual periods being observed to cycle with the moon. Learning to forecast events and behavior that are likely to result from human biology and actions enhanced the ability of humans to both better adapt to their environment and to affect it.

The growing capacity to affect the environment made human self-forecasting even more important, whether individually or in groups. While human self-awareness may have originally not been much more advanced than that for apes, the evolutionary impetus for further development of consciousness and self-forecasting from introductory stages was created by the increased control over the environmental situation that it spurred. In particular, advances in conscious self-awareness improved forecasting, including of others through empathy, thereby enabling greater adaptation to the environment that was ever more greatly impacted by humans as their conscious forecasting abilities grew. Improvements in humans' conscious forecasting abilities thus increased the impact(s) they had on the future environment and therefore magnified the evolutionary incentives to increase those forecasting abilities even more, leading to a cycle of ever greater conscious forecasting abilities and control over the environment over many generations of human beings.

Language developed as people attempted to explain the complex concepts associated with forecasting to others such as their offspring. Systems of symbolic oral sounds likely began as primitively as consciousness itself, but such communication between human beings had to grow in complexity with the expansion of the ability to forecast that had to be explained to children and others. Since language helps in labeling stimuli, it thereby aids in categorizing the environment for more efficient reactions to it, and even talking to oneself without uttering a sound, as apparently occurs quite frequently and usefully in most people, is beneficial for such understanding of reality (Robson, 2010d). Thus, the self-forecasting that spawned consciousness also promoted the development of language in human groups that created a more organized form of conscious thinking through the very categorization of self-aware states with the symbols of sounds called words. Consciousness is therefore typically perceived in language form, as in talking to oneself.

One's entire self and conscious mind is a product of the environment and relationships to others. Signals from the environment and others are constantly being detected by the brain, but many of these are inhibited by the brain's thalamus from entering into conscious awareness. Since the conscious can process only a finite amount of information to engage in its decision-making function, the forecasting that developed to improve this decision-making process necessitates the thalamus filtering, which helps keep out information deemed irrelevant for conscious focus on a particular task. Conscious thoughts or concentration thereby affects what information is processed, as well as categorizes the information received into a usable comprehensible form.

On the macro level of conscious perception, information from the five classical human senses is categorized to make particular combinations of electromagnetic fields seem like solid objects. This characteristic of conscious thought evolved from the usefulness of interpreting the electromagnetic signals sent by other electromagnetic fields and received by the connected electromagnetic fields that make up the individual organism. Visual and audio communication evolved to enhance cooperation and coherence between living beings and their environment that included other organisms.

In the beginning, all organisms communicated via light waves (de Quincy, 2005), at both the cellular and DNA level and across individual organisms (Narby, 1999). It may even be possible for the subconscious to evaluate the electromagnetic emissions of inanimate things (Peat, 2005). Everyone and everything emit radioactive waves (Segre, 2002), and these waves interact with each other. The radioactive waves and positrons created in the process of transforming protons into neutrons are caused by the periodic decay of all matter that also create neutrinos (Close, 2009), which may be part of the invisible communication process. Neutrinos, which may actually stretch out across distances equaling billions of light years (*NewScientist*, 2009zzd), are neutrally charged particles that pass through other matter without effect (Close, 2009). However, these particles could potentially be detected in some fashion by the human body, such as by the pineal gland, thus potentially enabling psychic powers. Since each individual also emits a significant amount of antimatter such as the positrons in the process of decay (Gefter, 2009b), any psychic sixth sense might be subconsciously categorizing the impacts of the electromagnetic waves of such antimatter that is matter travelling backwards in time to create feelings of premonition, although most psychic powers may involve emissions and detection of neutrinos and light photons.

Each individual organism may therefore be affected by the radioactive and light emissions of others, as well as possibly be able to deliberately affect others by the organism's own emissions. The periodicity of those electromagnetic emissions could conceivably be controlled in an informational manner by the pineal gland, although not all information enters into consciousness or is totally controlled by a human's conscious thoughts. However, the fact that organisms are affected by and can affect unseen psychic forces (McTaggart, 2008) and the fact that decay occurs only as a probabilistic function that is affected by the temperature of the radiating matter (Childress, 1999) provide evidence of such effects and potential controlling influence that may be harnessed by conscious will.

Conty (2002) suggests that a prehistoric ritual of entering and leaving the center of a labyrinth can help liberate the mind into understanding the creative aspect of consciousness that affects all things. However, it seems possible to understand consciousness without any elaborate rituals. A conscious understanding of consciousness can enhance the coherence of human beings with their environment.

A. Forecasting, Imagination, and Hope

As the mind sees repeated events that are related, it starts to forecast and speculate. In fact, the entire brain has been conceptualized as "a probability machine that constantly makes predictions about the world and then updates them based on what it senses" (Huang, 2008). "Humans have a fascination with the future" (Brooks, 2010d) that may merely reflect their unique self-awareness that developed out of the usefulness of forecasting for decision-making.

The human subconscious itself does evaluate probabilities (Stanovich, 2004), as is necessary in the environment of uncertainty that is ruled by quantum theory. In fact, since every particle in the universe exists in a specific place at a particular time only with a certain probability (Capra, 1999), the evolution of the ability to evaluate probabilities seems quite natural.

Some primitive predicting is evidenced even in rodents like rats (Begley, 2007d), which also have the KLK8 gene mutation for memory and learning (*NewScientist*, 2007j). However, that gene may have evolved separately in mice to enable their more limited form of forecasting, and human self-predicting and consciousness have enabled a huge expansion in their own version of the forecasting advantage in reacting and learning.

Predicting led, in humans, to conscious forecasting of themselves possibly beginning with women seeing their reproductive cycles being related to moon cycles (Shlain, 2003). The tendency of women's menstrual cycles to synchronize via pheromones emitted in their sweat (Bernstein and Nash, 2008) and to occur with approximately equal frequency with that of the moon (Shlain, 2003) may actually have prompted recognition of the first patterns that initiated group forecasting and consciousness. Rats, whose brains are wired for both empathy and problem-solving based on some type of at least subconscious probability predictions, never progressed to this step of self-forecasting individually or in groups, possibly because rodent menstrual cycles are not tied to an observable object like the moon, and hence rats may not consciously think.

The portion of the brain involved in reflective learning from experience that is called the medial prefrontal cortex, which exists in humans as well as other organisms like monkeys, is also utilized for learning from forecasting through imagined self-experience of different future scenarios (Fox, 2008). Learning from self-prediction, which seems to have motivated the development of consciousness, may be what makes humans significantly different from other organisms. Larger amounts of "grey matter" in the right anterior prefrontal cortex are associated with more introspective people who feel more confident after making conscious choices, as is consistent with a hypothesis that the development of this part of the brain is required for the particular type of conscious awareness of which human beings are capable (*NewScientist,* 2010aaf).

Conscious self-prediction combined with categorization and labeling also enable a person to understand and forecast others better by enhancing the comprehension of the individual's "own motivations, emotions, and behavior" (Burling, 2004). This trait derives from a form of empathy (Nettle, 2007). Much of empathy involves projecting one's own preferences on others in order to deduce their reaction to a situation from a forecast of what one's own self would do (Brunstein, 2008).

Complex predicting in an uncertain world leads to humans imagining different scenarios that might unfold and making preparations for dealing with those possibilities. The rather unique human imagination (Kenneally, 2008b) therefore probably resulted from those complex forecasted scenarios that enable humans to try to avoid undesirable situations and, instead, attain desired outcomes. The fact that quantum reality is characterized by complex mathematics that incorporates imaginary numbers (Penrose, 2004) may symbolize a deeper significance and meaning for imagination and its usefulness in adapting to the world.

Much of the brain's energy, including most of the amount utilized when a person is not focused on a particular task, is directed toward day-dreaming in the medial prefrontal cortex, which enables simulating future scenarios and revisiting past events for potential posting in long-term memory via the hippocampus (Fox, 2008). The hippocampus creates brain waves at different gamma frequencies that strengthen neural connections more when the neurons in that area of the brain fire more rapidly and thereby enable the formation of easier conscious recall (Thomson, 2010a). It thereby permits learning that results in the formation of new neural connections (Fox, 2008). Both "subjective experience" and "meaning", whose neural basis continues to perplex experts (Rowe, 2008), appear to result from these brain activities.

In humans at least, forecasting also led to the development of hope. Hope creates pleasant feelings that offset depression, and it tends to spring up eternally in humans as forlorn hopes are replaced with new ones (Hirschhorn, 2008). As such, it seems to reflect some positive probability of a better future and the anticipated more pleasant situation such predictions might bring. The mere imagining of the future can bring some reinforcing pleasure to the human brain (Klein, 2003). In that sense, the act of forecasting can be self-reinforcing within the brain itself, as well as enhance decision-making abilities. In fact, as explained in Chapter 11, the conscious has developed its own method of providing rewards to itself for what it perceives to be good forecasts and decisions that turn out to be correct.

Self-forecasting, imagination, wishful thinking, the power of belief, and consciousness may all be integrated in some fashion. For instance, the tradition of making both predictions and wishes by the light of a flame, such as on birthdays (Knight, 1999), may reflect an origin for forecasting and wishful thinking at nightly campfires used to keep predators at bay, stay warm, and cook to facilitate digestion. Certainly, the seemingly magical properties of fire would have promoted a belief in it that could have enabled wishful thinking to have the power of positive thinking, as is explained in Chapter 16. Rewards to such predictive thinking would have certainly encouraged further development of forecasting and imagined scenarios. Consciousness would have easily grown out of such predictions of the future, as optimal forecasting requires self-forecasting that in turn promotes a reflective self-awareness to make better predictions about oneself. The beginnings of the unique human powers of consciousness may therefore have actually begun shortly after the harnessing of fire 1.8 million years ago.

Consciousness developed to enable better predictions about oneself and therefore about the future in general. Self-conscious thought can be largely turned off by focusing on something, even just one's own breathing process, and the result is living for that experience in the present (Capra, 1999). Doing so can result in a blissful loss of sense of oneself and a union with all else (Gynn and Wright, 2008) that enable relaxing the separated self-awareness involved in individual conscious decision-making. While such moments without forecasting and conscious decision-making can provide happiness, living strictly for the nirvana of the present tends to be consciously unfulfilling. The conscious evolved in humans to enable them to make better decisions by forecasting the future, and it is driven to carry out that function. The conscious developed to make better decisions using forecasts, and so humans, unlike apes and other animals, naturally spend much of their time "living" in the future.

Despite its powerful and mysterious properties, the process of consciousness remains based in simple "if-then" logic that is applied to information inputs received from the senses and emotions. The information that is allowed to filter into consciousness is in part determined by the conscious focusing attention and thereby directing processes and actions that are carried out subconsciously. While some decision-making and forecasting can be conducted without consciousness, self-forecasting requires a self-awareness that typically includes a categorization process for the sensual perceptions and emotions the person experiences. Categorization of sensual perceptions and emotions, and therefore self-awareness, is enhanced by language and communication with others.

B. Communication, Chanting, Hypnosis, and Music

Teaching others, especially the young humans who have an especially long learning development process (Bernstein and Nash, 2008), about forecasts, predicting, and other complex concepts probably initiated the development of human oral communication and language. The fact that the area of the brain involved in planning is also active when people make symbolic gestures, which probably preceded speech (Coghlan, 2008a), is consistent with this hypothesis. Since conscious thinking is possible without language (Burling, 2004), human language probably developed originally as a tool to communicate ideas already known by the communicator as opposed to being a medium for discovering new concepts (*NewScientist*, 2007z).

Although training in tool use is possible via copying that doesn't require anything more than visual observation, teaching complex concepts does require communication (Burling, 2004). Some human learning does occur by simple copying, but humans evolved to be able to use the "mirror" portion of that brain to do much more, including predicting the intentions of others (Slack, 2007). Psychologists have suggested "that the ability to infer the mental states of others may underlie language" (Ananthaswamy, 2009a). This capacity to recognize others' feelings is enabled in the anterior insula and frontal operculum even when reading, hearing, or seeing a story (*NewScientist,* 2008zw) and seems to be enhanced by language (*NewScientist,* 2009zzn). These facts support a hypothesis that one of the incentives for teaching was empathy, which caused people to want to help others, especially the young and inexperienced, to obtain their own desirable outcomes. However, the development of language was also motivated to assist others in understanding how to cooperate better in group activities and goals.

Human communication may have started with hand and visual signals but probably evolved to vocal noises at least partially in order free up the hands for other tasks such as carrying, making, and using tools (*NewScientist*, 2007h). According to Burling (2004), people still communicate emotions more with gestures and visual signals than with spoken words. For instance, unless inhibited by society, as in some Asian and military cultures, humans tend to use more innate facial expressions when in the presence of others than when alone to relay their emotions, and children can recognize facial signals after only three months of life (Bernstein and Nash, 2008). The eyes evolved to display inner feelings (*NewScientist*, 2007e), as seems natural since the eyes are part of the brain (Peat, 2005) as well as essential in the very emotionally communicative experience of romantic love at first sight.

Initial verbal communication, which may have begun as much as 1.9 million years ago (McClenon, 2002), likely consisted initially of short differentiated sounds for each different concept and event being described (Burling, 2004). The earliest, fairly complete spoken languages probably appeared in the last half million years (McClenon, 2002). These forms of communication were initially more tonal in nature, insofar as the meaning of words was heavily influenced by the spoken tone, as continues to be the case for many Asian languages today. The primitive sounds voiced by chimps to communicate with each other today (*NewScientist*, 2006b) may not be much dissimilar to the earliest ones utilized by humans. Tonal languages are easier for children to learn, especially because they apparently don't require two fairly recent gene mutations, ASPM and microcephalin, which are prevalent in European populations with less tonal languages (Schultz, 2007b), thereby providing some evidence of the development of language being partially motivated to teach the young.

Language itself is learned via copying that utilizes the mirroring portion of the brain (Thomson, 2008) that may have something to do with empathy (*NewScientist,* 2009zx). Burling (2004) provides lots of evidence on how language evolved, including from gestures that may initially have been only natural instinctive behavior which other animals like bonobos also evolved to recognize, understand, and use to communicate with each other. Some evidence of early verbal communication is also evident in chimps, which combine various sounds, including one for a type of predator from which they optimally flee, when they want the entire group to move on for any reason (*NewScientist,* 2006b).

Visual communication was later replaced by sounds because noises could be used when the eyes are optimally focused elsewhere or where vision is blocked. A FOXP2 gene has been found to be crucial in learning verbal communication, as it permits memory and reproduction of sounds in both humans and animals (Coghlan, 2007b), and birds use a similar form of this gene to learn to sing properly, (Yong, 2008).

Teaching about the future, especially an uncertain one, seems to require much more complex communication than is possible with mere copying or even simple gestures and sounds. As humans grew into larger groups it became easier to communicate with sentences of individual words than to teach a large number of short differentiated sounds for each event (*NewScientist*, 2007zf). People still categorize groups of concepts and events in order to be able to facilitate recall and analysis of more of them within the context of the finite working memory of human consciousness, that represents the conscious brain activities which process information recalled and that tend to operate using whatever language or information categorizing system that an individual knows (Bernstein and Nash, 2008). Language may be the most efficient means of categorizing bits of information and is therefore selected as the encoding form for most memories and conscious thoughts.

Since some organisms appear to have a conscious working memory (Coghlan, 2008a), and since some even have a short-term memory that is superior to humans' normal capacity (Hooper, 2007b), the limited mental abilities of other organisms may at least partially stem from their lack of a language to categorize things and concepts. Such categorization enables consideration and processing of an enormous body of information within the limited working memory of consciousness (Bernstein and Nash, 2008).

Tests show that humans' memory and recall of the words of languages today operate in a manner similar to internet search engines (*NewScientist,* 2007zj), which themselves are being enhanced with improved categorization and information processing capabilities (Giles, 2008b). Memory of an item or event is initiated by a specific network of brain cells being initially excited, and this same set of neurons fire again together when that particular item or event is remembered (Marshall, 2008a). Although the number of connections or synapses between brain cells is large enough to permit memory of an almost infinite quantity of things, humans develop methods of categorizing remembered items that facilitate conscious recall upon demand since only a limited number of ideas can enter consciousness at any time without creating mass confusion (Marshall, 2008a). Some of these limits may exist because too many ideas create overlapping neural firings in the coherent activity of conscious thought that would lead to chaotic confusion, although it's also possible that some creative inspiration arise from brief intervals of such chaos that aren't meaningfully perceived in detail despite the implications of them being consciously recognized in subconsciously processed form via emotions or intuitive thoughts.

While language appears to be useful in the memory and recall process (Bernstein and Nash, 2008), the emotion-driven amygdala remains especially useful for enhanced storage of memories related to strong feelings (Marshall, 2008a). The amygdala helps in categorizing memories, including for determining whether they are pleasant or unpleasant based on the brain's release of internally produced opiates like to the amygdala's receptors (Esch and Stefano, 2005a). While chimps have been found to have better short-term memories than humans (Hooper, 2007b), they haven't developed the search program that humans have, possibly because their ability to categorize long-term memories for easy recall may be impaired by their lack of language capacity.

People utilize much of their brain capacity for evaluation, and especially exceptional abilities in memory in some humans are actually associated with inferior skills of analysis (Gigerenzer, 2007). The reason may be that too many connected memory neurons firing too easily or often swamps and therefore confuses the neural firings related to evaluation. The brain's process of categorization that seems to occur in the prefrontal cortex results in many memories being forgotten, or consciously recalled only with great difficulty, if they are either specified consciously to be low priority or not consciously experienced (Marshall, 2008a). Events that induce larger emotional reactions tend to be remembered better unless there is some inhibition of the connection between the amygdala, where many emotions are felt, and the hippocampus, where memories are temporarily stored (*NewScientist,* 2008u).

In order to permit the appropriate cooperative activities to optimally take advantage of a recognition of cycles in groups of humans, such as hiding from predators in trees when the moon phase indicated the approach of women's menstrual cycles, would have required some communication like language. Explaining the complex associations between things, events, and the related appropriate actions required more than mere teaching by example. Even a relatively simple forecasting pattern like the relation between menstrual and lunar phases would have been difficult without some language that showed a relationship between objects or events.

A form of sign language in the form of pointing to objects and waving of the hands to indicate verbs may have preceded verbal language. However, sounds emitted while making each hand signal due to the instinctive emotions it prompted would have eventually led to those noises being identified with the visual symbols, thus resulting in the first words. Analyzing and predicting cycles, as well as determining the optimal course of action given such forecasts, would have been rewarded from a survival standpoint, and explaining such predictions and analysis would have likely been reinforced both socially and sexually, thereby causing the evolution of ever greater forecasting, analytical, and communication skills.

Such analysis and forecasting eventually led to deep universal questions on where humanity and everything are heading long-term (Urban, 2007). The question "where is everything going?" can be translated as "wo fuehrt alles?" in German, which utilizes an almost identical sound structure "wofuer alles?", which means "what is the purpose of everything?" The language similarity for the two questions may provide some insights as to the reason for the widespread desire of humans to know the meaning or purpose of everything.

Language, especially those like Native American ones that categorize things and concepts into less compartments, is considered to be a mere vibrating entry into an alliance with the vibrations of the universe (Peat, 2005). As previously mentioned, it is likely that language originally developed with a strong emphasis on tone that was also related to gestures (Burling, 2004). The very vibrations of spoken words have such a very important impact on the meaning of verbal communication in older languages that some believe such sounds represent a form of communicating with the entire cosmos (Peat, 2005).

The frequency of vibrations of all people and things, as well as of the sound waves of particular words or chants, may in fact be picked up and evaluated by the pineal gland to detect resonance with oneself (Strassman, 2001). This process enables spoken language to communicate more than the mere contextual word meanings that are consciously understood. Initially, the various tones were probably evoked by different emotions, thus enabling communication of feelings that could encompass deductions about the forecasted environment and the appropriate actions therein.

The actual individual sounds used by humans for communicating orally were probably rather simple and repetitive, however. People in modern times remain strongly influenced by focused repetition and hypnotic suggestion (DeLaurence, 1902). It seems likely that this human characteristic reflects the initial methods of communication that were designed for affecting others as much as possible through the dissemination of information in the form of sounds that reflected emotions. People today listening to a story exhibit electromagnetic brain waves similar to those of the reader, with greater comprehension of the story for those whose brain waves overlap the most with the reader (*NewScientist*, 2010ar). Thus, communication of emotions seems to remain important even with today's complex languages, as brain waves are strongly impacted by the emotions.

Hypnotic repetition as in chanting represents a powerful component of many religions (Underhill, 1965), and hypnotic chanting has been hypothesized to be the precursor of language (McClenon, 2002). When hypnotic repetition of sounds is used to induce a meditative trance, there is an activation of the pineal gland, which is known to be engaged in producing the hormone melatonin (*NewScientist*, 2007zl). Melatonin is involved in the repairing of neural connections and brain healing (Erb and Erb, 2003) and apparently assists in purging unnecessary or less useful linkings between brain cells using chemicals similar to those involved in the immune system's destruction of diseases (Trivedi, 2008). It also stimulates endorphins and oxytocin in the brain (Gynn and Wright, 2008), thereby being beneficial, soothing, and useful for social bonding. Melatonin released by the pineal gland during hypnotized states may server a similar function of pleasantly relaxing the brain to facilitate strengthening the connections between neurons that enable carrying out suggestions made under hypnosis.

The pineal gland tends to be more calcified and less "open" in most individuals as they approach adulthood in today's society that inhibits its conscious use, but its activation may be enhanced by electromagnetic waves as well as iron-rich diets that are related to the gland's iron components (Stray, 2005). DMT, which provides blissful and often illusory feelings of trusting unity that also seem to be enabled by the pineal gland (Strassman, 2001), and which could have evolved to magnify the pleasant effects that motivate surrender to the common cause, may also be produced by the pineal gland. These effects, which may have something to do with a continued desire for a trusting connection to others and a communal consciousness that has been fragmented by the development of selfish individual consciousness, could go a long way toward explaining the power of hypnosis. In any event, given that group chanting of individual words have a hypnotic effect in promoting group focus among humans (Howell, 1986), such sounds would also have usefully influenced humans to unite for important common tasks such as defending against more powerful predators.

Chanting may likely have developed different tones which combined to form music. It may have later developed into an association with instruments like drums which may have served a supporting or substitute role for the hypnotic communication of desired emotional states. For instance, while chanting to hypnotically focus groups of humans into having the courage to face predators in groups, weaker members of the group could have facilitated the process with vocal sounds and other noises they made by hitting wood to enable hypnotic effects without requiring the active defenders to engage in diversionary noise making themselves. In any event, since hypnotic repetition in the form of music such as with drums is also an important component of many religions (Underwood, 1965), music may have developed with chanting as a form of group hypnosis via sounds.

The fact that music aids in memory (*NewScientist*, 2010x) may be related to its original use in associating hypnotic sounds with the relevant group actions. Music, which has been hypothesized to have "developed to cement human relationships" (*NewScientist*, 2009zzb), would have certainly promoted such cooperative activities. Music has indeed been found to be genetically associated with the production of vasopressin that promotes social bonding, love, and altruism (*NewScientist*, 2009zzb).

Since chanting represents a form of music, it may also have led to the development of singing and other musical sounds evoked frequently by humans. Music that sounds most like that created via cave echoes is perceived to be the most "otherworldly" (Levitin, 2008). That phenomenon may provide hints as to some of the earliest physical origins of musical communication and its usefulness in "otherworldly" trips that may have been traveled via hypnotic group chanting in caves. In any event, since each music note represents a blend of hundreds of sound frequencies (McTaggart, 2008), simple chants with different musical pitches could communicate a wide range of different meanings and have significantly different hypnotic and other effects, especially when combined into songs.

Some scientists such as Darwin have actually suggested that oral language began with songs (*NewScientist*, 2008j), as would be consistent with other species using songs simply to woo mates (Kenneally, 2008a). This hypothesis is consistent with the fact that music creates pleasant feelings via dopamine or neural excitations in the same areas of the brain as do "rewards of food, sex, and psychoactive drugs" (*NewScientist,* 2008i). It is also consistent with the first languages being tonal ones and with language itself being preceded by chanting that my have later been utilized in the emotionally charged process of human courtship.

Since musical ability is concentrated in a separate part of the brain from language ability, and since children tend to learn musical skills after speech, it is possible that group chanting was the primary initial factor prompting the development of the symbolic communication associated with language (McClenon, 2002). However, chanting is not that much different from singing and can represent singing if the tone is varied. As a result, it is quite possible that chanting initiated the development of oral communication skills, but singing or chanting with more tone variation enhanced its further evolution. The fact that the older languages utilized more tonal symbols (Burling, 2004) is consistent with this hypothesis, which also implies a more holistic view of language that utilizes the entire brain and integrates more than the mere rational logic of the left side of the brain.

Spoken languages today, including Western ones, retain a musical nature although the concentration on the meaning of the words spoken leads to them not being perceived as such (*NewScientist,* 2008zzz). The ability to pick up spoken sounds from background noise actually depends on attuning one's brainwaves to those of the speech, and musicians have a greater ability to do so (*NewScientist*, 2009zzzzg) possibly because their musical ability has accustomed them to better synchronize their brainwaves to tones. Part of the power of group hypnosis by repetitive tones or chanting may involve the synchronization of brainwaves that it may induce, thereby facilitating cooperation. Romantic music could have a similar effect in putting lovers on the same wavelength.

Musical tones may also have once been useful for communicating forecasting. In particular, the complex concepts of prediction within the context of an uncertain and ambiguous future may be difficult to convey with mere toneless words. Human forecasting eventually culminated in the deeper questions of where humanity is heading and the meaning of life, and attempts to answer such questions are often thought to be more easily communicated artistically such as via musical tones. The Native American belief that speech represents vibrations with deep cosmological meaning (Peat, 2005) implies that musical vibrations may indeed be useful for answering deep questions..

For instance, music is most appealing when its tones reflect mathematical functions of frequencies that share the same neural pathway to activation of neurons with the brain's auditory thalamus (Harmer, 2008). The fact that the thalamus is actually the filter to the human cortex (Bernstein and Nash, 2008), which is where consciousness is centered, may provide clues on the origin of tones and forecasting. In particular, the coherence between the thalamus and musical tones could be related to a common origin for consciousness and music as well as language, with all three deriving from early self-forecasting efforts that had to be communicated for maximum group cooperation. The hypnotic chanting to do so evolved to easily pass through the thalamus filter in order to enable the effects to instantaneously permeate all in the group to face dangers like predators without questioning priorities or doubts. It has been found that today some people's brainwaves can predict those of others by a short-time period as during the reading of a story (*NewScientist*, 2010ar), perhaps indicating the integration between language and forecasting in group consciousness that has developed so far it enables subconscious prediction of communicated thoughts before they are spoken.

Since different tones or "accents" attract mates from different groupings with different tones and accents at least for whales (Kenneally, 2008a), singing could have also originally had another evolutionary purpose for its development. In particular, as with other organisms, variation in voice tones could have helped people to attract partners from other groups or clans with different singing accents that would have likely enhanced both genetic diversity and inter-group cooperation. More details on that phenomenon are provided in *NewScientist* (2007s), which explains that even birds learn new song dialects in a few years that make them less attracted to older birds in the same area tweeting the old songs. The fact that children raised in Israeli kibbutzes aren't attracted to each other (Bernstein and Nash, 2008) may stem in part from the familiarity of their voice tones, as well as the familiarity of their sight.

Different verbal tones, accents, sounds, and languages developed among humans in different groups because these forms of communication evolved for efficiency purposes in terms of time and specificity of the environment to have no connection to the original word signals (Burling, 2004). Nevertheless, with the exception of some unique grammar among Native Americans who don't always employ a subject or noun because of a culture that encourages a more fluid and accurate perception of things (Peat, 2005), the grammatical syntax remains similar across most languages despite the wide differences in the meanings of individual words (Burling, 2004). This similarity facilitates communication across clans whose languages evolved differently in different environments. Native American, who were separated from the rest of humanity by oceans for many millennia, were never influenced to develop the same grammatical framework because they had little or no contact across the great watery divide.

Communication between clans was promoted in ancient times by the custom of individuals of at least one sex entering into another clan after puberty, with such customs having evolved to promote diversity in the gene pool and inhibit the biological weaknesses in the offspring of incestual relations (Howells, 1986). For this cultural tradition to function most effectively, the speech capacities of humans needed to be able to adapt to different dialects and languages. Given that learning two languages in early childhood comes quite easily and actually enhances humans' mental and social capacities (Bernstein and Nash, 2008), it seems likely that humans evolved in an environment of peaceful harmony among neighboring clans with intermarried loved ones.

Intertribal mating, which diversifies the gene pool and avoids incestuous problems, as opposed to intertribal warfare, which kills off the fit, would make humanity more genetically robust. Since individual human personality traits are normally not significantly affected by the environment during pre-puberty years (Nettle, 2007) but are strongly influenced by adolescent peer pressure (Pinker, 2003), it seems likely that the movement of mates into other clans occurred at puberty.

Regardless, the development of language itself would have been very useful in promoting cooperation among humans and purging those with inadequate "agreeableness" genes since information about an individual harming others or the group would have spread rapidly among people and thereby led to ostracizing that could even culminate in an exiling which might have been fatal (Nettle, 2007). Humans' evaluation of themselves, i.e., their self-esteem, is heavily dependent on the opinions of others (Bernstein and Nash, 2008), and language can help communicate such important views. Thus, language would have facilitated reinforcement of behavior that was considered good for the group as a whole via "gossip" and the development of a reputation among other humans. This advantage of language is in addition to the analytical and forecasting it permits through both spreading knowledge and creating a structure for analysis and forecasting that was more powerful than instinctive intuition.

Written languages may have begun with cave paintings 30,000 years ago, and markings that likely represented mathematical symbols began 15,000 years ago, but modern written languages started about 10,000 years ago (Gynn and Wright, 2008). These scripts were originally pictorial, symbolizing more directly and intuitively what is meant (Shlain, 1998). The initial written languages may have been knots or labyrinths that were used for counting but that also symbolized deep spiritual concepts as opposed to "material" things which are mere illusions (Conty, 2002) or symbols that categorize the characteristics of everything. In ancient times, writing had a magical connotation that carried special powers (Knight, 1999), as is consistent with its importance in spiritual symbolism.

Reading normally consists of visual stimuli enabling perception of the shapes of individual written symbols or letters that are rapidly processed subconsciously in the left side of the brain into a whole image and meaning just as many other visual stimuli, such as faces, are recognized in the process of interpreting the external environment (Dehaene, 2009). The fact that reading a language involves the perception of fragmented stimuli that are processed subconsciously into a holistic form which can be consciously interpreted implies that language itself represents a natural development of the capacity of organisms for awareness and conscious self-awareness. Since the area of the brain involved in reading "originally evolved for the visual activity needed to track animals" (Flanagan, 2009c), language is merely a further development in the direction of optimizing behavior within the constraints of the existing environment. The changes in the FOXP2 gene that are thought to have enabled language in human ancestors a half million years ago and appear to control over a hundred other genes that differentiate human brains from those of chimps (Callaway, 2009e) seem to represent the biological foundation by which this development occurred.

Within the context of the group consciousness characterizing prehistoric human societies until 35,000 years ago, the development of more advanced methods of communication between humans reflects a desire to communicate better with the other parts of oneself that is integrated with other human beings. Psychic and other more ancient forms of communicating feelings enable emotional group understanding between humans that sometimes isn't as precise for logical analytical comprehension as symbolic words.

However, by 35,000 years ago females had learned the art of deception, such as via artful masks and other make-up designed to make themselves sexually more desirable, especially when they were older. Having already so manipulated their looks for probably tens of thousands of years, they then figured out how to use language in a manipulative manner that resulted in conscious deceit. Once conscious dishonesty via language disturbed the coherent communication process between humans, the group consciousness became fragmented into individual consciousnesses. While individuals divided from the human whole are motivated to act for selfish benefits and create a culture of deception and selfish exploitation, the innate irrepressible human desire for group consciousness motivates people to honesty for the mutually loving, trusting relationships for which they evolved to attain maximum happiness. Despite the widespread use of languages today to deceive, it seems unlikely it could have been utilized for that purpose prior to the ending of group consciousness 35,000 years ago.

In contrast, Miller (2001) has hypothesized that language partially developed out of an ever-escalating war between the sexes involving deception and evaluating deception in order for men to get sex and women to get love. However, the entire concept of deceptive communication across individuals appears to be inconsistent with the fact of human group consciousness and the necessity of human cohesion in prehistoric times. The mere concept of deception seems to be so unnatural that many young children literally don't know how to smile and can't even do so upon command unless they are truly happy (Burling, 2004). Native Americans have long tended to believe one's spoken word binds entire self and spirit (Hamilton, 2005).

Since words do permit people to learn to understand each other better and therefore enable greater empathy, it appears much more likely that language developed out of a desire for better bonding and group cooperation, as well as to enhance communication with respect to forecasting. It seems probable that inter-human deception, at least on the scale that exists today, grew out of the type of lies which characterize hierarchical societies like matriarchies and patriarchies that are based on falsehoods designed to induce people to obey their "superiors". As explained previously, deceptive religions with superior beings first appeared about 30,000 year ago in order to help manipulate people into accepting what were likely the first human societies with hierarchical structures. Prior to that time, honesty coincided with equality, group consciousness, and cohesion of purpose. Since people are actually better able to read lies through strictly hearing words than via a combination of words and sight (Wiseman, 2007), it seems likely that the first dishonesty occurred via mere words, such as during blind obedience to manipulative hypnotic matriarchal suggestions. Tens of thousands of years of experience with such cold lies has enabled people to develop better defenses against such cold lies unless they are combined with other persuasive stimuli.

While patriarchal conquerors replaced the dishonest matriarchs thousands of years ago, they have imposed their own set of religious and other lies. Evidence "that the acceptance of 'counterfactual beliefs' increases with age" (Dow, 2004) provides evidence in favor of a hypothesis that people are conditioned by today's patriarchal society to live with a set of deceptive assumptions but aren't predisposed to believe in those lies which are inconsistent with reality but which are at the foundation of patriarchy.

Language has actually been found to reduce creativity today, as it involves more activity in the logical left side of the brain (de Lange, 2010), and part of the reason for this detrimental effect may relate to rational attempts to perceive not only the meaning of statements symbols but also their validity. In addition, by generalizing concepts in terms of word symbols, language also omits some information and analysis that is more effectively evaluated utilizing subconscious processing that creates intuition and feelings that enter into consciousness and can enhance forecasting and decision making.

C. Consciousness, Forecasting, and Language

As already explained, consciousness largely arose from the mental processing involved in forecasting oneself, either as a member of a group in group consciousness or as an individual in a group of one in individual consciousness. The minimum level of brainwave frequency necessary for consciousness to exist may be fairly high because self-awareness requires synchronization in the neural firings of the sensory and other sectors of the brain that normally fire at different speeds, with such coherency permitting the integration of all such information into a single unified experience (Hamzelou, 2010b). The neural firings also have to be fast enough to allow the logical reflection on the experience that characterizes consciousness, thereby enabling forecasting and decision making based on such predictions as opposed to automated subconscious reactions. However, when the brain operates on the gamma frequencies that exceed those of the beta wavelengths necessary for normal conscious thought, the experience flows too quickly for conscious reflection, thus potentially making more reliance on subconscious reactions necessary in order to keep up with the fluid situations, such as rapid physical action.

As previously mentioned, consciousness developed with forecasting based on an awareness of the self in any situation, and humans needed to learn to communicate such information in order to enable coordination in their cooperative societies. They thus developed more advanced means of communication, including teaching, talking, learning, listening, and translating for meaning. They also learned to communicate with themselves and teach themselves in language form. Since language also appears to be an efficient method of storing, retrieving, and processing information in verbal sounds, the forecasting and conscious decision-making tasks related to them evolved to be performed in language form.

One's conscious talking to oneself in a language seems to have developed out of such using speech to teach others the art of forecasting and other useful complex analysis. Talking to oneself often involves the conscious trying to summarize in words what subconscious intuition, urges, perceptions, and feelings indicate in order to try and best address those issues with actions. Such internal dialogue can also consist of questioning and analyzing thoughts and ideas for actions in a verbal form that permits not only evaluation of consciously recognized pros and cons but also subconscious intuitive reactions to each silent statement.

People's mode of conscious thinking is therefore influenced by the language they speak. For instance, the Chinese language promotes a more suggestive and intuitive way of thinking that is more inclined to influence others than to develop a well-specified rational train of thought (Capra, 1999). Native American languages, which tend to emphasize actions instead of objects and thereby enable concentration on processes and relationships (Peat, 2005), are at least somewhat similar to the Chinese way of thinking. The Chinese language, which has many words whose employment as a noun, verb, or adjective depends only on "the emotional content of the sentence", and which reflects the Chinese philosophy that tends to emphasize actions as opposed to objects (Capra, 1999), may therefore be more in tune with reality in that respect than Western languages that are more focused on nouns or objects (Peat, 2005). Both the Chinese and Native American languages may reflect the origin of language as a method to induce actions in others as opposed to a method for logical thinking about fragments of things in isolation.

It is interesting that there is ongoing research into enabling people to communicate with each other by thoughts (*NewScientist,* 2008zzm). Perhaps, the effectiveness of first impressions in predicting personality that is more accurate when in the physical presence of someone than exists from just seeing a face (Wiseman and Jenkins, 2009) is one aspect of that communication. People may very well utilize electromagnetic patterns in the brain to form intuitive impressions, which result from translated signals from others by some consciously unperceived means, and subconscious sensing of electromagnetic waves and their quantum entanglement may be the means for doing so (Radin, 2006).

Einstein himself indicated that people's feeling of separateness is an illusion of human consciousness itself (McTaggart, 2008), and that view seems to reflect the nature of reality insofar as a perception of a totally separate self that is created by individualistic consciousness is inconsistent with the frequent communication, contact, and entanglement between the electromagnetic fields of people that integrate them into more of a whole.

Chapter 15
Individual Consciousness and the Relationship with the Subconscious

While the nature and even the location of the "mind" continues to be the subject of widespread dispute among scientists and others (Flanagan, 2009a), the characteristics of both the conscious and the subconscious portions of brain processing become clear when one integrates an understanding of their relationship with evolution. The mind as a whole is a changing set of coherent neural firings in the brain that evolved from a subconscious cause-and-effect analytical tool to help enable humanity as a whole to make decisions on survival and reproduction. This processing system developed through linear logic to eventually become aware of itself in order to enable it to better forecast the repercussions of its decisions and thereby improve decision-making. In particular, predicting responses in the environment longer-term required forecasting not only the direct impacts of decisions but also the effects on the self whose future decisions would affect the situation in the future. A conscious awareness of relevant information about the self facilitated such self-predictions.

The conscious mind is essentially this self-awareness that logically reflects on itself and everything else for the purpose of forecasting the self's future possible environments to enable better decision-making. The insula portion of the brain is the component responsible for the awareness of the senses and emotions (*NewScientist*, 2010as) while the prefrontal cortex is the part of the brain that is involved in the logical processing necessary for using information on the self-awareness (Hamzelou, 2010b) to forecast and make decisions.

The human conscious and subconscious mind became ever more powerful as the environment was increasingly affected by the actions of humanity, as such self-induced impacts require ever more complex thinking than was possible by the simple linear cause-and-effect analysis of other organisms. As humans' influence on their environment grew with their minds, self-forecasting became ever more important for them with respect to evolutionary adaptation to that changing environment. While the foundation of conscious analysis remains based on the same original Bayesian logic of all brains that enables learning, the human mind evolved the ability to conduct far more complex analysis by building self-forecasting onto this simple algorithmic system that is involved in subconscious processing.

The limited capacity of the conscious, however, has also required programming of the subconscious to carry out much of the multi-dimensional analysis. In addition, while the root rational logic of the conscious can't be controlled by the environment or others, it is strongly motivated via pleasure chemicals in the brain that provide pleasant feelings of happiness and self-fulfillment to serve the goal of enhancing the survival and reproduction of all humans, including the individual self. In addition, informational inputs to the conscious mind about the environment and subconscious emotions themselves certainly can affect conscious deductions, and the rational program called consciousness enables it to surrender control to subconscious tendencies and the environment whenever it deems doing so to be optimal.

The subconscious, which merely reacts to stimuli via neural responses, many of which were programmed in early childhood by the parental environment without ever being subjected to any analytical questioning, communicates to the conscious via feelings and is focused on the present moment, whereas the conscious evaluates and dreams about the future (Lipton, 2008). When inconsistencies exist between conscious thought and subconscious processing or feelings, the conscious is motivated to rationalize the differences (Gynn and Wright, 2008). This rationalization process can lead to the conscious overriding the subconscious, thinking of a reason to obey the subconscious, or some combination of those two possibilities. For instance, when there is a contradiction between logic and emotions in déjà vu feelings where a person remembers previous similar environmental stimuli but can't identify their past context, the conscious may rationalize perception of familiarity that doesn't seem justified as resulting from a dream as opposed to a prior experience whose details simply aren't remembered (Phillips, 2009a).

While the conscious can overcome subconscious feelings and tendencies with focus, reasoning with subconscious emotions is sometimes counterproductive, as frequent battles between the conscious and subconscious can result in neurological disorders (Lipton, 2008). In the case of conscious attempts to permanently repress innate human drives like for sex, the subconscious urges will only resurface in a different form that can result in especially undesirable behavior.

As explained in a prior chapter, human consciousness was originally a group one, where individual opinions of oneself were formed by the group, and decisions were naturally made in the best interests all the individuals in it that include oneself. Individual consciousness evolved out of forecasting oneself separately from that of the group, as it thereby developed opportunities for making better independent decisions for oneself. Nevertheless, the individual consciousness that developed out of this system remains highly motivated by subconscious feelings to act in the best interests of the group. Although group opinions therefore continue to mold each individual's own conscious thought, the diversity of experiences and opinions, including not only from humans via direct contact but also via media like books, permits each person to mold his or her own thoughts more independently.

The development of individual consciousness also enhances individuals' ability to take advantage of others for their own benefit. While internal brain chemicals, empathy and the risk of spiteful reactions create strong incentives not to engage in exploitation, expanded individualized consciousness has been increasing the incidence of selfish behavior as it creates feelings that are more independent of the group. Increasing selfishness has also increased incentives for expanded individual consciousness in order to prevent others from taking advantage of one's cooperative behavior. Tens of thousands of years of rule by exploitative matriarchal and patriarchal rulers have enhanced this process of individualization, which makes exploitation more difficult. While individualized thinking has weakened group consciousness, the evidence in Chapter 13 indicates that there remain strong enough innate tendencies in human beings to stop and even reverse this process which is so disruptive to humanity that it creates strong rational motives to return to some form of individualized group consciousness.

Group self-awareness and forecasting may have been partially related to the harnessing of fire and the appearance of the homo erectus ancestors of human beings 1.8 million years ago, and fire may also have been instrumental in enabling the move toward more individualized forms of consciousness that could have come with the evolution of homo sapiens about 200,000 years ago. In particular, the fact that human ancestors began to use fire to forge better tools about 200,000 years ago and that this process enabled "modern humans to conquer the world" (*NewScientist,* 2009zzzd) implies that the transformation of consciousness into an individualized one may have also been related to a further development of human use of fire. In particular, such technology no longer made cooperation crucial to survival, and so more individualistic people and groups with less empathetic incentives for cooperation could evolve without endangering survival. Increases in individualism might have even been spurred in this new environment that was more secure from predator attacks at any time as it might have spawned greater creativity that would enable further technological advances. At some point, the newest technology might have even been selfishly held secret by more individualistic people, thereby enabling them to exploit others. However, given the continued strong level of empathy and egalitarian traits in human beings today, and given the evidence on the timing of the origin of one-way worship, it is unlikely that the actual development of exploitative hierarchies occurred before the first matriarchal religions were established 35,000 years ago that were probably at least partially based on some secretive technological "magic".

Individual consciousness developed especially strongly in the last few thousand years as societies based on a more general 2-class society based on sexual groupings were replaced by patriarchal cultures based on conquest that created an individualized hierarchy consisting originally of conquering rulers and that resulted in special individualized treatment of particular elite individuals (Baring and Cashford, 1993). To the extent that matriarchies were also hierarchical with respect to having elite priestesses and female rulers, the development of individual consciousness was no doubt strongly spurred at least in some people by those prehistoric societies as well.

However, each individual's feeling of "self" continues to develop out of the image that everyone forms of oneself through memories of experiences, preferences, and other characteristics of an individual (Flanagan, 2009b). The individualized self is just a person's reflective self-awareness or conscious opinion of oneself. This self-view that is formed by coherent systems of brain cell firings is strongly impacted by others' conscious opinions, which are also influenced by the views of everyone else, and so a group consciousness continues to exist in human beings. The individual conscious that is a component of the group does represent a separate set of neural firings from others, but many factors contribute to greater coherency in the brainwaves and thus in opinions. At any rate, the individual consciousness terminates as a living whole at death along with the body even though fragments of the conscious will may persist in the form of psychic waves which persist forever.

Within the context of the feeling of self having evolved for the purpose of helping humans to self-forecast, as part of a group or as an individual therein, each person's conscious has enormous capacity to observe itself, including the individual's own behavior, emotions, and memories. Much of the information utilized in developing the perception of the self is derived from subconscious processing.

The individual mind consists of both a conscious, which is the person's command center that has final authority for decision-making, and a subconscious, which largely consists of automated processing that reacts to external and internal stimuli, including directives from consciousness itself. The conscious self effectively evolved in humans to supplement, and have final authority over, the subconscious decision-making process that exists in most organisms. The function of consciousness is to enable customized decision-making in light of probably future actions and reactions, including by oneself. That conscious dictator is highly motivated to make all decisions subject to its overriding purpose of serving the human species as a whole.

An individual is not directly conscious of the processes of the subconscious, which is involved in coordinating the more automatic actions of the human body and undertaking instinctual acts (Bernstein and Nash, 2008). Because it is not conscious, the subconscious and its activities seem quite complex and mysterious, when in actual fact the subconscious merely applies simple but effective rules of thumb extremely rapidly to any situation at hand (Gigerenzer, 2007). In the face of incomplete information, the subconscious algorithms tend to be more efficient, as the conscious can't possibly think of all the possible things that can happen and their probabilities to conduct adequate analysis. As a result, the conscious often is unaware of why actions are taken and is left trying to figure them out or rationalize them (Lawton, 2010), as the reason or motivation for actions and decision-making left to the subconscious are communicated to the conscious only through vague intuitive feelings.

Nevertheless, there is nothing magical about subconscious algorithms and their effectiveness. In fact, many are actually established consciously but are later consciously forgotten once the task has been assigned to the subconscious to free the conscious and conscious memory to address new problems and issues. Most people recognize at least intuitively from experience, that simple rules are often most effective for many applications that can be left for subconscious application. While the subconscious can process an enormous amount of information very quickly to make reasonably good decisions, its very simple rules make it easily manipulatable by others. Since people's decisions seem to be significantly impacted by totally irrelevant stimuli in the environment (*NewScientist,* 2010ai), it seems that the subconscious can even be manipulated by the environment itself.

Consciousness permits more complex decision-making. Conscious experience consists of information that has been unconsciously generated and then determined by the individual's subconscious to be made available for conscious evaluation, as in conscious forecasting of one's own actions and future. For instance, things that are consciously sensed by humans, via the eyes, nose, ears, etc., are generated by the subconscious mind, with the inputs from the senses being processed by the subconscious first before being sent into consciousness through the thalamus filter (Bernstein and Nash, 2008).

Simple rules are applied subconsciously for both processing and screening out information that is not of high enough relative priority to be consciously considered. Those rules are often changed through both subconscious learning and conscious focus. Conscious focus essentially consists of setting a priority goal or rule for the entire brain and then instructing the subconscious to do whatever it thinks necessary to carry out the instruction(s).

On the other hand, dreams and even awake imagination are largely generated internally by the subconscious. Dreams tend to be in images (McClenon, 2002) because the subconscious didn't originally evolve to communicate verbally and instead typically employs the basic sense most frequently used by humans, i.e., sight to portray the information as comprehensibly as it can in a form that can be generalized to many circumstances independent of actual words spoken. Dreams normally can't be remembered by the conscious, but they can be recalled when the mind is in an altered state, such as shortly after waking while the brain is moving from subconscious to conscious thought (Gynn and Wright, 2008). Dreaming serves many purposes, but an important one is to present intuitive feelings to the conscious that the subconscious has figured out using its own efficient algorithmic logic.

During dreams, three brain sectors, the left amygdala, the insula, and the midline that are responsible for emotions and the motivating reward system are especially active, but the prefrontal cortex that is responsible for rational self-reflection is very subdued (Hamzelou, 2010b). Since the insula is involved in being aware of the senses and the emotions, thereby creating an overall sense of self (*NewScientist*, 2010as), it appears that dreamers are aware of the dreams while they are simulated. However, people don't consciously analyze or reflect on their dream experience, at least partially to avoid conscious interruptions in the free flow of subconscious thoughts. Dreaming thereby permits imagining future emotional events without conscious interference with respect to the likelihood of them occurring, thereby helping the dreamer to be well prepared for future possibilities, including emotionally charged ones. Since the prefrontal cortex is activated when imagining things while awake (Fox, 2008), the lack of rational reflection in dreaming may be the major characteristic that differentiates it from conscious imagination.

"Lucid" dreaming, which involves being self-aware of the dreaming state, may be somewhere in between, insofar as the lucid dreams are characterized by more activity in the prefrontal cortex, which retrieves and analyzes information in the working memory of the brain (Hamzelou, 2010b). Those dreams can be recalled as a result, whereas dreams during deep sleep aren't normally directed for storage as memories at least partially because the neural firings involved in creating a perception of the self in them aren't connected by the neural firings associated with conscious reflection or analysis.

Dreams and imagination both seem to be highly related to the process of forecasting and problem solving, especially with respect to simulating different possible courses of events, actions, and likely consequences. There is even evidence that rats dream while sleeping, with the dreams including reenactment of puzzles with which they were faced with while awake (Eagleman, 2008). Human dreams involving guns have been found to help improve scores on "shoot 'em up" video games (*NewScientist*, 2009zzzx), and there is some evidence that human sleep enables people to learn through the dream experiences (Callaway, 2009d). Simulations of possible future events and a person's reactions to them are often part of the dreaming process, and such simulations can provide information to enhance conscious decision-making through subconscious intuition. This process is different from, but can complement, conscious day dreaming that utilizes more rational reflective logic to evaluate decision making within the context of forecasted future situations.

Creativity and dealing with new situations can be enhanced by dreams (Gynn and Wright, 2008), which develop various situations that can be analogous to ones faced or will possibly be faced in the future. While dreams can sometimes be consciously remembered, they more often are utilized to help the subconscious develop better rules to achieve its tasks, including by providing intuitive or gut feelings resulting from future situations that are similar to the dreamed simulations that are not normally subject to conscious recall. Dreaming also permits the subconscious to train the neural firings associated with tasks assigned to be carried out in the situations simulated during sleep.

Many external inputs to the subconscious besides dreams are never consciously perceived but do affect human beings, such as hormonal pheromones that are emitted in sweat (Williams, 2008b). Pheromones have been documented to be emitted by humans to create unconscious attraction to members of the opposite sex via a direct connection between the nose and the subconscious portion of the brain responsible for sex (Milinski, 2006). These consciously undetectable smells cause people to be attracted to those with different genes for their own immune system to provide for better defense against more diseases in any resulting offspring that have the more diverse gene combination (Potts, 2008). In addition, pheromones, which also affect the mating instincts of other organisms, appear to be useful for inhibiting sex between different species (*NewScientist*, 2010ag). Pheromones can also provide other signals, such as to enabling people to deduce fear in other humans (Williams, 2008b).

Other smells not consciously recognized can also affect people. For instance, the scent of women's tears significantly decreases males' testosterone levels even though its smell isn't consciously distinguishable from saline solutions that have no impact, with this effect possibly evolving because tears are an indicator of female fertility (*NewScientist*, 2011b). In particular, tears might indicate reduced chances of a woman being fertile, as females may be more prone to cry when they are menstruating, when they are already pregnant, or when they are in menopause. An automated turning off of the male sex drive in the presence of a sad woman might also have evolved because it might have indicated times of danger or shortages when more offspring would have been a counterproductive burden to humanity.

Other examples of external effects on the subconscious that are not consciously perceived include electromagnetic waves from stellar bodies like the sun and moon that are picked up by the pineal gland which controls hormone releases (Stray, 2005). Photon waves that people emit to possibly enable subconscious communication between each others' DNA (Narby, 1999), electromagnetic waves from the earth that human dowsers and other organisms apparently can detect (Stray, 2005), radioactivity that everything produces (Segre, 2002), and quantum entanglement effects from anything that has ever been in contact with anything else (Aczil, 2002) are just some other examples of phenomena that can affect human behavior subconsciously. Even others' emotions may be read from the electromagnetic waves reflected off others' sweat (*Newscientist*, 2008y), as well as from photon emissions. Humans are even able to subconsciously learn to recognize the lack of nutrition in artificial sweetners (Fox, 2009c). These consciously imperceptible inputs, many of which may be very difficult for most people to comprehend or even believe, are apparently processed by the subconscious and then utilized directly for subconscious action or for signals sent into consciousness via dreams, intuition, etc. Communication of subconsciously perceived information is often relayed to the conscious in symbolic or generalizable form such as via the emotions that reflect a summary of signals from the entire body.

Many subconscious signals sent via emotions may be experienced as gut feelings or intuition. Since decision-making has been found to be extraordinarily difficult, if not impossible, for people without emotions (*NewScientist*, 2008i), it seems that important subconscious recommendations based on simple rules and summarized by feelings are being routed to the conscious at all times. Many emotions derive from signals sent by other cells in the body (Lipton, 2008), and this information can also be useful in making decisions.

Since it is difficult for the conscious to prioritize and value outcomes without emotional direction (Klein, 2003), it seems likely that the subconscious also enables the conscious to attach relative emotional values to various actions and potential outcomes. The fact that people suffering from psychological paralysis of portions of their body have been found to activate the right orbitofrontal cortex area of the brain responsible for inhibition of emotions when they try to move the functional body part (Jones, 2009b) is consistent with a hypothesis that emotional inputs do influence decision-making but they can be overridden by conscious thought. The conscious can ignore emotional signals, but it does so at its own risk, as when a paralyzed person tries to move a limb that has been subconsciously learned by past failed attempts to lead to failure. It may be optimal to at least reconsider reasons why a proposed action should or shouldn't be undertaken when a gut feeling indicates not doing what is consciously determined.

However, emotions can also be manipulated to trick humans into making decisions that may be suboptimal, and, perhaps as a result, people with autism are less subject to such manipulation (*NewScientist*, 2008zzp). Autistic people are actually better able to focus on a particular problem without being distracted by signals that are irrelevant to the problem at hand but that are relevant for reflecting emotional values which are vital in many situations, especially social ones. Autistic people thus sometimes have abilities which others don't have that can contribute to the welfare of human groups, thus explaining why this trait wasn't eliminated by evolution and continues today in some human beings

However, groups of only autistic people wouldn't have great survival capacity by themselves, and so trying to mimic autistic people or become autistic in an attempt to gain some of their special powers represents a pathological direction for humanity. As opposed to trying to act like autistic people to eliminate the emotions necessary for decision-making to avoid being tricked into wrong decisions, it might be optimal to learn to better evaluate emotional signals. In particular, the conscious can deduce from knowledge, experience, and logic to disregard intuitive inputs that seem to be manipulative and counterproductive, including by reacting to such inhumane influences with human spite.

Studies have found that people with high levels of intelligence as measured by IQ scores can be poor decision-makers when they don't employ their analytical abilities sufficiently (Bond, 2009e). One cause of this problem may be their poor application of their intelligence to the evaluation of emotional signals that do provide useful information for the better decision-makers. Even the forecasting that is so important for decision-making requires a balancing between analytical and intuitive thinking (Loye, 2000), which are the tools the conscious has available to make decisions in a process of applying logic to informational inputs that include feelings generated by the subconscious.

While emotions may be needed to make the value judgments required to weigh the pros and cons of most decisions, the conscious can provide analytical judgments that provide alternative measures of both the probable outcome of taking any particular action but also its relative value. Conscious rationalization of subconscious intuition therefore serves as a means of determining if there is a good reason for the subconscious thoughts or if they should be modified. Conscious direction can also overwhelm subconscious feelings totally with alternative neural firings, thereby making it possible to prevent manipulation by the external environment or any type of poor subconscious decision-making.

The conscious can thus learn to deal with subconscious tendencies that are useless or counterproductive to optimal decision-making. For instance, there is evidence that people's voting behavior is influenced by home team sporting outcomes, with victories being associated with more votes for the incumbent (*NewScientist*, 2010an), implying that the subconscious feelings which influence decision making can be influenced by irrelevant events that reflect an individual's overall situation. The happiness from a sporting event that can carry over into satisfaction with a government leader who has nothing to do with that event can be inhibited by the mere conscious recognition of this fact, which can enable deducing the subconscious signal as biased.

Knowledge like people's decisions being significantly affected by irrelevant aspects of their physical environment, such as occurs when a harder chair for a negotiating person causes the individual to be more rigid on changing offered prices in a car purchase decision when (*NewScientist*, 2010ai), can be used to consciously reevaluate any initial decision in light of the potential bias caused by the relative pleasantness of irrelevant aspects of the environment. For instance, since the subconscious may empathetically attribute one's own feelings to others in making a decision, such as assuming the other side of a negotiation is empathetically perceived to be rigid because a person is just experiencing rigidity by sitting in a hard chair, it may only be necessary to take one's own circumstances into consideration to avoid irrelevant aspects of that environment to adjust one's intuitive attributions to others.

In any event, the fact that emotions and stress sometimes lead to the brain sending signals to the stomach and nearby adrenaline glands (Ludwig, 2004) is consistent with people experiencing intuition in the "gut". In addition, many emotional signals consciously experienced as intuition can be based on communication to the subconscious from the stomach, which summarizes the many signals it receives from all the cells that can include their detection of particular electromagnetic emissions of others. Many intuitive feelings may also stem directly from the pineal gland's processing of electromagnetic waves received by it, including both from the external environment and from the individual's own cells. All intuitive feelings, whether processed by the pineal gland or the gut, may be caused by subconsciously learned rules with respect to past experiences when the gut or pineal gland has received similar information. There could also be some subconscious processing of more material memories for this purpose.

The exact cause of the subconscious emotional signals may potentially be pinpointed more exactly by conscious recognition of specific cues, such as related to irrelevant aspects of the physical environment and hunger. In any event, by evaluating the subconscious feelings stemming from its own processing of all information using algorithms based on past similar experiences, the conscious can add some useful rational analysis of the reason for the summarized emotions that can help determine their relevance to decision-making and thereby even permit directives to the subconscious to ignore such signals in the future. Recalling memories of prior such feelings and their situational context may help here.

People's conscious memories themselves are merely the recall of bits and pieces of information about events and their general context that are consciously reconstructed to form a whole that is plausible but not necessarily accurate (Marshall, 2007). Much of conscious experience is represented by just putting into words the feedback from memory recall, from the senses, and from the chemicals in the brain like endorphins that create pleasure (Bernstein and Nash, 2008). Conscious decisions are therefore also based on incomplete information and therefore not necessarily optimal even in cases when logical thinking with full information might be more effective than simple subconscious algorithms.

Although conscious rationality apparently is useful enough for evolution to have given it final authority in making decision, the incredible power of the subconscious should not be underestimated. For instance, the algorithms necessary to solve the rubric's cube puzzle in less than 10 seconds, as some people are somehow able to do, are as yet something that today's enormous computing power still can't figure out (Palmer, 2008). As with machines themselves that can sometimes develop their own algorithms better than people consciously can (Buchanan, 2008d), many tasks are optimally delegated to the subconscious via an internal conscious order to achieve a particular goal with little or no conscious direction on how to do it.

Recall of the past and other information passing through the thalamus filter does enable evaluation of it for some particular task(s) of the conscious. The primary function of the conscious seem to be to set rules, guidelines, and priorities for the subconscious to follow and to determine when and whether focus should be directed on a particular activity or goal for maximum efficiency in an individual's operations. The conscious is needed to initially figure out matters that are either too complex or too tricky for simple subconscious decision rules without direction, but subconscious processing is generally more efficient once optimal actions or thoughts have been learned for repeated applications in similar situations (Bond, 2009b).

A. Division of Labor and Communication Between the Conscious and Subconscious

Although the subconscious may hold and process an enormous amount of information and memories that can't possibly be brought into the finite information-processing capacity of the consciousness (Gigerenzer, 2007), conscious thought has developed in order to permit complex analysis of what information is brought into consciousness. The coherent neural firings across the brain during conscious thought does enable drawing logical deductions based on various types and sources of information, but the very harmony in neural firings restricts the number of separate pieces of information that can be evaluated. Too many distinct items may cause a conflict in neural firings that leads to blurred confusion and therefore limit capacity of the working memory of the conscious. For optimal use of that capacity, information is often brought for self-aware reflections in the form of symbols that categorize things, and language is an important tool for categorization. Mathematics and the related numeric symbols can also assist in some situations. Depending on the task the conscious directs itself toward evaluating, it can send directions to the subconscious to indicate what information is to be allowed into conscious thought by focus on some category of information, as well as by setting priorities on information flow under various circumstances.

The information that is brought into consciousness for evaluation of any issue is itself subject to partial conscious control through data categorization and rules defined for the benefit of the subconscious information supplier. As previously mentioned, the focus that permits conscious concentration on only information relevant to a particular problem is an important component of intelligence (Motluk, 2007), and so conscious focus is vital for controlling the flow of information as well as for delegating tasks for subconscious processing.

The subconscious operates based on innate characteristics and habitual processes learned from past experiences in the environment and past conscious directions if there is no focused conscious direction in the present. In lieu of conscious direction, the subconscious can also communicate to the conscious its own interpretation of information, problems, and decisions, including solutions based on its own evaluation of both learned and innate rules applied to its vast store of information. Since much of the subconscious communication with the conscious occurs via dreams (Bernstein and Nash, 2008) as well as intuitive feelings, it is often in a form that is unclear and disorganized due to the relatively underdeveloped communicative abilities of the subconscious, which originally evolved to provide signals based on innate or learned subconscious algorithms for generalized application by more primitive decision-making brains. In particular, much of the human subconscious, like the processing of information by the pineal gland and gut, evolved before humans learned languages, and so it often communicates only in images and simulated sensory perceptions, such as in sequential stories that run like silent movies without much language use, although the conscious can imply language from them.

The imagined events that occur during dreams are similar to those experienced consciously by those under the influence of injected DMT, which itself can be responsible for dreams and meditation states of ecstasy (Strassman, 2001). A person may be quite conscious during dreams, including during the most blissful portion of sleep (Warren, 2007), but the individual may not remember the experience because feedback about it is not provided to the memory portion of the brain. Remembering dreams might overload the conscious working memory with excessive information that is irrelevant and that is instead designed for processing by the subconscious in the form of algorithms and intuitive feelings which develop out of the simulated events in the dreams.

Dreams themselves may represent indications of unresolved problems that the subconscious triggers temporarily into consciousness with simple but often deep symbolism that can lead to solutions as well as self-healing (Strassman, 2001). This sleep process, which isn't normally remembered, may later result in intuitive feelings or hunches. Normal outside stimuli are cut off from the thinking portion of the brain during deep sleep (Warren 2007), and so such sleep enhances the brain's capacity for addressing matters of concern in a refreshing manner without distracting external events.

The fact that noadrenaline and serotonin production in the brain is largely shut down during dreaming seems to indicate that it is then possible to re-experience events without excessive stress (Young, 2009a). In particular, because noadrenaline and serotonin inhibit stress (Klein, 2003), their absence during sleep is consistent with stress-related chemicals like cortisol not preparing the body for fight or flight reactions that would normally stress the body. The stress-reducing hormones are therefore unnecessary then. While many events imagined during dreaming would be wild enough to be very stressful if truly experienced, sleep doesn't stress the body perhaps because of a disconnect between the part of the brain responsible for dreaming and the part of the brain involved in controlling body actions at that time. Nonetheless, the rapid eye movements that occur during the dream state (Hamzelou, 2010b) indicates that the visual sense remains connected during dreaming, as is consistent with most dreams being experienced as a silent movie. Even those dreams with remembered language may reflect a connection between the visual perceptions and the verbal utterances that are later deduced to be related to the situations and actions.

Imagining future events, which may be enhanced by memory recall through neural activity in the hippocampus (Geddes, 2009a), and which are certainly based on remembered information, does seem to take place during sleep without the more extreme concern that conscious imagining might invoke. Such imagining during sleep can therefore help prepare a person for future possibilities through simulation of future situations that permit analysis of different possible actions and reactions of oneself and others without overly stressing the body.

Sleep itself is induced by a rise in the amount of adenosine in the brain that is produced by astrocyte neurons during awake hours and that suppresses activity in the brain cells of the cortex (*NewScientist*, 2010o), which is responsible for conscious reflection. Sleep therefore provides a means of allowing repair of the brain and the rest of the body with less energy (Young, 2008b) and without conscious interference. Such repairs are necessary not only because of any damage incurred over the day but also because of new learning that motivates changes in physiology, especially in the neural connections in the brain. Since basic cognitive functions like memory are enhanced by just a few minutes of sleep (*NewScientist*, 2008i), many brain repairs can apparently be preformed rather quickly when there are no interruptions by self-reflective thoughts.

In addition to the hormone melatonin produced by the pineal gland during dark hours for repair purposes, DMT may be released into the brain during deep sleep, with experiments indicating that many people under the influence of DMT experience a feeling of being tested and modified by reptilian, bee-like, and other weird forms that seems like an alien abduction (Strassman, 2001). This "dream" under the influence of DMT may reflect a symbolic attempt by the mind to indicate in a consciously comprehensible manner the complex process of self-healing that goes on during sleep and other times when DMT is active in the brain.

It is possible that hallucinating chemicals like DMT and certain electrical impulses are sometimes also released while people are awake. The self-reflection processing capacity necessary for categorization is then still functioning, thereby enabling conscious memory of them. Within this context, it is interesting that research has discovered "that when specific frequencies" of electromagnetic waves "are directed into the hippocampus area of the back brain, a large percentage of subjects reported abductions by unidentified flying objects (UFOs), out-of-body experiences, and a wide range of altered states of consciousness, including 'Union with God'" (Smith, 2006). The left side of the brain typically perceives incomprehensible thoughts or "mystical" feelings emanating from the emotional right side of the brain as being caused by external beings in its process of attempting to explain all experiences via logical influences (Gynn and Wright, 2008). The perceived experience of UFOs and aliens has for over a century also been induced in wake people whose perceptions of unnatural mental experiences change to fit their level of technology and culture (Gooch, 2007), and since people tended to have perceptions of ghosts, demons, and angels in earlier times with less technology (Gynn and Wright, 2008).

Sleep appears to help the brain process memories, with the more emotional experiences being far better remembered (Young, 2009a). However, the emotional content of dreams typically is restricted to help develop subconscious algorithms and intuition because the logical self-reflection process that takes place in the prefrontal cortex and that can categorize things for conscious recall has been turned off during sleep. Without that connection, the associations with dreamed events are strictly subconscious, thereby inhibiting bringing the events of dreams into conscious working memory. Dreams are remembered typically when they cause a person to awaken suddenly from them to enable entry into conscious reflection. Such remembered imagined events may represent attempts by the subconscious to communicate unresolved problems and inconsistencies for conscious evaluation.

Being largely devoid of any sense of outside perceptions during the deep dream state (Warren, 2007), hallucinations, an out-of-body feeling, and altered states that often exist in people under the influence of DMT (Strassman, 2001) might be quite normal during sleep. These "trips", like dreams in general, would help the brain evaluate different future possibilities and the feasibility of occurrence, thus helping with forecasting as well as preparing the brain subconsciously for action in various eventualities by rehearsing brain connections upon perception of some stimuli. However, such "trips" would normally be forgotten in most dreams, whose memory might clutter up the brain with unnecessary and generally incomprehensible information designed for general subconscious use for symbolically similar eventualities.

While DMT seems to be a more natural "high", other hallucinogens including LSD can also create similar perceptions (Nichols and Chemel, 2006) that can provide healing meaning to humans. Since inadequate sleep results in people feeling more stressed, becoming more depressed, and less able to consciously control their negative emotions in the amygdala (Young, 2009a), it is possible that the wildly imaginative nature of dreams is partly designed to assist humans in not taking too seriously their fears and other primitive emotions that they inherited from more primitive life-forms with far more limited brain capacity. The deepest interval of sleep, when there is rapid eye movement (REM), results in dreams that are more emotional and scary than non-REM dreams (Young, 2009c), implying that REM dreams are especially useful for dealing with terrifying situations and fear itself. Their most important function, however, may be to prepare the subconscious for optimizing reactions to such situations that relieve the need for stressful conscious analysis of the appropriate behavior in the situations simulated during sleep.

The fact that electrical stimulation of the hippocampus results in "a widening of emotional meaning, such that things not considered as significant would be perceived as meaningful, with hallucinations perceived as extremely real" (Smith, 2006), provides some especially useful insights on the functioning of the mind. The hippocampus is involved in categorizing memories according to time and place through emotions, conscious thought, or revisiting past events in the process of daydreaming. This area of the brain is also connected to the process of recognizing actual past events. The hippocampus seems to be fooled into believing the reality of hallucinations triggered by both electromagnetic waves and DMT when they are administered while a person is awake because the experiences haven't been generated by a conscious direction for daydreaming that is remembered as being imagined, and so the conscious deduces them as real. In particular, whereas daydreamed events are generally not considered important enough to be easily remembered, externally administered DMT or artificial electrical stimulation of the hippocampus while awake cause hallucinatory experiences to be recalled as true experiences because of the lack of an alternative explanation for the perception when the memories were formed. In contrast, while asleep and dreaming under the influence of internally generated DMT, there is no conscious memory whatsoever of imagined events because the conscious self-reflection necessary to categorize the events for conscious recall has been completely shut off at that time.

Like perceptions brought on by internally or externally administered DMT itself, the hallucinations induced by electrical stimulation of the hippocampus at specific frequencies may be caused by prompting more random connections between the electrical brain cell firings involved in memory than are normally allowed by the hippocampus. While sleeping, this electrical activity results in wildly imagined events that are simulated without conscious reflection of their possible occurrence in reality, and so there is nothing for the conscious to recall as significant. During daydreaming, the conscious recognizes the imagined nature of simulated events, and so they are also generally relegated to have little lasting emotional impact. On the other hand, artificial electrical stimulation of the hippocampus and DMT in an awakened state results in perceptions based on chaotically scattered bits of information which can be bizarre enough to provoke strong, self-reflective emotional responses that can't be explained, and they are therefore recorded in memory as real enough to greatly affect the person's conscious view of things.

Electrical stimulation of the hippocampus could even spark the internal release of DMT in the brain that might alone cause the abnormal remembered perceptions. In particular, such electricity may prompt the neurons in that area of the brain to register a strong emotion in consciousness that can't be explained with normal neural connections, and so the brain automatically supplies DMT to enable new and wild connections between memories of many different sorts to enable creative reactions to the abnormal experience. A new categorization of thoughts, memories, and perceptions can result that may even reflect reality more closely than the normal categorization process. In particular, normal conscious perceptions optimally adapt to ignore a great deal of irrelevant information in order to enable survival in a usual environment of normal experiences.

Given that far more detail is perceived even by the traditional five senses than is normally registered consciously or remembered but is indeed detected by some people (Gynn and Wright, 2008), DMT and electrical stimulation of the brain could even conceivably result in improved perceptions of more actual phenomena that permit greater understanding of microscopic reality. Babies' brains may actually experience true reality more closely than adults because they haven't yet learned to categorize information into artificial symbols that result in consciously ignoring information that is normally irrelevant to human survival.

The process of generating exceptional insights into problems which can't be resolved with standard modes of evaluating perceived relevant information seems to involve thinking consciously about a problem that can't be solved with mere conscious thought and then putting it aside until a solution can pop into consciousness (Miller, 2010). This diversion of conscious focus for other activities may enable the subconscious to access and process ignored information using its own algorithmic logic, with its conclusions later delivered for conscious evaluation in dreams, intuition, and other sets of neural signals.

In any event, an individual's subconscious, i.e., the portion of the brain and its logic about which one is not conscious, can powerfully affect a person's actions and the conscious itself via instincts, intuition, feelings, dreams, and control of information flow. However, despite the incredible capacities of subconscious processing, and despite the fact that they control almost all brain activity (Frith, 2008), the subconscious is far from infallible. For instance, it can have its simple rule-following algorithms manipulated by others able to predict them from past patterns in the behavior they cause. Perhaps as a result, the conscious has evolved to have final decision-making powers. Although the subconscious was the predecessor of the conscious and can act to carry out most tasks alone, even automatic functions like breathing can be controlled by the conscious. Athletes are also able to program their subconscious physical movement abilities by just consciously thinking about them or rehearsing them in their minds (Bernstein and Nash, 2008).

Conscious focus can swamp the robotic inclinations set by the neural firings of the subconscious and thereby change subconscious inclinations and actions. The conscious can, especially via repetition of focus or self-statements, cause undesired subconscious tendencies to dissipate as the associated neural connections in the subconscious are unlearned by being repeatedly subdued by neurons relating to self-reflective thought out-firing them. Neural firings that lead to undesirable thoughts or actions may eventually stop firing altogether even without further conscious focus after initial repression by strong-willed concentration.

However, redirecting unwanted tendencies, desires, or feelings tend to be easier and more successful than repressing them. In particular, the train of neural firings associated with stimuli that prompt undesirable thoughts or urges can be interrupted at some point between the subconscious perception of the stimuli and the decision to act undesirably on them when neural firings causing the desired reactions are consciously directed to out-fire them at the selected point. The neural firings associated with the redirected wanted tendencies eventually become habitual and automatically swamp the undesirable ones.

Habits represent a sort of averaged uncollapsed probability wave function that may be broken by conscious intent which collapses the wave on an alternative point. That collapsing of the wave function can be accomplished via self-hypnotic concentration on an alternative event that excludes the possibility of the habitual average. Such focus also enables collapsing the probability wave functions of other people and the environment, thereby maximizing the likelihood of experiencing that other event which occurs only within the context of the system of the interrelated probability wave functions of all things. This exercise of concentrated will power can be incredibly powerful and can't be negated by the wills of others as long as a person's intentions are consistent with human nature, as will be explained in Chapter 16.

In any event, the conscious receives information from the subconscious and then attempts to figure out what orders to give to it subsequently. Orders to the subconscious can include continuing as before, as well as consciously setting new rules for behavior, information filtering, and processing. The conscious can also direct the subconscious to follow the suggestion of others. Thus, the subconscious and all its various modules are made to follow the general procedures set by the conscious mind (Stanovich, 2004). Even when the subconscious appears to operate independently at times or be under the control of others, it does so only because of conscious direction or lack thereof.

Subconscious rules which have been set that are inconsistent with other established subconscious algorithms, or with satisfaction of innate needs and desires, can lead to recurring messages of conflict in intuition, emotions, and dreams. Indecision is a natural result. The subconscious leaves the final decision up to the conscious portion of the brain on how or whether to resolve those conflicts and continues according to pre-existing tendencies only if the conscious is unwilling to make a decision to change to resolve the inconsistencies.

Even seemingly forgotten traumatic events which have resulted in certain seemingly uncontrollable subconscious behavior, such as fear in some environments, can be changed by just bringing the memory of the traumatic experience into consciousness (*NewScientist,* 2008u). The conscious can then logically categorize the memory as a special situation that shouldn't be generalized to many similar but irrelevant circumstances for which the undesired reactions can then be easily recognized as inappropriate. The strong evidence of the frequent success of this procedure confirms that the conscious can rewire most subconscious tendencies, which often enough overgeneralize responses to stimuli.

However, there may be some characteristics of human nature that can't actually be rewired. For instance, the feelings of unhappiness and emptiness that inevitably result from attempts to rewire the brain to act strictly for selfishly utilitarian purposes may not be re-programmable, as human consciousness and brains evolved to serve humanity as a whole. As will be discussed more later, human happiness and self-fulfillment have been programmed by evolution to depend on serving that higher purpose, and a desire for physical, emotional, and spiritual union with others is too innate to consciously eliminate.

Individuals can change most rules for their subconscious at will, but it can also consciously set none at all. The subconscious can even be set to override consciousness in case of certain events, and it instinctively is pre-programmed by genes to do so in situations dangerous to self-preservation. Except for the basic evolved need to have an intent to mutually serve humanity as a whole, rules set or pre-programmed for the subconscious can be reprogrammed.

The human subconscious typically operates most efficiently using one-dimensional decision-making insofar as it makes selections between alternatives employing just one criterion deemed to be the most important for making the decision, as occurs among other organisms as well (Gigerenzer, 2007). While faster, the subconscious operates in a somewhat rigid rule-based form, and so the capacity to be able to change some subconscious rules is useful from an evolutionary point of view. Consciousness evolved at least partially for the purpose of monitoring subconscious reactions to the environment to improve the optimality of the algorithms employed. The flexibility of being able to use conscious decision-making and logic to reprogram the subconscious through the conscious is especially useful given the fact that life and the universe are ever changing.

While conscious interventions in subconscious tendencies, both with respect to resisting and changing them, are designed to enhance decision-making, the very process of evaluating behavior that is subconsciously directed can detract from the effectiveness of that behavior. Nevertheless, the conscious slowing and changing of subconscious rule-following certainly do inhibit manipulation by making behavior less predictable, as can be an important trait in the face of selfish other human beings who are able to forecast behavior from past patterns. The unpredictability of human beings also makes it difficult for other organisms to evolve traits over long periods of time that would enable them to take advantage of automated behavior.

On the other hand, an over-reliance on the logical left side of the brain leads to an overconfident, mechanical, and fragmented "take on reality" that inhibits insights into the interconnected nature of things (Flanagan, 2009d). Intuitive inputs into conscious decision-making from subconscious processing that can more efficiently evaluate and react to the incredible complexity of such relationships are therefore rational.

The subconscious can much more quickly process a far greater amount of information, especially about a person's own senses, and so it is normally left with as many tasks as possible without conscious interference, or at least as long as the individual feels good subconscious vibes about it doing so. Although the subconscious has very limited mathematical capacities, typically being unable to count over four, its simple rules of thumb often are superior to the most advanced statistical analysis when there is a possibility of past spurious correlations or accidental associations with less important factors (Gigerenzer, 2007). The subconscious is continuously thinking up new theories or rules for itself even without conscious intervention (Burling, 2004). It frequently even sets and engages in sequential decision-making via the anterior fronto-median cortex of the brain when the criterion deemed to be most important doesn't permit selection (Gigerenzer, 2007).

The subconscious can also signal that it believes it best to leave the decision up to logical conscious processing. The conscious can even set the rules for such signaling to be based on detection of particular tangible or even invisible stimuli like electromagnetic waves incoherent with a particular activity or situation. The incredibly powerful but non-controlling nature of the subconscious serves strictly as a useful tool for conscious direction in perpetuating humanity.

It's especially important in social situations to delegate more decision making to the subconscious, which can then process the enormous amount of invisible electromagnetic and other information that enables optimal behavior in relationships with other human beings. Some imbibe alcohol in order to help free their subconscious from conscious interference, thereby often allowing them to function better in social situations.

However, the same effect without the distorting influence of alcohol on perceptions can also be attained for social relationships by mere conscious focus on a general goal, such as making friends. The subconscious can then operate freely within this guideline without conscious interference in the form of slow evaluation of whether anything is socially acceptable or not. For instance, the "old adage" of "be yourself" may be optimal for making friends because it avoids conscious interference in the efficient subconscious processes necessary to really be humanly connected.

With respect to winning "mating" friends, the goal would have to be a mutual surrender that is consistent with human nature and its overriding purpose of serving humanity as a whole. For instance, the subconscious can be given instructions to provide flirting signals subtly indicating an offer of "love" to the partner that can be withdrawn if they are negatively received.

In any event, consciously learning about others and what others like can help in any social situation and in attaining goals related thereto. Showing sufficient interest to find out or ask can be very productive here. Conscious evaluation of that information can enable drawing people together, but the rapid reactions needed in an ongoing social encounter typically require more reliance on subconscious processing. Of course, the subconscious must be directed toward a mutual affection for the best chances of success, as obtaining social contacts for strictly selfish purposes typically results in negative signals or "bad vibes" that can be subconsciously perceived by the targeted "friends" in the form of an intuitive dislike and even be sensed as manipulative. Counterproductive spiteful reactions by others may therefore even result from such intentions.

Thus, selecting appropriate attitude guidelines or goals for the subconscious to follow enables better adaptation to many environments than consciously trying to evaluate the situation. Specific rules are less effective than general guidelines especially in social situations because of the uncertainty and complexity of other human beings, who act rather unpredictably due to their own human nature. Behavior in these situations that are too complicated and fluid for conscious evaluation are typically best left up to subconscious guidelines that permit rapid automated adaptation to a quickly changing environment. However, the overall guidelines or patterns can optimally be modified in light of new information.

The process of largely surrendering decision-making to the subconscious to follow a particular pattern of behavior typically involves focusing on a personality prototype specified by the conscious. Such a system can include hero worship, whereby individuals act as they think some known hero would act (Kharkhordin, 1999). Since even apes' tendency to look at those most revered in a society (Phillips, 2008a), it may be that hero reverence, or at least a desire to copy, is a natural tendency of primates in general.

An imaged personality may be learned via impressions from others as well as from idealized, exaggerated or even fictitious representations of individuals that are often promoted in stories and films. Different images can be combined to form new types as well as varied to be dependent on the situational context.

In any event, following a personality image permits the subconscious to more quickly determine appropriate actions. Since the subconscious can then simply prompt actions as it perceives that the imaged individual would in all specific situations which are encountered. Adopting a different personality actually causes the brain's overall electrical pattern to change distinctively (Gooch, 2007), and the personality selected can be determined at conscious will for different situations when appropriate (Gynn and Wright, 2008). It is even possible to have the subconscious adopt a specified personality until general or specific changes in the environment require either adoption of a new personality prototype or conscious intervention for determination of subsequent action.

While a person can change the individual's own personality image to fit different environments, some degree of consistency may be easier and better for purposes of meeting the expectations of particular social relationships. However, for optimality in running on the automatic pilot of the subconscious, it's important not to try to project an image to others, since attempts at image projection leads to feelings of self-consciousness that inhibit the natural flow of the subconscious. Each person needs to establish an attitude or personality prototype for the individual's own subconscious to follow without any conscious worrying about what others think for the subconscious to be most effective, as the conscious evaluation involved in trying to be self-aware of others' opinions is too time-consuming and disconcerting for subconscious efficiency. In addition, an individual basing actions on the opinions and expectations of others enables manipulation of the person, as well as possible undesirable perceptions by others as a manipulative artificiality.

Although many subconscious tendencies and characteristics are learned from others and their reactive opinions, those inclinations can be changed. It is especially important that perceptions of some intolerance by the elite's robotic slaves for a person's selected personality and the goals the individual seeks to achieve to be either ignored or treated with contempt to prevent self-conscious defensiveness that disrupts the natural subconscious flow. If any reaction at all to such subhumans is deemed optimal by the subconscious, the response might sometimes optimally be directed toward political or philosophical issues. Explaining an individual's religion to be humanity's purpose to "make love; not war" as opposed to the "make war; not love" philosophy that Christianity promotes for the benefit of the Christian elite may be especially appropriate.

Modifying a particular personality type that is somewhat affected by genes can take a significant amount of effort (Gigerenzer, 2007), and so the choice of the overall image may optimally reflect a style most similar to the basic personality most suited to an individual's inherited genetic composition. However, evidence such as that indicating a person's personality type is influenced by the perceived sexual attractiveness of the individual, with the more attractive being more extrovert (*NewScientist,* 2010af), by itself implies that any seemingly inherited personality tendencies are interactively determined with the environment. While personality tendencies are heavily affected by heredity (Giles, 2008a), they are far from deterministically set by an individual's genes. Each person has the capacity to employ many different personalities (Gooch, 2007), and it is possible to simply concentrate consciously on the image that a person considers appropriate for any particular situation.

Regardless of the personality pattern selected, playing a particular imaged part as in a play allows the subconscious to function most efficiently in the vast majority of environments. Social situations especially lend themselves to role-playing not only because they are invariably complex and fluid but also because they require the subconscious to process an incredible amount of information, much of which isn't even consciously perceptible. The conscious mind with its limited working memory really can't handle the extremely large number of variables and their complex interactions involved. The subconscious neural connections evolved under conditions of no conscious intervention, and so their effectiveness requires following them without conscious interruption that are optimally avoided, extreme circumstances such as violations of laws that can lead to penalties and incarceration notwithstanding. Being oneself in one's customized image also facilitates a complete trust and friendship with the necessary "chemistry" because it shows that it isn't just a product of rational "manipulation".

Despite the multitudes of different categories of personalities, there seem to be two types humans have learned to adopt. One is the playful social type that evolved for courtship and social relationships, and another is a professional sort that evolved for work-related activities such as seeking and obtaining food and water or defending against predators in prehistoric times. In humane societies, these personalties can be merged for playful use throughout life. While today's sick society that is founded on cruelty to human beings encourages the development of those two separate personalities for outward show, basic human nature can remain integrated by seeing them as mere adaptations to the environment, just as fighting lions required a different type of action than courting a mate in prehistoric times that could also serve to protect and win love. A huge number of different variations and mixes of various types of personality are possible, and happiness can be enhanced by the knowledge that each person has the freedom to choose and change the individual's self-image at any time.

One personality that is promoted strongly by U.S. culture is the slave's doctrine that combines the propaganda of just "work hard" with "if you're so smart, why ain't you rich?" This philosophy widely followed by Americans helps ensure that the non-ruling people blamc themselves for their slave-like predicament imposed by the capitalist system that needs to inhibit human spite since it seeks only machines to exploit. The U.S. philosophy itself affects conscious thought by focusing people's attention on work and making money. It thereby distracts the conscious from more natural and effective means of meeting human needs and desires that would eventually cause a spiteful overthrow of the cruelty and inhumanity of the system. A far better philosophy is to adapt the German adage of "work smart, not hard," that might be useful in all activities.

Even without a particular conscious image or hero personality being projected, people do exhibit patterns of thought and behavior that develop into personalities. Different personalities in different contexts are frequently created for the same person because the mind perceives an environmental situation where a particular pattern of thought and behavior is believed appropriate (Carter, 2008). Each human's self thus naturally differs in different situations (LeDoux, 2002).

Multiple personalities in the same person are sometimes even completely unaware of, or disconnected in the brain from, each other in extreme cases (Carter, 2008). The ability of people in altered states of minds, such as under hypnosis, to imagine being people long dead (Gooch, 2007) may also reflect such acting that can even seem real through empathy. As Morrell (2005) indicates, people become what they project. On the other hand, without conscious selection of an image or images, a person's pattern of behavior is determined by the environment that includes manipulative people and societies.

The media in today's modern age also provide role models through movies and television shows that can relay messages consciously and subconsciously about socially acceptable behavior, just as stories and myths did in the past. It has been shown that reading fictional stories enhances empathy and social skills, and television and movies may do the same, at least partly by identifying with the characters, indicating how one should act in different situations, and improving social reasoning (Oatley, 2008). Media images, including via films that are known to be fictitious, are so powerful that they often result in a combination of smiling and teary eyes in viewers upon seeing a happy resolution for the hero(es). The latter mixed emotion may reflect the happiness from the viewers feeling natural human empathy, which permits imagining the joy of the star(s) while the viewers also feel sadness because equal glee is not experienced in reality by the spectators, who in a humane culture would live in ecstasy.

In any event, selection of media images should be judiciously evaluated due to many their projected role models being designed to manipulate people into robotic slavery or exploitation. In addition, television with its focus on idealized role models, whom most can't possibly emulate for genetic or money reasons, can result in feelings of comparative unworthiness and helplessness that cause depression (Klein, 2003). The resulting negative emotions create stress that has very negative health effects (Lipton, 2008). The problem is further exasperated by the frustration created by television advertising of products that many can't afford and that don't provide any lasting happiness or meaningful relief from negative emotions even when they are consumed (Klein, 2003). In fact, the payment for such consumption items can increase unhappiness and feelings of emptiness because of regret from the recognition that the money paid didn't buy self-fulfillment or anything of sufficient value to be worth the price.

Some advertising has very direct negative effects on physical health. For instance, television advertising of chemical products for medicinal cures to particular health problems has an anti-placebo or docebo effect that increases the incidence of the advertised diseases through the power of negative thinking (Lipton, 2008). The advertised drugs themselves are also typically not much more effective in treating the "promoted" diseases than a "useless" placebo would be. In addition, the taking of medicines can have a very adverse impact through warnings about their side effects that become self-fulfilling properties (*NewScientist*, 2009zs). Death induced by application of "modern" medical care has risen to be the leading cause of mortality in the United States, although that fact may be at least partly because chemical drugs treat symptoms of diseases while largely ignoring the cause of diseases, virtually all of which have been linked to stress (Lipton, 2008). Thus, television, which promotes only such solutions to problems, may indeed by deadly for human beings, although lifeless corporate businesses do enhance their image through such brainwashing tools that help them maximize their profits without regard to the effect on people.

Given the extensive manipulation by the mass media that heavily employs psychology, deception, and purchased role models or experts to manipulate and exploit others (Miller, 2008a), caution should always be employed when selecting a particular personality image not observed in real life. Actual historical characters may be more appropriate role models. In addition, instead of utilizing other personalities as prototypes, it can often be optimal to instruct the subconscious to adopt the attitude that one has oneself employed in successful situations in the past that seem appropriate.

While a human being typically operates most effectively with conscious concentration on an image or role to be played for purposes of attaining a consciously set goal in any particular situation, it is also possible to have the subconscious ready to switch to the selection of another personality type upon perception of specific adverse reactions or changes in the environment. However, the power of subconscious processing is enhanced greatly if one doesn't consciously question the suitability of a selected personality at the time of its application, as such conscious doubts interfere with the unconstrained subconscious operation necessary to accomplish whatever tasks it can as efficiently as possible.

A failure to make a conscious choice of one's own image results in one's subconscious largely acting out an image desired or expected by others, including the media. The power of hypnosis itself works at least partially on role-playing to serve the suggestion(s) of the hypnotist. Except in cases of self-hypnosis, the result can be robotic slavery to the suggestion(s) or suggested image(s) set by the media or others. Whereas playing a role because of the expectations, desires, and hypnotic suggestions of others may have been beneficial to the group as a whole in the prehistoric days of group consciousness and mutual group hypnosis, such surrendering of control in the sick contemporary societies that motivate selfish fragmentation result inevitably in manipulation and exploitation to the detriment of oneself and humanity as a whole.

The subconscious is often strongly influenced by what a person is taught while a child (Lipton, 2008). As a result, Christian beliefs typically have a very powerful impact on people in the Western world (Lovelock, 2009). Evolution enabled strong ingraining of information learned as children because it helped kids coherently exist in environments similar to that of their caretakers that typically turned out to be similar for the next generation. This evolved human characteristic designed to perpetuate humanity has in the last few thousand years been hijacked by artificial patriarchal religions that have declared war on human nature and are thus the pathological enemy of humanity. As such, Christian brainwashing of children is extremely cruel and extraordinarily detrimental to their mental health. Through an ingrained belief in and fear of hell for failure to worship a fictitious deity, patriarchal religions like Christianity motivate people throughout their lives to consciously fight their human nature by suppressing innate subconscious drives and redirecting the natural purpose of human beings to serve humanity into a slavery to "superior" beings.

Thus, while the subconscious operates best without conscious interference, it often optimally needs direction by the conscious, or one will be directed by the environment in a counterproductively mindless stimulus-response fashion that is easily exploitable. Even without manipulation by others, failure to provide conscious direction results in a person acting like a chicken with its head cut off. Today, a failure to develop one's own conscious direction based on a conscious rational analysis of the meaning and purpose of one's own life almost invariably results in a robotic existence for the purpose of being mercilessly manipulated and exploited strictly for the benefit of other selfish individuals. Recent research findings indicating that human activity follows a habitual pattern not much different from that of animals and insects (Buchanan, 2010) implies that humans are still strongly affected by inclinations programmed into organisms over hundreds of millions of years.

On the other hand, with the knowledge that each person evolved to serve humanity as a whole, each individual can always change any undesired activity by acting according to that simple guideline. The personality style that may be designated for this purpose can be whatever the person perceives a human being of the future would want. Oneself projected empathetically to be living in a future humane society is a particularly easy image to create, and one can imagine that one is traveling back into the past to try to contribute toward the creation of such a blissful environment. Such a projection may actually not be too far from reality, as will be explained later. If one is ever unsure how to act or react in a given situation, one can ask oneself what a true human being would do. An individual thereby harnesses all of the tremendous subconscious and conscious power of human nature by effectively adopting a personality role model where one plays oneself.

B. Conscious Control of the Perceptions of Reality and "Psychic" Information

Although the subconscious actually determines what is allowed into consciousness (Douglas, 2007), the conscious has virtually complete power to program the subconscious to allow into consciousness whatever the conscious wishes. One can therefore truly create one's own reality consciously (Morrell, 2005). However, the conscious remains strongly motivated to construct a reality that contributes to the long-term benefit of humanity as a whole, as adverse feelings of emptiness occur upon attempting to manufacture a hedonistic reality that serves only oneself.

The conscious sets the rules for the subconscious with respect to both actions and perceptions that are allowed into reflective self-awareness. The capacity of the conscious to change the subconscious when it perceives that the environment warrants it, may be a key to the evolutionary survival and thriving of humanity

Conscious programming of the subconscious and its own rule-following process can also include a suppression of verbal consciousness and logic in order to permit better access to sixth senses, intuition, and the subconscious (Begley, 2007a). It is actually possible to better experience some subconscious knowledge by deliberately silencing the conscious voice and letting the subconscious take over completely, as the Taoists recommend (Capra, 1999). A Ouija board, if not consciously controlled, can also be used to obtain useful intuitive deductions from the subconscious (Lawton, 2007). Accessing the subconscious in such a fashion may actually represent attainment of a superconscious, which Morrell (2005) indicates is a phenomenon that integrates the entire mind and other information picked up by the sixth sense.

The altered state of consciousness that is experienced when the sense of self is given up by suppressing activity in the left frontal lobe of the brain does seem to permit tapping into a larger information field that may include psychic use of forces such as light, electromagnetic, and quantum waves not normally detectable (McTaggart, 2008). The process by which such forces can affect one's own brain may involve merely opening up and relaxing the brain's information filters to enable subconsciously provided information on external electromagnetic and other waves to affect the frequencies of one's own brain neural firings, thereby empowering one's subconscious, feelings, and intuition.

Given that matter itself is not really solid but more of a process of electromagnetic waves or vibrating fields of divided nothingness, the usefulness of the human capacity to modify simple subconscious rules that determine behavior and perceptions without conscious interference may be enhanced via methods of harnessing the laws of quantum mechanics. Besides an improved ability to develop and apply newer technology, a change in the way one interprets and reacts to internal and external stimuli can enable greater "psychic" powers. For instance, one can actually consciously feel good and bad vibrations as a result of coherence and incoherence, respectively, with all the other electromagnetic waves.

The vibrations of the external environment do form human beings, and aboriginal peoples attend more to such forces than people in modern cultures who rely less on their intuition (Lipton, 2008). Good and bad vibes are apparently perceived subconsciously and felt consciously as imprecise intuition. One can consciously try to enjoy and magnify good vibes while attempting to avoid bad vibes. This process can occur by simply moving toward environments that have a good intuitive feel to them while avoiding environments that intuitively create bad feelings.

Nevertheless, one can actually affect the vibrations in the external environment, as well as in the internal environment of a person's own body, through conscious internal direction of emotions (Braden, 2007). As a result, it is also possible to try to make bad vibrations in the environment coherent with one's conscious will via the control of those emotions. Emotions have an effect on the environment whether they are consciously directed or not, and there is a hypothesis that individual conscious regulation of the emotions will be the next big step in the evolution of human beings (Braden, 2009), thereby greater enabling control of any situation.

The biology of all organisms is affected by invisible forces that include microwaves, radio frequencies, and consciously undetectable frequencies, as well as by visible light spectra (Lipton, 2008). Such forces and the extraordinary power of the quantum entanglement of microscopic particles across space have an impact on individual cells (Braden, 2007). Cell activity may be generalized as a process of either opening up to positive stimuli to enable growth or closing up to avoid negative stimuli, and this process is affected by the conscious outlook that, for instance, shuts down the immune system during times of stress (Lipton, 2008). For example, negative emotions affect over one thousand different biochemical processes in the body in a manner that adversely impacts a person's health (Braden, 2009).

Matter without life is also affected by such powers, and quantum entanglement actually enables conscious intentions to have a discernable effect on both animate and inanimate objects across both time and space (McTaggart, 2008). The shattering of glass via singing in harmonic tendencies that magnify the intrinsic vibrations of the glass is just one example of seemingly mysterious forces at work (Lipton, 2008). Merely setting frequencies for the human voice vibrations to resonate harmoniously with the wavelengths of the receiving glass causes the shattering (Robson, 2010a), thereby providing an indication of the incredible power which human beings can harness via thoughts and actions that are coherent with the forces and nature of the universe.

The effects of thoughts and actions that are in harmony with nature might be even greater on live organisms and conscious human beings because they represent systems of vibrations that are operating with greater coherence so that resonating with them impacts the whole organism that in turn affects other matter and organisms. While both animate and inanimate matter are made of the same oscillating electrical fields caused by divisions of nothing, live cells and organisms represent organized systems of vibrations that are already at higher levels of resonance, and so resonating the frequency of one's own vibrations with theirs enables greater flow through to other matter.

A harmonious communal influence thus has a more powerful influence on the entire environment external to one's own system of vibrations. Choosing to resonate with other live systems doesn't imply, however, enslavement to them, as the power created by one's focused resonance can be directed to whatever one wishes in a manner that influences others more than oneself is impacted. For maximum power, the innate tendency to make a choice between two alternatives that leads to dividing the world into forces of "good" and "evil" must be redirected into harmony with the nature of the universe, so that the natural resonance with human nature in the quest of advancing humanity is considered "good" while patriarchal religions like Christianity and systems like capitalism that have declared war on human nature are categorized as "evil". The christians are the new lions and killer bacteria of today that prey on human beings to perpetuate an unnatural, ignorant elite, many of whom may be descendants of cruel barbarian conquerors who are long dead but who are worshipped in patriarchal religions like Christianity as "deities".

Despite such adverse influences that have such a powerfully negative impact on many people's subconscious, even the sickest mind created by Christian lies can be changed. The fact that both external and internally emitted external electromagnetic waves that are invisible can be controlled via emotionally confident, informed focus facilitates an understanding of how relatively easy it is to consciously change one's own ingrained beliefs and tendencies. Just as killer bacteria can be destroyed or made impotent and even transformed into cooperative species, so can the vermin of Christianity be purged or redirected for the good of humanity.

While the systems of electromagnetic fields that interact with each other and the environment in ways that can be readily measured today have been researched by biology and chemistry much more than the "psychic" external forces have been, the more invisible electromagnetic forces are indeed quite powerful and can be controlled (Lipton, 2008). It is only necessary to induce a conscious intent to do so, and the subconscious can do the trick, or learn to be able to do so through its own process of trial and error.

The subconscious itself not only affects to the environment but also reacts to it. Although the conscious has the power to act as the omnipotent commander for the subconscious, a failure to exercise that free will results in the subconscious being more readily controlled by external forces. For instance, the subconscious can, in lieu of direction from the conscious, act according to rules and suggestions set by others. For instance, many people let others or religions establish the behavioral guidelines for their subconscious and even enable external control of the direction of their own conscious thought. They thereby transform themselves into zombie slaves that would subconsciously appreciate being woken up and freed. Some people allow their thoughts and actions to be determined by genetically predisposed tendencies and desires, as well as by obsolete logic such as infantile behavior and ways of thinking learned automatically during childhood, and they too are equivalent to robots without conscious self-direction.

The subconscious controls the vast majority of all human activities, and it undertakes so many tasks that it sometimes seems to control people (Douglas, 2007). Even when the conscious has established the framework for subconscious processing, the time and reason for establishment of a subconscious tendency or rule is often not remembered or known. As a result, consciousness sometimes merely acts to provide justification for what the subconscious already has decided to do and done when conscious direction for the action isn't provided (Phillips, 2006). In particular, being unconscious of subconscious "logic" or rules, which the conscious may have actually set in the past but not remember, the actual reasons for actions determined by the subconscious can only be deduced by the conscious. Sometimes those deductions are invalid.

Nevertheless, despite the power and usefulness of the subconscious, a person is not at all a slave to it or its habits. It is quite certain that the subconscious doesn't control or manipulate conscious thought for any purpose except for its evolved task of advancing humanity. It merely follows orders of the conscious for that reason, while the conscious merely answers to itself for its own direction which is strongly influenced by feelings of self-fulfillment when it isn't acting selfishly or otherwise deviating from its evolved purpose. While the subconscious has an enormous influence via fulfilling automatic preset rules and filtering in information to the conscious, individuals can always consciously reset those internal rules, although it is easier to so without negative side effects and fetishes if subconscious tendencies are redirected toward ones coherent with innate human nature.

Subconscious rules, both for actions and for filtering in information to the conscious, can be consciously changed even if the conscious is unaware exactly what those rules are because it doesn't remember setting them or they were set in childhood. It is only necessary to establish a set of actions that conflict with them. Much less will power is required to maintain those alternatives if they serve humanity as a whole, as they thereby provide the greatest reinforcement of all in the form of feelings of self-fulfillment from having done what each person was born to do. The conscious mind can even override reflexes and all innate desires, including for nourishment and sex, as long as its underlying intent is to benefit of humanity as a whole. Thus, despite the mutual feedback between the conscious and subconscious, the conscious is 100% in control, albeit subject to the resurfacing of any subconscious tendencies, such as the sexual drive, when the conscious will power is weakened by establishing higher goals than serving humanity.

While the conscious can overrule old habits and the subconscious itself, it can only do so with effort. In particular, habits and addictions can be broken with sufficient self-will or conscious direction to change. An individual normally may have only a limited amount of "will power" each day, and excessive use earlier in the day reduces the amount that can be expended later, although the consumption of foods or liquids with lots of glucose or sugar can help build those resources back up (Phillips, 2008b). In any event, one can always summon the necessary will power to change a habit or addiction with enough belief that one can.

Breaking a habit initially requires "will power" being activated in the right frontal lobes of the brain that utilizes a great deal of finite brain energy, but this process can be facilitated by establishing pre-set rules to follow that are more powerful if they are written down (Phillips, 2008b). Setting rules to follow seems to involve the conscious reprogramming of the subconscious. They also make it easier for the conscious to decide to override any urge to violate them without having to evaluate whether doing so is optimal or not. They thereby conserve the limited energy and capacity of the conscious, which can suffer from overload or burn-out if it is utilized excessively.

The brain has a remarkable degree of plasticity with respect to its ability to change or rewire itself, even in the face of injury and disease, including consciously, and even inadequate learning in early years of life can be completely reprogrammed (Bernstein and Nash, 2008). However, healing "bad" or inadequate brain connections takes a great deal of effort from which children can be spared by playful, loving custodians. Parents of children can thereby also increase the likelihood of "good" brain connections since people faced with expending too much effort to have healthy brain connections may decide it's not worth it. They may even be unable to recognize unhealthy tendencies as such. The widespread evidence that parenting behavior does affect children's personalities (Bernstein and Nash, 2008) is consistent with this conjecture. People's behavior is also heavily influenced by genes and peer opinions (Pinkner, 2003), especially after puberty, but in the final analysis children simply grow up to have minds of their own, as most parents are well aware.

Although humans can use the conscious to reprogram the brain and the simple rule-based behavior of the subconscious, it's easiest to just stay the course if a person is comfortable with matters. A failure to make any decision to reprogram the subconscious results in an individual following old habits and addictions.

Human behaviors tend to be motivated by pleasure chemicals in the mind, such as endorphins for useful stimuli and dopamine in anticipation of rewards. These incentives evolved to usefully direct brain functions and actions. However, they can also lead to unwanted habits and addictions due to distortions in the brain's system of expectations that sometimes reward anticipation of a habit with dopamine even though it no longer provides any endorphin reward, especially with individuals who are unsatisfied with their lives (Klein, 2003). Without the satisfaction of the basic human need for self-fulfillment that only emanates from attempts to serve humanity as a whole, people can become addicted to habitual behaviors, selfish hedonism, narcotics, fears, and perverse worshipping rituals.

Since there is a limit to the amount of dopamine that the brain can produce and absorb, the need to increase dopamine levels to keep a pleasure from being boring can't be internally met to make it irresistible. In addition, if habitual behavior isn't followed by endorphins released upon the actual satisfaction of a human need whose frustration motivates the dopamine incentives to carry out the habit, internally created dopamine eventually stops being produced at the habitual activity.

Even hamburgers can be addictive (Gynn and Wright, 2008), but they don't die a natural death because they provide some needed nourishment. However, any addiction can be replaced by more desirable or healthy habits if desired. Unwanted habits die faster when there is something that provides more benefits. Since the highest lasting pleasure comes from attempts to serve humanity, breaking any addiction may be far easier if it is motivated by that purpose. For instance, some stimuli such as the taste of some nutritious foods like broccoli isn't intrinsic but instead derives from some bad memories, but previously aversive stimuli can be manipulated to be pleasurable (*NewScientist,* 2009s), especially if doing so enhances human life and thus the ability to serve humanity. As a result, it is possible to habituate people to like even healthy stimuli like hot peppers that they previously found repugnant (Klein, 2003). It apparently is also possible to reduce the appetite for desirable food by merely thinking about eating it (*NewScientist,* 2010aaaj).

Humans, as well as even animals, have a natural tendency to habituate to any pleasure or addiction for which they experience ever less happiness or contentment until it disappears (Klein, 2003), as long as their lives aren't totally empty and devoid of purpose. Without rewarding feelings of self-fulfillment for serving humanity as a whole, hedonism can easily arise that seeks to replace the emptiness with some pleasure. Religious worships provides the same type of "fix" through an absurd surrendering love to a fiction that is supposed to offer some absurdly anticipated reward in an fairytale afterlife. Many forms of hedonism and religious worship provide some momentary relief from the feeling of worthlessness that the conscious experiences for not carrying out the task for which it evolved, but they certainly are counterproductive to human health and happiness long-term. Some externally administered narcotics would actually kill a person if continually increased to keep the high from being boring, as there is a limit to how much the human body and brain can absorb. A conscious recognition of each person's higher purpose to serve humanity is the best and most natural way to remove such any undesired addiction, as well as to avoid the feelings of empty worthlessness that come from committing treason against oneself and one's human nature by surrendering to something other than humanity as a whole.

A feeling of self-satisfaction or self-fulfillment that is inconsistent with a particular habit apparently creates sufficient dopamine to reduce or eliminate the incentives created by the unwanted habit's dopamine. As a result, the feeling of self-fulfillment from serving humanity that evolved in individual human beings over millions of years to enable humanity to cooperate and survive an otherwise overwhelming environment is sufficient to eliminate habits inconsistent with that innate drive of human nature.

Many addictions result from sexual and emotional frustration that is rampant in today's cruel societies which impose morals designed to create the battle between the sexes and divide people away from their innately shared communality. Without the natural ecstasy of such harmony, people seek pleasures in other forms. However, just imagining the real possibility of a group hypnosis and sexual resonance among even a small number of human beings can help divert a person away from an addiction or fetish and back toward their human nature, which they must follow in order to be truly happy. Thinking about such fantasies can also increase their chance of becoming reality.

Surrender to one's own human nature in service to humanity as a whole can't take love away from loved ones, since human nature innately prompts mutual love in service to the innate higher purpose. Such a surrender to humanity also represents the best strategy in the battle against humans' vicious enemies such as christianity, which is an extremely perverse form of fetishism, and capitalism, which represents an especially extreme form of selfishness. In contrast, surrendering to a fetish or selfish perversion of human nature represents a retreat from that fight against inhumane enemies, which are absolutely merciless in torturing their surrendering prisoners of a selfish fetish or a fictitious worshipped deity. Even if it costs those enemies some resources to deal with the prisoners, the submission of human beings to the fetish monstrosities of Christianity and capitalism greatly enhances the motivation of the enemies' zombie servants to continue as obedient slaves, as does the subsequent torture. Regardless, surrender by a human being to a fetish or selfish addiction reduces the strength of humanity's army and negatively impacts its morale. Such knowledge that perversions of human nature represent treachery to humanity can help inhibit both the formation and continuance of addictions and fetishes.

It can also be useful to recognize that addictions as well as worship are just fetish substitutes for natural human pleasures, such as a mutual loving sexual relationship which is naturally more desirable. Even a fantasy of a future, hoped-for fulfillment of a desire for mutual love is more satisfying long-term than any fetish. Addictions, worship, and other fetishes are by their nature de-humanizing, as are any thoughts of indulging in them that can only bring hell on earth for a human being. It can also be useful in breaking an addiction to recognize that all undesirable fetish addictions of any type only make a person a slave to the "pusher" which are despicable characters who themselves addicted to the money fetish.

Serving any fetish, including money, typically results from an unhealthy repression of human nature. The value system for basic human needs and desires like for food and sex are processed in the same area of the brain as for objects that are identified with them (Lawton, 2010), implying that such objects have become mere fetish substitutes, as with a dog that has been trained to identify a bell with food. While an inability to satisfy human needs can also contribute to the development of an addiction to a fetish, the positive feeling of self-satisfaction resulting from trying to serve humanity as a whole as each person evolved to do can more than offset the allure of any fetish or unnatural pleasure.

Research indicates that individuals prone to addictive behavior often don't change the counterproductive actions or thoughts even after it has been made clear to them that modifying them can increase the probability of other rewards (*NewScientist,* 2008l). The cause may be that the chance and utility of such other rewards may be too low to offset the pleasure gained from the habit. People addicted to counterproductive behavior appear to value the benefits from it more highly than an improved possibility of an alternative reward, especially if that alternative reinforcement remains uncertain. Benefits from additive behavior can include "highs" generated internally in the brain or with imbibed drugs, which can seem more rewarding than the chances of another pleasure, especially if there are risks of failure in attempting to attain the alternative that can lead to negative feelings of frustration and depressive inadequacy. Prior failures to attain an alternative reward also contribute to a lack of confidence that reduces the chance of success, thereby further decreasing the perceived benefit from changing a habit conscious reason indicates should be broken.

People may also perceive the cost of rewiring their brains to exceed the benefits from any enhanced chances of other rewards that might result. Rewiring the brain requires not only conscious effort, focus, and self-confidence to enact the change but also an admission to oneself that one was previously wrong, which itself can inhibit the necessary self-confidence to overcome the disease. Past failed attempts to stop an addiction can also cause a person to think the individual isn't strong enough to end it, thereby making any further effort to do so seem futile.

However, the very positive lasting feeling of self-fulfillment that comes from carrying out the purpose for which each human being was born is sufficient to overcome any undesired "habit" that a person perceives is inconsistent with serving humanity as a whole. In particular, human nature evolved to make it possible to the great pleasure of self-satisfaction just from trying to carry out this purpose, and this benefit alone can be enough to overcome any addiction. Thus, it may be the natural human cure for breaking all habits.

Morrell (2005) has indicated that constant conscious repetition of commands to the subconscious can overcome subconscious resistance to change and removal of unwanted habits. She indicates that knowledge of why there are subconscious tendencies which seem to be inconsistent with what a person consciously wishes can help redirect those tendencies, especially with respect to those developed due to childhood experiences that are no longer useful or appropriate. Many subconscious tendencies and wishes were created out of experiences that are no longer relevant, and behavior subconsciously programmed during childhood can be especially counterproductive to an adult. Even some innate subconscious tendencies, such as the ability of pregnant women to more susceptible to reading other people's emotions expressed in their faces (*NewScientist*, 2009zzzzk) and the capacity of females at any time to be more sensitive to anything their fingers feel (*NewScientist*, 2009zzzzs) can be reprogrammed if they don't feel those inclinations are useful today.

In any event, to control a person's own subconscious, the individual must consciously give it irrevocable orders. Issuing commands to the subconscious merely involves concentration, which can be accompanied by consciously repeating a rule or guideline to oneself silently. The physiological effect of such orders is to overwhelm the old connected neuron firings in the subconscious, resulting in a future weakening of those old connections, which may disappear altogether upon sufficient repetition of the commands.

At the same time that old neural connections in the subconscious are eliminated, new ones are created by conscious commands. These new firing connections become ever stronger as a person repeats or concentrates on the new rules more, and their strengthening helps weaken the undesirable, inconsistent ones further, as incoherence between groups of neural firings leads to the weaker ones dying out as the brain naturally focuses on just one coherent set of thoughts. In addition, application of new rules for the subconscious to act in various situations, including in daydream simulations, also helps to weaken the old system of neural firings that can thereby be eventually disconnected altogether.

Self-directing a person's own subconscious keeps others from directing it through suggestions and other means. There is no need to even attempt to explain or rationalize a command to the subconscious, as reasoning with the robotic subconscious is futile. It just accepts the order supplied by the conscious neural firings that cause the subconscious neurons to fire as commanded. However, conscious reasoning may be required in order to maintain the will power behind the command with more neural firings when necessary, as conscious doubts about the correctness of the order can weaken the resolve. Consciously establishing plans for meeting innate needs and desires in ways that are consistent with the order may be essential for the effectiveness of this process.

Conscious thought can also resolve conflicts between various subconscious rules and priorities in general. For instance, the determination of whether the human trait of spite should be felt and what spiteful reactions should be undertaken is one of the most complex considerations that the conscious must weigh. This decision-making process, which causes humans to develop very early in life their own sense of justice, must always be weighed against the likelihood of counter-spite in others. Such evaluations require forecasting of both oneself and others, thus necessitating an application of rational logic to the empathized self-awareness of oneself and others in the future. This type of analysis may not be effectively possible with simple subconscious algorithms, at least partially because of cultural shifts in the last 35,000 years.

Because the conscious can command the subconscious to always try to detect exploitative behavior, it can be left to automated behavior for many tasks until a signal for possible manipulation is given for conscious evaluation. The conscious can then immediately interrupt subconscious processing that is perceived to be manipulated, and conscious intervention can then occur safely. Since spiteful actions penalize the manipulator to inhibit further exploitation by that individual as well as by others who are thereby taught by example as opposed to direct experience but may also prompt counter-spite, some rational analysis may be warranted before a knee-jerk decision rule is applied. Spite and counter-spite, which can be especially counterproductive between partners or desired mates, may actually have significantly contributed to the evolutionary growth of the human brain that requires extraordinary abilities to deal with such complexities within the context of serving both individual and group goals.

The conscious is therefore typically given the opportunity by the subconscious to make the final determination on whether to act immediately with spite to a perceived unjust situation. In particular, subconscious signals to the conscious that the body is prepared for spiteful reactions can be consciously directed into automatic relaxation if conscious thought perceives doing so is optimal. It is also possible to consciously evaluate any situation further before choosing to undertake a spiteful response, or postpone spite to a future time. The conscious can also program the subconscious to react in a spiteful manner according to preset rules that may not rationally seem to be in a person's personal interest if the individual consciously chooses to do so.

In addition, a person can consciously even influence what a person perceives is unjust and whether a person feels spite. If the intent is to develop some sort of long-term reputation of not being exploitable and thus inhibit attempts to do so, the usefulness of a reputation for feeling spite in cases of perceived exploitation and reacting spitefully to them should be evaluated relative to the costs associated with creating and maintaining such a reputation. This analysis could include the advantages of pride and positive feelings of self-satisfaction that come with actions coherent with one's human nature, but it might also be judged to apply only in certain future circumstances. A realization that one is thereby helping humanity was also once an important consideration here, as may explain the rewarding feelings that naturally arise from spite. However, in the context of today's societies where selfish behavior is rampant because their very foundation of is based on unjust exploitation, spiteful efforts may be totally futile. By anticipating the better feeling that can come with strategic applications of spite in the future, a person can control innate automated reactions and redirect for some more productive cause like replacing the contemporary inhumane culture.

Identical conscious thought processes also work for other behavior that derives partly from innate human urges. For instance, actions undertaken to respond to sexual urges can be directed from undesired fetishes in the present into behavior that maximizes the chances of normal human sexual intercourse in the future, especially if a person recognizes doing so helps one's future self and all of humanity.

Thoughts and tendencies programmed by brainwashing as children or otherwise can also be changed by the conscious. The fact that the conscious operates at a higher frequency than the subconscious thoughts that were "learned" in childhood or under hypnotic suggestion indicates that they too can be eliminated through conscious command. The continued firing of the conscious to hate any undesired "learning" can actually remove the connected subconscious neural firings quite quickly because of the chemical punishment those cells receive as a result of such conscious reprimands that are exhibited in neural out-firing the subconscious neurons. It only takes a conscious will to do so that is strengthened by the knowledge of the ability to do so. Habits that were once reinforced in some fashion but that the conscious no longer sees as productive can be similarly eliminated. The higher firing frequency of consciousness ensures that the conscious negative reinforcement of the subconscious thoughts operating at lower frequencies occur often enough to ensure discouragement of them in only a brief focused self-reflective thought that will take place at the same time as the undesired firings of neurons because of the more rapid frequency of consciousness. Continued discouragement ensures that the undesired neural connections are completely eliminated.

Every undesirable thought is thereby discouraged with conscious firings that provide the subconscious connections with the negative reinforcement which eventually redirects them to the desired subconscious neural connections with ones that are positively rewarded by conscious thought. A simple recognition of this fact and a desire to change alone can do the trick, although undesired feelings and behavior can be eliminated more quickly with a stronger will that results in more frequent and more powerful conscious application of this reward/punishment system.

The conscious should optimally redirect any undesirable subconscious feelings toward behavior and desires that the conscious perceives as useful to humanity. As long as one is serving one's innate human tendency to try to benefit humanity, even failed attempts to serve humanity successfully can be enjoyable and rewarding. Collapsing back to bad habits due to failure or perceptions of very small chances of success are thereby inhibited by the pleasure of trying to what one was born to do. In addition, as will be explained in Chapter 16, the likelihood of success is also increased dramatically if one applies the goal of doing one's best to serve humanity that also create a feeling of happiness and self-fulfillment.

When rules, image, or reputation that one's subconscious has or has been given to follow are inconsistent with other ones, the conscious has to make a decision, and this decision-making process interrupts the efficient flow of thoughts in the subconscious. To avoid the indecisiveness and ineffective action that can result from the distracting lack of focus, it is useful to adjust the various decision-making criteria of the subconscious so that the rules are not in conflict with one another. For this purpose, it's important that the guidelines that are set for the subconscious are consistent with the overriding purpose of each human being.

Giving orders to one's subconscious is a form of talking to oneself. While talking with oneself is often just the conscious arguing two different viewpoints in an attempt to logically think through to the optimal one, it can also represent communication with one's subconscious, or at least with one's conscious perception of what the subconscious would say in a conversation if the subconscious could communicate via language.

Within this context, berating oneself, or one's perception of the subconscious actions, for its stupidity in following orders can often occur. The criticism can also be of the conscious criticizing itself. While evaluating actions and decisions is useful for learning purposes for the future, such criticism is unhealthy if it is judgmental in any way or causes any feeling of guilt, regret, or blame. While self-evaluation can be useful, it's absurd to fight any innate component of one's mind or oneself as a whole. It's bound to be far more healthy to perceive errors in judgment and action with good humor.

Blaming the subconscious for anything is especially absurd and counterproductive since it is merely an automated program, partially written by the conscious, to serve humanity as a whole. The subconscious exists strictly to assist and follow orders, so. The subconscious and all the rules or orders it follows are either pre-programmed genetically, set by the conscious, or allowed by the conscious to be set by others. Via dreams, gut feelings, and other means, the subconscious often tries to provide feedback hints to the conscious to modify orders or subconscious tendencies that are leading to undesirable behavior, such as those which result in conflicting directions for the subconscious. Despite any internal dialogue imagined by the conscious, the subconscious and conscious are on the same side if they are unified in carrying out each person's higher purpose.

Although the full extent to which the existing patriarchal society and religion has perverted human behavior is not yet completely known, it is safe to assume that any mistakes or embarrassing past behavior were caused by the patriarchal influence and only need to be purged by conscious will. Recognition of this fact enables maximum focus on optimal behavior, as opposed to a excessive concentration on avoidance of improper actions that can lead to a person being paralyzed by fear of regret that can never be justified in the current environment of patriarchal tyranny.

Many mistakes and embarrassing behavior are caused by brainwashing by today's cruel cultures that start their disgusting indoctrination with young children. The limited conscious activity that infants exhibit causes most of the inhumane programming at that age to be the total responsibility of the environment, which includes most importantly the parents and other assigned "caretakers" in the contemporary societies. Since infants have no ability to consciously question the programming, their subconsciouses are effectively coded before they have any ability to stop the obscene torture of their minds. However, as people age, innate subconscious tendencies motivate them to change inhumane tendencies, and individuals can indeed figure out what subconscious tendencies needs to be eliminated, even if they don't remember the actual programming or the exact cause of it, especially if they receive help in recognizing them. Hypnotic influences are similar and can be similarly purged even if the causes aren't remembered. Such de-brainwashing helps one become a true human being and serve oneself as well as humanity as a whole.

Humans have optimally evolved to adapt to the environment for the good of all, including the individual self, and so the true blame for any errant judgment on an individual's part rests solely with the current society that was put in place by patriarchal psychopaths. Nevertheless, it can sometimes be unhealthy to cast excessive guilt on society or others for unpleasant situations, as such recasting of blame only provides an excuse for not striving for a better situation. A feeling of powerlessness can even result that eliminates the incentive to change. To avoid such relapses back into slavery to the enemies of human nature, it can help to recognize merely attempting to serve humanity makes a contribution to that higher purpose by serving as an example as well as by any effects it has in weakening the Christian and other enemies of human nature.

Humor is also very useful in this situation. As with one's own mistakes, finding the humor in the errant ways of others that lead to the unjust societies and disharmonious environment can lead to an outlook that is both healthier and more productive toward solving problems. This attitude can also direct hate against inhumane culture and not the cruel elite robots or their brainwashed slaves. Such a perspective can also enhance the ability to resonate with the human nature of all other people that thereby maximizes the chances of a successful fight against their cruel artificial programming.

While it can be useful to evaluate entire societies, as well as different people and entire groups of people, such analysis might optimally be focused on providing inputs to enable the conscious and subconscious to operate jointly with the intrinsic human nature of others to try to improve the situation. The conscious can force the subconscious to carry out this task by focusing on a particular goal as opposed to casting judgment or blame. When the overriding goal is to serve humanity in an enjoyable quest that permits the bliss of experiencing many of the infinite wonders of the universe created by divisions of nothing that one thereby helps to co-vibrate, harmony reigns between the subconscious, conscious, and superconscious that maximizes efficiency, happiness, and self-fulfillment. Humanity is also thereby best served in a natural process to which all human beings are innately driven.

C. Dealing with Needs and Desires

One can always let needs and wants control oneself and one's subconscious if one wants them to. One only needs to believe that one's needs and desires control oneself for that belief to become reality.

Desires such as for food and sex are promoted by chemicals in the brain like endorphins that create pleasant feelings at experiencing certain stimuli (Klein, 2003). Desires are strongly influenced by one particular gene that affects the amount of pleasure felt via the level of neural firings which take place when the need or desire has been satisfied, and this level varies across people to make satisfaction of needs or desires more important to some individuals (Nettle, 2007). Eating good food alone can actually result in the release of more pleasant endorphins than externally consumed opium (Klein, 2003).

Desires tend to arise when there are insufficient pleasure chemicals in some areas of the brain. In addition, some chemicals like serotonin, oxytocin, vasopressin, testosterone, and noadrenaline create tendencies to want to engage in specific types of activities. Those chemicals apparently prompt certain subconscious neural firings that either translate into automated responses or are transmitted to the conscious for evaluative decision-making. If repeated signals of a specified sort in certain circumstances generally lead to unsuccessful or counterproductive actions, the conscious can cause them to be ignored in a process that eventually indicates to the thalamus to filter out such unwanted messages in such situations. This suppression results in a lack of reinforcement of the subconscious neural firings associated with the signals, thereby causing them to discontinue. There may even be reactions in the brain or in the glands that then even suppress the production of the chemicals that led to the undesired neural firings.

Nevertheless, the conscious exists to make decisions, including on prioritizing the fulfillment of desires and attainment of goals, and so it can't simply suppress all urges and emotions because an undesirable feeling of uselessness for itself results when its rationality has no reason to exist. At the opposite extreme, the conscious can't just focus on satisfying strictly personal desires because it is then rationally prompted to find no meaning in the self, thereby resulting in an undesirable feeling of nihilistic emptiness. Only by recognizing the purpose for which consciousness evolved is it possible to consciously prioritize goals and satisfaction of desires in a way that maximizes one's happy self-contentment with oneself, as only dedication to serving humanity as a whole can lead to a rewarding feeling of self-satisfaction and self-fulfillment.

Many believe that humans are driven to seek to satisfy Maslow's hierarchy of five human needs, which are, in order of priority, are physiological needs, security, belongingness, esteem, and self-actualization (Bernstein and Nash, 2008). However, a large number of people skip satisfaction of some of the lower needs because of greater value or probability of attainment with respect to satisfaction of some higher level needs, such as the primary innate goal of humans to serve humanity as a whole. People are driven by adverse feelings of emptiness if they don't carry out their evolved purpose.

As a result, satisfying any or all of Maslow's goals is likely to bring happiness and feelings of self-fulfillment only to the extent they are consistent with serving humanity. For instance, studies have indicated that people who strive for and attain all the goals aren't any happier unless they emphasize goals that are most consistent with their innate nature. In particular, good relationships with other human are positively correlated with happiness, but meeting goals of esteem and self-actualization through money or selfish goals tend to bring feelings of emptiness and unhappiness.

Maslow's set of hypothesized needs represent merely a means to enable one to serve humanity better. An innate urge to fulfill those needs or desires arises in all people from time to time, but decisions on whether to postpone satisfaction of them and what goals should be set to try to fulfill is left upon to conscious decision-making. A person can also consciously change priorities for trying to satisfy desires or attain goals toward that end. The particular goal upon which focus is optimally directed in any moment optimally depends on the circumstances as opposed to any reflexive subconscious prioritizing, as when there is a higher chance of an attempt to achieve a specific goal of meeting a particular need being successful with limited effort. Any overriding purpose set by a person can also influence the prioritization of goals and desires. For instance, a minimum level of some physiological and security needs must be met in order to be able to serve humanity in the future, but people can be motivated to sacrifice their own lives for a cause.

Many soldiers and martyrs actually die for their self-actualization goals even though their particular goals may not be consistent with those of humanity. However, most people today are too confused by religious lies and lack of information about the purpose of humanity to find a self-actualization goal that brings true fulfillment or happiness. As a result, most concentrate on the lower needs or desires, especially with respect to sex because of the strong worldwide repression of sex drives. Nevertheless, when false or inadequate direction causes a person to habitually prioritize goals above those of serving humanity, feelings of emptiness and unhappiness typically result even when many or all other needs and desires are satisfied.

Many human needs can be, and often are, postponed for later satisfaction when there may be better opportunities for meeting them, as when the environment seems more favorable for satisfying other needs. For instance, the sex drive can be, and often is, controlled by the conscious, which can inhibit it, especially when trying to have sex now is likely to result in negative repercussions for meeting future sexual and other human desires. Postponing satisfaction of a need or desire does permit satisfaction of other needs, such as eating or working to make enough money to buy food to eat, thereby creating better chances for satisfaction of future needs and desires.

Some people may make the relative importance of human needs dependent upon the cultural norms of the society in which they live. Others may place little or no value on meeting social norms. For example, some spiritualists value only ideals and internally generated feelings, and so they unnaturally try to ignore more primitive desires including social acceptance. On the other hand, a complete, consciously directed suppression of lower needs and desires can result in perversions and even death.

A good example of an individual concentrating on a higher goal is one who believes a return to a human society is optimal for humanity and who might therefore naturally place negative value on esteem and belongingness in a hierarchical culture. In particular, a true humanist might perceive "normal people" in contemporary societies as unnaturally selfish robots, and "friendship' with such inanimate objects can't satisfy desires for social acceptance by real human beings. Humanists might even be willing to risk satisfaction of personal physiological and security needs in order to serve humanity as they see fit. They can even try to help transform the brainwashed slaves of today into human beings and thereby create the opportunity for more human comradeship.

Nevertheless, even for humanists, natural human desires arise, and a human being is motivated to try to satisfy them as part of the goal of serving humanity. Acting according to a person's human nature is certainly one component of this process that includes trying to satisfy the individual's own desires. Desires that a person has very little chance of satisfying and that aren't enjoyed merely in striving for them can cause frustration which can even intensify the desire into a slavery to it, as can setting any goals whose attainment is unlikely.

To avoid such unpleasant situations, a person can establish an objective of just trying to do the best that the individual can in an effort to attain the goal. Such goal setting also avoids the doubts and regrets associated with having selected the objective and having failed to achieve it. The unquestioning focus and belief necessary for optimal activity is thereby attained, thus maximizing the probability of achieving a particular outcome.

Nevertheless, while focusing in any moment on any one such particular goal is efficient, it can be useful for flexibility in activity to be prepared to switch priorities when various conditions arise. Those situations can be pre-specified to avoid frequent evaluations by the conscious that divert efficient attention on the current primary goal and task. Maintaining the overriding purpose of serving humanity as a whole can help ensure both coherence in thoughts and a happy self-fulfillment regardless of failure in goal setting or achievement.

The subconscious appears generally able to subjectively evaluate efficiently the probability of satisfaction of a need or desire for any given action. It then weighs the likelihood and value of successful satisfaction against the amount of effort expended and the risks or costs of failure. Information on innate abilities and past experiences as well as rules given to it by the conscious are inputs to the evaluation process.

A conscious analysis of the probabilities, rewards, and costs or risks of alternative courses of actions can help an individual make decisions, as well as change subconscious tendencies that seem to lead to undesired outcomes, such as addictions to drugs, etc. To eliminate undesired behavior, it can be helpful first to try and consciously think logically on the reason for it. Values and probabilities of event outcomes can be assigned that justify the subconscious tendency toward the targeted behavior or addiction, and conscious biases, over-optimism, and rationality can thereby be factored out. With the source of the problem thus found, alternative actions, behavior, attitudes, and knowledge can be creatively thought up, investigated, and evaluated until the values and probabilities are changed sufficiently to make the undesired behavior relatively unattractive.

For instance, the value received for a given habit can be contrasted with the infinitely more valuable human activities of serving humanity as a whole while experiencing and exploring more wonders of the universe, including the possible ecstasy of mutual love. Mutual love, or an entertaining search for this emotional bliss in lieu of a desired partner, is innately far more enjoyable than any other experience. As a result, merely undertaking new actions and experiencing new things, including a blissful connection with humanity and the universe as a whole, can help in breaking any unwanted habit which reduces the degree or chance of mutual love.

Although drugs like alcohol, cocaine, opium, and nicotine can create desired feelings of happiness and even useful behavior, such as by decreasing inhibitions and by promoting more self-initiative that the feelings of happiness alone help create for a time, their positive effects tend to decline with increased use, whereas novelties always provide reinforcing food for the brain (Klein, 2003). This fact makes it easier to see the benefits of alternative modes of action that create more natural and useful highs. Frustration in new endeavors can be avoided by merely setting goals to try to do the best one can in any novel situation without the attainment of any particular goal being necessary in order to have the conscious to be satisfied with the thought and decision-making processes of the brain that it controls.

Alcohol, narcotics, and cigarettes seem to have originated over ten thousand years ago, and the core motivation may have been, and may be, to return to the bliss of egalitarian group consciousness that was eliminated with matriarchal hierarchies. However, they are only temporary fixes, and only serving humanity can bring lasting happiness. Many may use drugs to try to become more enlightened, but they become useless and counterproductive once one knows the true meaning of life, the universe, and everything. In addition, because enlightenment requires rational understanding and the natural human feelings that existed for millions of years before humans used drugs, they could actually inhibit understanding the purpose of human life.

Recognition of facts about the extraordinary plasticity of the brain can be very useful in breaking habits and subordinating satisfaction of desires to lower levels of priority. For instance, it has been documented that even an individual's subconscious actions, such as directing the level and direction of blood flow, can be controlled by hypnotic focus to enable seemingly miraculous events like walking on hot coals without pain or blistering (McClenon, 2002). Opioids are naturally released at the experience of sufficient stress, providing one possible reason for the needles poked into a person in acupuncture to reduce pain (Klein, 2003), and so any stress associated with breaking an addiction may be more than offset by experiencing pleasant chemicals, especially with enough belief in oneself and one's will power. Given that chiropractors appear to measure and affect nerve tension in a manner that seems consistent with quantum physics (Lipton, 2008), it is possible that treatment by acupuncture may also exert its effects through the quantum entanglement of microscopic particles in the body. Events that seem mysterious according to classical Newtonian physics are actually consistent with the more accurate quantum laws of nature that seem to be reflected in the ancient spiritual views of peoples like the Native American Hopi (Braden, 2007).

Because sexual desires are heavily repressed by today's cultures, the urge for satisfaction of this particular human desire merits special consideration. Since sexual desires can be affected by the cortex thinking portion of the brain as opposed to being a mere mechanical spinal or thalamus reflex (Rathaus, Nevid and Fichner-Rathaus, 2008), sexual behavior seems to be under the control of the individual conscious. To exert such control, as is often necessary in today's contemporary cultures that so unnaturally suppress human sexuality, it is sufficient to understand the nature of the human sexual drive and its evolved purpose to promote the continued existence of humanity.

As long as one doesn't try to completely suppress the sexual drive permanently, perceive sex as a need, feel an obligation to propagate one's genes, or feel obligated by social pressure to engage in sexual intercourse, sexual desire serves as a very useful tool for humanity as a whole. Sex draws people together and enhances social bonding and health, besides procreating offspring when desired. However, a person can also live quite happily without it, as long as the urge to try to satisfy it isn't completely repressed.

Complete elimination of naturally evolved desires, such as for sex and nourishment may not even be possible long-term without redirection into perversions. As previously mentioned, those who easily could overcome natural desires for nourishment and procreation died out without offspring long ago. While the human body has adapted to live without food and sex for extended periods when the external environment requires it, internally repressed desires cause psychological problems like depersonalization, obsessive-compulsive behavior, hypochondria, phobias, and hysteria (Gooch, 2007). The latter "disease", which actually used to be treated with "leeches on the labia", "was once a fashion statement, so strongly linked was it with breeding and social superiority" (Shetty, 2009), thus providing an indication of the connection of bizarre repression of women's sex drive to social mores.

Permanent conscious self-repression of sexual drives is especially counterproductive because the subconscious will generally cause this natural human desire to resurface in some bizarre form. When sexual activity with others is externally repressed, relief from frustration and stress remains possible with masturbation as long as there is no internal suppression of the natural desire for sex and related fantasies that may be more likely realized as a result of imagining such thoughts.

Besides providing stress relief, fantasies help a person consciously evaluate what is feasible and how. In particular, it's difficult, if not impossible, to get excited about something that's not going to happen, and the subconscious works in fantasies to see how an event can happen. The lower level of excitement about something with a low probability of occurrence motivates thoughts of other ways to satisfy a particular desire. Fantasies thus provide the conscious with subconsciously processed information as to whether and how a particular desired outcome can be achieved. They also enable a conscious evaluation of whether the wish is worth it.

In contrast, a conscious decision to repress a natural desire like natural human sex for religious reasons, or due to the persuasion and pressure of others, often results in unnatural perversions and violence (Reich, 1936). In particular, a philosophy can lead to fantasies of fulfilling the desire in some undesired fashion, and some carry out perverse fantasies as a result.

Perverted fantasies, including diseases associated with one-way worship, can result from any internal repression of desires that are caused by the artificial repression of sexuality by contemporary cultures. Such warped fantasies indicate the need to eliminate any internal repression of natural desires, regardless of the cause. For instance, it's important not to allow the disgustingly prudish societies of today to cause one to pretend to oneself that one doesn't want sex. Despite the cultural taboos against human sexuality and bodies, people recognize subconsciously the desirability of sex and the beauty of human bodies, and so it is possible to engage in enthusiastic light-hearted subtlety when making efforts for the social bonding involved in sex. Such exaggerated play-acting can be part of the courtship culture at the same time that the benign ulterior purpose is also clearly evident. To avoid perversions, it's important to remember the real motive and not to repress it even when sick perverts a person rebuff or reprimand a person for acting like a human being.

The sex drive itself stems from the hypothalamus and doesn't appear to be consciously controllable (Klein, 2003). Except for a refractory period after orgasm (Rathaus, Nevid, and Rathaus, 2008), the signals from the hypothalamus to consider ways to have sex may be fairly persistent, especially after longer periods without sex. Because sexual excitement takes place in other sectors of the brain (Graziottin, 2004), these signals themselves do not represent sexual excitement but merely a desire for sexual stimulation. While the sex drive may not be consciously controllable, the actions taken in response to that drive are indeed affected by conscious decisions. Even preying mantises can modify their sexual activities in reaction to observations in the environment (Brooks, 2010c), and the human sex drive can lead to sexual excitement in any particular moment only if it is uninhibited by the conscious.

The brain's thalamus filter often can inhibit sex drive signals from entering into consciousness when a person's conscious is focused or pre-occupied with other tasks. On the other hand, if the conscious becomes aware of or focuses on a stimulus that a person finds sexually exciting, it is difficult to repress the thought of sex. While the conscious has enormous capacity to temporarily repress thoughts of sex for social, cultural, religious, or other reasons (Graziottin, 2004), attempts to completely suppress sex drive signals forever seem to intensify their allure. As opposed to self-repression of a sexual drive, which eventually resurfaces in fetish or perverse forms, merely redirecting or postponing an unwanted signal for sex to a more appropriate time, place, and/or partner is natural and healthy. Knight (1999) has advised neither denying nor worshipping sex but instead accepting sexual desire as a part of each person's divine being, using it to make life in the individual's physical body a pleasant and creative experience.

The urge for sex drive is similar to the drive for sufficient food and water, except that sexual intercourse with others can be postponed indefinitely whereas the need for nutrition can't be. Postponement of sexual desires, however, may only be feasible if there are fantasies of sex with others that have some chance of occurrence in the future and that can be imagined in the present during masturbation. Attempts to permanently self-repress the sex drive or realistic fantasies of it may only lead to the urge resurfacing in perversions.

Since trying to self-repress truly human sexual desires typically leads to them burning even stronger and often causes perverse thoughts at the very least, it's important to maintain such desires and direct any frustrated desires into activities that create the opportunity of satisfying them. For instance, making oneself as attractive as possible, going out to places where possible desired mates might hang out, flirting, and courting are all means of increasing the chance of sexual activity. In spite of the obscene repression of sex by contemporary cultures, every individual lusts for it with a physically and socially attractive person. As a result, even in a perverse society that is designed to frustrate sexual desires, as most patriarchal ones are, there is always a chance that emotionally satisfying sexual activity will indeed happen. If a person is unable to focus and resonate enough with human nature to actually experience an emotionally and sexually ecstatic relationship, the individual will no doubt be able to do so in parallel universes. An understanding of internal and external psychic forces that enable the power and self-confidence associated with such focused emotional coherence with others will be explained in Chapter 16.

In prehistoric times, merely serving humanity better naturally guaranteed more sexual opportunities. Doing so today can also enhance a person's attractiveness today. In addition, if the individual internally establishes top priority on the natural goal of helping humanity thrive, as appears to have evolved to make human beings experience happy self-fulfillment, the drive for personal sexual pleasure may only occur when it is subconsciously perceived to be optimal for purposes of striving to make the person's best contribution toward the higher goal. In such cases, there is no need to consciously suppress the sexual goal to keep unfulfillable urges from occurring, as the subconscious will discontinue them based on the overriding purpose that it is innately attuned to follow. In particular, a subconscious algorithm can be set to make the sexual desire secondary to serving humanity goal as a whole, with sexual activity itself being specified as just one of the possible opportunities for doing so, as it should be.

Although humans have been programmed by evolution to want to experience the bliss of mutual love, each individual has the right to choose when and with whom the person surrenders. When a person doesn't have an acceptable willing partner, there is a drive to try to do something to find such a mate. Although the desire for sexual release will likely arise if some potential or temptation for fulfillment is perceived, the sexual drive remains a weaker force than the evolved tendency to want to serve humanity as a whole. Sex and sexual desire represent just one aspect of human nature and succumbing to that desire is just one means of serving humanity.

The more intense a desire, the more likely a person will in fact think of acceptable ways to experience them. As previously mentioned, lust can help provide natural fantasies in the meantime. Given female capacity to experience multiple orgasms (Bernstein and Nash, 2008), one way for the male to productively direct a frustrated desire involves trying to burn ever harder for longer during fantasized sexual intercourse and to have a potential female partner come to more orgasms during that process before experiencing an orgasm inside her.

Sexual desire itself had to evolve to be powerful enough to motivate any selfless actions that may be needed to obtain a sexual partner. Similarly, evolution had to evolve emotions like love to be strong enough to overcome rationalizing away social behavior (Stefano and Esch, 2005). The natural human desire for both sex and love together is strong enough to bring partners together in emotional and sexual ecstasy despite all the cultural influences today that divide people into selfish entities and that frustrate the drives for sex as well as love and thereby artificially separate them.

For example, frustrated sexual desire for a particular partner has always motivated individuals to do all sorts of things for satisfaction of that desire, which once upon a time brought empathetic feeling from the desired partner. However, in today's world of indoctrination and motivation to selfishness, it more typically brings only exploitation and frustrated humiliation. Sexual excitement brings some kind of high, created at least partially via just pleasure chemicals produced internally in the brain like morphine (Esch and Stefano, 2005a). Sexual excitement might also bring pleasure through the ecstasy of DMT which may, together with oxytocin, enhance feelings of trusting unity. While such chemical reactions evolved to promote a surrendering mutual love and egalitarian bonding, the hierarchical cultures of today inhibit the coherent but diverse equality necessary for that harmonious bliss.

The exploitation inherent in today's religious and social systems results in people consciously inhibiting the expression of the pleasure chemicals associated with mutual surrender to a larger whole. Nevertheless, there remains an innate human desire for complete mutual surrender to a partner that can be especially strong when a sexual desire does present itself for consideration. Repression of that desire to unify completely with a mate leads to all sorts of fetishes like surrendering worship to enable uninhibited experiencing of the natural human ecstasy associated with mutual love that is difficult in cultures that program people to exploit each other instead of love one another. An unnatural one-way worship of the bodies of the opposite sex, or even the far more disgusting perversion involved in the worship of fictitious deities, are rampant in systems that encourage repression of human desires. Perverting the natural human emotions involved in intimate egalitarian relationships into superficial fetish worship is often facilitated by brainwashing human beings into perverts dependent on and believing in superior beings.

While the normal sexual stimuli for men are signs of women of child-bearing age, and while the normal sexual stimuli for women are signs of sexually excited males (Bergner, 2009), each individual may have variations in those natural stimuli due to cultural, religious, and experiential conditioning that has caused other stimuli to be sexually exciting, including fetishes. Although the brain can learn to be excited by fetishes, especially if an orgasm becomes identified with them due to past experience, reprogramming the brain connections back to the normal sexual stimuli that are genetically innate is facilitated by the fact that the brain was set up for such normal connections.

For people programmed from childhood to believe in christianity, the natural co-surrender to a mate is replaced by one-way worship and servitude of a fictitious male deity. The satisfaction of human desires is often perceived to exist in Heaven after surrendering to a man-made god and serving that entity in some perverse fashion on earth.

Because of the sick cruel society of today, actual satisfaction of human sexual urges can be problematic. Since most people in contemporary cultures don't know what they really want innately as human beings, a process of convincing a desired mate to overcome his or her programming and thereby enable mutual surrender is necessary. While many have already surrendered their "souls" to a fictitious cruel god, they have thereby condemned themselves to a self-imposed hell, and the resulting eternal suffering for their human bodies creates the motivation to repossess their human nature by replacing the lies of the fictitious god with actual knowledge in their minds.

The fact that even women who are programmed to consciously believe in Christian morals and lies remain innately excited sexually by anyone who is physically and emotionally passionate about them, if only subconsciously, improves the chances of mutual surrender in societies filled with the robots of the cruel religion. Rescuing females from the enslavement to a sadistic patriarchal doctrine is uncontrollably alluring to them subconsciously, even if the chains of their brainwashing seem extremely strong and difficult to break when confronted with their enslaved robotic conscious. The same holds for informed women since males are innately hard-wired to want mutual surrender to their mates' desires.

Recognition that the christian doctrine is a fairy tale that is filled with absurdities and holes in consistency and that is designed to divert people's human nature from its evolved task of serving humanity can help individuals overcome the cruel brainwashing implemented to force slavery to a "holey" hierarchy. An improved understanding of the innate human desire for the blissful mutual surrender to a loving partner and humanity in general can also assist in eliminating artificial programming by a cruel religion.

Sexual desire seems to come up as a periodic impulse that is put for consideration by the conscious when it isn't engaged in tasks upon which the individual has previously decided, or which is deemed more important, such as for purposes of satisfying future desires. For instance, concentrating on a job or work in order to live and to have a good enough reputation to attract the opposite sex can override consideration of urges to fulfill desires immediately. Overriding immediate satisfaction of sexual desire is even easier with a recognition of the greater importance to human happiness of the self-fulfillment involved in serving humanity according to each person's own unique human nature that includes seeking and potentially finding a human mate for mutual surrender.

The idea of fantasizing about an ever-changing, hotter, and more appreciative passion of satisfying the sexual desires of a trusted mate can help one practice in one's mind for the possibility and focus on it. Here human passion is the key, since some idealized perfect form of sex is by definition unhuman and therefore unattractive, unexciting, boring, and even disgusting. The strict missionary position with no foreplay that is so crudely recommended by the religious fundamentalists and other sick perverts is just one example of such unattractive behavior. For those men passionately enjoying providing oral stimulation to females, it's not only possible to have one's cake and eat it too, but that's often what the females naturally want subconsciously. "Ladies first" in starting an orgasmic union may actually be most exciting to both partners. Attempting to suppress fantasies of such natural passions can only result in perversions.

In any event, fantasies can assist a person in deprogramming the sick societal inhibitions against sex and make an individual ever more ready for the bliss of compassionate mutual love and empathetic orgasms that all humans are capable and worthy of giving and receiving. Merely focusing on such thoughts increases the chances of experiencing them and is far more exciting than any perverse thoughts or acts. All perversions can therefore naturally fade away eventually even if natural human passion is never actually realized in the parallel universe one experiences.

Prior to anyone surrendering unconditionally in today's inhumane societies, however, probing a partner to determine willingness and ability to do the same is necessary to avoid exploitation. While surrender to other humans in general is a naturally evolved human trait that enables leaving the self behind and creates very positive feelings in humans, conscious manipulation or abuse of that tendency by others can make it a very miserable experience. The human mutuality that normally would come naturally with surrender by another must be assured to avoid the exploitation by a selfish brainwashed robot.

While a striving for unity through sexual desire is a naturally evolved human trait, so is a spiteful loss of interest in an exploitative mate, and so a desire for a purely selfish individual is naturally redirected toward another who is not. For that natural reaction to occur, however, it may be necessary to refuse to self-repress the desire for sexual, emotional, and spiritual unity with at least some human being, including possibly with another whose enslaving brainwashing has been removed with one's own help. With that freedom of thought experienced, it becomes easy to realize that a purely sexual relationship with a programmed robot wrapped in a human body is about as alluring as mating with a rubber doll.

In cases of brainwashed christians who are too possessed with fear of a fictitious hell to be anything but robotic slaves to a fictitious god for the exploitation by others, the eternal bliss of mutual surrender isn't possible. Despite christianity being designed for patriarchal rulers, many women are among the most fanatical adherents of that religion, at least partially out of that fear of hell after death. Many females have even been programmed to see it as their moral obligation to try to convince others, especially their boyfriends and husbands, to follow the same life of christian slavery. Their disgusting and very selfish intent of making others experience the same hell on earth that they suffer is motivated by their futile attempt to maximize their perceived chance of entering a fictitious heaven after death. If women can't be helped to recognize that the chance of Christian heaven or hell occurring after death is zero, or if it isn't possible to wake up the brainwashed zombies in some other fashion, exploitation of them is the only humane treatment, since they aren't "alive" human beings, and since they will otherwise be exploited by monsters.

These facts actually make breaking through the artificial programming easier than otherwise, thereby enabling a possible rapid de-programming of the religious cults like Christianity that pervade the world's societies today, if people can be made consciously aware of them. Greater knowledge of these facts increases the chances for all of replacing the patriarchal hell with Heaven on earth. It can also be reassuring for the informed to recognize that the more enlightened one is, the stronger the psychic waves to which all people are innately receptive, especially if the intent of them is humanly benevolent and therefore not reflected back spitefully. One enlightened person's psychic waves can dominate all the patriarchal ones which are extraordinarily weak because of their programmed artificiality that is incoherent with human nature and the universe. In fact, Christian psychic emissions may have no impact at all on individuals acting in harmony with their human nature other than as random flying light photons. The ruling patriarchs are so uninformed that they don't know how to harness psychic forces adequately. Even if they did, they would recognize that psychic powers can be consciously exercised only by people acting coherently with human nature that would prohibit the hierarchal systems that enable their rule, one-way worship, sexual repression, and other obscene perversions of christianity.

It is likely that the ruling elites have so totally transformed themselves into selfish machines that they are completely impotent psychically, having no control of their electromagnetic emissions. Their pineal glands are likely to be totally calcified, and their external emissions become more like waste excretions, to be manipulated like dung fertilizer by enlightened human beings. The hierarchal societies they have created are so perverse that not only do they suffer the emptiness associated with a lack of self-fulfillment but even their own sexual desires are frequently not fully satisfied, and so they are easily manipulated.

Sometimes the self-repressed sexual desires of the patriarchs and their slaves can cause them to overindulge in satisfying some other desire to try to compensate for the missing dopamine reward. Some perverse behavior like overeating can even increase frustration of sexual desires. For instance, overeating due to repression of the sex drive makes a person fatter and less attractive to potential partners.

When the frustration of a particular desire is caused by external factors, depression often results (Bernstein and Nash, 2008). Widespread prudery in a society is a major cause of such external repression and problems (Reich, 1936). Cultural suppression of potential mates' sex drives can result in an individual giving up on finding a partner, thereby further reducing the chances of sexual satisfaction and magnifying the depression, even if a person doesn't self-repress the individual's own sexual urges.

Externally forbidding something tends to only make people want it ore (*NewScientist,* 2009x), but the effect may be even stronger and more frustrating with respect to the innate human sex drive. Feelings of helplessness typically result from sexual frustration, as does a tendency to blame oneself that creates a feeling of worthlessness and thereby magnifies the negative feeling (Klein, 2003). A humane society couldn't condemn people to such depressed states, but people learn to exist in the hierarchical societies of today that effectively treat human beings worse than dogs.

For the enlightened, it can be useful to recognize that, despite the cruelty of contemporary societies which are truly responsible for virtually all depression, what a person wants will be attained, at least in a future parallel universe, and perhaps in the universe the individual gets to experience if the person focuses hard enough. It can actually be fun to fantasize about achieving a certain desired experience even if an individual doesn't focus enough to actually experience it. As previously indicated, fantasizing about fulfilling a particular desire can actually help attain it, or at least recognize the effort needed to experience realization of it and determine whether it is worth it or not. Fantasizing about the past can also help a person evaluate outcomes that the individual may wish to try to achieve in the future. Unlike the need for food, the desire for a loving sexual partner can be satisfied for life with mere enjoyable attempts to find a desirable mate and fantasies about potential ones, especially if an individual sets a higher priority on serving the purpose for which humans evolved, i.e., serving humanity as a whole.

While the plasticity of the brain allows most everything (Bernstein and Nash, 2008), the enormous pleasure, happiness, procreation potential for spreading genes, and motivation that sexual desire provides makes permanent self-repression of the sex drive something that is not only potentially impossible but also consciously undesirable. The impetus to sexual excitement can be modified within the brains of invalids (Rathaus, Nevid, and Fichner-Rathaus, 2008), but it can't be suppressed permanently with conscious thought alone. As previously mentioned, humans have a tendency to want what they currently don't have, and that fact is especially true for satisfaction of natural human desires. Human nature pushes people to strive to satisfy desires and to achieve goals.

Since it's not healthy to repress natural urges, seeing every goal as a challenge and to change goals frequently, especially if frustrated, represents a way to adapt to the adverse environments created by contemporary societies. When a major goal, such as mutually passionate love, is achieved, it's not a final equilibrium but instead naturally results in developing new goals. In the case of some major attained goals, like a mutual romantic relationship, striving to maintain the achieved objective can be quite a goal in and of itself. Doing so long-term may actually require mutual surrender to the higher purpose of serving humanity.

It is extraordinarily difficult to postpone attempts to achieve a goal or satisfy a desire if putting it off increases the chances of satisfying it in the future. Doing so, and even deciding whether to do so, however, is far easier if an individual establishes the higher purpose of serving humanity as a whole. In addition, establishing that higher purpose also enable a value system that can enable postpone satisfaction of any goals or desires if doing so is perceived to serve that higher purpose.

Regardless, better means and chances of attaining desires with less undesirable outcomes, behavior, fetishes, and addictions can certainly be thought up in any environment, including in the sick contemporary one that selfish rulers have created for humanity. Natural desires can thereby be channeled into consciously desirable behavior. It is useful to remember that any fetish or religious worship of anything reduces or even eliminates the chance of real life experiences, and television shows and virtual reality do the same, even if they can provide some temporary diversion from miserable or boring circumstances. Such diversions aren't nearly as good as even the hope or chance of real human experiences. Human fetishes of any nature can easily be exploited by others and should normally result eventually in human spite when it is realized they are being manipulated. For instance, going crazy for trading female bodies is a natural male trait that innately creates desire in real women, but females brainwashed into exploiting male sex drives for the maximum utility and monetary value they can obtain transforms them into lifeless robots that themselves are mere fetishes as unappealing as blow-up dolls.

Many of people's desires today such as for money are merely a means to an end as opposed to being intrinsic to human nature, and some are just Pavlovian responses to stimuli associated with satisfaction of a need or wish (Nettle, 2007). For instance, the same part of men's brains lights up when they look at erotic pictures of women as when they take financial risks (*NewScientist,* 2008x). Both money and pornography are mere Pavlovian fetishes trained or "conditioned" into people to facilitate their manipulation. It's useful to be aware of the fact that letting a person's desires and needs rule, especially with respect to fetishes, can result in greater control by others of the path that the individual follows. While innate desires can't be totally self-repressed, they must be self-controlled and directed for the higher purpose of human beings, or they will be exploited by the inhumane monsters created by contemporary cultures.

Any signal that leads to a desire to engage in some unwanted habit or addiction should be recognized as something similar to an unnatural and counterproductive fetish. Such a fetish for a fetish represents a further separation from a real human desire, and seeking satisfaction of a fetish's fetish is an extra layer removed from natural lust, thereby making it an even greater perversion of human nature. To eliminate fetishism or an obsession for a fetish of any type, a person can prepare, train, and discipline the mind to focus on something else when the signal or fetish provokes an unwanted Pavlovian response. Understanding that an individual is not one of Pavlov's manipulated dogs that can be controlled by such fetishes makes it easier to eliminate the connection to the unwanted behavior.

It's also important not to fall into the trap of testing out whether one can control the addiction to a fetish by taking a step toward engaging in the behavior, as that thought might proceed to the next step and then just a little more, until it happens all over again. Such actions effectively represent attempts to apply the Pavlovian theory with respect to oneself, and they are therefore dehumanizing as they effectively cause one to become one of Pavlov's animals.

Of course, one is not a powerless dog, but, to ensure that fact, it's generally best not to think about any addiction or obsession, much less test one's resistance to it. If the thought of an undesired addiction or obsession even arises, repeating to oneself that the brain's neural connections which prompted one to have the addiction are being written over helps the conscious neurons to out-fired by other neural connections or thoughts. The very thought that neural connections causing a fetish are being written over by the thoughts helps because repeating the thought provides the confidence and positive thinking to weaken and eventually stop the system of neural firings causing the fetish. The process will lead to the eventual elimination of the old connections altogether as long as one's own conscious repeats that fact whenever the thought of it arises. However, the thought of testing out whether unwanted connections are truly dead causes them to strengthen by re-firing if they were dying, or to even reawaken from the dead and regrow if they were completely dormant previously. Such tests are therefore generally very counterproductive toward eliminating unwanted habits.

With respect to some undesirable habits or behaviors that are prompted by certain stimuli, it's useful to just consciously program the mind to purge the very thought of the habit or behavior upon any receiving signals. Such stimuli represent fetishes and can include sights, sounds, feelings of any kind, a desire, and any particular behavior that solicits a tendency to want to engage in the unwanted habit or addiction. A curiosity as to whether one has sufficient self-discipline to resist various stimuli is also a fetish signal, and any combination of fetish indicators are optimally ignored, regardless of whether they are internally, externally, or accidentally generated. The disgusting thought of essentially being Pavlov's slave dog, into which a fetish essentially transforms oneself spiritually, can help terminate any neural firing connections associated with acting on such signals or testing oneself like a hound.

However, the drive to satisfy basic human needs and desires, such as for nutrition, temperature control, and sex, that are controlled through feelings of hunger, cold, and sexual desire via the hypothalamus (Bernstein and Nash, 2008) cannot be so easily controlled. The hypothalamus and the striatum, both of which appear to be activated when they are not satiated with sufficient amounts of various chemicals such as dopamine (Arnow, Desmond, Banner, Glover, Solomon, Polan, Lue, and Atlas, 2002), seem to send their signals into consciousness to make that decision-making process aware of all innate human needs and desires which merit attention because they haven't been sufficiently addressed recently.

It is interesting that a lighting up of the striatum is associated with rewarding behavior such as self-penalizing actions motivated by spite (Whitefield, 2009). This fact implies that spite is a drive similar to that of other human needs and desires insofar as the stratum creates urges for satiation when the conscious perceives injustice. It can therefore be deduced that the drive for spite can be consciously affected by changing opinions with respect to what is fair, possibly indicating that the drive for other human needs can also be redirected by changing conscious beliefs. While beliefs may not be able to eliminate human needs, they can certainly affect their prioritization for satiation, as well as the timing, frequency, and manner of meeting those needs. A belief system coherent with overall innate human nature, such as one based on the facts that humans evolved to promote the continued existence of humanity, would be optimal for that purpose. Such a belief system can enable postponement of satisfaction of desires for food, water, and sex when a person perceives that doing so is optimal for humanity as a whole.

In any event, focusing attention away from feelings of hunger, thirst, cold, sexual desire, and spite can temporarily remove them from consciousness, especially if plans exist for their satisfaction in the future at a time perceived to be better for humanity. Since attempts to completely self-repress innate desires merely causes their redirection into fetishes, it is naturally human to formulate and implement plans to maximize the chance of satisfying unfulfilled needs and desires. Failure to satisfy some human needs, such as hunger and thirst, can eventually result in death, but frustration of sexual desires tends to merely result in a greater obsession with sex that is dangerous only if it is self-repressed and thereby leads to the development of fetish replacements for it that create extremely perverted desires and acts.

Since the frustration, stress, and perversions associated with frustrated desires and urges can be avoided by postponing indulgence in them into the future instead of repressing them, it can be useful to focus attention away from such matters by engaging in activities that are currently rewarding. Both work and leisure, as well as mere knowledge enhancement, can be utilized to divert focus away from unfulfilled needs and desires. Most people have naturally learned to engage in sex only when it is appropriate, as is naturally healthy as long as there isn't a permanent self-repression that leads to unnatural perversions.

The same principle holds true for undesirable emotions, including unrequited love. Emotions should, like urges and desires, be postponed to a future time when they might be more useful, appropriate, or welcome. Imagination can be useful here, and imagined love with a potential future mate can also be combined with sexual fantasies, thereby integrating the natural human desire for a mutual surrender between mates, both emotionally and sexually.

Postponement of emotions via imagining their future realization can be useful with respect to any feeling. For example, instead of immediate expressions of anger at the feeling of spite, the emotion can be delayed to a possible future time when there is a better opportunity for it to have a more desirable effect.

Depressive feelings themselves can be reduced by minimizing the amount of time one spends consciously comparing one's own situation to that of others. It's also useful to reduce the frequency of contrasting one's circumstances in the present to past situations to avoid feeling anger against oneself. Comparisons by nature imply a hierarchy that is unnatural to humanity. Even within the existing hierarchy, it's important to see life as an enjoyable game or challenging battle to legally replace the cruel hierarchies with humane egalitarian systems. In this war, believing in any of the propaganda and lies of the inhumane cultures is treasonous to humanity.

D. Mind Control

Acceptance of suggestions, whether conscious or not, enables control of one's mind by the suggesting entity and is a form of hypnosis. Hypnotic suggestions can occur not only via induction, as typically happens by repetitive remarks that lead to a relaxed state which make the hypnotized person more susceptible to suggested remarks, but they are also possible through suggestions while awake that usually have a gradual effect which is increased with repetition of the suggestion (McClenon, 2002). Persuading people to accept suggestions has been found to be positively related to the simplicity of the suggestion, the perceived self-interest of the person in following the suggestion, congruity of the suggestion, confidence of the suggesting agent, and empathy between the parties (Bond, 2010).

Hypnotic suggestions can even be received subliminally through media, such as occurs when the conscious attention of a person is distracted by some aspect of the medium at the same time that visual stimuli are subconsciously registered in the individual's brain, and these subliminal suggestions can affect behavior without the person ever having any conscious memory or knowledge of the stimuli (*NewScientist*, 2009p). In the latter case, the information apparently passes to the subconscious without conscious questioning by the thalamus filter, thereby resulting in the subconscious being programmed with neural connections that are subsequently communicated intuitively via intuitive feelings to influence behavior.

Brand name preference due to positive subconscious recognition from advertisements is a good example of such subhuman processes that influence buying decisions. However, as with all intuitive feelings, whether having been induced by suggestions and hypnosis or not, a person's conscious can override them, especially when the individual is consciously aware of any manipulative origination for them, or when the person consciously evaluates other information as may be more motivated for important decision-making tasks.

Hypnosis often leads to behavior that is similar to playing a particular part as an actor that can result in the subconscious causing a person to live and experience that part (Gooch, 2007). This process only needs direction from a trusted hypnotizing agent, whether an external one or via self-hypnosis, to activate. For instance, some behavior can be simply programmed into people, such as walking more slowly after reading a particular article about old age (Wiseman, 2007). Hypnosis can also lead to significant changes in perceptions, such as in seeing sounds, as apparently occurs by opening up connections in the brain between the senses that had long ago been suppressed in most people (Robson, 2009b). Such artificially induced connections or reconnections are generally useless or even counterproductive, as they create too much irrelevant noise, and so those interconnected neuron firings eventually disappear if they aren't reinforced with biologically rewarding chemicals like dopamine. Similarly, hypnotic suggestions for behavior that doesn't lead to a positive consequence are eventually extinguished.

There are varying types and degrees of hypnosis, but hypnotic control largely consists of getting an individual to follow the suggestions of the hypnotist with little or no conscious questioning or interference. In particular, a person must be persuaded to trust the hypnotist or hypnotic medium for control to be obtained. Hypnosis occurs when a subject allows at least a partial surrendering of himself or herself to the hypnotist (Urban, 2007), who can thereby gain varying degrees of control of the subject through direct access to the subject's subconscious. The human desire for interconnection and coherence with others and the universe, as is incorporated in a desire to mutually surrender as one in group hypnosis, may be one of the traits that has helped make humans hypnotizable.

An origin for speech via hypnotic chanting in prehistoric times of coherent group consciousness may contribute to the power of verbal statements on hypnotized people. The natural inclination to believe spoken words may also stem from that beginning when empathy caused people to only chant and say things that were useful to others and the group as a whole. The ability to consciously analyze statements likely developed thereafter when people were confronted with conflicting remarks or chants, or when the verbal signals turned out to be unnecessary or counterproductive. However, most statements were probably accepted automatically as factual until the development of deception and dishonesty that began with the exploitative matriarchies 35,000 years ago.

A chanting origin for speech in cultures of group consciousness and honesty may help to explain the natural human tendency to want to follow another's suggestion. In such prehistoric cultures, suggestions were always made to benefit others, not to manipulate them for purely selfish reasons. To avoid misleading others, people likely provided their suggestions along with an indication of their perception of their reliability, perhaps originally through the degree of tonal urgency in chant-initiating sounds.

Because a suggestion took effort and thought to make, there was an implicit obligation in the group consciousness of most of prehistory to utilize the suggesting person's mental work. Following a suggestion was a sign of oneness, and failure to do so might insult the suggesting person, who might not be willing to provide helpful suggestions in the future. Any group consciousness among friends remains subject to the same tendency even today.

The brain continues to find it relaxing to just follow suggestions, and a conscious effort at skepticism is needed to inhibit that tendency. Following a suggestion reduces the amount of additional mental processing time expended to solve a problem, thereby economizing on thinking resources. However, following suggestions blindly in today's societies that encourage exploitation, manipulation, and deception can be suicidal. In a capitalist society where people are encouraged to act in their own best interests and are therefore motivated to make suggestions that may be of limited value and even be useless or counterproductive to the individual receiving the suggestion, the suggestions of others have to be analyzed for potential manipulative intent, as well as for poor quality due to ignorance of the suggesting person who may also be manipulated or misinformed.

The degree of trust placed in a suggesting entity depends on a subconscious and sometimes conscious weighing of the relative importance of the goal for which the suggestion is made and the opportunity cost associated with expending the effort required to rationally evaluate a suggestion or make one's own decision. The analysis also optimally takes into consideration the benefits and costs to the suggesting entity from engaging in manipulative deception which may later be detected and punished in some way. In particular, people making suggestions at least subconsciously recognize the risk of spite, as well as the loss of any friendship, unless their recommendations are made with benevolent intent. There is therefore a natural inclination to trust a suggestion in lieu of other information, and people often trust total strangers, the television, or a brand name for unimportant consumer purchases. While the analysis necessary to determine whether a recommendation should be trusted can be done totally subconsciously, the conscious can always set rules for the subconscious and intervene when it perceives it optimal to do so.

Humans have a capacity for detecting potentially biased suggestions that can be evaluated both subconsciously and consciously. For instance, sensory perceptions like tone of voice and facial expressions can be analyzed subconsciously at the same time that other information is pondered to determine whether a signal of intuitive suspicion should be sent by the subconscious to the conscious mind, which can evaluate further the motivation, reputation, and expertise of the suggesting agent. A component of this analysis involves a recognition that manipulation results in the loss of desirable trust and friendship that may have some positive utility to a potential manipulator, thereby inhibiting deceit and exploitation especially if an individual shows signs of spiteful responses in the face of such behavior such as with respect to a reputation for being spiteful.

In the case of complete trust and focus on the hypnotic agent, the conscious inhibitory firings are completely shut off, resulting in hypnotic sleep and maximum suggestibility. In particular, hypnosis can be used to motivate a subject to enter a dream-like state whereby the subject is cut off from sensory perceptions and conscious thought. In this state, the subject can be easily influenced to act as the hypnotist instructs. This control may be maximized not only with the attainment of the subject's trust but also via the use of persuasion, drugs, deceit, offers of rewards, and threats of punishment. The subject can also be persuaded to reprogram his/her own subconscious, especially with respect to having a particular reaction, like a behavior, desire, or feeling, in response to a particular stimulus programmed by the hypnotist. The signals themselves can be feelings or perceptions. Pleasant feelings can even be programed into a hypnotized subject upon following the commands of a hypnotist, thereby providing reinforcement for obedience. The hypnotist can also have the subject not remember the programming, since the conscious is cut off from the memory portion of the brain during dreamlike hypnotic states, thereby making it more difficult to consciously realize the control that is exerted, much less change it.

In cases where there is unity of purpose between the hypnotizing agent and the hypnotized subject, suggestive influence can last forever, as in mutual love that involves mutual hypnosis. In that form, a unity of purpose exists to maintain the ecstasy of the balanced love that is further maximized by being shared through empathy which further promotes a desire to mutually please.

The power of suggestion without a deep hypnotic trance works in a manner similar to a trance-like state except with varying degrees of shutting down rational analysis of statements. A partial or complete termination of conscious questioning, doubt, and analysis is dependent on a person trusting the entity making the suggestion. Whether the hypnotizing entity is another person, a group, or even a medium like television, the hypnotic power of a suggestion is more powerful when the suggesting entity is more trusted.

Tarot card reading and other prophesizing "magic" work in a fashion similar to a suggestion, but the advice is provided by random events such as those at least inhibits manipulation by others. A random choice of card or other event can even result in suggestions that can be self-fulfilling if one believes in them, just as the suggestions of others and hypnotists can be. The choice of card(s) or other sign(s) may be random, but once selected, the power of positive thinking and belief can influence their occurrence if they are believed, as in any hypnotic effects. However, as with any suggestion, signals of a magical sort affect one's future only to the extent one's faith in them causes one to incorporate them into one's subconscious.

Contrary to the allegations that following hypnotic suggestions involves nothing but role-playing, it has been well documented that hypnosis has strong biological and physiological effects (McClenon, 2002). Since repetition of suggestions typically increases their power, it seems likely the hypnotic process involves the creation of neural connections which can be strengthened through the repeated firings of those connected neurons that are induced by repetitive suggestion. Because of trust in the hypnotizing agent, those neural firings are not very inhibited by conscious evaluation, nor are they later subjected to an automatic conscious reevaluation when the firings are repeated if they were never associated with any conscious doubts, memories, or associations with other stimuli. Hypnosis tends to be more powerful and long-lasting if the hypnotized person is motivated to repeat the suggestions, as when the individual believes, or wants to believe, the suggestion. Belief thereby constantly strengthens the neural connections that lead to following the suggestion by not only permitting them without interfering doubts and conflicting neural firings but actually reinforcing them with the conscious thoughts involved in believing in them.

Some suggestions can be entered into the subconscious as rules to be followed no matter what, but they can always be broken by conscious intervention in any event if they appear to be leading to counterproductive behavior. Even with suggestions entered directly into the subconscious without any conscious memory under a hypnotic trance, it's only necessary to evaluate the signals that are leading to a particular behavior to change the subconscious rules that are followed upon such signaling.

In the case of exploitative hypnotism for the selfish benefit of the hypnotist, spite results, thereby creating a desire to eliminate or reverse any tendency to follow suggestions. A hatred of the suggesting agent for breaking the individual's trust can even be dangerous to the exploitative hypnotist due to the possibility of revenge. Like the mutual love, group consciousness, and trust that create hypnotic effects, the effects of hypnotized suggestions of any sort can be eliminated by merely deciding to do so. Even in prehistory, behavior based on suggestions often had to be purged or changed upon the death of others with whom a person had shared a group love, and so a natural human ability to ignore some suggestions that don't seem appropriate had to evolve. Such connections were often broken anyway when there arose another hypnotizing influence, including from oneself and yet others.

The easiest way to break an unwanted influence or suggestion is to place top priority on trying to serve humanity as a whole in one's own way. Any influences or suggestions inconsistent with that purpose are automatically purged through that order to the subconscious. Setting this purpose in life above all else also motivates self-love as an individualized component of the love of humanity, thereby permitting self-hypnosis.

To break hypnotic programming of any type, it is merely necessary to consciously focus on instructing the subconscious to eliminate the connection between a stimulus and the reaction. On the other hand, if hypnotized behavior develops into habits that are reinforcing short-term, then just breaking the connection between the signal and the behavior isn't enough. It is also necessary to break the habit. This process requires seeing that the habit, whether acting weird, having weird desires, alcoholism, etc., brings insufficient benefits long-term compared to other behavior and that there is no further net gain from the habitual behavior no matter how much pleasure or benefits it has brought in the past. Breaking a habit can provide pleasure in the form of learning not only that it can be broken but how. This pleasure can be greatly magnified to more than offset any habit if there is the self-fulfillment that comes from committing oneself to trying to serve humanity as a whole in a way that the person chooses.

Even if a hypnotist programs a victim to react upon a signal that is not consciously remembered, the victim can, upon recognizing some hypnotic control, instruct the subconscious to have all signals and rules consciously re-evaluated for maintenance or rejection. The latter process may be a sort of spiritual rebirth that leads to an elimination of old habits and subconscious tendencies that are no longer wanted. It can often even remove undesirable behavior and beliefs programmed into oneself before the age of six when the hypnotic state of theta waves dominated.

Breaking hypnotized commands requires instructing the subconscious to place a lower priority on any reward from obedience. For instance, the special type of attention and perceived love for slaves compares negatively to self-reinforcing behavior which is consistent with human nature and which therefore provides a greater reward in the form of a feeling of self-fulfillment from being a human being as opposed to a subhuman servant. Acting according to a person's own human nature can also increase the chances of satisfying other true human desires as opposed to fetish substitutes that lead to only temporary pleasure and are pathological long-term. For instance, winning praise through some professional accomplishments can replace a stage hypnotist's suggestions to act absurdly as a means to obtain desired attention, since professional accomplishments bring more positive attention and economic or other rewards than acting out the hypnotist's suggestions that can be inconsistent with the more desirable form of attraction.

A person can program the individual's own brain to have each detection of even a minor undesirable connection between neurons lead to an immediate breaking of at least that connection at the same time that another stronger connection grows or is reborn within the brain. The replacement connection can be whatever the individual wishes, including merely laughing at the folly of the prior connection. Even without conscious detection of an undesirable neural connection, actions, or thoughts that are inconsistent with desired goals can be swamped by interconnected electrical firings relating to desired actions and intent, thereby causing the disappearance of the undesirable thoughts that are no longer reward chemically in the brain. However, it is important that a connection to the overriding innate goal of human beings to serve humanity as a whole be made at least implicitly for all thoughts to avoid a relapse resulting from lack of complete coherence in the brain with innate subconscious tendencies.

It can be useful to recognize that each person creates all the chemicals and electrical connections in the individual's own brain and has complete control over them, including any pleasure chemicals, DMT, and melatonin involved in hypnosis. One only has to believe this fact take complete command of oneself.

For instance, the subconscious can be reprogrammed via self-hypnosis (Hewitt, 1996), which merely involves suggesting to oneself. Self-hypnosis represents a form of group consciousness that unifies the conscious and subconscious. It requires sufficient faith in oneself to enable one to follow through on one's own suggestions without conscious questioning or inhibitions. Self-hypnosis consists of a group of one person, and it may therefore optimally apply the most primitive form of group consciousness to the fragmented consciousness of today that divides not only individuals but even each person's conscious from the subconscious. By integrating the subconscious tendencies for group consciousness among all people with the individual consciousness, a desire for a larger unity with all human minds should follow naturally.

A deeper and potentially more powerful self-hypnosis can be achieved via recordings of oneself using the same tone and words used to hypnotize others, such as to relax, imagine being in a relaxing place, become sleepy, fall asleep, obey the hypnotist's instructions, and then awaken (Hewitt 1996). An altered state of self-hypnosis is also possible via meditation (McClenon, 2002). Meditation itself can be induced by concentrating on some needed action, which typically is breathing but can also be some activity such as archery that one has learned so well that one merely lets the subconscious take full control, enabling a person to just marvel at that and become unified with the activity of the subconscious (Capra, 1999).

Although meditative hypnosis in any form can be used for mere relaxation and enjoyment by allowing the subconscious to integrate better with the conscious, self-hypnosis remains a form of the conscious directing the subconscious. For maximum effectiveness, there has to be a coherency between the conscious directions and the innately irrevocable drive for self-fulfillment through trying to serve humanity as a whole.

Self-hypnotized suggestions that are so deeply imbedded through conscious repetition that they are ingrained into subconscious connections of neural firings enable the subconscious to operate efficiently without subsequent conscious intervention. Without such suggestions, one's conscious may question and inhibit every action, leading to indecision and inefficient subconscious operation. On the other hand, letting one's subconscious go without direction can lead to either acting like a chicken with its head cut off or slavery to the suggestions of others. While such obedience can result in very powerful acts due to the power of the subconscious when operating without conscious questioning and inhibitions, the subconscious has no self-direction. The subconscious just operates efficiently according to the last order(s) or thought(s) received by it that a person either chooses consciously or lets be chosen by others and the environment.

For self-hypnosis to work, it's necessary for a person to believe in what the individual consciously instructs. Such a belief doesn't require deification of one's self or even necessitate a belief in everything one says or thinks. Instead, a mere recognition that one is in control of one's own brain is sufficient. A belief in one's own opinion on a matter can be strengthened by one's own knowledge on that subject, as can come from science.

Whereas hypnosis represents a form of submission to others, self-hypnosis is a submission to one's own self. Mere silent conscious repeating of rules for the subconscious to follow represents a form of self-hypnosis.

Because self-hypnosis enables better integration between the activities of the subconscious and conscious, the subconscious knows better when to ask for conscious decision-making, the conscious knows better when to let go for uninterruptible, efficient operation of the subconscious. While meditation can be especially useful for this integration, mere enlightenment about the meaning of life, the universe, and everything can do the trick of both letting go and maintaining self-control in the service of humanity.

Self-control, including of one's subconscious, merely requires one to give the orders and decide on one's own thoughts for oneself. These orders may be a form of self-hypnosis which can have maximum strength because there can be no questioning the benevolence of their intent. While commands to oneself can lead to failures, there is no frustration or loss of confidence as long as one's intent is to try to serve humanity as a whole, as such attempted service has to be in the way each individual decides is best to maximize the creative power of human nature. Learning from one's failures enhances that collective creativity as well as provides a source of entertaining humor.

While a person always has the power to ignore the suggestions of others, habitual tendencies, or unwanted urges, self-control can be facilitated in the face of needs, wants, and external attempts to influence when one sets a goal or behavior that one deems more important like serving humanity as a whole as a human being. Training the subconscious to serve humanity is more successful if negative feelings like frustration are avoided by merely attempting to achieve the goal and humorously enjoying all failed outcomes.

One short-term goal that is always achievable but yet never boring is to merely be more aware of the moment and the experience of it all (Klein, 2003). Many Eastern religions advocate as much (Capra, 1999). However, permanently maintaining such goals can transfer oneself into an impotent or manipulated robot or vegetable, and an unpleasant feeling of meaningless worthinessness can easily arise. It can be more satisfying to try to achieve a more worthwhile goal for which the mere attempt to achieve it may create a very pleasant feeling of self-satisfaction.

Although frustration, failure, and pessimism with respect to a new goal can lead to a return to prior habits or desires, focus and self-repetition of the need to change will overcome such tendencies if the effort to try to attain the objective is perceived to be enjoyable and self-fulfilling. In particular, just striving for a goal, such as helping humanity as a whole, that is recognized as worthwhile even when not successful, can create a sufficient feeling of pride and self-fulfillment to overcome any undesired thought. Unwanted habits and fetish desires can thereby be eliminated. Anything suggested by a hypnotist or others can be unprogrammed by another hypnotizing agent, including oneself.

Deprogramming is especially facilitated when it is an important matter, thereby motivating greater will power that prompts stronger and more frequent counter-commands. Ending the effects of childhood brainwashing is also easier when one recognizes that one can indeed de-program and program one's own brain, no matter what other hypnotic brainwashing has been applied by anyone.

To motivate such reprogramming, it can be useful to think logically to figure out that following a certain suggestion doesn't lead to the result that was stated it should. For instance, purely selfless obedience generally brings others to contemptuous disgust instead of positive attention, and so following the suggestions or commands of others can be counterproductive, by causing contempt instead of affection. Suggestions may also be more easily broken when they are recognized to be based on false assumptions, such as an assumption that one has to be exploited to attain goals. Coherence in goals and brainwaves is necessary for a successful accomplishment of tasks that is self-fulfilling.

Losing interest in some particular goal related to suggested thoughts and actions can also help motivate reprogramming, but innate drives for human needs and drives can't be eliminated permanently. They can only be redirected for the purpose of serving humanity as a whole in a way that the individual decides is best. Doing so enables each person to self-direct innate urges instead of having them manipulated by others.

To de-program undesired programming of any kind, it is only necessary to essentially write over earlier programming by having neural connections fire more powerfully than the originally programmed ones. Conscious firing of other neurons more powerfully will eventually silence neural connections that were externally suggested or that resulted from bad habits. Klein (2003) suggests trying new behavior or thoughts that will be more enjoyable because human nature finds repeated actions ever less pleasurable but reinforces novel experiences with chemicals like dopamine in the brain.

Pleasure chemicals, which are released more strongly in anticipation of a desired event than when the situation is actually realized, merely cause electromagnetic activity that reinforces thoughts and actions as long as there isn't conscious interference. Conscious focus to remove undesirable thoughts leads to electromagnetic activity in the brain incoherent with the connected subconscious brain cell firings of the unwanted thoughts that apparently disrupts the chemical reactions which are causing a repeat of the connected electromagnetic neural firings involved in the undesired tendencies. Connected firings seem to require repeated dosages of dopamine for continuation, and conscious thoughts can stop such reinforcement simply by a confident will to do so. Since the human brain evolved to promote human nature, natural human behavior that occurred in the distant past will always provide more self-fulfillment and thrills than artificial behavior based on "modern" technology. This fact can be readily seen by comparing mutual human sexual pleasure with sex with blow-up dolls and contrasting the artificial "highs" of harvested or manufactured narcotics with love.

While focus on alternatives that represent very pleasurable natural human behavior, such as mutual human sexual pleasure, is the optimal route toward breaking unwanted activity or thoughts, a subconscious tendency to intensify and redirect desires if they are internally repressed can sometimes make it difficult with that process alone to eliminate the neural connections causing undesirable thoughts. If one continues to feel obligated to do something that one doesn't consciously want to perform, perhaps due to hypnosis, childhood brainwashing, habit, or addiction, it can be useful to imagine actually carrying it out in some parallel universe to the fullest extent all the way to the end and beyond, in order to see the absolute meaninglessness to which the activity leads. Once the behavior is identified with an adverse feeling of emptiness, the neural connections to carry out the activity should weaken and disappear if one recognizes that a satisfying feeling of self-fulfillment can only be derived from serving humanity as a human being. Nevertheless, it's important not to repeat, renew, or test such perverse fantasizing, because otherwise the fantasies themselves can become yet another habit to be broken.

Recognition of the fact that any actual experiences aren't as enjoyable as the anticipation of them can further weaken tempting but undesired thoughts, since a single imagining of them has already provided the maximum pleasure they can supply. In particular, knowledge that anticipating a desired event provides more satisfaction than the actual experience can by itself aid in removing a fetish. Any further imagining or experiencing of the fetishes thereby becomes increasingly boring and less satisfying. It may also help to realize that discontinuation of being a slave to a particular habit, thought, pleasure, suggestion, ideology, brainwashing, or unnatural desire enables one to act as a human, instead of as a boring lifeless robot.

Merely seeking human pleasures in the process of serving humanity provides the highest highs, including through realistic human fantasies alone. As will be explained in more detail in Chapter 16, fantasies of natural human pleasures can also maximize the likelihood of their occurrence, thereby providing further motivation for such thoughts when a person is trying to determine what to ponder.

It is important to emphasize that while each person is free to direct the complex actions of the individual's own subconscious, the conscious has evolved to be driven toward doing so for the benefit of humanity as a whole. Orders to the subconscious inconsistent with that very purpose of consciousness create negative conscious feelings of emptiness as well as incoherence with innate subconscious tendencies. The result if not only unhappiness but ineffectiveness in controlling the subconscious, whose innate inclination to serve humanity may only resurface in some bizarre form.

Although one can alternatively let the environment and others direct it by choosing to enslave oneself to one's desires or to the suggestions of others, such lack of will is contrary to human nature. While the subconscious is incredibly adaptive and powerful, it can also be manipulated quite easily. For instance, brain cells extracted from rats, and placed in nutrient solutions, naturally grow connections and seemingly seek direction from electrical stimulation that can be programmed by a computer to direct a robot (Marks, 2008). It's only necessary to be aware of the fact that one's conscious is in charge of such a powerful and obedient servant to ensure that someone else, or some other program, doesn't control it.

While external electromagnetic waves, including from the earth itself, can affect human minds (Stray, 2005), people can also change the frequency of their own brain's electrical activity by mere intent (McTaggart, 2008). A mere recognition of faults in one's own brainwaves and thought processes appears to result in changes in neural firing frequencies and perceptions (Thomson, 2010a). Humans, who themselves also emit electromagnetic waves, are, in the final analysis, "in charge of their own minds" (Silva, 2002).

Moreover, there is irrefutable evidence that humans can, with sufficient focused intent, affect the electromagnetic waves of other people and things (McTaggart, 2008), Human beings with concentrated emotions, like those occurring around Princess Diana's death and 9/11/01, can even affect the earth's own electromagnetic field to an extent that is significantly detectable by satellites (Braden, 2009). Being composed of electromagnetic waves themselves, humans both affect and are affected by those of others and other things.

External electromagnetic stimuli, induced by an electromagnetic coil held over an individual's head to create a weak electric current, can be used to treat mental problems like depression (*NewScientist,* 2008zzt) and schizophrenia (Thomson, 2010a). However, such forces do not have a completely overwhelming effect. They only promote tendencies in those who don't have sufficient interest, desire, or will to determine their own fate. Existing technology can't even measure electromagnetic impulses from the head precisely enough to obtain both spatial and temporal information in the brain (*NewScientist,* 2009zg), much less totally control them. The quantum facts of not being able to detect both the position and speed of particles simultaneously may be one reason humans can always overcome any external controlling force to collapse their wave functions on their own selected point(s).

As a result, although it has been found possible to use observed patterns in the neural firings in a person's brain upon the existence of specific thoughts so that the individual can willfully cause changes in a computer screen's output (*NewScientist*, 2010aat), it would seem impossible to fully manipulate a person's brain against the individual's will. In fact, with sufficient will power and psychic knowledge, the brain can adjust firing connections to adapt to any counterproductive external influences. Complete control of an individual's thoughts is therefore impossible through external application of such forces.

Even if more electromagnetic control of the mind could be harnessed in some fashion, it would be necessary to be able to control all human brain processes, since human nature has evolved so that the mind is very plastic in firing connections, chemical reinforcement, and quantum communication through neural microtubules and photon entanglement. A targeted victim of electrical or other mind control can indeed learn to redirect or even block out all effects, electro, signals, hypnotized, or otherwise, that are culturally artificial, inconsistent with human nature, or interfere with natural human urges. It merely requires a recognition that one can do so by rewiring neural connections and firings without repressing natural human needs, desires, and drives. In particular, given that the source of such electromagnetic waves can be detected by the body subconsciously, the brain can learn to recognize the external effects and either signal their nature to the conscious, such as via intuition, or have their effects completely overridden with neural connections that can be initially set subconsciously or consciously.

Nevertheless, invisible electromagnetic forces from natural sources do affect psychological tendencies on an ongoing basis that can lead to measurable average effects across groups of people, and so they must be taken into consideration. For instance, the stock market has been shown to be affected by lunar cycles that have been found to have an effect on average in tests on past data and may be related to the electromagnetic effects of the moon's changing but cyclical orbital position around the earth (Yuan, Zheng, and Zhu, 2006). However, because such effects are induced externally, they may not continue in the future if people willfully choose to react against or overcome the forces. In particular, people can use conscious thought to override the normal neural firings in the brain resulting from external forces, especially if they are both aware of the impacts and have an incentive to change the impact of them. The effects of the moon cycle on stock market prices creates beneficial opportunities for those aware of and able to exploit the effect of those forces on unaware humans (Murphy, 2007), and human beings are innately motivated to rewire their brain cell firings to take advantage of such phenomena.

If useful for their survival, it is quite likely that humanity will eventually adapt to overcome all external forces, as each individual is motivated to do so unless killed by them. Adaptation can occur consciously rather quickly, especially if it is felt that there is human exploitation by others that can cause feelings of spite. Individuals can order their subconscious to filter out any external effects just by giving the order, which the subconscious can and must obey irregardless of the effects of external forces.

The human mind can alter effects created by the environment through mere rationally based orders from the conscious to the subconscious to offset negative external influences on neural connections, including those created by external hypnotists and radio waves of various frequencies. In fact, human consciousness may represent a tool that evolved at least partially for both programming the subconscious and directing, emitting, and deflecting radio or electromagnetic waves from eliciting behavior that is incoherent with human existence. Even evolved human reactions which may have been useful in some past environments but which are no longer are perceived to be useful can be reprogrammed.

The large range of electromagnetic brain wavelengths between 0.5Hz and 125Hz that reflect differing firing frequencies of neurons and alternate states of consciousness, including highly focused ones at the highest frequencies, can be controlled by the conscious to serve its direction without any change in external inputs. Internal inducement of changes in those frequencies provides an indication of the overwhelming power of the brain in comparison to environmental factors. Attempts to control another's brain via electromagnetic waves can't affect one's own neural firing frequency if one consciously refuses to resonate with such incoherent external forces.

Focus is useful for ensuring that external factors don't affect one's own behavior. Concentrating on a task provides direction to the subconscious to ignore all but relevant information. When one feels a potential external attempt at influence, it can be useful to automatically concentrate more closely on the particular task(s) in which one is currently engaged, or preparing to engage. The attempted unwanted influence should thereby be easily ignored consciously, although alternative possible courses of action can still be evaluated subconsciously, including with respect to any diversionary information that is indeed relevant or might justify a change in focus. To the extent that otherwise negative influences thereby enhance focus and cause some evaluation of alternative actions, they can actually have a positive impact.

To avoid following unwanted suggestions, including those received during childhood or similarly impressionable and hypnotic states that a person may not even remember, it may only be necessary to think about the power of consciousness, which has complete control over the subconscious and whose suggestions have to be accepted by the subconscious as orders as opposed to mere suggestions. Conscious self-commands are automatically designed to override any suggestion or even orders from others whether remembered or not. It is only necessary to believe in the factual power of one's own conscious thoughts, as otherwise one may allow suggestions and lies to the contrary to weaken conscious directions, thereby causing the conscious itself to be influenced by subconscious signals based on past habits or suggestions of others.

The power of consciousness over oneself is thus virtually unlimited as long as one doesn't allow one's recognition of this fact to be weakened by doubts that may stem both from the environment as well as from strong subconscious urges. This environment includes others' statements and even beliefs that serving a cruel fictitious deity is in people's best interests. Many who hold such views are actually acting strictly out of self-interest, as they believe in their deity only because they think by doing so they will get a after death, as well as possible rewards while living. The suggestions and "psychic" waves relating to the felt beliefs in a lie by Christians and other similarly brainwashed religious zealots, however, cannot serve humanity as human nature evolved to be grounded in truth in order to maximize the power of human beings that is based on the quantum reality of the universe. The only urge that can't be ignored is the purpose for which human consciousness evolved, i.e., to serve humanity as an equal but different member of that species.

Many other people believe strictly in a rational selfishness that can lead to hedonism, which they then promote in order to develop a sort of group consciousness of fellow hedonists that is naturally more enjoyable than being alone. Many of these individuals also believe in some fictitious deity out of ignorance and fear of penalties in an afterlife in which they have been indoctrinated to think is at least possible if not likely. Since selfish hedonism can't bring true lasting happiness without some higher purpose, they may have further incentives to believe in religious lies in order to establish this hedonist reason for being. However, pure selfishness, even when supported by fellow hedonists and motivated as the will of a make-believe deity, is eventually and often punished by their own innate human nature with adverse feelings of uselessness. While such people may find some temporary relief from this agonizing meaninglessness by being with and helping others like fellow hedonists, their lives remain misdirected. No amount of pleasures, friends, and token helping of others can allow them to escape the emptiness of their lives. Recognition of these facts enables a person to confidently ignore or redirect those for the benefit of humanity with unequivocal conviction.

The beliefs in selfishness or religious lies that are brainwashed into people starting at a very early age inhibit people from acting as their human ancestors did with respect to acting communally for the benefit of all. They are then a disgrace to their heroic predecessors.

Brainwashed people today may attain some of the benefits of communality when they form into subgroups of humanity, such as via the nationalism that is widely promoted in most countries. However, these subgroupies or patriots remain separated from the whole, and no amount of group identification or patriotism can keep them happy. Because the existence of separateness drives them to feel superior to others outside the group, hierarchies typically result that create disharmony that is unnatural to human beings. Such incoherence between people's wills and desires not only limits their power and happiness, but it often even leads to spite, violence, and war, especially from those not in their subgroup. It therefore usually causes extraordinary misery.

Religious lies, selfishness, and subgroupism are inconsistent with human nature, and any beliefs in them weaken the power and ecstasy of communal human resonance. Besides creating disharmony between human beings, such faith in inhumane lies creates internal incoherence within the individuals with those beliefs. Inhumane beliefs fragment the mind and thereby weaken the resonance between the conscious and subconscious. Besides creating feelings of emptiness, uselessness, and pointlessness, a failure to think and act as humans evolved to do disrupts the potency of subconscious processing and can even cause natural subconscious urges to be redirected into perverse forms.

Inhumane beliefs and views of others must be ignored to avoid catching their disease. There can be no feeling of social obligation to follow their advice or entertain the possibility that their inhumane opinions might be valid. Those afflicted with the diseases like Christianity and pure rationalism or hedonism have been transformed into programmed robots by the beliefs that have been ingrained into them, and their suggestions and views generally don't provide any useful direction. While the innate human nature of those others could cause them to make meaningful suggestions, and while their opinions on a few matters might occasionally have some degree of validity due to random chance, there is a far higher probability that their views are motivated by a desire to exploit for their own benefit and thus counterproductive to others and humanity as a whole. The fact that socially better connected people tend to treat other humans as objects without any empathy (Fox, 2010) provides both evidence and a useful perspective on this issue. As a result, when in doubt, it is often optimal to totally ignore the advice from others since it is more like to lead to be counterproductive than useful for those following their own human nature.

While it is often necessary to disregard the opinions and perceived feelings of people afflicted with diseases like Christianity and rationalism or hedonism, it can sometimes be useful to make token efforts to show the consistency between some of their random or subconsciously humane views to be the same as one's own in order to make temporary alliances with these robots. However, social ostracization and rejection by these programmed zombies should be taken as a hilarious complement to the beauty of one's human nature that can shame these sickos into seeking a cure for their disease.

Individuals can only attain maximum power, happiness, and feelings of self-fulfillment from serving humanity, as setting this higher purpose for all actions and thoughts automatically harmonizes the innate human natures of all individuals. Even in today's world where so many are inhumanely brainwashed into fragmented parts of the whole, the bliss and power of such communality can be attained with just the conviction that all present and future human beings are being subconsciously driven to serve humanity even if they don't consciously realize it, just as their ancestors were.

For an individual to obtain maximum resonance with people's innate communality, it is vital to completely purge any internal views and thoughts inconsistent with human nature. Even a person alone can accomplish this goal which harmonizes the individual's thoughts and deeds and thereby unifies the subconscious and conscious for maximum internal power. Such coherence eliminates the need to decide whether subconscious urges or conscious rationality should be followed. The unified operation of the conscious and subconscious for the purpose for which they evolved doesn't require making much more than a self-directive to do so, as long as there is a belief that the command will be followed. Once human control over oneself has been obtained, one's influence over the external environment becomes extremely potent, as the internal harmony resonates with the subconsciouses of all other human beings. Both internal and external power is thus maximized when one acts consistently with human nature and its overriding goal of serving humanity.

To exercise such power, it is always useful to attempt to coherently use all external factors, as can be done via increased focus, and to never hope for or rely on luck. Redirecting undesirable external forces while resonating with coherent ones can help one to control the environment instead of being controlled by it. Since all human beings are equipped from birth in the art of persuasion (Bond, 2010), anyone can accomplish this mission through mere suggestions as well as via self-hypnosis, "psychic" powers, and actual physical actions that are intended to benefit humanity as a whole.

E. Consciousness as Part of the Whole

The universe is made up of probabilities, but events aren't totally random. A person can influence those probabilities with sufficient focus and will power. While information about the chances of events shouldn't be ignored, since they indicate the amount of will power necessary to successfully overcome the odds, a person should never believe that probabilities are unchangeable destinies that the individual is enslaved to experience impotently.

Somehow someway, humans have evolved to seek to connect with and serve others and the universe, and not just because of neural pleasure firings. It may be more of a selfless reverence that is mutual. How else can one explain the fact that people sacrifice their lives for others or for a belief in how to help others and improve the universe? Perhaps this characteristic partially results from humans having had to serve the greater good of the group and humanity in general in order to survive. They literally couldn't live without each other. The evolved trait of human communality may explain why humans became the dominant force on the earth, as human nature may be more consistent with the structure and nature of the universe than other organisms without that capacity to serve higher goals.

Most who have experienced subconscious knowledge feel that we are all connected and one, and this hypothesis is consistent with quantum mechanics (Peat, 2005). In quantum physics, sound waves can be thought of as phonon particles just as light waves can be thought of as photon particles (*NewScientist*, 2007r). Both are mere vibrational waves of far different frequencies (Szego, 2002), which themselves vary for different types of sounds (Capra, 1999) and light (Szego, 2002). Everything perceived, however, is merely the brain's categorization of interconnected fields, such as electromagnetic ones (Capra, 1999). These interrelated fields normally exist in the form of waves of potential that can be collapsed into individual points on the wave in the form of a particle when they interact with other waves in ways that exclude the possibility of other potential points on their waves (Penrose, 2004).

Nothing people observe or experience is real, and even "objective" observations affect humans' perception of reality (Brooks, 2007b). According to Merali (2007f), there is no such thing as objective detachment from the environment, as is proven by experiments in quantum physics which indicate that measurement affects measured particles. The measurer effectively interacts with the measured object because the measurement collapses the probability wave function of that electromagnetic field to a single state, thereby eliminating other potential states or averaging waves of them. While the environment may randomly choose what information to report, i.e., what state on the probability wave function will be observed, that observation does affect the past, present, and future probability wave function of the observed electromagnetic field.

For instance, Begley (2007f) indicates that experiments have shown it's not tangible effects associated with measurement that cause a single photon particle to go through only one slit when observed, as a shutter that is opened just after the photon goes through one of two holes in a screen keeps the photon from going through both holes simultaneously, just as it would do if the shutter remained closed. In particular, a different shadow on a subsequent screen that it hits is observed when the photon goes through just one hole (i.e., when the observation shutter is open) than when it goes through both holes (i.e., when the observation shutter is closed). The implication is that the photon goes through one hole in one parallel universe and through the other in another universe, and one sees the average blurring of both universes unless one focuses on observing through which hole the photon passes and hence which parallel universe one experiences or observes.

Thus, what happens in more than one universe affects what is subsequently observed, as long as one doesn't now or in the future definitely know which potential past universe did happen. Quantum uncertainty may even make it impossible to ever deduce the exact set of past universes that had to occur for any particular present moment to exist. The sum of all potential past universes that haven't been excluded from connection to the present or certain future moments therefore does affect what is experienced, as in a smoothed wave.

Nonetheless, anything that happens now does affect things that happened in connected past universes. Some events now can collapse past occurrences into points of certainty, while only the probability of occurrence of other past events is affected. When only the probability of a past event is impacted, people living in those past universes can still change the chance of the occurrence by consciously exercising greater will power that can then affect the future. The past, present, and future are thus determined jointly by all human beings in all time periods.

The probability wave functions of the present are affected by the future probability wave functions that can be connected to them. Since there are alternatives sets of unconnected parallel future universes that are possible, each set's effect on the present is determined not only by the probabilities established by that set of connected parallel universes but also by the relative probability that the given mutually exclusive set will happen compared to alternative sets that remain connected to the present situation. Actions taken in the present, however, affect the probability of those alternative sets occurring, and so the probability of each future set is influenced by each individual in the present. The probabilities of both the present and future are therefore jointly determined. As is consistent with the reality of quantum physics, there is no one-way causality in the time dimension as the future and present can affect the past just as the past affects the present and future.

Humans' learning to forecast that may have been initiated by observing a connection between women's procreative cycle and the moon contributed to their ability to have a greater effect across the time dimension. By forecasting in the present moment different probabilities for future occurrences, humans have a greater impact on what future universe can be experienced. Instead of merely experiencing whatever happens, they can affect the probability and even certainty of future events. The self-awareness that was motivated by the optimality of self-forecasting in a world where human actions increasingly impact the future environments permitted the innate logical deductive capacities of brains in general to be applied to that information. This process enabled awareness of both the information and the deductive capacity itself that combined are consciousness, which can collapse probability wave functions in the past, present, and future on points it selects.

The very self-forecasting that led to people's consciousness and sense of self that is separate from other things prompted humans to see themselves in relation to other objects as well as being interconnected with them. Humans learned to not only forecast their own actions in the future but to predict the corresponding effects those actions would have on the environment that would in turn affect their own actions in an infinite feedback loop that has interconnectivity as its foundation. Whereas the communality of all is normally smoothed into a wave of average possibilities, humans can collapse the waves on points they choose. Separate consciousness merely creates particular forms of connectivity between probability wave functions that collapse some on particular points to achieve the intent determined by that consciousness. The points of collapse have to be consistent with the waves of potential of all else, and so the power is limited by the set of feasible points on the interrelated waves that exist.

Separate consciousness thus enables powers like a deity, and the further development of separate consciousness enables greater powers to collapse the wave functions of everything on desired feasible points. More individualized consciousnesses effectively create individual deities as long as they continue to resonate with the probability wave functions of the wills of all other individual deities. Each individual can collapse those wave functions on the points the person chooses and thereby allow the individual to experience the set of feasible universes that the person wishes.

The power of human deities, however, remains dependent on humans maintaining their human nature, which gives them humans' power to control the environment by resonating with other human beings in an evolutionary process that made a unified group consciousness of diverse human beings the highest ecstasy that all at least subconsciously seek. This blissful feeling of being one in a mutual surrender to present or future human beings can only motivate such deities to resonate in bliss with other diverse deities for a meaningful existence. The only alternative is a boring, meaningless emptiness that is experienced regardless of the amount of wealth or power one has.

Human nature thus makes a single omnipotent deity in the future impossible, as such a deity would immediately recognize its own pathological emptiness and die of infinite boredom and meaninglessness. It would destroy all past, present, and future events with it to create an impotent nothingness whose existence is proven to be impossible because of everything existing today. An impossible, worthless, useless nothingness is analogous to the microscopic facts of the meaninglessness of pure nothing that isn't attracted to anything, that experiences only nothingness, and that is totally impotent. Such nihilism couldn't be willed by anyone acting according to their human nature that evolved to promote the survival of humanity, and anyone without the powerful human nature has little power to will anything, much less something that is inconsistent with the wills of human beings.

In addition, future humans, who likely will have greater knowledge and thus more potent will power than those of the present as long as they keep their human nature, will have to resonate more with other humans, including across time, to attain their potency, as explained in more detail in the next chapter. Humans may be deities, but they are communal ones that rule jointly with all past, present and future human beings. This deduction on the possible directions for the universe long-term can assist human beings in understanding better their own meaning and purpose, as well as the probabilities of future occurrences that both affect themselves and that they themselves impact with their conscious choices.

Human forecasting farther and farther into the future has long led to the question of where the universe is going and the ultimate meaning or purpose of everything. This search for meaning has roots in the basic tendencies of even the most primitive organisms. In particular, because events in the environment are often correlated with subsequent events of positive or negative value to organisms and their survival, the brains of all living organisms associate various stimuli as signals for possible subsequent rewards, and their subconscious minds relate self-behavior in different environments to greater chances of satisfaction of organisms' needs and desires. This process of efficient linear decision-making is encouraged by evolution to be innate. The identification of signals with subsequent rewards is the basic source of the yearning to seek meaning in everything (Klein, 2003), but this drive seems to be a conscious one only in humans, who seek to deduce the effects of their actions on their future selves as is determined interactively with the external environment. Knowledge of a purpose for everything including themselves enhances humans' ability to forecast the environment and themselves, as well as choose actions in the present that contribute to this purpose and thus have meaning or meaningful implications toward reaching that goal.

Given the basic reality of interconnectivity between everything that is recognized to be an absolute truth due to entanglement and evidence of the holistic properties of matter, interconnectivity must be a core component of the answer to the ultimate question with respect to meaning. The human desire to seek an answer to the question of their own purpose therefore naturally results in a yearning for greater interconnectivity as a possible source for the final answer that was initially prompted by an observed interconnectivity between one aspect of human biology and astrological events.

The human desire to unite with the universe may also have to do with the fact that everything evolved from the original whole nothingness to the quantum reality of electromagnetic waves that are interconnected with everything else. This unperceived but scientifically proven reality of waves of attraction then makes a second fractal layer of unseen reality consisting of chemical atoms and compounds such as DNA. That second sphere of reality in turn creates a third layer that humans can actually perceive with their five senses. Those senses merely see, hear, and feel electromagnetic force fields in a categorized form. Although human brains have evolved to perceive many such fields as solid, since they act with a force that typically has a practical effect as if they were relatively impenetrable, people deep down in their quantum selves want to experience the basic reality of oneness and interconnectivity. Perhaps the human characteristic of wanting to connect with and serve humanity as a whole by optimally resonating with the universe through their analytical forecasting is what allowed them to be the dominant life form on earth. In particular, their nature is more harmonious with the underlying reality than any other living organism. Interestingly enough, some people under the influence of hallucinogenic drugs that are naturally produced by the human body not only experience more empathy and receptivity to all else but actually perceive pure light as the true reality and the perceived layer of reality to be mere "shadows and plays of light" (Strassman, 2001).

One implication of these phenomena is that each individual is indeed one with the universe, and that the differentiation of selves and the compartmentalization of things are mere illusions. The very physical patterns of this universe, which is only one of the infinite number of realities with different sets of integrated characteristics or laws that have evolved and that will continue to be created from the infinite number of paths which will all occur, can themselves evolve, and all the possibilities and "parallel realities" had to "meld together to yield" the current universe in order for it to exist (Davies, 2007).

People's consciousness is the experience of the universe created by the neural firings in their brains that are prompted by a combination of sensual stimuli, emotions, and deductive reasoning. Each moment is a creation out of infinite possibilities that allows human consciousness. In many possibilities, a person dies or doesn't live, and so people aren't conscious of those parallel universes. Given the rapid advances in technology that supply a 50% chance of occurrence of eternal life even for 60-year olds today (Kurzweil and Grossman, 2004), humans may very well live forever in some universes. Such possibilities are the only ones an existing human being will ever be conscious of in the far future. However, there will still be many other parallel universes of which humans are unconscious (Merali, 2007i) because they don't exist in them.

F. Conclusions on Consciousness

The conscious developed out of simple logical forecasting processes that were applied to self-prediction. Consciousness is therefore based on a genetically ingrained logic system. That system is dependent on information provided to it by the emotions and senses that the subconscious allows to filter through the thalamus.

The code for the conscious is essentially based on the logic of "if A, then B". The information the conscious receives from prior experiences, including both from the subconscious and itself, is utilized by that system of logic to set goals as well as strategies and tactics to attain those goals. While this basic framework for the conscious applies to all humans, different genes, different experiences, and some randomness in thought processes result in people consciously thinking differently from others. As will be explained in Chapter 16, at least some of the stochasticity is internally affected by human will power, which supplies individual free choice through the self-reflective thought in decision-making that enables individual selection of the set of feasible universes that the individual will experience. The collapsing of wave functions is accomplished through the brainwaves of consciousness that initiate a cascading process that begins with internal wave collapses via microtubule openings in neurons.

Since the decision-making, forecasting, and self-forecasting that produced the conscious evolved to permit better decision-making, the conscious naturally was given dictatorial powers over the subconscious. However, the exercise of those powers may require greater and more repeated firings of the neural connections associated with some commands, such as suppression of desires to meet the needs for human life like nutrition. The logical forecasting, decision-making, and wave collapsing of the conscious is based on the information which it receives from the subconscious and which varies depending on an individual's experiences that themselves differ both stochastically and for other reasons such as genetic ones. The conscious, however, also provides rules to the subconscious for determining what information is filtered to the conscious. The conscious self-awareness which is partially formed by subconscious signals to it, thus has power over those signals and subconscious processing in general. Although that power is limited somewhat by the genetic capacities of the brain, those capacities evolved to permit a great deal of flexibility or free choice.

The conscious is kind of like a dictator to its subject, interactively instructing it in rules to follow and changing or deprogramming bad rules, especially those suggested from outside. The subconscious can be reprogrammed to a large degree, but some things are far harder to completely change.

Many subconscious tendencies are innate or purely genetic, while many others have been developed at least in part by the past environment, including the opinions of and treatment by others. They also can continue to be affected by the present and future environments. However, the conscious is in full control at all times to overrule any and all subconscious tendencies, whether innate or due to past, present, or future environments, or some combination thereof. It does so by having the conscious neurons electrically out-fire the subconscious ones.

The conscious thereby influences everything about a person's brain. Even emotions such as love, spite, and hate can be eliminated in many situations by training the neural firings associated with them via out-firing them with conscious thoughts to do so. For instance, upon seeing a lion behind a zoo cage, the subconscious neural firings that prompt signals to the heart to beat faster and prepare for a dangerous situation, thereby resulting in feedback to the brain in the form of fear that is consciously felt, can be negatively reinforced by conscious suppression so that the reflex of rapid heart beat and related experience of fear doesn't exist for the caged lion. Similarly, desires such as hunger for food and sex can be similarly impacted under most circumstances. A natural wish to surrender to some group consciousness can also be suppressed, but there seems to be an innate negative reinforcement of the brain in the form of a feeling of empty meaninglessness when there is a repression of the desire to serve humanity as a whole.

The conscious fully controls subconscious inputs to it by the filtering process that lets information, thoughts, and fantasies into consciousness, providing rewarding feedback to the subconscious neural firing connections that provide desired information flows and negative reinforcement for connections that supply information deemed to be irrelevant or counterproductive to a conscious purpose with a higher priority. The conscious thus needs to have a goal of higher priority in order to have sufficient "will power" to control subconscious signals for lower priority goals like sex. Pursuing goals inconsistent with the evolved human purpose of serving humanity as a whole, however, leads to negative reinforcement of the brain in the form of worthlessness that is eventually pathological because it leads to the logical deduction of self-destruction because of lack of value or meaning for itself.

Even if the conscious tires of exercising its will, it remains in charge as long as it never doubts the fact that it is in charge. The conscious is an innate human program that evolved to be based on rational logic to enable humanity to survive and reproduce, and this program can't be controlled by the environment or others, unless a person chooses to permit external control. Many people do surrender to let the environment or subconscious tendencies to take control sometimes or even always.

The conscious can also choose to surrender to other individuals or groups of people, as is a natural human tendency in mutual surrender, such as when the individual serves the group that serves the individual as well. However, one-way surrender is inhibited by human nature that has evolved for equality and isn't designed for the artificial inferiority, hierarchy, fetishes, and exploitation inherent in any one-way surrender to an individual, group, fictitious deity, or absurd mystical force. The conscious is naturally programmed to only allow tendencies that advance humanity as a group of equal individuals mutually serving each other.

In the current environment that encourages, promotes, and indoctrinates people with slavery to selfishness and servitude to "superiors", many are unaware of their innately egalitarian nature, much less the complete power their conscious has to totally control their own brains. As a result, they allow the "superiors" to manipulate and control them through various processes that restrict and ration the satisfaction of human needs and desires. The masses may feel they have no choice except to accept their position as inferior slaves competing to serve their superiors in order to receive reinforcement in the form of satisfying their various desires at various times.

However, human beings can conscious control their desires as well as seek alternative ways of satisfying them if they recognize a higher purpose that is consistent with evolved human nature, which drives each person toward a completely mutual surrendering love between people. The service to humanity and to all human beings includes first and foremost serving oneself as a member of that species. Because of the pleasant feeling of self-fulfillment that comes from just trying to serve all humanity in the way each individual believes is optimal, serving oneself optimally requires setting that purpose as the overriding goal that governs all actions, just as human ancestors did for millions of years.

By setting an overriding goal of serving humanity as a whole, the conscious and subconscious can be set to operate in a unified fashion for the purpose for which they evolved. Subconscious signals will be sent for letting it operate without interference when that is optimal The subconscious will signal to the conscious when it needs interfering decisions or evaluations of actions that can be made quickly to allow control to revert back to the subconscious when that is optimal. Since the subconscious is involved in any processing of "psychic phenomena in the environment, doing so also enables a unified consciousness that can evaluate and act appropriately on signals from others, including inhumane ones. Mere concentration on the overriding goal of each human being when in conscious doubt as to the appropriate action permits the most efficient and effective process.

The conscious might even optimally set a role model as a human being of the future with traits that each individual believes would exist based on existing prototypes, imagination, and combinations thereof. The personalities could freely change for different situations, as well as be modified using the standard Bayesian brain learning process that characterize the entire brain system, to enable maximum adaptability to the fluid environments. Over 50 trillion powerful minicomputers in the form of the cells in each individual stand ready in harmonious cooperation to serve the purpose of humanity, and the power of such coherent processing is virtually infinite given the quantum nature of the universe and human beings, as explained in the following chapter.

www.ingramcontent.com/pod-product-compliance
Ingram Content Group UK Ltd.
Pitfield, Milton Keynes, MK11 3LW, UK
UKHW020130290926
15471UKWH00003B/401

9 788883 980688